ROUTLEDGE HANDBOOK OF ENVIRONMENTAL POLICY

This Handbook provides a state-of-the-art review of research on environmental policy and governance.

The *Routledge Handbook of Environmental Policy* has a strong focus on new problem structures – a perspective that emphasizes the preconditions and processes of environmental policymaking – and a comparative approach that covers all levels of local, national, and global policymaking. The volume examines the different conditions under which environmental policymaking takes place in different regions of the world and tracks the theoretical, conceptual, and empirical developments that have been made in recent years. It also highlights emerging areas where new and/or additional research and reflection are warranted. Divided into four key parts, the accessible structure and the nature of the contributions allow the reader to quickly find a concise expert review on topics that are most likely to arise in the course of conducting research or developing policy, and to obtain a broad, reliable survey of what is presently known about the subject.

The resulting compendium is an essential resource for students, scholars, and policymakers working in this vital field.

Helge Jörgens is Associate Professor of Public Policy at Iscte – University Institute of Lisbon and Integrated Researcher at CIES-Centre for Research and Studies in Sociology, Portugal.

Christoph Knill is Full Professor of Political Science at Ludwig-Maximilians-Universität (LMU) Munich, Germany.

Yves Steinebach is Associate Professor of Public Policy and Public Administration at the University of Oslo, Norway.

"This collection of concisely written articles by a group of leading European and international scholars provides a compelling introduction to important environmental policy themes from both historical and contemporaneous perspectives. Key analytical concepts and paradigms which have guided environmental policymaking over the past decades are discussed, factors affecting environmental policy performance and environmental policy change are introduced, and the challenges that will need to be overcome to achieve transformations of energy systems and urban environments are considered. A broad range of topics, including geoengineering, indigenous knowledge in climate policy, the role of litigation, and new forms of environmental activism, are covered. The book will be of great use to teachers, students, and practitioners looking for a solid overview of central environmental policy concepts, research questions, and empirical approaches."

Miranda Schreurs, *Professor of Environmental and Climate Policy, Technical University of Munich*

"This fascinating book provides a wide-ranging overview of advances in the study of environmental policy over the past few decades. The editors are to be commended for their selection of cutting-edge topics and of top-rate scholars to cover them. A great volume that provides an overview of key topics in environmental policy and highlights recent advances in scholarship."

James Meadowcroft, *Professor at the School of Public Policy, Carleton University, Ottawa, Canada*

ROUTLEDGE HANDBOOK OF ENVIRONMENTAL POLICY

Edited by
Helge Jörgens, Christoph Knill, and Yves Steinebach

Cover image: © Gandee Vasan / Getty Images

First published 2023
by Routledge
4 Park Square, Milton Park, Abingdon, Oxon OX14 4RN

and by Routledge
605 Third Avenue, New York, NY 10158

Routledge is an imprint of the Taylor & Francis Group, an informa business

British Library Cataloguing-in-Publication Data
A catalogue record for this book is available from the British Library

ISBN: 978-0-367-48992-2 (hbk)
ISBN: 978-1-032-50311-0 (pbk)
ISBN: 978-1-003-04384-3 (ebk)

DOI: 10.4324/9781003043843

Typeset in Bembo
by SPi Technologies India Pvt Ltd (Straive)

The Open Access version of Chapter 24 was funded by European Research Council.

CONTENTS

CONTRIBUTORS

Mikael Skou Andersen is Professor of Environmental Policy Analysis at Aarhus University, Denmark.

Christian Aschenbrenner is Doctoral Researcher at LMU Munich, Germany.

Michael Böcher is Professor of Political Science and Sustainable Development at Otto von Guericke University Magdeburg, Germany.

Fabio Bothner currently works as a climate protection manager. Previously, he was Lecturer at the University of Bamberg and at the FernUniversität Hagen, Germany.

Aron Buzogány is Assistant Professor of Political Science at the University of Natural Resources and Life Sciences, Vienna (BOKU).

Benjamin Cashore is Li Ka Shing Professor in Public Management and Director, Institute for Environment and Sustainability (IES) at the Lee Kuan Yew School of Public Policy, National University of Singapore.

Catherine Chen is Postdoctoral Researcher at LMU Munich, Germany.

Andreas Duit is Professor of Political Science at Stockholm University, Sweden.

Xavier Fernandéz-i-Marín is Ramón y Cajal Fellow at the University of Barcelona, Spain.

Doris Fuchs is Professor of International Relations and Sustainable Development and Speaker of the Center for Interdisciplinary Sustainability Research at the University of Münster, Germany.

Jana Gheuens is PhD Researcher at the Brussels School of Governance and the Vrije Universiteit Brussels, Belgium.

Anat Gofen is Associate Professor with the Federmann School of Public Policy, Hebrew University of Jerusalem, Israel.

Alexandra Goritz is PhD Candidate at Freie Universität Berlin, Germany.

Michael Howlett is Burnaby Mountain Professor and Canada Research Chair (Tier 1) at Simon Fraser University in Burnaby, BC, Canada.

Dave Huitema is Professor of Public Administration and Policy at Wageningen University, the Netherlands.

Detlef Jahn is Professor Emeritus of Comparative Politics at the University of Greifswald, Germany, and Alumnus of the European University Institute (Florence), Italy.

Martin Jänicke is Professor Emeritus of Comparative Politics at Freie Universität Berlin and Affiliate Scholar at the Institute for Advanced Sustainability Studies (IASS) in Potsdam, Germany.

Andrew Jordan is Professor of Environmental Policy at the University of East Anglia, the United Kingdom.

Helge Jörgens is Professor of Public Policy at Iscte – University Institute of Lisbon and Integrated Researcher at CIES - Centre for Research and Studies in Sociology, Portugal.

Kirsten Jörgensen is Senior Lecturer of Political Science at Freie Universität Berlin, Germany.

Kristine Kern is Professor and Head of the Research Group Urban Sustainability Transformations at the Leibniz Institute for Research on Society and Space in Erkner, Germany. She is also affiliated to Åbo Akademi University in Turku, Finland.

Lisa Klagges is Research Associate and PhD Student at the Institute of Political Science and Communication Studies at the University of Greifswald, Germany.

Christoph Knill is Full Professor of Political Science at LMU Munich, Germany.

Nina Kolleck is full professor at the University of Potsdam, Germany.

Duncan Liefferink is Associate Professor in the Environmental Governance and Politics group at the Institute for Management Research, Radboud University Nijmegen, the Netherlands.

Sylvia Lorek is Chair of the Sustainable Europe Research Institute, Germany.

Pia Mamut is Postdoctoral Researcher of Political Science at the University of Münster, Germany.

Ina Möller is Assistant Professor at the Environmental Policy Group of Wageningen University, the Netherlands.

Ishani Mukherjee is Associate Professor of Public Policy and Lee Kong Chian Fellow at Singapore Management University (SMU).

Sreeja Nair is Assistant Professor (Public Policy) at the Lee Kuan Yew School of Public Policy, National University of Singapore.

Anica Rossmoeller is Doctoral Student of Political Science at the University of Münster, Germany.

Karoline S. Rogge is Professor of Sustainability Innovation and Policy at the Science Policy Research Unit (SPRU) at the University of Sussex, UK, and Deputy Head of the Competence Center Policy and Society at the Fraunhofer Institute for Systems and Innovation Research ISI, Germany.

Raúl Romero is Lecturer of Environmental Sociology at the National Autonomous University of Mexico.

Claudia Ros is a student of Social Anthropology at the National Autonomous University of Mexico.

Barbara Saerbeck works for Agora Energiewende (Germany) as Senior Associate Key Questions in Berlin, Germany. She is also Lecturer at the Berlin School of Economics and Law.

Simon Schaub is Research Fellow at the Institute of Political Science at Heidelberg University, Germany.

Patrick Scherhaufer is Researcher and Senior Lecturer at the Institute of Forest, Environmental, and Natural Resource Policy at the University of Natural Resources and Life Sciences, Vienna (BOKU).

Kei Schmidt is a recent MSc graduate (Environmental and Energy Policy) of the Department of Social Sciences at Michigan Technological University and is now Regulatory Specialist with the US Army Corps of Engineers, USA.

Paul-Philipp Schnase is Student Research Assistant at the Chair of Policy Research and Environmental Politics at Fernuniversität in Hagen, Germany.

Jonas J. Schoenefeld is Research Scientist at the Institute for Housing and Environment in Darmstadt, Germany and Visiting Researcher at the Tyndall Centre for Climate Change Research, School of Environmental Sciences, University of East Anglia, in Norwich, United Kingdom.

Johannes Schuster is Postdoctoral Researcher at Leipzig University, Germany.

Israel Solorio is Associate Professor of Public Administration at the National Autonomous University of Mexico.

Qi Song is PhD Candidate within the EMPOCI project at the Science Policy Research Unit, University of Sussex, UK.

Detlef F. Sprinz is Professor at the University of Potsdam, Germany, Senior Fellow at PIK – Potsdam Institute for Climate Impact, Potsdam, Germany, and Affiliated Researcher, ESSCA School of Management, Angers, France.

Christina Steinbacher is Doctoral Researcher and Research Fellow at the Chair of Empirical Theory of Politics at LMU Munich, Germany.

Yves Steinebach is Associate Professor of Public Policy and Public Administration at the University of Oslo, Norway.

Paul Tobin is Senior Lecturer in Politics at the University of Manchester, UK.

Annette Elisabeth Töller is Professor of Policy Research and Environmental Politics at FernUniversität in Hagen.

Diarmuid Torney is Associate Professor in the School of Law and Government at Dublin City University, Ireland. He is Co-director of the DCU Centre for Climate and Society and Programme Chair of DCU's MSc in Climate Change: Policy, Media and Society.

Jale Tosun is Professor of Political Science at the Institute of Political Science at Heidelberg University, Germany.

Maria Julia Trombetta is Associate Professor in International Relations and International Security at the University of Nottingham Ningbo, China.

Mareike Well is PhD Candidate at Freie Universität Berlin, Germany.

Adam M. Wellstead is Professor of Public Policy in the Department of Social Sciences at Michigan Technological University, USA.

Rüdiger K.W. Wurzel is Professor of Comparative European Politics and Jean Monnet Chair in European Union Studies in the School Politics and International Studies at the University of Hull, UK.

ACKNOWLEDGEMENTS

We thank Annabelle Harris at Routledge for her suggestion to develop this Handbook and Jyotsna Gurung for her clear guidance and support throughout the whole process. Our Handbook brings together the contributions of many of the leading scholars in the field of environmental policy analysis. We would like to express our sincere gratitude to all of them – it was a pleasure working with you! We also thank Inês Rocha Trindade, Daniela Rodrigues, and Sara Canha at Iscte – University Institute of Lisbon for their support in developing the concept for this Handbook, as well as Leonie Köhler at Ludwig Maximilian University of Munich for her invaluable help in preparing the final manuscript.

1
INTRODUCTION
A Research Agenda for Environmental Policy Analysis – Past, Present, and Future

Helge Jörgens, Christoph Knill, and Yves Steinebach

1.1 The Rise of Environmental Policy within Political Science

This Handbook provides a state-of-the-art review of research on environmental policy and governance. Since the 1960s, environmental policy has become a central topic of national and international policy-making. Today, environmental policy is among the most important, and certainly most dynamic, policy domains, with a wide range of topics and subfields such as biodiversity conservation, air pollution control, water protection, waste management, and, above all, the fight against climate change and the transition towards a carbon-free economy.

Along with this development, the study of environmental policy gradually evolved as a central issue in various subfields of Political Science, including Public Policy, Comparative Politics, and International Relations. Since the late 1970s, research on environmental policy has constantly evolved into a major field of public policy analysis. Empirical work on topics of environmental policy-making is published not only in specialized journals such as *Environmental Politics* (since 1992), the *Journal of Environmental Policy & Planning* (since 1999), or *Global Environmental Politics* (since 2000), but constitutes a key topic in all major journals in the fields of Comparative Politics and Public Policy. A broad literature has developed that studies the making of environmental policies across different stages of the policy cycle, including problem definition, agenda-setting, decision-making, implementation, and evaluation. At the same time, scholars have analyzed various factors that determine national variation in environmental policy ambitions, including the role of political parties, social movements, and political institutions. Significant advances have been made in assessing cross-national interdependencies and drivers that affect the diffusion and convergence of environmental policies across national borders. Related to these developments, scholars of International Relations have studied the emergence and impact of international environmental regimes and organizations. More recently, students have turned to investigating the role of the bureaucratic bodies of these organizations and their influence on environmental policy beyond the nation state.

This general trend has been reinforced by the rapid increase of public and scholarly attention to climate policy and the corresponding rise in scholarly activities and publications dedicated to this topic, especially in the forms of journal articles, monographs, edited collections, and the

DOI: 10.4324/9781003043843-1

so-called "grey" literature.[1] Environmental policy has also received much and growing attention in graduate level programs in Environmental Policy and Politics, Environmental Studies, Political Science, Public Administration, and International Relations.

1.2 State of the Art

At this point in time, a general review and assessment of existing knowledge and future research priorities is highly warranted. Yet, although various handbooks and edited volumes exist in the field of Environmental Policy and Politics, most of these differ from this volume in at least one of three ways. First, most of the more recent handbooks emphasize the global dimension of environmental policy and politics (e.g. Dauvergne, 2012; Falkner, 2013; Harris, 2014; Kalfagianni et al., 2020). This is due to the predominance of the climate change issue but neglects the ongoing relevance of the nation state and its domestic policies, as well as environmental policies at the regional and sub-national levels, in all areas of environmental policy-making. Second, many of these books are mainly concerned with the politics side, lacking a systematic examination of environmental problems, policies, and processes (e.g. Schelly and Banerjee, 2018). Third, most of the handbooks focusing on the policy side of environmental governance were published ten or more years ago (e.g. Meijer, 2010; Dauvergne, 2012; Wijen et al., 2012). They are often characterized by a compartmentalized structure, focusing very much on the traditional areas of environmental policy-making, such as nature, air, water, soil, and waste, or study environmental policy from the perspective of distinctive academic subfields (e.g. Dauvergne, 2012; Wijen et al., 2012; Bäckstrand and Lövbrand, 2015).

This Handbook presents an up-to-date state-of-the-art review of environmental policy analysis with a strong focus on emerging topics and new problem structures. It emphasizes the preconditions and processes of environmental policy-making, and takes on a comparative approach that covers all levels of policy-making – local, national, and global – and different parts of the world. In short, this Handbook fills an important gap in a vibrant and rapidly growing literature.

1.3 Structure of the Book

The resulting compendium is a go-to guide for established and newly interested researchers, for students, as well as for governments and policy-making entities. It presents theoretical, conceptual, and empirical developments that have been made in recent years and highlights emerging areas where new research and reflection are warranted. This structure and the nature of the contributions will allow the reader to quickly find a concise expert review on topics that are most likely to come up when conducting research or designing environmental policies. Moreover, it provides a reliable and quick survey of what is presently known about problems, processes, and achievements in the field of Environmental Policy. In light of these different objectives, the book is structured into four broader parts.

The first part of the book focuses on important theories, paradigms, and analytical concepts of environmental policy-making. Apart from a concise overview chapter on the emergence and constitution of environmental policy as a policy domain, this part provides various contributions that present essential concepts that have been developed and applied to capture major analytical approaches and paradigmatic shifts in the study of environmental policy. This includes the fundamental differentiation between environmental outputs, outcomes, and impacts as well as conceptual chapters on the environmental state, polycentric governance, and ecological modernization. All contributions cut across different subfields and areas of Environmental Policy. Table 1.1 briefly summarizes the contributions of the Handbook's first building block.

Table 1.1 Analytical concepts and paradigms in environmental policy analysis

Chapter	*Main arguments and findings*
Emergence and Development of the Environmental Policy Field (Chapter 2, Böcher)	• Environmental policy has established itself as a policy field differentiated by the four constituent elements of problems, institutions, actors, and measures. • Environmental problems have evolved from visible to "wicked" or even "superwicked" problems. • Specific environmental policy actors such as NGOs and green parties have emerged and are influencing environmental policies. • Environmental policy is a well-established policy field, but new conflicts may arise due to the climate crisis which also might overshadow other important environmental problems like biodiversity.
Environmental Policy Outputs, Outcomes, and Impacts (Chapter 3, Steinebach)	• The number of environmental policies in place has more or less constantly grown over the last decades. • The policy instruments have moved from a hierarchical approach to more flexible and less prescriptive forms of intervention over time. • All environmental policy instruments have their strengths and weaknesses and thus must be combined to achieve the highest level of policy effectiveness. • Environmental policies vary in their effectiveness depending on their policy design and the level of resource provision for policy implementation.
The Environmental State (Chapter 4, Duit)	• Renewed focus on the environmental state is needed in order to address ongoing processes of global environmental change. • This requires a model of the political economy of the environmental state as distinct from that of the welfare state. • More research is needed to empirically assess the drivers and consequences of environmental state expansion.
Polycentric Governance (Chapter 5, Jordan and Huitema)	• Polycentric thinking challenges simplifications in environmental governance, including the idea that "global problems" require "global solutions". • It is a useful means to account for the rapidly changing contours of governance particularly in dynamic area such as climate change. • The state, even if not completely hollowed out by austerity and captured by neoliberal forces, is not capable of and does not have the legitimacy to govern complex issues such as climate change alone.
Ecological Modernization and Beyond (Chapter 6, Jänicke and Jörgens)	• Ecological modernization is the policy-driven innovation and diffusion of resource efficient clean technologies. • As a political strategy, ecological modernization attempts to reconcile economic growth with environmental preservation by making products, processes, and services more eco-efficient. • Ecological modernization policies are characterized by a high degree of political feasibility, which results from their market compatibility as well as economic and social co-benefits. • In recent years the gap between environmental goals and actual improvements has widened and ecological modernization seems to be reaching its limits. • To overcome these weaknesses it is necessary to proceed to structural solutions. The authors propose the term *ecological modernization 2.0* to describe this advanced strategy.

The book's second part is dedicated to the factors affecting environmental performance. The focus is on various aspects that can contribute to or prevent environmental policies from achieving their intended effects. We start this assessment with a broad overview chapter on the determinants of environmental performance. Following on from that, we engage in a more detailed discussion of specific factors. The second part of this Handbook includes chapters on the role of environmental administrations, policy integration, as well as policy implementation and evaluation. In addition to these factors that essentially relate to different stages of the policy cycle, a number of chapters focus on cross-cutting determinants of environmental performance, including international environmental bureaucracies, lawsuits by environmental non-governmental organizations, the inclusion of indigenous knowledge and expertise in policy-making, as well as the science–policy interface (see Table 1.2).

Table 1.2 Determinants of environmental policy performance

Chapter	*Main arguments and findings*
Determinants of Performance in National Environmental Policies (Chapter 7, Jahn and Klagges)	• Environmental performance is a controversial concept that needs to be carefully operationalized and requires contextualized analysis. • The challenge of studying political determinants is that there are many non-political aspects that have an impact on performance, such as economic, geographic, and climatic conditions. • The best-known political aspect is the positive effect of consensual structures on environmental performance. • National environmental performance is strongly determined by international factors, such as international environmental agreements, and the integration into supranational organizations – especially in the EU.
Bureaucracy and Environmental Policy (Chapter 8, Knill and Steinebach)	• Bureaucracies are clearly underrated with regard to their importance for environmental matters. • Effective bureaucracies are crucial for the proper functioning of environmental policies throughout the different policy stages. Independent and well-equipped administrations *not* only produce better-designed environmental policies, they are also faster and more effective in applying them. • The ability of bureaucracies to produce and implement environmental policies depends on different factors such as the analytical capacities of the people working in the administration, their capability to overcome both inner- and outer-institutional boundaries, and the bureaucracies' independence (autonomy) from direct political intervention. • Despite their importance, environmental administrations are under "siege" in modern democracies as their organizational capacities do *not* keep pace with the massive expansion of implementation tasks.
Analytical Perspectives on Environmental Policy Integration (Chapter 9, Steinbacher)	• Research on environmental policy integration (EPI) suffers from a lack of systematization. • The chapter offers four novel analytical perspectives on EPI distinguishing between (1) horizontal and vertical EPI and (2) policy substance versus policy process orientations. • Horizontal EPI strives to remedy the cross-sectoral challenge of environmental policy-making, while vertical EPI seeks to manage complexity and to reduce the present EPI implementation gap.

(*Continued*)

Table 1.2 (Continued)

Chapter	*Main arguments and findings*
Environmental Policy Implementation (Chapter 10, Tosun and Schaub)	• To better understand the challenges of policy implementation it is rewarding to adopt an analytical lens that differentiates between processes, including how policies are designed, which implementation structure is chosen, how agencies make decisions, how target groups react to a public policy, and what the intended impact of a policy is. • Politics can influence the process of policy design and produce policy outputs that are imperfect. Likewise, agenda-setting dynamics and lacking political or societal support may result in policy outputs. • Implementation structures lacking mechanisms to facilitate horizontal or vertical coordination, and a division of competence among ministries, can hinder effective policy implementation. • Constraints on government agencies and the nature of the policy targets can have an impact on how the stipulations of a policy are translated into more concrete policy instruments and instrument settings. • Target groups vary in their characteristics, actions, and needs, which also affect policy implementation.
Environmental Policy Evaluation (Chapter 11, Schoenefeld)	• Environmental policy evaluation has grown substantially around the world. An often-repeated hope is that it will provide the necessary knowledge to drive continuous policy improvement. • Yet, such improvements through evaluation face important challenges. These challenges include high levels of complexity, a lack of cumulative knowledge-building based on evaluations, and a range of methodological questions. • These challenges have triggered an active discussion about how to respond, including ideas for how to generate cumulative knowledge, how to better conceptualize the role of evaluation in public policy processes, how to govern evaluation itself, and how to work towards a common set of environmental policy evaluation guidelines.
International Public Administrations in Environmental Governance (Chapter 12, Jörgens et al.)	• International public administrations (IPAs), defined as the secretariats or bureaucratic bodies of international organizations, have become important, albeit often-neglected actors in international environmental governance. • There is consensus among scholars that IPAs are best conceptualized as (partially) autonomous actors with a potential to influence international environmental policy processes and outputs. • IPAs rely on their formal and organizational autonomy, their expertise and procedural knowledge, and their central position in issue-specific information flows to advocate for their own policy ideas and preferences. They make use of the complex multi-level and multi-actor structure of the international system to create support for their preferred policy options. • Traditionally, IPA influence is often hidden as they prefer to act "behind the scenes". More recently, secretariats have taken on a more proactive role, openly advocating their policy preferences and bringing supportive non-state or sub-national actors into multilateral negotiations. • IPA research has moved from case studies to comprehensive mappings of the position and centrality of IPAs in issue-specific policy networks, using social network analysis (SNA).

(*Continued*)

Table 1.2 (Continued)

Chapter	*Main arguments and findings*
The Role of Litigation of Environmental Non-Governmental Organizations in Environmental Politics and Policy (Chapter 13, Töller et al.)	• While legal tradition shaped the implementation of the Aarhus Convention in Convention states, Court of Justice of the European Union (CJEU) case law forced several EU member states to extend their standing rights for ENGOs. • For about half of ENGOs the right to take legal action is part of their strategy. Suing activities display regional and organizational patterns and depend particularly on staff, the intensity of political activity, and whether associations receive public money. • Lawsuits filed by ENGOs can improve not only the application of environmental laws, but also environmental quality – under the condition that the law sets unambivalent rules that are clearly not being complied with. • The European Commission's outsourced enforcement strategy appears risky, given the high variation in standing rights and overall efficiency of legal systems as well as the heterogeneity in environmental movement strength.
Indigenous and Local Knowledge in Environmental Decision Making: The Case of Climate Change (Chapter 14, Solorio et al.)	• This contribution argues that the inclusion of indigenous and local knowledge in environmental decision making is hampered by a three-layered explanation: a distributive, a procedural, and an epistemic injustice. • The distributive (in)justice refers to the way in which indigenous peoples carry the burden of climate change impacts and policies while rarely receiving the direct benefits. • The procedural (in)justice refers to the exclusion of indigenous peoples from the main decision centers of climate policy. • The epistemic (in)justice captures the way in which climate policy overshadows the indigenous people's capacity to communicate its own knowledge and way of relating to the environment.
The Science–Policy Interface and Evidence-Based Policymaking in Environmental Policy (Chapter 15, Wellstead et al.)	• There is a growth in scholarship developing taxonomies of barriers and drivers of evidence-based policy-making. • Most of the barriers to evidence in environmental policy centers on the lack of information or uncertainty by policy-makers. • Ambiguity related barriers are often overlooked. Policy process concepts can help explain the challenges associated with the science–policy interface. • Policy-process-related causal mechanisms improve the understanding of evidence-based policy-making. • Environmental-based policy innovation labs are a promising avenue for evidence-based policy.

In part three of the Handbook, the analytical interest shifts to the study of patterns and determinants of environmental policy change. We begin with an overview that captures the general patterns of policy change at the aggregate level and identifies the major drivers and consequences of policy growth in the environmental domain. Two subsequent chapters then focus more on cross-national and international drivers of environmental policy change, including the analysis of environmental pioneers and laggards as well as the dynamics of policy diffusion and policy convergence. These assessments are complemented by contributions on the factors affecting the choice and change of environmental policy designs and the impact of changing frames through the securitization of environmental policies. The part concludes with a chapter on patterns of environmental policy change in South-East Asia and thus sheds light on environmental policy-making in a highly dynamic geographic region (see Table 1.3).

Table 1.3 Environmental policy change

Chapter	*Main arguments and findings*
Policy Change and Policy Accumulation in the Environmental Domain (Chapter 16, Knill)	• The phenomenon of environmental policy accumulation is defined by the constant growth of environmental policy portfolios; i.e. the number of policy targets and policy instruments is growing over time. • Causes of policy accumulation are political incentives for policy overproduction as well as international factors (globalization and international policy harmonization). • Policy accumulation might come with negative side effects, in particular policy complexity and the overburdening of implementation bodies.
Leaders, Pioneers, and Followers in Environmental Governance (Chapter 17, Liefferink et al.)	• While leaders actively seek to attract followers, pioneers focus mainly on stringent internal policies. • Leaders and pioneers may exert structural, entrepreneurial, cognitive, or exemplary leadership, or combinations of these types. • Leadership implies followership. The initial focus on states as leaders has been gradually widened to include the role of followers. • Leadership in and by the Global South has so far remained largely unexplored.
Convergence and Diffusion of Environmental Policies (Chapter 18, Knill et al.)	• Policy diffusion and policy convergence capture related yet different phenomena. • Diffusion and convergence are driven by a number of factors, in particular international interdependencies between countries that emerge from cooperation and communication at the level of international and supranational organizations as well economic interlinkages in globalized markets. • To analyze the effects of convergence and diffusion on environmental problem-solving, this chapter captures (1) the spread of different types of environmental policy targets and instruments and (2) changes in the similarity of national environmental policy portfolios over time. • The findings indicate that diffusion and convergence should generally help to strengthen the capacity of national governments to address environmental problems.
Policy Design for Sustainable Energy and the Interplay of Procedural and Substantive Policy Instruments (Chapter 19, Mukherjee)	• Contemporary research in the policy sciences places *effectiveness* as the central goal of policy design. • This emphasis permeates both micro-level design considerations for specific policy calibrations, as well as more meso-level policy tool mixes. • Effective instrument design, therefore, augments the task of looking at individual tools and considers them as tool "compounds" that comprise substantive *and* procedural means which interact through the process of designing tools and subsequent tool calibrations. • In line with the growing literature on policy design and multi-component policy means, this chapter illustrates the notion of such substantive-procedural design compounds, by comparing what is known about the formulation of three classes of energy policies: renewable energy targets or quotas; feed-in-tariffs; and net metering or smart grids.

(*Continued*)

Table 1.3 (Continued)

Chapter	*Main arguments and findings*
Securitization, Climate Change, and Energy (Chapter 20, Trombetta)	• Addressing climate security and energy security are not just about identifying a set of objective threats and developing policies to deal with them. • The chapter emphasizes the discursive construction of climate and energy security, which identifies threats to be considered, the entities to be protected, and the means to be employed. • Climate change and energy security are deeply related; however, they have been constructed as two distinct discourses. This creates tension and increases insecurity. • The chapter analyses the evolution of climate and energy security discourses and shows how they started to be integrated. • Considering climate and energy security as part of the same discourse has implications for security practices and for climate and energy policies.
Environmental Policy Dynamics in Southeast Asia: Two Steps Forward, One Step Back (Chapter 21, Nair et al.)	• Southeast Asia is a historically important bioregion in terms of environmental indicators such as biodiversity, freshwater resources, and species' richness and distribution. • It is also a region that faces intense pressure from population growth, industrial expansion, environmental change, and the subsequent land and forest degradation. • Drawing on the theoretical work on policy dynamics and change this chapter discusses how policy-makers are dealing with these changes, by focusing on policy formulation, choice of instruments, capacities, and policy networks, supported by examples from the region. • There is a pattern of "two steps forward and one step back" in the region as there has been a gradual extension of interest and capacity in the environment, better polices, and implementation/enforcement. • However, there are still many problems in low-capacity states (Cambodia, Laos) and backsliders (Myanmar, the Philippines, Thailand), and continuing ongoing problems in Malaysia and Indonesia.

The fourth and final part of this book concentrates on more fundamental – paradigmatic – challenges related to the transformation of environmental policy-making. This includes contributions discussing the need for a longer time horizon in environmental policy-making, the role of local policy-making in multilevel governance constellations, and policy mixes facilitating the shift towards renewable energies. Further chapters capture the paradigmatic shift towards sufficiency in environmental governance and the transformative impact on environmental policy-making that might emerge from the new environmental and climate movement. In addition to the challenges associated with mitigating the key threats, there is a growing discussion on the potential for climate change adaptation and related policy implications which are discussed in a chapter on the governance of geo-engineering. Finally, important paradigmatic challenges, especially for global climate policy, arise from specific problems and governance approaches in autocratic regimes and in fast-growing economies. Both topics are addressed in individual chapters. The book concludes with a discussion of future research challenges and captures the broader role of environmental policy in advancing theory-building and analytical concepts in political science and its various subdisciplines (see Table 1.4).

Table 1.4 Transformation of environmental policies: paradigmatic challenges

Chapter	*Main arguments and findings*
The Challenge of Long-term Environmental Policy (Chapter 22, Sprinz)	• By definition, long-term environmental policy (LoPo) challenges are difficult to solve. • Policy options to cope with LoPo encompass institutional design, information, dis-/incentives, as well as regulation and enforcement. • Predicting LoPo choices, coping with time inconsistency, and assessing the effectiveness of LoPo policies are among three challenges that merit dedicated future LoPo research.
Cities and Urban Transformations in Multi-Level Climate Governance (Chapter 23, Kern)	• Since the Rio Conference in 1992 international organizations have supported sustainable development and climate policy at the local level. • Leading cities have become players in global climate policy, but local climate action is not a panacea. • As "ordinary cities" have less capacities than leading cities, they need support by national governments. • Urban experiments may not lead to sustainability transformations but to "projectification". • The decarbonization of cities and towns requires the scaling of urban experiments within, beyond, and across cities.
'Policy Mixes for Addressing Environmental Challenges (Chapter 24, Rogge and Song)	• The chapter introduces an extended conceptual framework for policy mix research. • Two analytical approaches for delineating complex policy mixes are proposed. • These approaches are illustrated by the example of global climate change for informing policy mix design.
Fifty Shades of Sufficiency: Semantic Confusion and No Policy (Chapter 25, Fuchs et al.)	• This chapter explores the meaning of sufficiency in environmental policy discourse. • Delineating the variance in definitions used and the associated effects for evaluating policies with respect to their sufficiency focus. • These difficulties are illustrated with policy examples in the consumption areas of mobility, food, and housing. • The current lack of real sufficiency-oriented policies are highlighted.
The New Climate Movement: Organization, Strategy, and Consequences (Chapter 26, Buzogány and Scherhaufer)	• The new climate movement has succeeded in mobilizing the masses and bringing protests and civil disobedience as relevant forms of resistance back into the public space. • There are significant distinctions in the profiles, narratives, and organizational forms of different groups of the new climate movement. • The new climate movement uses both legal and illegal forms of protest. • The activities of the new climate movement have consequences for public discourses, strategies of political parties, policies, and the activists themselves. • Studying the new climate movement is essential for public policy scholarship interested in how climate and energy policies emerge and to what effect.

(*Continued*)

Table 1.4 (Continued)

Chapter	*Main arguments and findings*
Geoengineering and Public Policy: Framing, Research, and Deployment (Chapter 27, Möller)	• Geoengineering is a contested concept that is usually associated with techno-scientific imaginaries of halting or reversing global warming. Public policy needs to be aware of the different meanings and intentions with which the term is used. • Geoengineering techniques are inherently anticipatory; they shape contemporary policy despite large uncertainties about whether or not they will ever exist. • To facilitate governance, it is helpful to think about how geoengineering techniques might differ in terms of the political organization of their implementation.
Environmental Policymaking in Authoritarian Countries (Chapter 28: Chen and Aschenbrenner)	• Understanding environmental policy-making in non-democracies. • Investigating the applicability of authoritarian environmentalism in different non-democracies. • Uncovering and explaining distinctive patterns of environmental policy-making in Singapore and Russia. • Empirically summarizing the characteristics of environmental policies passed in Singapore and Russia during the period 1970–2020.
Environmental Policy in Fast-Growing Economies: The Case of India (Chapter 29, Jörgensen)	• The growth-first paradigm and carbon lock-in, international environmental and climate governance, the North–South debate, and the equity paradigm have shaped environmental policy in India. The institutional structures of its political system, the strong role of the state, federalism and centralization, the judiciary, and democracy have each played a role. • India's vibrant civil society organizations (CSOs) and grassroots movements have won significant influence in mobilizing and campaigning for environmental protection, agenda-setting, and implementation. • On the one hand, India's democratic system offers an opportunity structure for environmental actors; on the other hand, its centralized and heavily bureaucratized state impedes environmental action.

Note

1 Grey literature includes policy briefs; reports by think tanks, national governments, and international organizations; and similar research-based but non-peer-reviewed publications.

References

Bäckstrand, Karin & Lövbrand, Eva (Eds.) (2015): *Research Handbook on Climate Governance*. Cheltenham: Edward Elgar.

Dauvergne, Peter (Ed.) (2012): *Handbook of Global Environmental Politics*. 2nd ed. Cheltenham: Edward Elgar.

Falkner, Robert (Ed.) (2013): *The Handbook of Global Climate and Environment Policy*. Malden: Wiley.

Harris, Paul G. (Ed.) (2014): *Routledge Handbook of Global Environmental Politics*. London: Routledge.

Kalfagianni, Agni, Fuchs, Doris, & Hayden, Anders (Eds.) (2020). *Routledge Handbook of Global Sustainability Governance*. London: Routledge.

Meijer, Johannes & der Berg, Arjan (Eds.) (2010): *Handbook of Environmental Policy*. New York: Nova Science Publishers.

Schelly, Chelsea, & Banerjee, Aparajita (Eds.) (2018): *Environmental Policy and the Pursuit of Sustainability*. London: Routledge.

Wijen, Frank, Zoeteman, Kees, Pieters, Jan, & van Seters, Paul (2012): *A Handbook of Globalisation and Environmental Policy: National Government Interventions in a Global Arena*. 2nd ed. Cheltenham: Edward Elgar.

PART I

Analytical Concepts and Paradigms in Environmental Policy Analysis

2
EMERGENCE AND DEVELOPMENT OF THE ENVIRONMENTAL POLICY FIELD

Michael Böcher

2.1 Introduction[1]

Environmental policy is considered a comparatively young policy field. It emerged in the 1960s and 1970s after it was recognised that the economic boom following World War II, especially in the strong industrialised countries, had led to negative effects on nature and the environmental goods of soil, air, and water (Böcher & Töller, 2012a). This chapter examines how environmental policy was constituted from a new policy field to an established policy area and how this political domain differentiated itself. The chapter proceeds as follows: first, the concept of "policy field" is defined in order to unfold an analytical framework for presenting the main stages of the development of environmental policy. Then, this framework is used to exemplify key landmarks in environmental policy development. At the end, the chapter's findings are summarised once again before I venture an outlook on further environmental policy development.

2.2 Definition of a "Policy Field"

In public policy analysis, certain sub-areas of policy are referred to as a policy field or policy domain. In an early definition, policy domain is referred to as the "components of the political system organized around substantive issues" (Burstein, 1991). The term "policy domain" is often used synonymously with terms such as "policy area" or "policy field" (Töller et al., 2021). Early definitions refer to a policy field pragmatically along the logic of political departments for which specific ministries exist. According to this definition, a policy field is "a substantively delimited area of regulations and programs, i.e., of policies, as they are normally organisationally combined in the area of responsibility of ministries or parliamentary committees" (Pappi & König, 1995, p. 111, own translation). These terms make it clear that there are political sub-areas that can be distinguished from one another in relation to certain social problems that are dealt with politically. Whereas in the past policy fields were primarily distinguished according to whether specific state institutions such as ministries existed to deal with them, definitions have evolved since the 2000s. More recently, research has focused on how such policy fields emerge

DOI: 10.4324/9781003043843-3

and differentiate, and what their constituent features are (Loer et al., 2015; Massey & Huitema, 2013, 2016; Töller et al., 2021). Thus, policy fields can be distinguished on the basis of which political problems they deal with and which actors and institutions exist that are specifically geared to certain political problems. I follow here a definition based on the heuristics of our own analytical framework, the "Political Process-inherent Dynamics Approach (PIDA)" (Berker & Böcher, 2022; Böcher & Töller, 2015), which understands political processes as inherently dynamic and identifies their central influencing factors as problems, actors, institutions, and measures, as well as situational aspects. According to PIDA, a policy field is defined as "a specific and long-term constellation of interrelated problems, actors, institutions and measures" (Böcher & Töller, 2012b, p. 4; Loer et al., 2015, p. 9; Töller et al., 2021). Based on this definition, I will clarify how it came about and how environmental policy developed into an established policy field. The development and differentiation of environmental policy will therefore be presented based on its components, which are derived from the PIDA framework: problems, institutions, actors, and measures.

2.3 Problems: Emergence of Environmental Policy as a Separate Policy Field

Constitutive of a policy field is the existence of specific policy problems that differ from problems dealt with by other existing policy fields. It is conceivable that the maturation of the policy field consists in a differentiation of the problems to be dealt with (Böcher & Töller, 2012b). After World War II, the first priority in many countries was to rebuild the economy, to put the population to work, and – in agricultural policy – to secure the supply of food. After an unprecedented economic boom, however, the ecological consequences of economic activity then became more and more visible in the 1960s and 1970s (Böcher & Töller, 2012a). Although there were earlier political activities dealing with nature conservation or the protection of water resources for example, broad social attention to environmental problems did not emerge until the 1960s. This is related to a more critical public, such as the student movement, which in the USA criticised the use of defoliant agents as weapons in Vietnam, for example (Morin et al., 2020, p. 8). In addition, in the 1960s there were some best-selling books dealing with environmental issues; a very well-known one is Rachel Carson's *Silent Spring* of 1962 (Carson, 1994). Concrete environment-related problems, such as air pollution, overexploitation of natural resources, and the pollution of rivers, in combination with a critical public and activities of scientists, led to an initial spark out of which emerged environmental policy, at first mainly in industrialised countries such as the USA, Sweden, or Japan (Jörgens, 1996). At this time, the first specific environmental regulations emerged in many countries, for example in the USA with the National Environmental Policy Act (NEPA) in 1970, the Clean Air Act in 1970, and the Water Pollution Control Act in 1972, which responded directly to environmental problems perceived at the time (Robertson, 2014, pp. 13–14). In this early phase of environmental policy, many of the problems were dealt with quite successfully which, on the one hand, could be witnessed by human sense and, on the other hand, could be well managed by regulatory instruments (e.g. pollution of water and air by establishing end-of-pipe solutions like emissions standards and filter technologies). However, the perception of the problem changed over time: on the one hand, situational events brought catastrophes to people's attention that had an influence on environmental policy; one only has to think of the nuclear reactor accidents in Chernobyl in 1986 and Fukushima in 2011. In addition, after environmental policy had been dealt with primarily on a national state level until the 1980s, environmental problems were increasingly perceived as transboundary or even global (Morin et al., 2020). In this way, the policy field of environmental policy was able to develop based on problems that had not previously been dealt with by established policies and were perceived as being so relevant as to require independent political action, before establishing itself as a result of differentiating newly perceived problem situations.

While environmental policy up to the 1980s was primarily shaped by the nation state, since then more and more environmental problems have become drivers of environmental policy change that transcends national boundaries; one need only think of the depletion of the ozone layer, the global loss of biodiversity, or climate change. Here, important multilateral environmental agreements emerged, like the Montreal Protocol on Substances that Deplete the Ozone Layer, the Convention on Biological Diversity, or the United Nations Framework Convention on Climate Change (UNFCCC) (Morin et al., 2020, pp. 18–24).

Environmental policy differentiated itself into an international field that could no longer be successfully pursued by national state policies alone. The problems became more global and also more complex. While traditional environmental policy was characterised by rather simple problems, which could be mastered by solutions such as standards for emission reduction or bans on certain hazardous substances, the environmental problems became more complex with the discovery of new problem situations.

In general, environmental policy is characterised by different types of problems (Böcher & Töller, 2012a, 2019; Pollex & Berker, 2022). Environmental goods are public goods, which means that the incentives for actors or states to behave in a non-environmentally friendly manner and to profit as free riders from the environmental activities of others are high. Moreover, some environmental problems are persistent, that is there is a very heterogeneous distribution of polluters and beneficiaries of environmental overuse and problems are transboundary and cannot simply be reduced to zero (Böcher & Töller, 2019, p. 100). The spatially uneven distribution of environmental problems also represents a special problem structure of environmental policy; for example, in the case of climate change, people in the countries of the Global South in particular are more at risk than people in the Global North, even though their contribution to climate change is much smaller, which raises questions of environmental justice (Dolšak & Prakash, 2022). In addition, there is the time dimension, which means that in environmental policy decisions regarding long-term effects have to be taken today, even though many environmental impacts will only be felt in the future which is a particular problem for political systems geared to short-term interests and electoral successes. The dilemma of environmental policy is thus that political decisions often have to be taken without the exact nature of the problems and the consequences of action being clear. Uncertainty about problems as well as about solutions is thus another feature of the environmental policy problem structure (Böcher & Töller, 2019, p. 100). Environmental policy is characterised by (scientific) uncertainties (Morin et al., 2020, p. 28), which means that different actors can draw on different expertise depending on their political interests (Böcher & Krott, 2016).

Due to this complex structure of environmental problems, today we speak of "wicked" problems (Roberts, 2000) or even "super wicked" problems. Both the problems themselves and the solutions are socially controversial, and in the case of super wicked problems, the time to deal with them is running out (Jain, 2019; Roberts, 2000). Climate change or the loss of global biodiversity currently represent global environmental problems that are considered "wicked" or "super wicked" because in both cases irreversible tipping points may be reached and time to act is running out (Wohlgezogen et al., 2020).

2.4 Institutionalisation

In addition to the discovery of problems that need to be solved politically and contribute to the formation and differentiation of a policy field, institutionalisation is an important aspect in the development and stabilisation of policy fields. Here, it is a matter of the state successively building up policy-field-specific institutional preconditions that exist permanently, even if governments change. Jänicke described this as part of "capacity building" in his pioneering studies on environmental policy development, making it an important prerequisite for successful environmental policy (Jänicke & Weidner, 1997). Jörgens identifies an institutionalisation that began in the industrialised countries in the late 1960s and, remarkably, continued until the 1990s (Jörgens, 1996). Here, national ministries of the environment as supreme enforcement authorities, national environmental agencies, panels of environmental experts to provide scientific policy advice on environmental policy, regular national environmental reporting, environmental framework laws, and a commitment to environmental protection in national constitutions were introduced (Jörgens, 1996, pp. 62–72). Institutionalisation, then, is about creating or enriching the rules underlying government policy and action with content relevant to environmental policy, given an understanding of institutions as rules that influence the political process in the sense of "polity". The active creation or initiation of ministries, authorities, and expert bodies, as well as the introduction of continuous reporting, is about ensuring that the state has permanent resources of the necessary knowledge and administrative capacities and responsibilities. In terms of institutionalisation, it was initially Japan, Sweden, and the USA that started to establish national institutions, before other industrialised countries began to establish national environmental agencies or national ministries of the environment in the wake of the first international conference on the environment in Stockholm in 1972 (Jörgens, 1996, pp. 79–80). This process continued until the beginning of the 1990s, when, after Western industrialised nations in particular had previously established corresponding ministries and offices, the Eastern European transition states created independent environmental ministries after the end of the cold war (Jörgens, 1996, p. 78). The fact that at the end of the 1980s there was a wave of new environmental ministries was also due to the Chernobyl nuclear disaster. Germany, for example, which until then had only a national Federal Environmental Agency that had existed since 1974, but not an independent environment ministry, reacted by creating one in 1986.

Figure 2.1 illustrates this institutionalisation based on the years of the foundation of environmental ministries. By 2008, environmental ministries had been established in 130 countries. The increase was initially large in the Western industrialised countries and Asia, before sub-Saharan African Countries, North African and Arabian states, and the Eastern European states and Russia as well as the Latin American states had established corresponding ministries. The reasons for the various waves can be described in terms of: the discovery of environmental problems at the end of the 1960s and the beginning of the 1970s, the first environmental conference in Stockholm in 1972, the end of the cold war in 1989/1990, and the United Nations Conference on Environment and Development in 1992.

2.5 Political Actors

For a policy field to become established, there must also be sector-specific actors dealing with the relevant issues. In environmental policy, state capacities were quickly built up. In environmental administrations and ministries, the number of state actors dealing specifically with environmental issues gradually increased. In addition, expert bodies play a major role, which

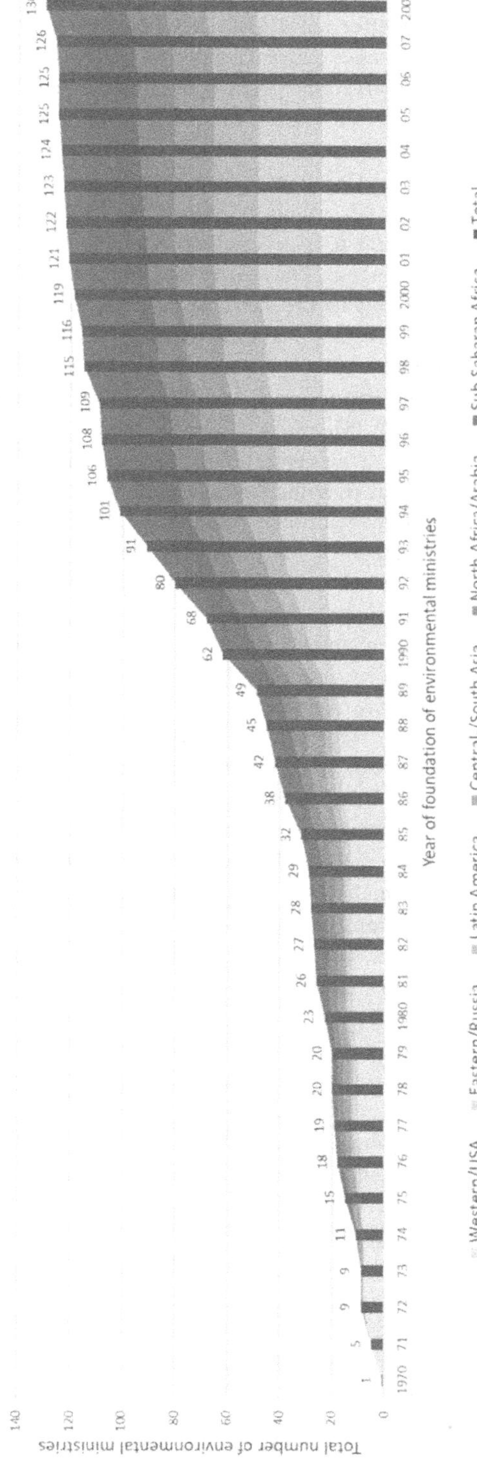

Figure 2.1 Institutionalisation of environmental ministries, 1970–2008.

Source: Own illustration using country group classification by Inglehart and Welzel (2010). Data sources: Aklin and Urpelainen (2014); Busch and Jörgens (2005a, 2005b); Povitkina et al. (2021).

also emerged with the establishment of environmental policy in the various countries, but also at higher and intergovernmental levels (Böcher & Töller, 2012a).

In the course of the emergence of environmental politics, activists founded new NGOs in the 1960s and 1970s. Although conservation organisations such as Birdlife International or the Sierra Club had existed earlier, some actors from these organisations founded the so-called second-generation NGOs (Morin et al., 2020, p. 131): this is where the now globally operating NGOs Friends of the Earth (1969) and Greenpeace (1971) emerged. These NGOs became known mainly through protest actions, for example Greenpeace was founded in the course of protests against US and French nuclear tests or whaling (Santese, 2020). In the 1980s, the activities of environmental NGOs changed. They increasingly became actors that built up counter-expertise to prevailing paradigms in the sense of policy advice, for example against nuclear power. At the same time, the strategies of already existing NGOs changed. Greenpeace or the World Wide Fund for Nature (WWF) also focused on providing expertise, but new, so-called third-generation NGOs were also founded, such as the Worldwatch Institute (1974) or the World Resources Institute (1982) (Morin et al., 2020, p. 132). These NGOs provided policy advice and maintained a close exchange with political actors (Morin et al., 2020). What all these NGOs have in common is that they increasingly differentiated their issue portfolios parallel to the development of environmental policy. In addition, there are large environmental NGOs that pursue a very broad range of topics and, at the same time, a differentiation of NGOs that pursue special topics (Carter, 2018, pp. 145–176). While in the beginning there were classic environmental issues such as anti-nuclear power or air pollution, these later expanded to include more general sustainability issues and, above all, climate protection. As a result of the intensified climate problem, completely new social movements also emerged that are less firmly organised than the classic environmental NGOs. In 2018, for example, the global "Fridays for Future" movement emerged in the wake of school strikes orchestrated by Swedish activist Greta Thunberg (de Moor et al., 2021). Fridays for Future (FFF) is primarily concerned with ensuring that the countries of the world implement the agreements of the 2015 Paris Climate Agreement and that the 1.5 degree target is achieved. In the course of the worldwide activities of FFF, other, more radical, climate groups such as "Extinction Rebellion" or the "Last Generation" emerged, which want to draw attention to the climate crisis with civil disobedience and more radical action (Gunningham, 2019).

A peculiarity of environmental politics is that along this policy field in many countries a separate – new – political party was formed: green parties emerged because, in the wake of the discovery of negative ecological effects of economic activity, new societal conflicts arose that could not be interpreted on the spectrum of classical left–right (labour vs. capital) cleavages (Dolezal, 2010). In addition, post-materialist milieus formed in many Western societies, which favoured the emergence and success of the aspiring social changes and "new politics" of green parties (Carter, 2018, p. 86; Jahn, 1993; Miller, 1991; Strenze, 2021). Green parties were initially founded out of the protest movements of the 1960s and 1970s, like the anti-nuclear or peace movements (Böcher & Töller, 2019). The first green parties came into being in New Zealand and Tasmania in 1972 (Carter, 2018, p. 86). The Green Party was exceptionally successful in West Germany and later in the Federal Republic of Germany, where the Greens were twice part of the federal government coalition. Green parties became members of national and sub-national parliaments or coalition governments in more than 30 countries (Carter, 2018; Grant & Tilley, 2019; Muller-Rommel, 2019). Carter (2018, p. 86) points out here, however, that the success of these parties should not obscure the fact that, with few exceptions, they have remained a phenomenon of Western liberal democracies in the Global North. In countries of the Global South, green parties hardly play a noteworthy role. In the countries where the Greens became successful, their electoral successes created pressure that ensured that other established parties also became more involved with environmental policy issues than before. This is well illustrated, for example, by Germany, where after the Greens' first electoral successes in the 1980s, other parties came under increasing pressure to develop independent environmental policy programmes (Böcher & Töller, 2012a).

2.6 Measures

The emergence and development of a policy field also involves the introduction of concrete measures to help regulate the problems specific to that field. In environmental policy, various instruments are often discussed, the differentiation of which is not only a typical feature of the establishment of policy fields in general, but of environmental policy in particular (see Chapter 8 in this volume on the evolution of diversity among environmental policy instruments). Policy instruments are used to influence collective action so that policy goals can be achieved (Böcher, 2012). For environmental policy, it can be stated that a number of alternative forms of political instruments were developed precisely in connection with environmental problems, which became more and more widespread and were found in many countries (Capano et al., 2020; Jordan et al., 2013; Pacheco-Vega, 2020; Salamon, 2002a; Wurzel et al., 2013). Especially in the early days of environmental policy, regulatory command-and-control approaches were used, for example by introducing limits and standards (Hahn, 2013). "Governments have used regulatory instruments since the onset of environmental policy making in industrialized countries in the 1970s" (Böcher, 2012, p. 14), and until the 1980s, such policy predominantly relied on regulatory instruments (Holzinger & Knill, 2004). These were seen as appropriate to the environmental problems identified at the time since they seem to be well suited when the goal is to reduce environmental impacts to zero, for example, hazardous substances like asbestos. However, many environmental problems such as CO_2 emissions cannot be "banned" in the short term but require instrumental solutions that help to reduce the problem in the long term. Therefore, regulatory approaches became the subject of much criticism because of frequent problems with implementation, resistance from industry in particular, inefficiency, and control problems. Environmental economists in particular pointed out that these instruments were not very flexible and addressed environmental problems at only a comparatively high cost (Hahn, 2013). Especially in environmental policy research, different types of instruments have been developed, their pros and cons discussed, and different typologies developed (Wurzel et al., 2013). A distinction is made between regulatory (command-and-control using hierarchy), economic (e.g. taxes, subsidies, and tradable permits using market-based mechanisms and price signals), cooperative (e.g. voluntary agreements), and informational (e.g. persuasion and education) instruments (Böcher, 2012). Environmental policy can even serve as a model for other policy areas, because a large number of different instruments have been developed and tested for environmental policy purposes on the basis of the instrument types mentioned above (Salamon, 2002b, p. 607). However, policy instruments rarely appear in their pure form. In environmental policy in particular, the use of hybrids, that is mixed forms of different instruments, and the combination of several instruments to solve an environmental policy problem, is the rule. New instruments are often added "on top" of existing regulations without abolishing existing instruments (see Chapter 16 in this volume). This can sometimes lead to instruments hindering rather than reinforcing each other, or to environmentally contradictory effects (Howlett, 2019; Jacob & Jörgens, 2011). Reasons for this are often path dependencies or interests and power of specific environmental policy actors who prefer certain instruments or resist new instrument alternatives (Berker & Böcher, 2022; Capano et al., 2020; Howlett, 2021).

Two aspects contribute to the establishment of the environmental policy field with regard to the use of instruments. First, there is the successive international spread of national environmental policy instruments that have been established in forerunner countries (Tews et al., 2003). Here, policy diffusion or policy transfer took place when individual countries adapted the environmental policy measures of other countries in the sense of "lesson drawing" (Busch et al., 2005; Tews et al., 2003). Pioneering studies by the Environmental Policy Research Centre at Freie Universität Berlin (Jänicke & Weidner, 1997) have successively documented this process of the global spread of key environmental regulations (for a more updated overview of current trends in the international diffusion and convergence of environmental policy instruments, see Chapter 18 in this volume).

Second, the regulatory scope of environmental policy expanded in the wake of new and changing problems. Thus, the range of environmental policy areas to be regulated became increasingly differentiated and larger (for a detailed assessment of this trend of environmental policy growth, see Chapter 16 in this volume). During early environmental policy it was, among other things, questions of waste management, emissions control (e.g. air pollution), and water protection that had to be regulated; later aspects such as recycling, climate protection, and climate adaptation or renewable energies were added (Böcher & Töller, 2012b, 2012a). Figure 2.2 illustrates this successive process of differentiation and expansion of environmental policy regulations in Germany.

Looking at the development of the environmental laws and regulations passed each year in Germany from 1969 to 2021 shows that a certain saturation sets in at some point. Fewer and fewer areas remain completely unregulated and the density of regulation increases (Böcher & Töller, 2012b, p. 9). In Figure 2.2, this is shown, for example, by the newly added areas of chemical regulation and renewable energy/climate protection. Therefore, the accumulated stock of laws must also be considered. If the laws and regulations are differentiated according to their regulatory areas, a clear diversification of the regulatory fields can be seen over time, which is mainly related to the definition of new problems (see Section 2.3). This is clearly shown in the figure, from the 1990s onward, where the previously clearly dominant field of "emissions control" is "caught up" by other regulatory fields such as waste management or renewable energy and climate protection: the "blind spots" (years in which no laws or ordinances were enacted in certain regulatory fields) decrease significantly. This development is representative of many other states. For example, the OECD Policy Instrument Database (PINE), which provides data on environmental policy instruments used in more than 80 countries, also shows a growing differentiation of the various environmental policy sub-areas that have become the subject of environmental regulation over time (OECD, 2017, p. 12).

2.7 Conclusion and Future Research

Based on the definition of a policy field according to PIDA, I have aimed to show how environmental policy has been developed since its beginnings into an established policy field in most countries of the world. For this purpose, some developments in terms of problems, institutions, political actors, and measures have been used as examples. In this way, it could be shown that environmental policy also came into being because environmental problems were discovered and recognised in societies as tasks to be solved politically. Whereas in the beginning it was mainly visible environmental problems that were tackled with classic end-of-the-pipe measures, environmental problems developed into wicked problems that can only be solved supranationally, like biodiversity or climate change.

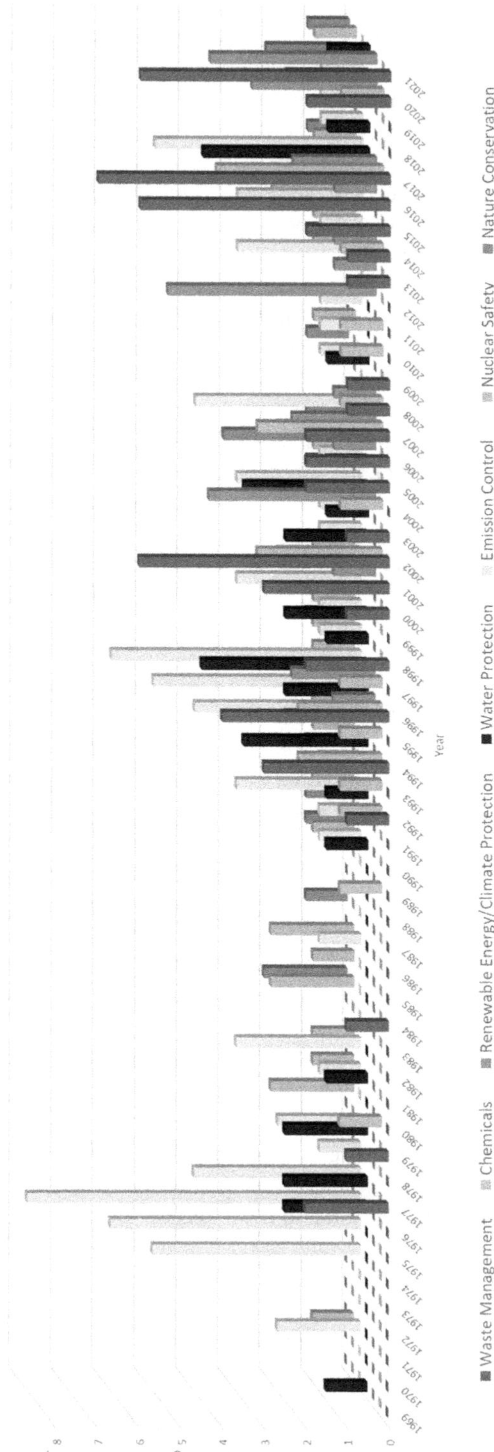

Figure 2.2 The development of environmental regulation in Germany in selected areas 1969–2021.

Source: Own calculations updated after Böcher & Töller 2012b.

States have to develop capacities to solve new kinds of problems. The institutionalisation of environmental policy, which can be observed worldwide in several waves from the 1970s, contributes to the establishment and longevity of environmental policy. This was exemplified by the establishment of environmental ministries, which took place in several phases in many countries around the world. The aspect of the "longevity" of a policy field is supported by this institutionalisation, even if there are also examples where environmental ministries or other environmental policy-related capacities were dismantled again due to a change in government as in Brazil (Menezes & Barbosa, 2021). Political actors that formed exclusively along the lines of environmental policy contributed to the establishment of such policy. They are NGOs such as Greenpeace or Friends of the Earth, which were newly formed, and later a whole number of NGOs and activists dealing with the climate issue, in addition to the state actors involved in environmental policy. In many countries even a separate party was formed, the Green Party, which in some countries entered parliament and even became part of a governing coalition.

In order to influence social action, the state uses measures of regulation. Here, environmental policy became a prime example of the use of all the different conceivable policy instruments between market, state, and voluntary agreements as well as information. The example of Germany shows that in the course of environmental policy development, there has been a growing number of environmental policy sub-areas that became the subject of state regulation – the blind spots became fewer and fewer, environmental policy became more differentiated and successively encompassed new problem areas or increasingly detailed regulations within existing measures.

Against the background of the definition of a policy field according to PIDA, these examples can help us to better understand how the interplay of environmental policy-specific problems, institutions, actors, and measures have resulted in a permanently lasting constellation, which has meanwhile established itself as environmental policy in most countries of the world and, moreover, has also developed internationally as a global policy field with its own institutions, regulations, actors, and measures such as international regimes. The development of environmental policy can be interpreted as a successively differentiating policy field, whose general existence today is no longer politically disputed (whereas its contents are!) – according to PIDA, environmental policy actually represents a constellation of interrelated problems, institutions, measures, and actors that is designed for the long term.

Up to now, environmental policy with its actors and institutions has been able to take up new problems such as climate change or the loss of global biodiversity. In fact, entirely new actors and institutions have emerged. The open question is whether climate protection will be the sole environmentally relevant problem that overshadows all other – still existing – problems to be dealt with in environmental policy. At present, almost only the climate crisis is being discussed at the global level, whereas even the associated biodiversity crisis or problems such as air pollution have a harder time being heard politically. In addition, it remains to be seen whether we will see greater polarisation and the emergence of new conflicts in the policy field in the future: these may arise due to radicalised climate activists, populist parties attempting to roll back environmental policy decisions, or, from a global perspective, between people in the Global South and Global North along new questions of global environmental justice for which the policy field has not yet come up with any satisfactory solutions. Nevertheless, there is some hope in the fact that environmental policy has so far always successively developed along its constituent elements of problems, institutions, actors, and measures with regard to new problem situations.

Box 2.1 Chapter Summary

- Policy fields represent a specific long-term constellation of interrelated problems, actors, institutions, and measures.
- Environmental policy has established itself as a policy field differentiated by the four constituent elements of problems, institutions, actors, and measures.
- Environmental problems have evolved from visible to "wicked" or even "super wicked" problems.
- Institutions such as environmental ministries have emerged in different waves in many countries in different regions of the world.
- Specific environmental policy actors such as NGOs and green parties have emerged and are influencing environmental policies, the latter especially in Western democracies.
- The development of environmental policy measures has led to the diffusion of specific policy instruments and the successive expansion of environmental regulations covering more and more environment-related aspects.
- Environmental policy is a well-established policy field, but new conflicts may arise due to the climate crisis which also might overshadow other important environmental problems like biodiversity.

Note

1 I would like to thank Nicolas Spohn for his support in researching the data and creating the graphics, and Lars Berker for valuable comments. I would also like to thank Sarah Schmitt for proofreading the manuscript. Moreover, I thank the editors for their extraordinary patience.

References

Aklin, M., & Urpelainen, J. (2014). The global spread of environmental ministries: domestic–international interactions. *International Studies Quarterly, 58*(4), 764–780.

Berker, L.E., & Böcher, M. (2022). Aviation policy instrument choice in Europe: high flying and crash landing? Understanding policy evolutions in the Netherlands and Germany. *Journal of Public Policy, 42*(3), 593–613. https://doi.org/10.1017/S0143814X22000034

Böcher, M. (2012). A theoretical framework for explaining the choice of instruments in environmental policy. *Forest Policy And Economics, 16*, 14–22.

Böcher, M., & Krott, M. (2016). *Science Makes the World Go Round. Successful Scientific Knowledge Transfer for the Environment*. Cham: Springer Nature.

Böcher, M., & Töller, A.E. (2012a). *Umweltpolitik in Deutschland: eine politikfeldanalytische Einführung*. VS Springer-Verlag.

Böcher, M., & Töller, A.E. (2012b). Reifung als taugliches Konzept zur Konzeptualisierung langfristigen Wandels von Politikfeldern? Überlegungen anhand des Politikfeldes Umweltpolitik. Paper presented to the *25th Conference of the German Association for Political Science (DVPW)*.

Böcher, M., & Töller, A.E. (2015). Inherent dynamics and chance as drivers in environmental policy? An Approach to Explaining Environmental Policy Decisions. *International Conference on Public Policy*, Milan. http://www.icpublicpolicy.org/conference/file/reponse/1434985890.pdf

Böcher, M., & Töller, A.E. (2019). *Umweltpolitik in Deutschland: Eine Politikfeldanalytische Einführung. 2., Komplett überarbeitete Auflage*. FernUniversität in Hagen.

Burstein, P. (1991). Policy domains: organization, culture, and policy outcomes. *Annual Review of Sociology, 17*, 327–350.

Busch, P.O., & Jörgens, H. (2005a). International patterns of environmental policy change and convergence. *European Environment, 15*(2), 80–101.
Busch, P.O., & Jörgens, H. (2005b). The international sources of policy convergence: explaining the spread of environmental policy innovations. *Journal of European Public Policy, 12*(5), 860–884.
Busch, P.O., Jörgens, H., & Tews, K. (2005). The global diffusion of regulatory instruments: the making of a new international environmental regime. *The Annals of the American Academy of Political and Social Science, 598*(1), 146–167.
Capano, G., Pritoni, A., & Vicentini, G. (2020). Do policy instruments matter? Governments' choice of policy mix and higher education performance in Western Europe. *Journal of Public Policy, 40*(3), 375–401. https://doi.org/10.1017/S0143814X19000047
Carson, R. (1994). *Silent spring*. 1962. Houghton Mifflin.
Carter, N. (2018). *The Politics of the Environment* (3 ed.). Cambridge University Press.
de Moor, J., De Vydt, M., Uba, K., & Wahlström, M. (2021). New kids on the block: taking stock of the recent cycle of climate activism. *Social Movement Studies, 20*(5), 619–625.
Dolezal, M. (2010). Exploring the stabilization of a political force: the social and attitudinal basis of green parties in the age of globalization. *West European Politics, 33*(3), 534–552. https://doi.org/10.1080/01402381003654569
Dolšak, N., & Prakash, A. (2022). Three faces of climate justice. *Annual Review of Political Science, 25*(1), 283–301. https://doi.org/10.1146/annurev-polisci-051120-125514
Grant, Z.P., & Tilley, J. (2019). Fertile soil: explaining variation in the success of green parties. *West European Politics, 42*(3), 495–516. https://doi.org/10.1080/01402382.2018.1521673
Gunningham, N. (2019). Averting climate catastrophe: environmental activism, extinction rebellion and coalitions of influence. *King's Law Journal, 30*(2), 194–202. https://doi.org/10.1080/09615768.2019.1645424
Hahn, R.W. (2013). *A Primer on Environmental Policy Design*. Taylor & Francis.
Holzinger, K., & Knill, C. (2004). Marktorientierte umweltpolitik—ökonomischer anspruch und politische wirklichkeit. In R. Czada & R. Zintl (Eds.), *Politik und Markt. Politische Vierteljahresschrift Sonderheft, 34*, 232–255.
Howlett, M. (2019). *Designing Public Policies: Principles and Instruments*. Routledge.
Howlett, M. (2021). Avoiding a panglossian policy science: the need to deal with the darkside of policy-maker and policy-taker behaviour. *Public Integrity, 24*(3), 306–318.
Inglehart, R., & Welzel, C. (2010). Changing mass priorities: the link between modernization and democracy. *Perspectives on Politics, 8*(2), 551–567.
Jacob, K., & Jörgens, H. (2011). *Wohin geht die Umweltpolitikanalyse?: Eine Forschungsagenda für ein erwachsen gewordenes Politikfeld*. FFU-Report 02-2011. Forschungszentrum für Umweltpolitik.
Jahn, D. (1993). The rise and decline of new politics and the Greens in Sweden and Germany: resource dependence and new social cleavages. *European Journal of Political Research, 24*(2), 177–194.
Jain, K. (2019). Climate change–a super wicked problem. *NOLEGEIN-Journal of Business Risk Management, 2*(2), 23–27.
Jänicke, M., & Weidner, H. (Eds.). (1997). *National Environmental Policies: A Comparative Study of Capacity-Building*. Springer Science & Business Media.
Jordan, A., Wurzel, R.K., & Zito, A.R. (2013). Still the century of 'new'environmental policy instruments? Exploring patterns of innovation and continuity. *Environmental Politics, 22*(1), 155–173. https://doi.org/10.1080/09644016.2013.755839
Jörgens, H. (1996). Die Institutionalisierung von Umweltpolitik im internationalen Vergleich. *Umweltpolitik der Industrieländer. Entwicklung–Bilanz–Erfolgsbedingungen*, 59–111.
Loer, K., Reiter, R., & Töller, A.E. (2015). Was ist ein Politikfeld und warum entsteht es? *dms–der moderne staat–Zeitschrift für Public Policy, Recht und Management, 8*(1), 5–6. https://doi.org/10.3224/dms.v8i1.19108
Massey, E., & Huitema, D. (2013). The emergence of climate change adaptation as a policy field: the case of England. *Regional Environmental Change, 13*(2), 341–352. https://doi.org/10.1007/s10113-012-0341-2
Massey, E., & Huitema, D. (2016). The emergence of climate change adaptation as a new field of public policy in Europe. *Regional Environmental Change, 16*(2), 553–564. https://doi.org/10.1007/s10113-015-0771-8
Menezes, R.G., & Barbosa Jr, R. (2021). Environmental governance under Bolsonaro: Dismantling institutions, curtailing participation, delegitimising opposition. *Zeitschrift für Vergleichende Politikwissenschaft, 15*(2), 229–247. https://doi.org/10.1007/s12286-021-00491-8

Miller, R. (1991). Postmaterialism and green party activists in New Zealand. *Political Science, 43*(2), 43–66.
Morin, J.-F., Orsini, A., & Jinnah, S. (2020). *Global Environmental Politics: Understanding the Governance of the Earth*. Oxford University Press, USA.
Muller-Rommel, F. (2019). *New Politics in Western Europe: The Rise and Success of Green Parties and Alternative Lists*. Routledge.
OECD. (2017). *Policy Instruments for the Environment*. Retrieved Nov 20, 2022, from http://oe.cd/pine.
Pacheco-Vega, R. (2020). Environmental regulation, governance, and policy instruments, 20 years after the stick, carrot, and sermon typology. *Journal of Environmental Policy & Planning, 22*(5), 620–635. https://doi.org/10.1080/1523908X.2020.1792862
Pappi, F.U., & König, T. (1995). Informationstausch in politischen Netzwerken. In D. Jansen & K. Schubert (Eds.), *Netzwerke und Politikproduktion. Konzepte, Methoden, Perspektiven* (pp. 111–131). Schüren.
Pollex, J., & Berker, L.E. (2022). Parties and their environmental problem perceptions—towards a more fundamental understanding of party positions in environmental politics. *Zeitschrift für Vergleichende Politikwissenschaft, 15*(4), 571–591. https://doi.org/10.1007/s12286-022-00515-x
Povitkina, M., Alvarado Pachon, N., & Mert Dall, C. (2021). *The Quality of Government Environmental Indicators Dataset, Version Sep21*. Retrieved Nov 20, 2022, from https://www.gu.se/en/quality-government
Roberts, N. (2000). Wicked problems and network approaches to resolution. *International Public Management Review, 1*(1), 1–19.
Robertson, M. (2014). *Sustainability Principles and Practice*. Routledge.
Salamon, L.M. (2002a). The tools approach and the new governance: conclusion and implications. In L. M. Salamon (Ed.), *The Tools of Government* (pp. 300–345). Oxford University Press.
Salamon, L. M. (Ed.). (2002b). *The Tools of Government: A Guide to the New Governance*. Oxford University Press.
Santese, A. (2020). Between pacifism and environmentalism: the history of greenpeace. *USAbroad–Journal of American History and Politics, 3*(1S), 107–115. https://doi.org/10.6092/issn.2611-2752/11648
Strenze, T. (2021). Value change in the Western world: The rise of materialism, post-materialism or both? *International Review of Sociology, 31*(3), 536–553. https://doi.org/10.1080/03906701.2021.1996761
Tews, K., Busch, P.O., & Jörgens, H. (2003). The diffusion of new environmental policy instruments 1. *European Journal of Political Research, 42*(4), 569–600.
Töller, A.E., Vogelpohl, T., Beer, K., & Böcher, M. (2021). Is bioeconomy policy a policy field? A conceptual framework and findings on the European Union and Germany. *Journal of Environmental Policy & Planning, 23*(2), 152–164. https://doi.org/10.1080/1523908X.2021.1893163
Wohlgezogen, F., McCabe, A., Osegowitsch, T., & Mol, J. (2020). The wicked problem of climate change and interdisciplinary research: tracking management scholarship's contribution. *Journal of Management & Organization, 26*(6), 1048–1072. https://doi.org/10.1017/jmo.2020.14
Wurzel, R.K., Zito, A.R., & Jordan, A.J. (2013). *Environmental Governance in Europe: A Comparative Analysis of the Use of New Environmental Policy Instruments*. Edward Elgar Publishing.

3
ENVIRONMENTAL POLICY OUTPUTS, OUTCOMES, AND IMPACTS

Yves Steinebach

3.1 Introduction

Environmental protection has become highly important among both the public and academia over recent decades. Scholarship has generated a significant body of research on the analysis of the determinants and performance of environmental policies. Scholars have analyzed why some governments tend to adopt more ambitious policy measures to tackle environmental degradation and climate change than others (Knill et al., 2010; Liefferink et al., 2009; Tobin 2017. Other contributions, in turn, have investigated what makes policies succeed or fail post-enactment and to what extent policy measures ultimately achieve their intended objectives (Limberg et al., 2021; Steinebach, 2022a). Given this wide range of issues covered, it becomes clear that comparative environmental policy research is a broad and highly differentiated field of study. This chapter discusses the literature on the determinants and performance of environmental policies according to their different research foci. As discussed in more detail below, there are essentially three dimensions that are typically employed when referring to environmental policies and that involve different research questions, analytical concepts, and theoretical explanations. These dimensions are policy outputs, policy outcomes, and policy impacts (Knill & Tosun, 2020: 25; John, 2011: 4). Policy outputs are the result of the policy-making process and are equivalent to policy decisions. Policy outcomes and impacts, in turn, are different consequences that follow on from these outputs. While policy outcomes focus on whether targeted actors (citizens, corporations, administrative entities, etc.) actually change and adapt their behavior in response to environmental policies, policy impacts capture whether the respective measures ultimately make a real-world difference such as the reduction of air and water pollution or of greenhouse gas emissions.

This chapter is structured as follows. The following three sections introduce the dimensions of policy outputs, outcomes, and impacts separately. The final section connects the different dimensions and discusses the shortcomings of the literature and avenues for future research.

DOI: 10.4324/9781003043843-4

3.2 Environmental Policy Outputs

Policy outputs are the direct result of the decision-making process. They represent what political institutions – including the parliament, government, and regulatory agenices – produce. This usually involves the adoption of environmental programs, laws, or regulations. Environmental policy outputs are thus defined by the content of a public policy as it is fixed in legal or administrative documents (Knill & Tosun, 2020: 25).

Scholars that focus on environmental policy outputs as the central variable of interest typically study whether countries differ in their instrumental approach to environmental matters (Tews et al., 2003), whether some institutional setups or political parties are faster and better than others in producing ambitious environmental policies (Liefferink et al., 2009; Tobin 2017), or whether countries become more equal or unequal in respect of their environmental policy portfolios over time (Holzinger et al. 2008a, 2008b).

A central challenge for the assessment of policy outputs is that policies substantially differ from one another, both regarding their stringency and the way they intend to change the behavior of citizens and businesses. A widely accepted typology in the public policy literature has been suggested by Hall (1993) who generally distinguishes between three 'components' of policy outputs (or policy change events in Hall's case). According to this typology, it is important to differentiate between (1) the broader goals ('paradigms') that guide governments, (2) the means ('instruments') they use, and (3) the exact calibration of the instruments applied.

Following this logic, the first aspect to look at is whether government deals with a given environmental problem at all. Here, the crucial question is whether government addresses, for instance, carbon dioxide (CO_2) emissions from industry or regulates the quality of surface water. The second dimension relates to the instruments used to address the target groups of a policy. In broad terms, these instruments can be either environmental regulations such as emission standards and limit values or more market-based instruments such as emission trading and eco taxes (Wurzel et al., 2013; Gunningham & Sinclair, 1999). Another instrument type that is frequently applied in the area of environmental policy is information-based instruments such as information disclosure on environmental practices and labeling schemes (Daugbjerg et al., 2014). The third, most narrowly specified, component of environmental policies is the calibration of the instruments applied. Depending on the exact instrument type, it defines, for instance, the exact emission limits or the level of tax burden. Moreover, the scope of a given policy may vary, for example when it applies only to newly licensed cars or industrial plants rather than to all vehicles or installations being registered in a country. Table 3.1 briefly summarizes the individual policy components that can be deduced from Hall's typology and that help to capture and compare environmental policy outputs and their ambitions.

In the literature, we find multiple applications and variations of Hall's typology. Knill et al. (2012), for instance, use the terms "policy density" and "policy intensity" which relate to the number of policy targets and instruments on the one hand, and their specific calibration on the other (for a comparable approach see Holzinger et al., 2008ba, 2008b; Gravey and Jordan, 2016; Lim & Duit, 2018). Schaffrin et al. (2015), in turn, rely on a more sophisticated approach to construct their "Index of Climate Policy Activity". The authors 'weight' the changes in policy instruments depending on a range of factors. These factors include, amongst others, whether the respective instruments come together with precise objectives, that is absolute emission reduction targets, and clear instructions on how and by whom the policies should be implemented and monitored. Following a similar logic, Burns et al. (2020) gauge the stringency of European Union (EU) environmental and climate policy proposals depending on whether the adopted measures involve binding targets/limits/standards and include credible monitoring and

Table 3.1 Components of environmental policy outputs

Component	*Description*	*Example*
Policy targets	Environmental issues that the government addresses	CO_2 emissions from industry or road transport
Policy instruments	Means that the government uses to address a given environmental issue	Regulations (emission standards, technological prescription); market-based instruments (green taxes, emission trading schemes); information (information disclosure, labeling scheme)
Instrument level (instrument calibration)	Defines the exact stringency of a given policy instrument	95 g CO_2 per km; 25 euros per ton of CO_2
Instrument scope (instrument calibration)	Defines the exact scope of a given policy instrument	Newly registered vehicles after the year 2020

enforcement mechanisms (see also Burns & Tobin, 2020). Moreover, researchers have relied on more data-driven or modeling approaches (Berrang-Ford et al., 2019). Steinebach et al. (2021), for instance, propose a measure of carbon pricing policy ambition that concentrates on the type of policy instruments used, the price of a ton of CO_2, and the share of national emissions covered.[1] These different aspects are ultimately merged into a single score of policy ambition using a measurement model that combines item response theory and factor analysis. Table 3.2 provides an overview of the different scholarly contributions on environmental policy outputs. It is sorted by the broadness (coverage) of the database used. The list does not claim to be complete.

But what is the aggregate knowledge on environmental policy outputs gained through the various contributions and approaches discussed above? First and foremost, we can note that more or less all countries around the globe have expanded their environmental policy portfolios, that is the number of policy targets addressed and the number of policy instruments used over time. Sommerer and Lim (2016), for instance, show that in both Western and non-Western countries the number of policy targets addressed more than tripled since the early 1980s. Limberg et al. (2021) make a similar observation with regard to the instruments used. In addition to this, governments have continuously tightened the provisions already in place, that is, further reduced the level of emissions allowed by law and expanded the scope of their policy measures (Steinebach, 2022a). The general upward trend in increasing environmental protection efforts slightly stagnated in the late 2000s, especially in EU member states. Steinebach and Knill (2017) found "an exceptionally long four-year period of almost complete regulatory inactivity" (p. 429) in clean air and water protection at the EU level after the outbreak of the financial crisis and the multiple challenges it created. Burns et al. (2019) qualify this observation by stating that the European Commission has not *stopped* producing environmental policies but reduced its policy ambitions (see also Pollex & Lenschow, 2020), Overall, however, environmental policies are practically never scaled back once they are established and this even under rather adverse (political) conditions. The years under the Trump administration in the United States and the Bolsonaro administration in Brazil seem to be an exception to this rule (Hejny, 2018; Bomberg, 2017; Barbosa et al., 2021).

With regard to the type of environmental policy instruments used, most governments have moved from a hierarchical approach to more flexible and less prescriptive forms of intervention over time (Gunningham & Holley, 2016). In the 1970s and 1980s, most environmental policy outputs took the form of environmental regulations. By the early 1990s, policy-makers started

Table 3.2 Scholarly contributions on environmental policy outputs and data availability

Authors	*Analytical dimensions*	*Coverage*	*Policy (sub)field*	*Data availability*
Sommerer and Lim (2016)	Targets (combination of distinct targets and instruments)	37 Western and non-Western countries, 1970–2010	Air, water, nature and biodiversity, noise, soil	Available upon request
Lim and Duit (2018)	Targets (combination of distinct targets and instruments)	25 OECD countries, 1975–2005	Air, water, noise, soil	Available upon request
Holzinger et al. (2008aa, 2008b)	Targets, instruments, instrument calibration (level and scope)	24 OECD countries, 1970–2000.	Air, water, soil	https://www.polver.uni-konstanz.de/holzinger/research/research-projects/enviromental-policy-convergence-in-europe-envipolcon/project-deliverables/
Knill et al. (2012)	Targets, instruments, instrument calibration (level and scope)	24 OECD countries, 1976–2003	Air	publicpolicy-knill.org
Limberg et al. (2021)	Targets and instruments	13 countries from 1980–2010	Air, water	publicpolicy-knill.org; https://ejpr.onlinelibrary.wiley.com/doi/10.1111/1475-6765.12406
Burns et al. (2019)	Instruments, instrument intensity (objectives, scope, integration, budget, budget, implementation, monitoring)	European Union level, 12 OECD countries and Romania	Air, climate, energy, industry, biodiversity (note: the policy areas covered vary by country)	Available upon request
Schmidt & Sewerin (2019)	Instruments, instrument intensity (objectives, scope, integration, budget, budget, implementation, monitoring), technology specificity	9 OECD countries, 1996–2014	Energy	https://www.sciencedirect.com/science/article/pii/S0048733318300702
Schaffrin et al. (2015)	Instruments, instrument intensity (objectives, scope, integration, budget, budget, implementation, monitoring)	3 OECD countries, 1998–2010	Air, climate	Available upon request (but also included in Schmidt & Sewerin, 2019)
Gravey and Jordan (2016)	Targets, instruments, instrument calibration (level and scope)	European Union level, 1992–2014	Air, chemicals, industry, nature and biodiversity, waste, water	Available upon request
Burns et al. (2020)	Policy intensity (existence of binding targets/limits/standards and monitoring and enforcement mechanisms)	European Union, 2004–2014	Air, climate, biotech, nature, noise, waste, water	https://iaeepresearch.wordpress.com/research-outputs/
Steinebach et al. (2021)	Instruments (carbon tax versus emission trading), instrument calibration (level, scope, other characteristics)	200 countries, 1990–2019	Climate (CO_2 emissions)	http://xavier-fim.net/publication/climate-policy-2020-who-puts-a-price-on-carbon-global-empirical-analysis-carbon-pricing-policies/

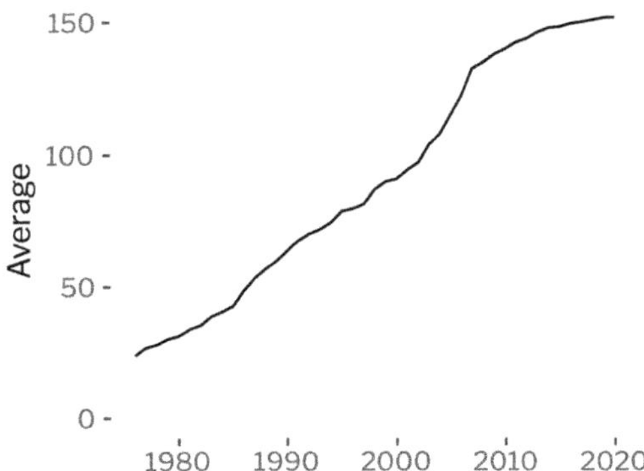

Figure 3.1 Average number of target-instrument combinations over time (21 OECD countries).

to make increasing use of market-based and voluntary initiatives. Despite the scientific "frenzy" (Jordan et al., 2005: 489) surrounding the use of 'new' environmental policy instruments, however, they have never eclipsed more traditional and hierarchical forms of environmental protection (Tews et al., 2003). Rather, governments have tended towards "regulatory pluralism" (Gunningham & Sinclair, 1999) and the use of so-called "instrument mixes" (Howlett and Rayner 2013) that aim at the same policy target but comprise multiple different policy instruments.

These broad trends in policy outputs can be illustrated by the following graphs. Figure 3.1 shows the average number of environmental policies (target–instrument combinations) in place for a sample of 21 OECD countries.[2] The environmental policy areas under scrutiny are air and water pollution as well as nature conservation. The figure reveals that the *average* national environmental policy portfolio has grown from less than 25 policy measures in the 1970s and 1980s to about 150 measures at the end of our investigation period. This equals a six-fold increase in the portfolio size for the sample of industrialized democracies.

Figure 3.2, in turn, shows the average number of policy instruments used per target over the same period and sample. It shows that in the early days of environmental policy, the underlying environmental issues were typically addressed with a rather limited tool set of about one instrument. Since the late 2000s, by contrast, each target is addressed *on average* by a mix of three different policy instruments.

3.3 Environmental Policy Outcomes

In the previous section, we intensively discussed the analytical dimension of environmental policy outputs. We now shift our focus to policy outcomes. Here, the central question is whether targeted actors change their behavior in response to environmental policies. This includes citizens and corporations (Weaver, 2014) as well as the administrative actors in charge of carrying out key implementation tasks such as local inspections and controls (Steinebach, 2022b). In the case of policies that are adopted at the international and supranational level, the outcome dimension might also involve whether domestic actors such as national governments take the necessary measures to transpose the respective policies in the national context (Bondarouk & Mastenbroek, 2018; Pollata and Newig, 2017). Depending on whether the focus is on the international or supranational institutions producing these policies or the national governments transposing them, the same policy actions are assessed once as outputs and once as outcomes.

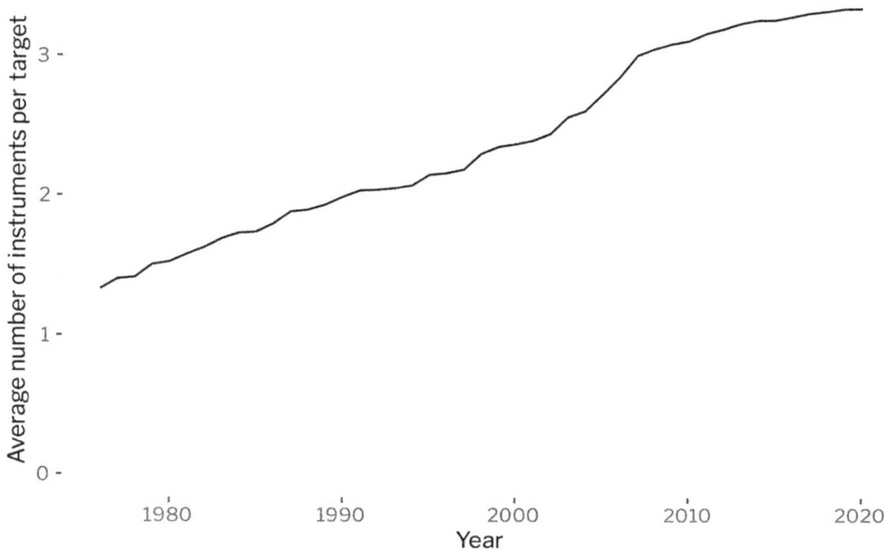

Figure 3.2 Average number of policy instruments used per target (21 OECD countries).

Scholars that focus on environmental policy outcomes as the central variable of interest typically examine whether different types of policy instruments vary in the way they change the policy addressees' behavior (Olejniczak et al., 2020), why individuals and businesses do or do not comply with environmental policies (May & Winter, 1999), and what it takes to make the respective policies work (Taylor et al., 2012).

In the context of policy outcomes, an important theoretical debate centers on the question of whether the choice of policy instruments ultimately makes a difference for the functioning of environmental policies (Somanthan et al., 2014). As discussed above, the literature broadly distinguishes between three instrument types. These are environmental regulations, economic or market-based instruments, and information-based instruments. According to Vedung (1998), these three instrument types can be described as 'sticks', 'carrots' and 'sermons' as they tend to change behavior through different mechanisms (see also Capano & Howlett, 2021). Tables 3.3 summarizes the different types of instruments used in environmental policy and presents their mode of action as well as their strengths and weaknesses.

Environmental regulations such as emission or technology standards are considered 'sticks' as they typically prescribe what is permitted (and what is not) and sanction non-compliant behavior. They thus function through the mechanism of 'command and control'. In consequence, a crucial prerequisite of regulations to exert the intended effect is that governments have the administrative capacities to make a credible threat that non-compliant behavior will be both detected and penalized (Schwartz, 2003). The central strength of environmental regulations is that they establish a quite straightforward cause–effect relationship and that, in consequence, the government can be relatively sure that the policies will achieve the envisaged behavioral changes on the side of the target group. A weakness, in turn, is that the target group has no real incentive to go beyond what is legally required from them (Baldwin et al., 2011; Gunningham et al., 2004).

Market-based instruments such as emission taxes or trading schemes are often presented as 'carrots' as they drive towards the expected behavior with economic promises. These economic

Table 3.3 Different types of instruments used in environmental policy

	Environmental regulations	*Market-based instruments*	*Information-based instruments*	*Nudges*
Mode of action	Sticks; command and control	Carrots; economic incentives	Sermons; mindful and responsible individuals	Psychological human traits
Examples	Emission standards; technological prescriptions	Taxes; emission trading scheme	Information campaign; labeling schemes	Opt-out clauses; deliberate labeling designs
Strength	Low uncertainty; straightforward cause–effect relationship	Great(er) flexibility for the target group	Contributes to the functioning of other instrument types	'Circumvent' human decision-making process
Weaknesses	Need for high enforcement capacities; no incentive for the target group to go beyond what is legally required	Require strong market competition and the existence of cheaper alternatives; potential to increase social inequalities	Are not particularly effective in the context of public good protection	No instrument in their own right; potentially 'undemocratic' and manipulative

incentives can be either positive or negative. Market-based instruments can increase the costs of the production or consumption of emission-intensive goods. Likewise, they can reward the use of environmentally friendly technologies. Market-based instruments thus alter the relative costs of goods and services, triggering (rationally acting) citizens and businesses to respond to price signals and, in this way, to change their actions towards a preferred outcome (Andersen, 1994; Nordhaus, 2007). The key benefit of market-based instruments is that they do *not* prescribe the use of specific technologies or emission reduction targets and thus provide greater flexibility to the target group in its approach to pollution management (Jenkins, 2014). Moreover, the respective production or consumption decisions taken always reflect the preferences of the involved individuals and therefore draw on information that is unlikely to be known (*ex ante*) by the regulators. Another aspect discussed in the literature is that market-based instruments are less dependent on the government's administrative capacities (Howlett et al., 2009). While this argument might be true for the direct comparison between environmental regulation and market-based instruments, it should *not* imply that market-based instruments can function without administrative capacities. As in the case of regulations, emission taxes or trading schemes need the establishment of administrative structures as well as the employment of (frontline) implementers that control the information citizens or plant operators submit (Steinebach & Limberg, 2021). A downside is that market-based instruments do only function at the 'margin'. In other words, only as long as more environmentally friendly alternatives are both available and cheaper, market-based instruments exert the intended effects (Hahn & Stavins, 1992: 465).[3] In addition, if not designed correctly, market-based instruments can have severe social consequences for parts of the population and provoke opposition (Povitkina et al., 2021). Thus, to receive broad-based public support, market-based instruments must overcome the perception that the burden imposed by the policies falls heaviest on people with the lowest incomes and the greatest need for polluting activities such as road transport (Beiser-McGrath & Bernauer, 2019). In fact, both market-based and regulatory policy instruments create costs for the target group. Yet, market-based instruments show more transparency than classical regulation, who must bear the main costs of environmental protection, and exactly how high these costs are.

Information-based instruments are typically referred to as 'sermons'. Information constitutes the least coercive of all instruments as there is no obligation to act on the details provided and individuals do *not* have to fear any sanction or financial losses. The central reason behind using information-based instruments is that human judgment is strongly based regarding knowledge and beliefs. Governments can use information campaigns or labeling schemes to inform citizens about the negative externalities of their actions (e.g. the effects on CO_2 emissions, animal welfare). The 'mode of action' of instrument-based instruments is thus that mindful and responsible individuals rethink and change their behavior patterns once confronted with additional and formerly unknown information. A strength of information is that it can be used as both an independent policy measure as well as a "metapolicy instrument that supports others" (Vedung, 1998: 48). Citizens, for instance, will find it easier to change their behavior in response to a green tax if they know about the environmental consequences of their consumption decisions. A weakness is that information has proven to function (surprisingly) well when the information provided directly affects the immediate self-interest and well-being of individuals (smoking, nutrition, etc.) but makes less of a difference when the information provided aims at the protection of public goods such as the state of the climate or the environment (Pal, 2014: 139). Moreover, the impact on behavior is strongly determined by the exact design of the information provision scheme (see discussion below on *nudging*). Haq and Weiss (2016), for instance, show that national labeling schemes on the CO_2 emission of passenger vehicles significantly differ from one another and that most of them could be improved by a better placement and readability of the provided details.

In recent years and given the limited success of the attempts at changing behavior by education and information, other ways to accomplish behavioral change have been explored. In this context, behavioral intervention, so-called nudges, have entered the toolbox of governments (Sunstein, 2011). In contrast to the instrument types discussed above, nudges do *not* change the options themselves nor the costs and benefits associated with them. Rather, they seek to 'nudge' individuals towards a preferred option through changes in the choice environment and exploiting psychological human traits such as the status quo bias (the tendency of individuals to prefer the current state over potential changes) or anchoring (the tendency of individuals to base their judgment on a particular reference point) (Carlsson et al., 2019). Ebeling and Lotz (2015), for instance, show that consumers are more likely to use green energy sources if they have to explicitly decide against this (opt-out) option. Likewise, Heinzle and Wüstenhagen (2012) find that consumers are less sensitive to the scale of the revised EU energy label (A+++ to D) compared to the old one (A to G) as the letter A functions as an anchor for consumers' judgment of energy efficiency. In consequence, the categories extended with plusses (A+, A++, A+++) are not perceived as real improvements regarding the devices' energy efficiency. This observation was also the reason why the latest reform of the EU energy label has rescaled the labels to the original A to G (European Commission, 2021). A benefit of nudges is that they 'circumvent' decision-making processes and thus more or less *directly* trigger behavioral changes. A limitation of nudges is that they are actually *not* an instrument in their own right but usually have to be incorporated in the design of existing instruments such as taxes or information-based instruments. They thus might suffer from the same (dis)advantages as their respective 'host' instrument. Moreover, there are concerns that nudges are manipulative and 'undemocratic' as they appeal to automatic cognitive biases and, in this way, bypass citizens' rational deliberation processes (Wilkinson, 2013).

Next to the choice of the (right) policy instrument, there also several other factors that determine whether the target group ultimately changes its behavior or not. By and large, Winter and May (2001) identify three reasons (motivations) for (non-)compliance when it comes to environmental policy provisions. The first reason is "calculated motivation" (Winter & May, 2001: 676) for compliance. Here, citizens and businesses will only follow a given environmental regulation when they conclude that the benefits of compliance exceed the costs of compliance. This cost–benefit calculation, in turn, is determined by the expected risk of getting 'caught' and the severity of punishments (fines, sanctions, etc.). The second reason is "social motivation" (Winter & May, 2001: 678). This type of motivation implies that those being targeted by a given regulation want to earn the respect of the people with whom they interact. The third reasons are "normative motivations" (Winter & May, 2001: 677). Individuals are more likely to accept and follow rules that they consider fair and reasonable.

The first two aspects – calculated and social motivations – are primarily influenced by environmental enforcement authorities and their actions. Only when the administration possesses sufficient resources to also engage in frequent inspections, citizens and businesses do consider the chance of getting caught and being is greater than the cost of compliance. The social motivation to comply, in turn, mainly depends on the way the administrators treat and interact with the target group (May & Wood, 2003; Nielsen, 2015). A formalized and professionalized 'enforcement style' can be expected to positively affect compliance but might also 'backfire' if the rules are perceived as too strict or demanding. In this latter scenario, a more flexible approach towards implementation and enforcement might lead to a greater social acceptance (Winter & May, 2001). The normative motivation to comply, in turn, is strongly determined by the exact policy design. Bernauer et al. (2020), for instance, find that the support for environmental policies is substantially reduced when there are too many exemptions to the regulations. Likewise, high levels of complexity and interactions between different policy measures have the potential to undermine the support for environmental policy (Adam et al., 2019; but see Fesenfeld, 2020).

3.4 Environmental Policy Impacts

Policy impacts focus on the extent to which policy decisions and their subsequent implementation and enforcement have actually brought about the expected results (Knill & Tosun, 2020: 26). The difference to the policy outcome dimension is that the focus is not (only) on whether the targeted actors have changed their behavior in response to policy outputs but also on whether the measures adopted ultimately make a real-world difference such as the reduction of air and water pollution or the improvement of environmental quality (Jahn, 2016).

There are several large-N studies that focus on environmental impacts (Fiorino, 2011). The vast majority of these studies uses politico-institutional features such as the regime type (*democracies versus autocracies*) (Li & Reuveny, 2006; Bättig & Bernauer, 2009; Povitkina, 2018), the type of government (*concentrated versus diffused governmental power*) (Wälti, 2004; Poloni-Staudinger, 2008), or the system of interest intermediation (*pluralism versus (neo-) corporatism*) (Scruggs, 1999; 2001) to explain differences in a country's environmental quality (see Chapter 7 by Jahn and Klagges). Whilst these analyses have done great work in improving our understanding of political and institutional drivers of environmental performance, they actually 'skip' the stages of policy formulation and implementation. In consequence, the studies can say but little about whether the observed differences are due to a better policy design and instrument choice or are caused by varying implementation capacities and procedures.

The vast majority of contributions that (really) use environmental policies as the central factors in the explanation of environmental changes typically deal with the influence of a specific policy measure in a single country (see, for instance, Cheng et al., 2017; Wang et al. 2019).

These works are definitely helpful to assess the impact of certain environmental measures in a given spatial context. Yet, they hardly produce generalizable results beyond the individual cases under scrutiny. There are only a few impact-oriented studies, in turn, that take on a more comprehensive perspective on the impacts of environmental policies and that are cross-sectional in scope. The central insight of these studies is that there is no unconditional or 'automatic' link between environmental policy outputs and impact (Knill et al., 2012; Nicolli et al., 2012; Sewerin, 2013). In other words, when a government produces (more) policy outputs, a country's environmental quality is not necessarily improved.

The literature identifies two factors that reduce the "slippage between [environmentally] regulatory policy outputs and impacts" (Knill et al., 2012: 441). Limberg et al. (2021) find that policies can indeed improve a country's environmental performance – but only if they are matched by a simultaneous expansion in administrative capacity. New policy measures must thus go hand in hand with additional capacities, that is the resources needed in applying, monitoring, and enforcing the respective policies, to make a real-world difference. Fernández-i-Marín et al. (2021), in turn, find that the effectiveness of environmental policy portfolios is higher if governments tend to develop 'tailored' policy solutions. Such solutions imply that governments do *not* use the ever-same instruments and instrument combinations across the various environmental problems they are dealing with but deliberately try to 'diversify' their instrument mixes.

Overall, it must be emphasized that scholarly contributions dealing with the real-world impact of policy measures do often find it difficult to establish causality. Environmental performance changes caused by factors other than existing policies are often incorrectly attributed to policies as impact. Conversely, a policy may have an impact but is then offset by other factors. Therefore, it is important to say that changes in environmental performance changes are not necessarily identical with impact, although it is often presented as such in everyday politics and debates. Therefore, it is not only difficult to attribute impact to policy design or to the quality of implementation processes though there is also a fundamental difficulty in identifying impacts as such in the first place.

3.5 Conclusion and Future Research

The central aim of this chapter has been to offer a systematic perspective on environmental policy research and the different aspects investigated by environmental policy scholarship. We have learned that the literature has made substantial progress in (1) mapping the different governmental responses to environmental issues (policy outputs); (2) identifying the aspects that make environmental policies succeed or fail in altering the target group's behavior (policy outcomes); and (3) assessing whether policies have ultimately managed to solve the underling environmental problem or not (policy impacts). Despite this progress made, however, there are still some shortcomings and thus room for improvement.

First, with regard to *policy outputs*, there is still a strong focus on the analysis and comparison of advanced industrialized democracies. Economically less developed countries or autocratic governments are typically less a part of comparative assessments (but see Sommerer & Lim, 2016; Tosun, 2013). This is surprising given that environmental issues matter all over the world. In the context of climate policy, autocracies such as China, Russia, or Saudi Arabia do play a particularly important role in combatting climate change (Eaton & Kostka, 2014). Future research on environmental policy outputs should thus move "beyond the usual suspects" and assess whether the knowledge gained on policy outputs also holds in other spatial and cultural contexts (Death, 2016).

Second, concerning the *policy outcome* dimension, governments and scholars face the challenge that environmental policy issues are addressed by an ever-growing number of different policy targets and instruments. While the relevance of positive and negative interactions between different elements within the policy mix is well-known and acknowledged in the literature (Howlett & Rayner, 2013), the dominant analytical focus is usually placed on a pair-wise comparison of different instrument combinations (Gunningham & Sinclair, 1999; Gunningham et al., 2004). Yet, Adam et al. (2018) show that the functioning of public policies is substantially different when there are three or four policy instruments addressing the same societal issue simultaneously. In consequence, and to reduce or prevent the unwanted consequences of "overcongested policy spaces" (Mulder, 2015: 154), scholars should invest more efforts in identifying the most effective environmental policy mixes including multiple instrument combinations.

Lastly, we need a dynamic perspective on *policy impacts*. A crucial insight from the research on policy impacts is that environmental policy measures can only make a real-world difference if the measures taken are also backed by administrative capacities (Limberg et al., 2021; VanDeveer & Dabelko, 2001). However, there is ample evidence that the connection also works the other way around, that is, that additional policy measures might easily overburden the administration and in this way lead to an overall worse rather than better policy performance (Limberg et al., 2021). Identifying the right balance between the environmental policy stock-up for implementation and the administrative capacities available as well as potential 'tipping points' seems highly important to ensure the long-term problem-solving capacity of the state in environmental and climate matters.

Box 3.1 Chapter Summary

- The number of environmental policies in place has more or less constantly grown over recent decades.
- Environmental policies are practically never scaled back once they are established, even under rather adverse (political) conditions.
- Policy instruments have moved from a hierarchical approach to more flexible and less prescriptive forms of intervention over time.
- All environmental policy instruments have their strengths and weaknesses, and thus must be combined to achieve the highest level of policy effectiveness.
- Environmental policies vary in their effectiveness depending on their policy design and the level of resource provision for policy implementation.

Notes

1 In the case of emission trading schemes, the authors also assess whether there is a national cap on greenhouse gas emissions, and if so, the annual percentage rate at which this cap is decreasing.
2 The data are not yet published but available on request. Please consult the author.
3 Obviously, marked-based instruments can also foster innovation. In the United States, for instance, carbon-sequestration tax incentives have massively increased the investment in carbon storage technologies (Anderson et al., 2021).

References

Adam, C., Hurka, S., Knill, C., & Steinebach, Y. (2019). *Policy Accumulation and the Democratic Responsiveness Trap*. Cambridge: Cambridge University Press.

Adam, C., Steinebach, Y., & Knill, C. (2018). "Neglected challenges to evidence-based policy-making: The problem of policy accumulation." *Policy Science*, 51, 269–290.

Anderson, J.L., Rode, D., Zhai, H., & Fischbeck, P. (2021). "A techno-economic assessment of carbon-sequestration tax incentives in the U.S. power sector." *International Journal of Greenhouse Gas Control*, 111, 1–16.

Andersen, M. S. (1994). *Governance by green taxes: Making pollution prevention pay*. Manchester: Manchester University Press.

Barbosa, L.G., Alves, G.M.A.S., & Grelle, C.E.V. (2021). "Actions against sustainability: Dismantling of the environmental policies in Brazil." *Land Use Policy*, 104, 1–4.

Baldwin, R., Cave, M., & Lodge, M. (2011). *Understanding Regulation: Theory, Strategy, and Practice*. Oxford: Oxford University Press.

Bättig, M.B., & Bernauer, T., (2009). "National institutions and global public goods: Are democracies more cooperative in climate change policy?" *International Organization*, 63(2), 281–308.

Beiser-McGrath, L.F., & Bernauer, T. (2019). "Could revenue recycling make effective carbon taxation politically feasible?" *Science Advances*, 5(9), 1–8.

Bernauer, T., Prakash, A., & Beiser-McGrath, L.F. (2020). "Do exemptions undermine environmental policy support? An experimental stress test on the odd-even road space rationing policy in India." *Regulation & Governance*, 14, 481–500.

Berrang-Ford, L., Biesbroek, R., Ford, J. D. et al. (2019). "Tracking global climate change adaptation among governments." *Nature Climate Change*, 9, 440–449.

Bomberg, E. (2017). "Environmental politics in the Trump era: An early assessment." *Environmental Politics*, 26(5), 956–963.

Bondarouk, E., & Mastenbroek, E. (2018). "Reconsidering EU compliance: Implementation performance in the field of environmental policy." *Environmental Policy & Governance*, 28, 15–27.

Burns, C., Eckersley, P., & Tobin, P. (2020). "EU environmental policy in times of crisis." *Journal of European Public Policy*, 27(1), 1–19.

Burns, C., & Tobin, P. (2020). "Crisis, climate change and comitology: Policy dismantling via the back-door?" *JCMS: Journal of Common Market Studies*, 58, 527–544.

Burns, C, Tobin, P, & Sewerin, S. (Eds.) (2019). *The Impact of the Economic Crisis on European Environmental Policy*. Oxford: Oxford University Press, 2018.

Capano, G., & Howlett, M. (2021). "Causal logics and mechanisms in policy design: How and why adopting a mechanistic perspective can improve policy design." *Public Policy and Administration*, 36(2), 141–162.

Carlsson, F., Gravert, C. A., Kurz, V., & Johansson-Stenman, O. (2019). *Nudging as an Environmental Policy Instrument* (April 25, 2019). CeCAR Working Paper Series No. 4, Available at SSRN: https://ssrn.com/abstract=3711946 or http://dx.doi.org/10.2139/ssrn.3711946

Cheng, Z., Li, L., & Liu, J. (2017). "The emissions reduction effect and technical progress effect of environmental regulation policy tools." *Journal of Cleaner Production*, 149, 191–205.

Daugbjerg, C., Smed, S., Andersen, L. M., & Schvartzman, Y. (2014). "Improving eco-labelling as an environmental policy instrument: Knowledge, trust and organic consumption." *Journal of Environmental Policy & Planning*, 16(4), 559–575.

Death, C. (2016). "Green states in Africa: beyond the usual suspects." *Environmental Politics*, 25(1), 116–135.

Eaton, S., & Kostka, G. (2014). "Authoritarian environmentalism undermined? Local leaders' time horizons and environmental policy implementation in China." *The China Quarterly*, 218, 359–380.

Ebeling, F., & Lotz, S. (2015). "Domestic uptake of green energy promoted by opt-out tariffs." *Nature Climate Change*, 5(9), 868–871.

European Commission. (2021). New EU energy labels applicable from 1 March 2021. Available at: https://ec.europa.eu/commission/presscorner/detail/en/ip_21_818

Fernández-i-Marín, X., Knill, C., & Steinebach, Y. (2021). "Studying policy design quality in comparative perspective." *American Political Science Review*, 115(3), 931–947.

Fesenfeld, L.P. (2020). The Effects of Policy Design Complexity on Public Support for Climate Policy. https://ssrn.com/abstract=3708920 or http://dx.doi.org/10.2139/ssrn.3708920

Fiorino, D.J. (2011). "Explaining national environmental performance: approaches, evidence, and implications." *Policy Science*, 44, 367–389.

Gravey, V., & Jordan, A. (2016). "Does the European Union have a reverse gear? Policy dismantling in a hyperconsensual polity." *Journal of European Public Policy*, 23(8), 1180–1198.

Gunningham, N., Grabosky, P.N., & Sinclair, D. (2004). *Smart Regulation: Designing Environmental Policy*. Oxford: Oxford University Press.

Gunningham, N., & Holley, C. (2016). "Next-generation environmental regulation: Law, regulation, and governance." *Annual Review of Law and Social Science*, 12(1), 273–293.

Gunningham, N., & Sinclair, D. (1999). "Regulatory pluralism: Designing policy mixes for environmental protection." *Law & Policy*, 21(1), 49–76

Hahn, Robert W., & Robert N. Stavins. (1992). "Economic incentives for environmental protection: Integrating theory and practice." *The American Economic Review*, 82(2), 464–468.

Hall, P.A. (1993). "Policy paradigms, social learning, and the state: The case of economic policymaking in Britain." *Comparative Politics*, 25(3), 275–296.

Haq, Gary, & Martin Weiss (2016). "CO2 labelling of passenger cars in Europe: Status, challenges, and future prospects." *Energy Policy*, 95, 324–335.

Heinzle, S.L., & Wüstenhagen, R. (2012). "Dynamic adjustment of eco-labeling schemes and consumer choice – The revision of the EU energy label as a missed opportunity?" Bus. Strat. Env., 21, 60–70.

Hejny, J. (2018). "The Trump Administration and environmental policy: Reagan redux?" *Journal of Environmental Studies and Sciences*, 8, 197–211.

Holzinger, K., Knill, C., & Arts, B. (2008b). *Environmental Policy Convergence in Europe: The Impact of International Institutions and Trade*. Cambridge: Cambridge University Press.

Holzinger, K., Knill, C., & Sommerer, T. (2008a). "Environmental policy convergence: The impact of international harmonization, transnational communication, and regulatory competition." *International Organization* 62(4), 553–587.

Howlett, M., & Rayner, J. (2013). "Patching vs packaging in policy formulation: Assessing policy portfolio design." *Politics and Governance*, 1(2), 170–182.

Jahn, D. (2016). *The Politics of Environmental Performance*. Cambridge, UK: Cambridge University Press.

Jenkins, J.D. (2014). "Political economy constraints on carbon pricing policies: What are the implications for economic efficiency, e nvironmental efficacy, and climate policy design?" *Energy Policy*, 69, 467–477.

John, P. *Making Policy Work*. London: Routledge, 2011.

Jordan, A., Wurzel, R.K.W. & Zito A. (2005). "The rise of 'New' Policy instruments in comparative perspective: Has governance eclipsed government?" *Political Studies*, 53(3), 477–496.

Knill, C., Debus, M., & Heichel, S. (2010). "Do parties matter in internationalised policy areas? The impact of political parties on environmental policy outputs in 18 OECD countries, 1970–2000." *European Journal of Political Research*, 49, 301–336.

Knill, C., Schulze, K. & Tosun, J. (2012). "Regulatory policy outputs and impacts: Exploring a complex relationship." *Regulation & Governance* 6(4), 427–444.

Knill, C., & Tosun, J. (2020). *Public Policy: A New Introduction*. London: Red Globe Press.

Li, Q., & Reuveny, R. (2006). "Democracy and environmental degradation." *International Studies Quarterly*, 50(4), 935–956.

Liefferink, D., Arts, B., Kamstra, J., & Ooijevaar, J. (2009). Leaders and laggards in environmental policy: A quantitative analysis of domestic policy outputs, *Journal of European Public Policy*, 16(5), 677–700.

Lim, S., & Duit, A. (2018). "Partisan politics, welfare states, and environmental policy outputs in the OECD countries, 1975–2005." *Regulation & Governance*, 12, 220–237.

Limberg, J., Steinebach, Y., Bayerlein, L., & Knill, C. (2021). "The more the better? Rule growth and policy impact from a macro perspective." *European Journal of Political Research*, 60, 438–454.

May, P., & Winter, S. (1999). "Regulatory enforcement and compliance: Examining Danish agro-environmental policy." *Journal of Policy Analysis and Management*, 18, 625–651.

May, P.J., & Wood, R.S. (2003). "At the regulatory front lines: Inspectors' enforcement styles and regulatory compliance." *Journal of Public Administration Research and Theory*, 13(2), 117–139.

Mulder, A.J. (2015). Interaction between EU instruments and member-state instruments: The end of CO_2 emissions trading in Europe? In: Gronwald, M., & Hintermann, B. (Eds.). *Emissions Trading as a Policy Instrument: Evaluation and Prospects*. Camdridge: MIT Press, pp. 149–180.

Nicolli, F., Mazzanti M., & Iafolla, V. (2012). "Waste dynamics, country heterogeneity and European environmental policy effectiveness." *Journal of Environmental Policy & Planning*, 14(4), 371–393.

Nielsen, V.L. (2015). "Law enforcement behaviour of regulatory inspectors. In understanding street-level bureaucracy" In: Hupe, P., Hill, P., & Buffat, A. (Eds). *Understanding Street-Level Bureaucracy*. Chicago: Chicago: Policy Press, pp. 115–131.

Nordhaus, W.D. (2007). "To tax or not to tax: Alternative approaches to slowing global warming." *Review of Environmental Economics and Policy*, 1(1), 26–26.

Olejniczak, K., Śliwowski, P., & Leeuw, F. (2020). "Comparing behavioral assumptions of policy tools: Framework for policy designers", *Journal of Comparative Policy Analysis: Research and Practice*, 22 6, 498–520.

Pal, Leslie. (2014). *Beyond policy analysis – Public issue management in turbulent times* (5th ed.), Toronto: Nelson Education.

Pollata, J.A.M., & Newig, J. (2017). "Policy implementation through multi-level governance: Analysing practical implementation of EU air quality directives in Germany. *Journal of European Public Policy*, 24(9), 1308–1327.

Pollex, J. & Lenschow, A. (2020). "Many faces of dismantling: Hiding policy change in non-legislative acts in EU environmental policy." Journal of European Public Policy 27(1), 20–40.

Poloni-Staudinger, L.M. (2008). Are consensus democracies more environmentally effective?, *Environmental Politics*, 17(3), 410–430.

Povitkina, M. (2018). "The limits of democracy in tackling climate change." *Environmental Politics*, 27(3), 411–432.

Povitkina, M., Jagers, S.V., Matti, S. & Martinsson, J. (2021). Why are carbon taxes unfair? Disentangling public perceptions of fairness, *Global Environmental Change*, 70(102356), 1–14.

Schaffrin, A., Sewerin, S., & Seubert, S. (2015). "Toward a comparative measure of climate policy output." *Policy Studies Journal*, 43, 257–282.

Schmidt, T.S., & Sewerin, S. (2019). "Measuring the temporal dynamics of policy mixes – An empirical analysis of renewable energy policy mixes' balance and design features in nine countries." *Research Policy*, 48(1), 1–13.

Schwartz, J. (2003). "The impact of state capacity on enforcement of environmental policies: The case of China." *The Journal of Environment & Development*, 12(1), 50–81.

Scruggs, L. (1999). "Institutions and environmental performance in seventeen Western democracies." *British Journal of Political Science*, 29(1), 1–31.

Scruggs, L. (2001). "Is there really a link between neo-corporatism and environmental performance? Updated evidence and new data for the 1980s and 1990s." *British Journal of Political Science*, 31(4), 686–692.

Sewerin, S. (2013). *Comparative Climate Politics: Patterns of Climate Policy Performance in Western Democracies. PhD Thesis*. https://kups.ub.uni-koeln.de/5513/1/Sewerin_Comparative_Climate_Politics.pdf

Somanathan, E., Sterner, T., & Sugiyama, T. (2014). National and sub-national policies and institutions. In: Edenhofer, O., Pichs-Madruga, R., Sokona, Y., Farahani, E., Kadner, S., & Seyboth, K. (Eds.), *Climate change 2014: Mitigation of climate change. Contribution of working group III to the fifth assessment report of the intergovernmental panel on climate change*. Cambridge: Cambridge University Press.

Sommerer, T., & Lim, S. (2016). "The environmental state as a model for the world? An analysis of policy repertoires in 37 countries." *Environmental Politics*, 25(1), 92–115.

Steinebach, Y. (2022a). "Instrument choice, implementation structures, and the effectiveness of environmental policies: A cross-national analysis." *Regulation & Governance*, 15, 225–242.

Steinebach, Y. (2022b). "Administrative traditions and the effectiveness of regulation." *Journal of European Public Policy*, online first, 1–20.

Steinebach, Y., Fernández-i-Marín, X., & Aschenbrenner, C. (2021). "Who puts a price on carbon, why and how? A global empirical analysis of carbon pricing policies." *Climate Policy*, 21(3), 277–289.

Steinebach, Y. & Limberg, J. (2021). "Implementing market mechanisms in the Paris era: The importance of bureaucratic capacity building for international climate policy", *Journal of European Public Policy*, online first, 1153–1168

Steinebach, Y., & Knill, C. (2017). Still an entrepreneur? The changing role of the European Commission in EU environmental policy-making, *Journal of European Public Policy*, 24(3), 429–446.

Sunstein, C.R. (2011). *Why Nudge?* New Haven: Yale University Press.

Taylor, C., Pollard, S., Rocks, S., & Angus, A. (2012). "Selecting policy instruments for better environmental regulation: A critique and future research Agenda." *Environmental Policy & Governance*, 22, 268–292.

Tews, K., Busch, P.-O., & Jörgens, H. (2003). "The diffusion of new environmental policy instruments." *European Journal of Political Research*, 42, 569–600.

Tobin, P. (2017). "Leaders and Laggards: Climate Policy Ambition in Developed States." *Global Environmental Politics*, 17(4): 28–47.

Tosun, J. (2013). *Environmental Policy Change in Emerging Market Democracies: Eastern Europe and Latin America Compared*. Toronto: Toronto University Press.

VanDeveer, S., & Dabelko G.D. (2001). "It's capacity, stupid: International assistance and national implementation." *Global Environmental Politics*, 1(2), 18–29.

Vedung, E. (1998). "Policy instruments: Typologies and theories" In: Bemelmans-Videc, M.L., Rist, R., & Vedung, E. (Eds). *Carrots, Sticks, and Sermons: Policy Instruments and Their Evaluation*, Transaction: New Brunswick, NJ, pp. 21–25.

Wälti, S. (2004). "How multilevel structures affect environmental policy in industrialized countries." *European Journal of Political Research*, 43(4), 599–634.

Wang, K., Yin, H., & Chen, Y. (2019). "The effect of environmental regulation on air quality: A study of new ambient air quality standards in China." *Journal of Cleaner Production, 215*, 268–279.

Weaver, R.K. (2014). "Compliance regimes and behavioral change." *Governance*, 27, 243–265.

Wilkinson, T.M. (2013). "Nudging and manipulation." *Political Studies*, 61(2), 341–355.

Winter, S.C., & May, P.J. (2001). "Motivation for compliance with environmental regulations." *Journal of Policy Analysis and Management*, 20, 675–698.

Wurzel, R.K.W., Zito, A.R., & Jordan, A.J. (2013). *Environmental Governance in Europe: A Comparative Analysis of the Use of New Environmental Policy Instruments*. Edward Elgar Publishing.

4
THE ENVIRONMENTAL STATE

Andreas Duit

4.1 Introduction

In the early 2000s, environmental policy and politics scholars started exploring how the state has responded to environmental problems. A core impetus for this research program was, at the time, the conviction that the state had, with only a few exceptions, been overlooked in contemporary debates on how to address ongoing processes of global environmental change (Paterson, 2016). The state's response to environmental problems was seen not only as a research gap to be filled, but also as a way of unlocking the potential of the state in providing more progressive and ambitious environmental mitigation efforts. In the two decades that ensued, a small but lively body of scholarly work has emerged.

In this chapter, my ambition is to first provide a brief overview of extant scholarly work on the environmental state. I will then go on to identify some blank spots in our current understanding of the environmental state, and finish off by suggesting a few outstanding research issues. More specifically, I develop two arguments about the future research agenda for environmental state research. The first argument is that environmental state scholars should reassess their theoretical starting points and instead develop a model of the political economy of the environmental state as distinct from that of the welfare state. The second argument is that the environmental state literature is at risk of getting bogged down in definitional and conceptual debates that are unlikely to be resolved without more empirical research, and that such empirical research should primarily focus on two key issues. The first issue is uncovering the mechanisms that determine the scope and consequences of state involvement in environmental matters – what are the drivers of the expansion of the environmental state? The second key issue is to determine if and when state environmental policy interventions are effective in mitigating environmental problems – does the environmental state matter?

4.2 A Brief Summary of Environmental State Research

The exact starting point of the evolution of a scholarly research agenda focusing on the relationship between the state and the environment is difficult to pinpoint. There is no single foundational text or seminal study, but rather a gradually growing interest in how state–environmental interactions are structured and, more importantly, related to the fight against environmental

DOI: 10.4324/9781003043843-5

degradation. Some early works that display an interest in the state's relationship to environmental problems include Martin Jänicke and Helmut Weidner's work on capacity building (Jänicke and Weidner, 1997; Weidner and Jänicke, 2002), Lennart J. Lundqvist's study of the Swedish case (2001), as well as a 2002 paper in which John Dryzek and co-authors argued that environmental conservation is emerging as a new core state imperative in modern states, and illustrated this process in a case study of the USA, Germany, and Norway (Dryzek et al., 2002).

The beginning of a more focused scholarly debate around the notion of the environmental state can be found in Robyn Eckersley's *The Green State: Rethinking Democracy and Sovereignty* (2004). Almost at the same point in time, John Barry and Robyn Eckersley published an edited volume *The State and the Global Ecological Crisis* (2005), which consisted of a collection of theoretical and empirical chapters on the role of the state in mitigating environmental problems. James Meadowcroft's contribution to this volume, a chapter with the title "From Welfare State to Ecostate" (Meadowcroft, 2005), along with some of his later writings (Meadowcroft, 2012; Gough and Meadowcroft, 2011), have been especially influential in shaping the subsequent scholarly debate. About a decade later, two edited volumes by Bäckstrand and Kronsell (2015) and myself (Duit, 2014) were published, followed by a special issue in *Environmental Politics* (Duit et al., 2016). The most recent major publication on the environmental state is another special issue in *Environmental Politics*, edited by Hausknost and Hammond (2020).

Although the environmental state literature does have a considerable theoretical slant, there is nevertheless a handful of empirical studies as well. The perhaps richest empirical study of the environmental state is Detlef Jahn's monograph *The Politics of Environmental Performance* (2016), in which the determinants of state involvement in environmental matters are investigated. Thomas Sommerer and Sijeong Lim's paper (2016) provides a comprehensive mapping of environmental states in a comparative perspective. Carl Death's book *The Green State in Africa* (2016) offers a rich exposé of how African states have created capacities for environmental management. In addition, the edited volumes by Bäckstrand and Kronsell (2015) and myself (Duit, 2014) contain several empirical chapters, although most of them are descriptive rather than inferential.

4.3 Environmental State, Green State, Ecological State, or Ecostate?

There are several different terms in the contemporary debate that are seemingly used to denote more or less the same underlying construct. For instance, Meadowcroft (2005) uses the term "ecostate", Eckersley (2004), Barry and Eckersley (2005), and Bäckstrand and Kronsell (2015), in turn, discuss the notion of a "green state", and Duit et al. (2016) and Hausknost and Hammond (2020) prefer the term "environmental state". However, a closer inspection reveals that these terms are in fact building on some fairly different assumptions with substantial analytical implications. The main dividing line goes between, on the one hand, *descriptive* conceptualizations that primarily seek to provide a foundation for empirical research, and on the other hand, *prescriptive* conceptualizations that are intended to include a normative and often forward-looking definition of what an environmentally sustainable state would (and should) look like (cf. Bäckstrand and Kronsell, 2015).

Robyn Eckersley's notion of the "Green State" (2004) is a good example of a more prescriptive conceptualization, in the sense that she explicitly sets out to identify what a state would look like and do if it were to have ecological sustainability as its core imperative. In contrast, the term "environmental state" is most commonly used in the descriptive sense to refer to those aspects of already existing states that have evolved in response to environmental problems. The environmental state concept does not include any assumptions about present and potential performance of the state and is simply a short-hand term for those functions of the modern state that are directly and sometimes indirectly involved in carrying out environmental policy. As such, it is a decidedly non-normative and non-teleological concept with the primary function of facilitating empirical research. In this vein, James Meadowcroft, Peter Feindt, and I defined the environmental state as "a state that possesses a significant set of institutions and practices dedicated to the management of the environment and societal– environmental interactions" (Duit et al., 2016, p. x).

It should be clear that environmental states in the descriptive sense abound among today's nations, but it is less certain that the type of state envisioned in the prescriptive sense will ever see the light of day. With very few exceptions, most contemporary states have developed organizational, distributional, ideational, and regulatory responses to environmental problems, albeit with large variations in scope, resources, and ambitions (Duit, 2016). However, the fact that all states are dealing with environmental problems does not necessarily mean that environmental management is at the heart of what a state does, as a central part of its *raison d'être*. Few would claim that any state comes close to being an ecostate in this sense, but a counter-argument is that it is doubtful whether the welfare state ever was the dominating justification for the modern state. In sum, while most environmental state scholars agree on certain features of their object of study, there is no universally accepted definition of what an environmental state is. However, to continue the analogy with the welfare state literature, it can be noted that there is, to my knowledge, no widely accepted or used definition of the welfare state either.

4.4 Where to Find the Environmental State

The environmental state can be given a temporal and geographical location. In terms of the point in time at which the environmental state first appeared, it is an established fact that most industrialized Western countries experienced an environmental awakening at the end of the 1960s and at the beginning of the 1970s. This awakening consisted in a "discovery" of environmental problems as a fundamental and systemic dysfunctionality of modern societies, as well as in a subsequent first wave of organizational and regulative responses, mainly aimed at addressing point-source emissions (Weidner and Jänicke, 2002). In other words, it was at this point in time that states started to use their organizational and regulative powers to produce environmental public goods (albeit on a very modest scale at first).

Identifying where in the world the "first" environmental state emerged is somewhat trickier, even though the literature has established various suggestions for environmental "pioneers" such as Sweden, the USA, and Japan (Lundqvist, 1980; Jänicke and Weidner, 1997; Weidner and Jänicke, 2002; Knill et al., 2012)

The focus in the literature on countries in the global north as pioneers of environmental policy does not mean that environmental states are non-existent in other parts of the world. Empirical studies on the global scale indeed find that environmental capacity has increased in all parts of the world, and industrializing countries often do more than just emulate green pioneers with a certain delay of adoption (Weidner and Jänicke, 2002). For instance, when examining 11 Central and Eastern European countries and 17 Latin American countries in five environmental policy areas, Jale Tosun finds an upward environmental regulatory trend in most countries in her sample (Tosun, 2013). Knill et al.'s comparative study of Mexico and Hungary also finds that both countries' national environmental regulations have become more stringent over time and this despite pressures from regulatory competition (Knill et al., 2008). In a similar vein, Sommerer and Lim find ample evidence of expanding environmental policy portfolios in non-Western countries, as well as declining growth of regulatory expansion in countries traditionally considered as environmental pioneers (Sommer and Lim, 2016). Carl Death's work on the "Green State in Africa" provides another in-depth illustration of how the state's involvement in environmental matters does not necessarily have to follow Western blueprints (Death, 2016). In sum, empirical studies on the global scale indeed find that environmental state capacity has increased in all parts of the world.

4.5 What Can the Environmental State Do?

A key dividing issue in scholarly debates on the environmental state is over its capacity to successfully address environmental problems. One strand of environmental state scholars holds guardedly positive views of the prospect engaging the state in environmental problem solving, often pointing to the perceived success of the welfare state in addressing social problems (Meadowcroft, 2005; Duit et al., 2016; Bäckstrand and Kronsell, 2015), with the implicit assumption that a similar application of state capacity is needed to address environmental problems. Much of these hopes are pinned on the state's unique position as a third-party regulator of the market economy, with the power of changing the rules governing market interactions (Christoff, 2005). Others point to the fact that only states have the organizational capacity to monitor the state of the environment, how market actors use and abuse natural systems, and to sustain the scientific research needed to understand society's impact on natural systems (Duit et al., 2016).

Other accounts have been skeptical about the prospects of the state ever playing a constructive part in addressing environmental externalities (Barry & Eckersley, 2005). There have been three main sources of skepticism. A first and most prominent objection has to do with how environmental issues tend to become systematically disregarded in favor of economic growth in a political system resting on the dual pillars of representative democracy and the market economy. This issue has been a key concern from the very beginning of the environmental debate in the late 1960s. Early writers such as William Ophuls advocated various forms of autocratic solutions to this problem, but more recent scholarship has explored the potential of deliberative and participatory alternatives to representative democracy (Lafferty & Meadowcroft, 1996; Ward, 2008; Dryzek and Pickering, 2018). Both approaches are based on the notion that liberal representative democracy has a limited capacity for solving environmental problems, which means that we have to go beyond this particular system of governance for solutions to the environmental crisis. In some renditions, going beyond liberal democracy entails far-ranging overhauls of the entire democratic system in order to be able to address environmental problems, whereas others argue that less drastic adjustments are preferable (Bäckstrand et al., 2010).

In a system of representative democracy and the market economy, the state has often been viewed as a rather passive entity that promotes the preferences of the most powerful interest groups. As industries generating environmental externalities also tend to be more resourceful, environmental issues are systematically disregarded in favor of policies and decisions promoting economic growth. In this vein, scholars in the so-called Treadmill of Production school essentially argue that existing environmental regulations are viewed as codifications of the best available technologies (BAT), which have been implemented because they do not pose any fundamental risk to the regulated industry in question while at the same time can be used to avoid criticism for not taking environmental responsibility (Buttel, 2004; Schnaiberg et al., 2002). Continuing a line of argumentation first advanced by the Treadmill of Production theory, Hausknost (2020) argues that there is a very distinct glass ceiling set by the market to how progressive the state can be in addressing environmental problems. The state in general, and the democratic state in particular, is intimately connected with the market economy and will therefore never advance policies that would endanger economic growth in a significant manner. Environmental reforms are achievable and sometimes even substantial, but a more fundamental reconfiguration of the state–market relationship is simply not possible, the argument goes.

A second, less-often-voiced argument against the state has to do with a perceived incompatibility between administrative rationality and the needs of the environment. This objection is often found in natural resource management studies, where it is argued that government agencies, by virtue of being bureaucratic and hierarchical organizations, tend to engage in so-called command-and-control management of environmental resources (Holling & Meffe, 1996) which runs counter to the behavior of most environmental resource systems. In addition, there have been concerns that an overly technocratic and legalistic green public administration risks alienating ordinary citizens, which in turn will preclude the formation of the necessary levels of legitimacy for various types of environmental policies (Paehlke & Torgerson, 2005). Proposed solutions to the problems associated with environmental public administration have often consisted of increased levels of public participation in environmental governance (Fischer, 1993), self-governance (Ostrom, 1990), as well as in blueprints for more polycentric (McGinnis, 2000) governance systems.

A third objection focuses on the supposed decline in state capacity that has occurred in the last two or three decades. According to this well-known narrative, the state has lost much of its power to the market, to sub-national governing bodies, and to international organizations, which in turn has affected the state's reform capacity in a negative way (Pierre, 2000). Scholars in the ecological modernization tradition tend to agree with this narrative in the sense that they view the state as more or less impotent in the face of global and lifestyle-related environmental problems which require governance of global supply chains and individual-level behavioral patterns (Mol and Spaargaren 2000). Instead, they turn to the market, the global community, and civic society for new forms of environmental regulation that do not rely on the state for successful implementation (Christoff, 2006; Mol et al., 2009)

4.6 Three Outstanding Questions about the Environmental State

Although the literature on the environmental state has expanded in recent years, there are still a number of central research questions that available studies have not been able to provide answers for. I suggest that these outstanding issues can be grouped under the headings of the how, why, and with what capacity the environmental state has evolved.

4.6.1 How?

A first key question is: *How have states responded to environmental problems?* This is largely a descriptive question in the sense that it is primarily concerned with the mapping and classification of state responses to environmental change. Some earlier work in the genre of comparative environmental governance can provide a starting point for such a mapping. A central finding from several studies (Jänicke and Weidner, 1997; Weidner and Jänicke, 2002; Holzinger et al., 2008) is compelling evidence for a regulatory expansion in the environmental policy area that has been ongoing since the beginning of the 1970s. Not only have most industrialized countries expanded their environmental policy portfolios in terms of the number of environmental problems for which there are policies, there is also a strong tendency toward more stringent policies (Holzinger et al., 2011). We also know from studies such as Jordan et al. (2003), Durant et al. (2004), Tews et al. (2003), and more recently Schulze (2021) that the type of policy instruments employed in environmental management have changed over time, from command-and-control policies carried out by government administrations and agencies towards more flexible, "soft", and network-based governance arrangements in which many different stakeholder groups are involved. Schultze's study of energy policy mixes offers a clear illustration of this trend: states are increasingly choosing soft over hard policy instruments to address environmental problems (Schulze 2021). Taken together, these studies suggest that the content of the policy portfolios of environmental states is changing, both in terms of policy instruments and policy problems.

Taking inspiration from established typologies in other areas of comparative research (e.g. Esping-Andersen, 1990; Hall and Soskice, 2003) a set of studies has aimed at uncovering similar typologies of environmental states in a comparative perspective (Dryzek et al., 2002; Jahn, 2016; Duit, 2016; Koch and Fritz, 2014; Zimmerman and Graziano, 2020). The assumption here is that environmental states can be expected to appear in clusters or types, rather than being ordered along a single dimension ranging from "laggards" to "pioneers" (as has been a leading assumption in the environmental performance literature).

The rationale for this assumption is that the environmental issue, to a certain extent, represents an external shock that hit most industrialized states about the same time in the late 1960s. As demonstrated in Lundqvist's classic study of clean air policies in the USA and Sweden (Lundqvist, 1980), this common shock had very different expressions when filtered through national political, geophysical, and economic contexts. This kind of interaction with national contextual factors is likely to be widespread and path-dependent in the sense that countries are likely to follow regulatory and organizational trajectories once embarked upon in early stages of each policy trajectory. Over time, new policy issues will then be filtered through not only the larger political context but also through the regulatory and organizational regimes already in place in the environmental area within the respective country. Evidence in support of different types of environmental states is, however, mixed. There is little convergence in the classification of environmental states found in the available studies (Dryzek et al., 2002; Jahn, 2016; Duit 2016; Koch and Fritz, 2014; Zimmerman and Graziano, 2020), and neither classification is able to generate distinct types of environmental states. However, as these studies have typically relied on fairly crude indicators and without allowing for temporal dynamics in typology formation, the final verdict on the existence of environmental state typologies should not be cast just yet.

4.6.2 Why?

The existing literature that views the modern state as next to powerless in mitigating environmental externalities leaves us puzzled by the growth of environmental regulatory outputs over the last several decades (Holzinger et al., 2008; Sommerer and Lim, 2016; Schulze 2021). In fact, there are even some instances in which the environmental state has extended its scope beyond the national level to regulate purely global environmental problems such as climate change, depletion of transboundary natural resources, and the spread of toxic substances. Why has the modern state, despite its tendency to be captured by business interests, its obsession with economic growth, and its confinement within national borders, nevertheless expanded its environmental responsibilities to cover an ever-increasing number of environmental problems both within and beyond national territories? The puzzle here concerns the combination of social, economic, and political mechanisms that determine the timing, scope, and composition of policies and institutions for addressing various environmental problems.

The literature suggests a number of possible structural factors that might explain variation in state responses, commonly measured as policy output (Fiorino, 2011, Cao et al., 2014). *Problem pressure* is an obvious explanation for a large class of end-of-pipe environmental problems that have their main effects on people living in the vicinity of plants and industries, but is less applicable as an explanation for diffuse-source pollutants such as greenhouse gases which have their effects located centuries and hemispheres away. The influence of *green social movements* (Dryzek et al., 2002; Rootes, 2013; Böhmelt et al., 2015) and *green parties* (Poguntke, 2002; Knill et al., 2010; Lundquist, 2022) on policy making is another possible explanatory factor. Research on *regulatory competition* (Vogel, 1997) and policy convergence and policy diffusion (Holzinger et al., 2011) suggests that trade and international organizations are important vectors for the spread of environmental policies and innovations. Finally, the changing structure of the *advanced market economy* (Christoff, 2006; Mol et al., 2009; Lundquist, 2021) as well as the rise of *post-materialistic values* (Inglehart, 1995) that follow in its wake have been suggested as explanations for environmental policy expansion. In sum, there is no shortage of hypotheses about what might explain regulatory expansion and contractions in the environmental area. There is, however, a dearth of studies that seek to assess the relative explanatory strength of these hypotheses, which means that a clear picture of the driving forces behind the environmental state is yet to emerge.

4.6.3 Capacity?

A third and final question about the environmental state is whether it has the necessary capacity for addressing the challenge of environmental problems. This is an issue that lies at the heart of the debate between skeptics and proponents of the environmental state. It mainly concerns the policy outcome side of the environmental state and is, therefore, best investigated by analyzing causal relationships between policy outputs and outcomes. The literature on environmental performance has offered some insights into the correlates between structural and institutional factors and environmental outcomes (Scruggs, 2003; Jahn, 2016), but has left the mechanisms generating these correlational patterns largely unexplored. Results from this line of research present a scattered and sometimes contradictory overall picture (Fiorino, 2011).

If one is to single out one reoccurring finding, it seems that corporatist schemes of interest representation and possibly a more consensual type of democracy have been beneficial for a higher level of environmental performance, at least for the reduction of point source emissions among Western industrialized countries (Jahn, 1998; Scruggs, 2003; Neumayer, 2003; Scruggs, 2003; Walti, 2004). However, the beneficial role of corporatism can be questioned in light of Matto Mildenberger's recent work on the notion of "double representation" – that both employers and employees are essentially representing fossil fuel interests – which leads to a situation in which both labor and capital are collaborating in resisting more ambitious climate policies (Mildenberger, 2020). There are also a number of studies that indicate linkages between higher levels of institutional quality (Povitkina and Matti, 2021), democratic quality (Von Stein, 2022; Povitkina and Jagers, 2022), and an improved environmental track record. A recent study by Limberg et al. (2021) offers some fresh insights about the role of institutional determinants of environmental policy outcomes: only when a state possesses sufficient administrative capacity is a more extensive environmental policy portfolio linked to better environmental performance.

The rather limited extent of this body of knowledge has multiple causes, such as restricted and varying selection of cases and unreliable and differing measures of environmental performance (Fiorino, 2011; Meadowcroft, 2014). But more than anything, its limitations are largely due to a combination of the absence of a coherent theoretical approach, poor data on policy outputs, and the application of statistical methods that disregard fundamental problems in correctly identifying the effect of state interventions (policies) on environmental outcomes. This is an area of central importance for the study of the environmental state: understanding if and when the state is able to break through the glass ceiling to solve environmental problems is really the key issue. If it turns out that the state is generally unable to provide effective environmental policies, as suggested by Hausknost (2020), Blüdhorn (2020), and others, there is little reason to pursue this research any further. But if there are signs that the state is sometimes able to act as a positive force in addressing environmental problems, then a central research priority should be to identify the scope conditions for when this is more likely to happen.

4.7 The Political Economy of the Environmental State

Much of the theorizing about the environmental state has been based on assumed parallels between the challenges that led to the emergence of welfare states in the late 19th and early 20th century, and the contemporary challenges presented by accelerating global environmental problems (Meadowcroft, 2005; Gough, 2016). Chief among these similarities is that both welfare states and environmental states are faced with the task of mitigating negative market externalities. Just as the welfare state took on gradually increasing responsibilities for mitigating the social and human costs of the market economy, Meadowcroft suggests that we have been witnessing a similar development within the environmental realm for the last four decades (Meadowcroft, 2012). Thus, one way of understanding the emergence of an environmental state is to see it as a functional requirement of the market economy – something that is needed for the continued functionality of the latter, but for which it is not possible to provide without interventions from non-market actors.

There are, however, important points of divergence between the environmental state and the welfare state (Gough, 2016). The first is the nature of the public good produced by the two states. The goods produced by the welfare state (e.g. free health care and education) typically have positive and fairly immediate effects on the well-being of large parts of the citizenry. This is not always the case for environmental goods. In fact, environmental problems beyond emissions that have localized and immediate health effects (i.e. first-generation point-source emissions) exhibit a set of characteristics that does not square well with the logic of representative democracy (Lafferty & Meadowcroft, 1996). Many environmental problems tend to display a substantial degree of geographical and temporal displacement of both causes and effects, and hence between policy costs and benefits. But the real challenge here is the fact that most contemporary governance systems are designed for addressing problems within the confinements of national borders over the short or medium term. We simply lack governance structures that would allow for the allocation of responsibility and the transferal of resources between generations and across distant places required to address things such as global warming and biodiversity loss.

Second, environmental problems tend to require that people give up some of their welfare in order to address them – they need to drive and fly less, recycle more, adhere to green building regulations, save water and energy, not eat delicious giant shrimps, and risk losing their jobs because of stricter environmental regulations or higher carbon taxes. Despite the fact that most people profess a deep concern for environmental degradation, reductions in short-term well-being nonetheless tends to be a too high an obstacle for passing ecologically necessary regulations and policies. Although a healthy environment is often described as a public good, Lafferty and Meadowcroft point out that much of what is labeled "environmental politics" is really a game of redistribution of costs and benefits between different groups in society (Lafferty & Meadowcroft, 1996).

Third, the very fact that most citizens, organizations, cooperations, and political parties agree that environmental issues are highly critical issues sets into motion the well-known logic of a valence issue (Green, 2007) – since everyone is in agreement about the importance of the environment no political actor has an incentive to politicize the issue. The emergence of green parties notwithstanding, most parties in industrialized countries profess a whole-hearted dedication to green values and the importance of saving the planet. The institutions of liberal representative democracy were once designed for reaching collectively binding decisions in contentious zero-sum issues between opposing societal interests. Somewhat paradoxically, the task of providing public goods for which there is widespread agreement seems to be much more difficult.

Fourth, environmental politics is a knowledge-intensive and expert-dominated field (Bäckstrand, 2004). Policy issues – typically in the form of newly discovered threats to the environment – usually emanate from academic researchers and experts, and they also play a dominating role in policy debates. In fact, many policy debates are primarily fought between different research camps rather than between different political actors. The hegemony of science in environmental discourse has experienced some challenges in recent environmental debates such as global warming and genetically modified organisms (GMO), in the sense that some political actors have taken positions that run counter to the consensus or majority view in the scientific community. For the present purpose, however, it is sufficient to note that the scientification of the environmental debate makes it difficult for political actors to act as policy entrepreneurs in the environmental area, mainly because they cannot control the selection of policy issues or the framing of such issues. This, in turn, makes the environmental area less fertile ground for anyone wishing to cultivate one's political capital.

In sum, although there are important lessons to be learned from the processes that gave rise to the welfare state, there are also some fundamental differences between these two states which means that the logic of the environmental state cannot simply be inferred from that of the welfare state (Gough and Meadowcroft, 2012). As a consequence, progress in environmental state research is probably best served by approaching the political economy of the environmental state as something independent of that of the welfare state.

4.8 Conclusion and Future Research

With a few exceptions, the scholarly debate on the environmental state has been theoretical and conceptual in nature. However, a continued search for a more complete definition of the environmental state, or a prolonged hypothetical debate about its ability to solve environmental problems, is unlikely to generate further scientific progress. What is lacking are empirically grounded accounts of what the state can and cannot be expected to do with regards to the environmental crisis. Nor are we in possession of a detailed understanding of what the state has done for the environment, and what the driving forces behind state action in environmental matters has been in a historical and comparative perspective.

A main task for future research is thus to develop an empirically informed program which can then serve as the foundation for a second wave of theorizing about the state. In doing so, outstanding empirical questions include, but are certainly not limited to, the three areas outlined above. A better understanding of the driving forces of state involvement in environmental matters and its problem-solving capacity is a prerequisite for determining the scope of the state as a vehicle for sustainability transformations. A more detailed image of the political economy of the environmental state will also allow for theorizing the environmental state *sui generis*, as something distinct from the welfare state, and governed by its own political economy.

Box 4.1 Chapter Summary

- We have reviewed and summarized the research on the environmental state.
- We have identified research gaps.
- More research is needed for understanding why the environmental state has expanded and the consequences for addressing environmental problems.
- Developing a better model of the political economy of the environmental state is a priority.

References

Bäckstrand, Karin. (2004). "Scientisation vs. Civic Expertise in Environmental Governance: Eco-Feminist, Eco-Modern and Post-Modern Responses." *Environmental Politics* 13(4): 695–714.

Bäckstrand, Karin, Khan, Jamil, Kronsell, Annica, & Lövbrand, Eva (Eds.) (2010). *Environmental Politics and Deliberative Democracy: Examining the Promise of New Modes of Governance*. Cheltenham: Edward Elgar.

Bäckstrand, Karin, & Kronsell, Annica (Eds.) (2015). *Rethinking the Green State: Environmental Governance Towards Climate and Sustainability Transitions*. London: Routledge

Barry, John, & Eckersley, Robyn. (2005). An Introduction To Reinstating the State. In J. Barry & R. Eckersley (Eds.), *The State and the Global Ecological Crisis*. Cambridge, MA: MIT Press.

Blühdorn, Ingolfur. 2020. "The Legitimation Crisis of Democracy: Emancipatory Politics, the Environmental State and the Glass Ceiling to Socio-Ecological Transformation." *Environmental Politics* 29(1): 38–57.

Böhmelt, Tobias, Thomas Bernauer, and Vally Koubi. (2015). "The Marginal Impact of ENGOs in Different Types of Democratic Systems." *European Political Science Review* 7(01): 93–118.

Buttel, Frederik H. (2004). The Treadmill of Production: An Appreciation, Assessment, and Agenda for Research. *Organization & Environment*, 17(3): 323–336.

Cao, Xun, Helen V. Milner, Aseem Prakash, and Hugh Ward. (2014). "Research Frontiers in Comparative and International Environmental Politics." *Comparative Political Studies* 47(3): 291–308.

Christoff, Peter (2005). "Out of Chaos, a Shining Star? Towards a Typology of Green States." In *The State and the Global Ecological Crisis*, (Eds.) John Barry and Robyn Eckersley. Cambridge, MA: MIT Press.

Christoff, Peter. (2006). Ecological modernization, Ecologial modernities. In P. H. G. Stephens & with J. B. and A. Dobson (Eds.), *Contemporary Environmental Politics. From Margins to Mainstream* (pp. 179–200). New York: Routledge.

Death, Carl (2016). *The Green State in Africa* New Haven: Yale University Press.

Dryzek, John, Hunold, Christian, Schlosberg, David, Downes, David, Hernes, Hans-Kristian (2002). "Environmental Transformation of the State: The USA, Norway, Germany and the UK." *Political Studies* 50(4): 659–682

Dryzek, John S, and Jonathan Pickering. (2018). *The Politics of the Anthropocene*. Oxford University Press.

Duit, Andreas (Ed.) (2014), *State and Environment. The Comparative Study of Environmental Governance*. Cambridge, MA: MIT Press.

Duit, Andreas. (2016). "The Four Faces of the Environmental State: Environmental Governance Regimes in 28 Countries." *Environmental Politics* 25(1): 69–91.

Duit, Andreas, Peter H. Feindt, and James Meadowcroft. (2016). "Greening Leviathan: The Rise of the Environmental State?" *Environmental Politics* 25(1): 1–23.

Durant, Robert F., Young-Pyoung Chun, Byungseob Kim, and Seongjong Lee. (2004). "Toward a New Governance Paradigm for Environmental and Natural Resources Management in the 21st Century?" *Administration & Society* 35(6): 643–682.

Eckersley, Robyn. (2004). *The Green State: Rethinking Democracy and Sovereignty*. Cambridge MA: MIT Press.

Esping-Andersen, Gösta. (1990). *The Three Worlds of Welfare Capitalism*. Cambridge: Polity Press.

Fiorino, Daniel J. (2011). "Explaining National Environmental Performance: Approaches, Evidence, and Implications." *Policy Sciences* 44(4): 367–389.

Fischer, Frank. (1993). Citizen Participation and the Democratization of Policy Expertise: From Theoretical Inquiry to Practical Cases. *Policy Sciences*, 26(3): 165–187.

Gough, Ian. (2016). "Welfare States and Environmental States: A Comparative Analysis." *Environmental Politics* 25(1): 24–47.

Gough, Ian, and James Meadowcroft. (2011)." Decarbonizing the Welfare State" in John S Dryzek, Richard B Norgaard, and David Schlosberg (Eds.) *The Oxford Handbook of Climate Change and Society*. Oxford University Press.

Green, Jane. (2007). When Voters and Parties Agree: Valence Issues and Party Competition. *Political Studies*, 55(3): 629–655.

Hall, Peter A., & Soskice, David. (2003). *Varieties of Capitalism: The Institutional Foundations of Comparative Advantage*. OUP Oxford. Retrieved from

Hausknost, Daniel. (2020). "The Environmental State and the Glass Ceiling of Transformation." *Environmental Politics* 29(1): 17–37.

Hausknost, Daniel, and Marit Hammond. (2020). "Beyond the Environmental State? The Political Prospects of a Sustainability Transformation." *Environmental Politics* 29(1): 1–16.

Holling, C. S., & Meffe, Gary. (1996). Command and Control and the Pathology of Natural Resource Management. *Conservation Biology*, 10(2), 328–337.

Holzinger, Katharina, Christoph Knill, and Thomas Sommerer. (2011). "Is There Convergence of National Environmental Policies? An Analysis of Policy Outputs in 24 OECD Countries." *Environmental Politics* 20(1): 20–41.

Holzinger, Katharina, Christoph Knill, and Thomas Sommerer. (2008). "Environmental Policy Convergence: The Impact of International Harmonization, Transnational Communication, and Regulatory Competition." *International Organization*, 62(4): 553–587.

Inglehart, Ronald. (1995). "Public Support for Environmental Protection: Objective Problems and Subjective Values in 43 Societies." *PS: Political Science and Politics* 28(1): 57–72.

Jahn, Detlef. (1998). "Environmental Performance and Policy Regimes: Explaining Variations in 18 OECD-Countries." *Policy Sciences* 31(2): 107–131.

Jahn, Detlef. (2016). *The Politics of Environmental Performance*. Cambridge: Cambridge University Press.

Jänicke, Martin, and Helmut Weidner. (1997). *National Environmental Policies. A Comparative Study of Capacity-Building*. Berlin: Springer.

Jordan, Andrew, Rudiger K.W. Wurzel, Anthony R. Zito (2003). New Instruments of Environmental Governance?: National Experiences and Prospects. London: Frank Cass Publishers.

Knill, Christoph, Marc Debus, and Stephan Heichel. (2010). "Do Parties Matter in Internationalised Policy Areas? The Impact of Political Parties on Environmental Policy Outputs in 18 OECD Countries, 1970-2000." *European Journal of Political Research* 49(3): 301–336.

Knill, Christoph, Schulze, Kai, and Jale Tosun (2012). "Regulatory Policy Outputs and Impacts: Exploring a Complex Relationship." *Regulation & Governance*, 6(4): 427–444.

Knill, Christoph, Jale Tosun, and Stephan Heichel. (2008). "Balancing Competitiveness and Conditionality: Environmental Policy-Making in Low-Regulating Countries." *Journal of European Public Policy* 15(7): 1019–1040.

Koch, Max, & Fritz, Martin. (2014). "Building the Eco-social State: Do Welfare Regimes Matter?" *Journal of Social Policy*, 43(4): 679–703.

Lafferty, William M., & Meadowcroft, James. (1996). *Democracy and the Environment. Problems and Prospects*. (W. M. Lafferty & J. Meadowcroft, Eds.). Edward Elgar.

Limberg, Julian, Yves Steinebach, Louisa Bayerlein, and Christoph Knill. (2021). "The More the Better? Rule Growth and Policy Impact from a Macro Perspective." *European Journal of Political Research* 60(2): 438–454.

Lundquist, Sanna. (2021). Explaining events of strong decoupling from CO2 and NOx emissions in the OECD 1994–2016. *Science of The Total Environment*, 793: 148390.

Lundquist, Sanna. (2022). "Do Parties Matter for Environmental Policy Stringency? Exploring the Program-to-Policy Link for Environmental Issues in 28 Countries 1990–2015." *Political Studies*: online first, 1–22.

Lundqvist, Lennart J. (1980). *The Hare and the Tortoise: Clean Air Policies in the United States and Sweden* Ann Arbor: Univ. of Mich. P.

Lundqvist, Lennart J. (2001). "A Green Fist in a Velvet Glove: The Ecological State and Sustainable Development." *Environmental Values* 10(4): 455–472.

McGinnis, Michael D. (2000). Polycentric Games and Institutions: Readings from the Workshop in Political Theory and Policy Analysis. Ann Arbour: Univerisity of Michigan Press.

Meadowcroft, James. (2005). "From Welfare State to Ecostate." In *The State and the Global Ecological Crisis*, eds. John Barry and Robyn Eckersley. Cambridge, MA: MIT Press, 3–24.

Meadowcroft, James. (2012). "Greening the State." In *Comparative Environmental Politics. Theory, Practice and Prospects.*, eds. Paul F Steinberg and Stacy D VanDeveer. Cambridge, MA: MIT Press, 63–87.

Meadowcroft, James. (2014). "Comparing Environmental Performance." In *State and Environment. The Comparative Study of Environmental Governance*, ed. Andreas Duit. Cambridge: MIT Press, 27–52.

Mildenberger, Matto. (2020). *Carbon Captured: How Business and Labor Control Climate Politics*. Cambridge, MA: MIT Press.

Mol, Arthur, Sonnenfeld, David and Spaargaren, Gert. (2009). The Ecological Modernisation Reader. Environmental Reform in Theory and Practice. London: Routledge.

Mol, Arthur, & Spaargaren, Gert (2000). Ecological Modernisation Theory in Debate: A Review. *Environmental Politics*, 9(1), 17–49.

Neumayer, Eric. (2003). "Are Left-Wing Party Strength and Corporatism Good for the Environment? Evidence from Panel Analysis of Air Pollution in OECD Countries." *Ecological Economics* 45: 203–220.

Ostrom, Elinor. (1990). *Governing the Commons: The Evolution of Institutions for Collective Action* (p. xviii, 280 p.). Cambridge [England]; New York: Cambridge University Press.

Paehlke, Roger., & Torgerson, Douglas. (2005). *Managing Leviathan: Environmental Politics and the Administrative State*. Toronto: University of Toronto Press.

Paterson, Matthew. (2016). "Political Economy of the Greening of the State" In eds. Teena Gabrielson, Cheryl Hall, John M. Meyer, and David Schlosberg (eds.) *The Oxford Handbook of Environmental Political Theory*. Oxford University Press.

Pierre, J. (2000). *Debating Governance -Authority, Steering, and Democracy*. Oxford: Oxford University Press.

Poguntke, Thomas. (2002). "Green Parties in National Governments: From Protest to Acquiescence?" *Environmental Politics* 11(1): 133–145.

Povitkina, Marina, and Sverker Carlsson Jagers. (2022). "Environmental Commitments in Different Types of Democracies: The Role of Liberal, Social-Liberal, and Deliberative Politics." *Global Environmental Change* 74: 102523.

Povitkina, Marina, and Simon Matti. (2021). "Quality of Government and Environmental Sustainability" eds. Andreas Bågenholm, Monika Bauhr, Marcia Grimes, and Bo Rothstein. *The Oxford Handbook of the Quality of Government*:

Rootes, Christopher. (2013). "Mobilising for the Environment: Parties, NGOs, and Movements." *Environmental Politics* 22(5): 701–705.

Schnaiberg, Allan, Pellow, David and Weinberg, Adam. (2002). The Treadmill of Production and the Environmental State. In A. P. J. Mol & F. H. Buttel (Eds.), *The Environmental State Under Pressure* (Vol. 10: pp. 15–32). Bingley, UK: Emerald Group Publishing Limited.

Schulze, Kai. (2021). "Policy Characteristics, Electoral Cycles, and the Partisan Politics of Climate Change." *Global Environmental Politics*, 21(2): 44–72.

Scruggs, Lyle. (2003). *Sustaining Abundance. Environmental Performance in Industrial Democracies*. Cambridge: Cambridge University Press.

Sommerer, Thomas, and Sijeong Lim. (2016). "The Environmental State as a Model for the World? An Analysis of Policy Repertoires in 37 Countries." *Environmental Politics* 25(1): 92–115.

Tews, Kerstin, Per Olof Busch, and Helge Jörgens. (2003). "The Diffusion of New Environmental Policy Instruments." *European Journal of Political Research* 42(4): 569–600.

Tosun, Jale. (2013). *Environmental Policy Change in Emerging Market Democracies: Eastern Europe and Latin America Compared*. Toronto, University of Toronto Press.

Vogel, David. (1997). *Trading Up: Consumer and Environmental Regulation in a Global Economy*. Cambridge MA: Harvard University Press.

Von Stein, Jana. (2022). "Democracy, Autocracy, and Everything in Between: How Domestic Institutions Affect Environmental Protection." *British Journal of Political Science*, 52(1): 1–19.

Walti, Sonja. (2004). "How Multilevel Structures Affect Environmental Policy in Industrialized Countries." *European Journal of Political Research* 43(4): 599–634.

Ward, H. (2008). Liberal Democracy and Sustainability. *Environmental Politics*, 17(3), 386–409.

Weidner, Helmut, and Martin Jänicke. (2002). *Capacity Building in National Environmental Policy. A Comparative Study of 17 Countries*. Berlin: Springer.

Zimmermann, Katharina, and Paolo Graziano. (2020). "Mapping Different Worlds of Eco-Welfare States." *Sustainability* 12(5): 1819.

5
POLYCENTRIC GOVERNANCE

Andrew Jordan and Dave Huitema

5.1 Introduction

Polycentric governance thinking represents a significant and growing strand of research in environmental social science. Polycentric governance systems are those in which 'political authority is dispersed to separately constituted bodies with overlapping jurisdictions that do not stand in hierarchical relationship to each other' (Skelcher, 2005: 89). 'Overlapping' means that the scope of the issues that are addressed by such bodies are not discrete. A related term – 'polycentric order' – was coined by Polanyi as long ago as the 1950s. This refers to the outcomes and processes of mutual adjustment between independent elements in a complex system (Polanyi, 1951). Ostrom et al. (1961) were, however, the first to use the term to describe the practical challenge of governing, in their case the delivery of public services in US cities.

Their work appeared during an era of rapid expansion of the federal government. As academics, they swam against the tide of thinking at the time (which was pro 'big government') by pleading for the self-governing capacity of (local) bodies, and their capacity to learn from each other (without one dominating the other). Indeed, Vincent Ostrom's early work implied that public services could be better delivered by different combinations of smaller bodies self-organising at different scales, and that actors (in his case, members of the public) would choose accordingly (McGinnis and Ostrom, 2011: 16). Hence, the type of services offered, levels of service provision, and responsible jurisdictions would be different for each public service. Throughout their long careers, Vincent and Elinor Ostrom continued to emphasize the importance of acting at the local level, and together with other polycentric thinkers, devoted significant time to understanding how local bodies could spontaneously coordinate – that is with little or no hierarchy (Aligica and Tarko, 2012: 242).

In summary, polycentric governance involves:

> multiple governing authorities at different scales ... Each unit within a polycentric system exercises considerable independence to make norms and rules within a specific domain (such as a family, a firm, a local government, a network of local governments, a state or province, a region, a national government, or an international regime).
>
> *(Ostrom, 2010a: 552)*

DOI: 10.4324/9781003043843-6

Crucially, 'independence' does not mean that units completely ignore each other; and polycentricity should not be equated with complete fragmentation and/or anarchy. In the early 1960s, Vincent Ostrom et al. (1961: 831) famously argued that:

> To the extent that [centres of decision making] take each other into account in competitive relationships, enter into various contractual and cooperative undertakings or have recourse to central mechanisms to resolve conflicts, the various jurisdictions … [they] may be said to function as a 'system'.

Since then, significant work has been conducted on environmental and sustainability issues using the lens of polycentric governance thinking (see for instance Sovacool et al., 2017; Berardo and Lubell, 2019; Thiel et al., 2019). One environmental issue which has attracted significantly more attention in recent years is that of climate change. It certainly fascinated Elinor Ostrom in the period just before her death in 2012 because it was generally considered to be a global issue, resolvable only by states acting in a coordinated manner at the international level. By the late 2000s, climate change governance had been well over 30 years in the making. But the product of all that activity – the international climate regime, centred on the 1992 United Nations Framework Convention on Climate Change (UNFCCC) – was heavily criticised for being too slow to produce results. Enter Elinor Ostrom, who powerfully argued that 'new' and more dynamic forms of governing climate change were not just possible and necessary, but were already appearing around, below, and to the side of the UNFCCC.

Her message was undoubtedly a positive one: not all forms of climate governance would have to be designed hierarchically by international negotiators meeting in UN fora. Many were in fact emerging spontaneously from the bottom up, producing a more polycentric pattern (Ostrom, 2010a, 2010b). Scholars (and practitioners) had, she claimed, become too fixated with the resolution of collective action dilemmas at the international level, ignoring empirical developments at the local level. Moreover,

> part of the problem is that 'the problem' [of climate change] has been framed so often as a global issue that local politicians and citizens sometimes cannot see that there are things that can be done at a local level that are important steps in the right direction.
> *(Ostrom, 2009: 15)*

In other words, her message was not only analytical but also unashamedly normative – local actions and initiatives were effective; they *should* be celebrated and, where possible, actively scaled up. In fact, she expected the many local initiatives to eventually connect and together inspire other actions, thus morphing into a system, providing collective solutions.

In order to understand why she advanced this seemingly counter-intuitive argument, we need to revisit the foundational assumptions and propositions of polycentric governance thinking. We will do this in the first half of this chapter. In the second half, we examine the extent to which they account for the emerging patterns and dynamics of climate governance. We conclude with a number of broader points about the value of a polycentric governance approach and identify priorities for future research and policy practice.

5.2 What Is Polycentric Governance?

Polycentric thinking should be viewed against the backdrop of a much larger debate about 'governance'. That word derives from the verb 'to govern'. Following Kooiman (1993), *governing* can be defined as directed behaviour, involving governmental and non-governmental actors, which is aimed at addressing a particular issue or problem. Governing is a process; it involves the creation of institutions – rules, organisations, and policies – that seek to stabilise behaviours, and it involves attention to the normative underpinnings of those institutions. By contrast, the term *governance* describes 'the patterns that emerge from the governing activities of social, political and administrative actors' (Kooiman, 1993: 2). Governance is not the same as *government*, which centres on the institutions and actions of the state. Thinking in terms of governance (and government) allows non-state actors such as businesses and non-governmental organisations to be brought into an analysis of societal steering (Lemos and Agrawal, 2006: 298). To quote Rosenau (1992: 4), governance:

> is a more encompassing phenomenon than government. It embraces governmental institutions, but it also subsumes informal, non governmental mechanisms ... whereby those persons and organizations within its purview move ahead, satisfy their need and fulfil their wants.

As noted in Section 5.1, the term 'polycentric' essentially means many centred. In that section we also noted that polycentric governance systems are those in which political authority is dispersed to separately constituted bodies with overlapping jurisdictions that do not stand in a hierarchical relationship to each other. The reference to 'separately constituted bodies' certainly places polycentric systems at the governance end of the government-to-governance continuum of governing types. However, the most important operative term in polycentric governance thinking is 'overlapping'. In that sense, polycentric systems share a number of important similarities with certain (so-called 'Type II'[1]) variants of multi-level governance, which comprise a complex, fluid patchwork of overlapping jurisdictions that mutate as demands for governance change in response to societal problems (McGinnis and Ostrom, 2011: 15).

Just as it is insightful to understand governance by relating it to government, we can better understand polycentric systems by comparing them to monocentric ones, that is governing systems controlled hierarchically by a single unitary body. Whilst no governance system is fully mono- or polycentric, they should be viewed as poles on a spectrum of governing types (Galaz et al. 2012: 22). That said, the debate on how to measure the level or degree of polycentricity within a particular governance system is far from resolved (Aligica and Tarko, 2012).

Finally, the literature on multi-level governance has been criticised for its tendency to describe rather than explain processes of governing (Jordan, 2008: 23). But over a number of years, polycentric thinkers have developed a set of quite specific propositions which arguably help not only to describe, but also to explain and ultimately actively govern polycentric systems. In the next section we introduce each one in turn and unpack its implications for the way in which climate change is or could be governed.

5.3 Polycentric Governance Theory: Central Propositions[2]

Some commentators have listed the positive and negative features of poly- and monocentric systems, but such lists struggle to explain why or how the features arise in the first place. And in fact, some have noted how polycentric thinkers tend to stress the positive aspects over the negative ones (Dorsch and Flaschland, 2017). In the following, we therefore draw on the work of Jordan et al. (2018a, 2018b) who have distilled polycentric thinking into five basic propositions.

5.3.1 Local Action: Governance Initiatives Are Likely to Take Off at a Local Level Through Processes of Self-Organisation

In many ways, this is the key proposition. Each individual actor in a polycentric system plots their own (local) actions, based on their own preferences, whilst responding to external stimuli. They are open to information about the experiences of others, and information about the consequences of their actions, both for themselves and for others. In response, they will adjust their behaviour (that is 'coordinate') with others.

5.3.2 Mutual Adjustment: Constituent Units Are Likely to Spontaneously Develop Collaborations with One Another, Producing More Trusting Interrelationships

Once constituent units have emerged, they will naturally interact with one another. Vincent Ostrom (1999: 57) even went as far as defining polycentric systems in such terms: they have "many elements", he wrote, "[which] are capable of making mutual adjustments for ordering their relationships with one another within a general system of rules where each element acts with independence of other elements". This explains why polycentric systems are often likened to complex adaptive systems (Tarko, 2017: 58): mutual adjustment is how actors adapt to changing external conditions, their actions in turn feeding back on other actors.

5.3.3 Experimentation: The Willingness and Capacity to Experiment Facilitates Governance Innovation and Learning

One of the main benefits of polycentric governance is that it facilitates – even encourages – actors to try out different approaches, which is known as natural experimentation. Over time, common methods emerge, so that the results of experiments in one setting actively inform those in other domains (via learning and scaling up).

5.3.4 The Importance of Trust: Trust Builds Up More Quickly When Units Are Able to Self-Organise, thus Increasing Collective Ambitions

In international political theory, the level of trust between state actors is assumed to be low. At a more local level, Vincent and Elinor Ostrom argued that things may work out rather differently. Trust, for example, may be in greater supply, born of (among other things) the greater likelihood of face-to-face interactions between actors. And when trust is more plentiful, the standard assumption of rational choice theory – that actors maximise their short-term interests – may not apply.

5.3.5 Overarching Rules: Local Initiatives Are Likely to Work Best When They Are Bound by a Set of Overarching Rules

Overarching rules are assumed to provide predictability, a means to settle disputes and reduce the level of discord between bodies to a manageable level. Their primary functions are to protect diversity (Proposition 1) and facilitate mutual adjustment (Proposition 2). However, their exact form and functioning are things that continue to excite debate amongst polycentric governance thinkers (Aligica and Tarko, 2012: 254ff.).

5.4 The Governance of Climate Change

In this chapter we have introduced the foundational assumptions and propositions of polycentric governance thinking, differentiating them from the related (but distinct) discussion of multi-level governance. In this section, we will examine the extent to which they account for the emerging patterns and dynamics of climate governance.

5.4.1 *Changes in the Locus and Focus of Governance*

Over the last 30 years the governance of climate change has altered significantly. At the 2015 UNFCCC climate summit held in Paris, world leaders changed the original emphasis on member states and international norm setting, and agreed to establish a more bottom-up system of governance through which states pledged to make emission reductions, then gradually ratchet them up as part of a process of ongoing assessment and review (Keohane and Oppenheimer, 2016). Crucially, the Paris Agreement also offered strong encouragement to existing and new climate actions by non-state and subnational actors (Hale, 2016), thus exemplifying a more general trend towards greater polycentricity. Today, there certainly appears to be more examples of greater polycentricity within the still relatively monocentric domain of state-led international policy-making. For example, more than 100 regional governments committed themselves to reducing emissions by at least 80 per cent by 2050, a target exceeding that of most sovereign states (Setzer and Nachmany, 2018).

More significantly for those advocating a polycentric approach, myriad governance initiatives have blossomed around, below, and to the side of the state-dominated UNFCCC process. Much of this 'groundswell' (Falkner, 2016) of activity is conventional in the sense that it links different forms of state-led governing (e.g. government-driven coalitions promoting carbon pricing such as the Carbon Pricing Leadership Coalition, or the European Commission collaborating with mayors in cities through the Covenant of Mayors for Climate and Energy). But it is also adopting novel, hybrid forms (e.g. international standards developed by non-state actors, or subnational governments collaborating across borders without the involvement of their national governments). If one moves outwards to explore the activities of non-state actors such as business, still more forms of governing come into view. These include voluntary commitments to reduce emissions by various leading companies, but also highly complex systems for monitoring and trading in emissions, and efforts to disclose the carbon risks for businesses and investors (for a fascinating analysis see Green, 2014). Many of the private initiatives are being steered by business, seemingly independent of state action. For example, the World Business Council on Sustainable Development coordinates Action 2020, an initiative to embed sustainability in business practices, as well as more sector-specific activities, such as the Cement Sustainability Initiative. To give another example, as part of the Science-Based Targets initiative, a partnership formed by the United Nations, and several business and environmental organisations, over 200 of the world's largest and most energy-intensive companies have voluntarily taken on 2050 reduction targets.

5.4.2 *How Polycentric Is Climate Change Governance?*

The governance of climate has never been and is unlikely ever to be fully monocentric, even if global governance initiatives by the UN are sometimes portrayed as undermining national sovereignty and replacing it with undemocratic global government. However, the simple matter of the fact is that there is no single world government capable of managing climate change across the globe. Over the last 30 years, the main international agreement – the UNFCCC – has, as noted above, moved in the direction of more spontaneous, bottom-up action; the Paris Agreement, for example, only commits member countries to make emission reduction pledges and then engage in revision and review activities. With hindsight, the 2008 Copenhagen conference ('COP 15') marked the high water mark of monocentric governing.

As the debate about international climate governance ground on, many actors did not, as Elinor Ostrom foretold, wait for the UNFCCC regime to push them to act: they took matters into their own hands. There is certainly not a monocentric system of governance in which a single, hierarchical unit structures the activities of all other units. Although the UNFCCC has established a common set of overarching norms and rules, its hierarchical steering power remains relatively limited. Rather, climate governance incorporates a variety of actors and institutions operating at multiple scales (Cole, 2015; Jordan et al., 2015). These include states and international organisations, but also civil society groups, companies, cities, and NGOs. Together, these actors have claimed the authority to address climate change in various ways, sometimes working alone, sometimes working in tandem through hybrid forms of governing. For example, hundreds of sports clubs and bodies across the world have joined a new UNFCCC-led initiative (Sports for Climate Action) through which they measure and report on their emissions, and gradually develop strategies and policies to reduce them to net zero by 2040.[3] The fashion industry (Fashion for Global Climate Action) has embarked upon a similar engagement initiative with the UNFCCC.[4]

Moreover, there are recognisable sub-domains of governing within the broader system of climate governance. These include international and national governance (centred on the annual Conference of the Parties to the UNFCCC), as well as transnational governance, which includes many forms of private governance involving businesses and industry associations. Interestingly, the degree of polycentricity varies across these sub-domains. For example, relatively unitary states such as the United Kingdom (UK) have adopted very long-term targets and strategies with independent advisory bodies to report on their fulfilment, and there are also more loosely coupled networks of national emissions trading systems, each with its own array of internal processes, emission reduction targets, and carbon prices. Even within an individual sub-domain, it is possible to observe a significant degree of internal variation, implying that the system as a whole is possibly 'doubly polycentric'. For example, the transnational climate governance sub-domain includes a wide variety of different initiatives, which are themselves unevenly distributed across the world (Bulkeley et al., 2014: 117–133). Some initiatives have a handful of members whereas others – e.g. city networks – have many hundreds. Some were initiated by states or state dominated organisations such as the EU, whereas others have emerged organically, from the bottom up.

Until the late 2000s, this steady proliferation of initiatives was widely perceived as a negative development – a 'fragmentation' of and possibly a distraction from international efforts (Biermann et al., 2009). Those who studied the new initiatives in more detail were more sanguine, seeing them as an alternative to a global regime that appeared to have become steadily more gridlocked (for example Hoffmann, 2011). Elinor Ostrom herself contributed to, and was open-minded about, the precise relationship between the various levels, units, and domains; she saw it as an empirical puzzle to be studied. If she were alive today, she would probably recognise the current pattern of governance as being inherently polycentric. The system has a hybrid and modular form in which the governance activities of states and a wide array of non-state actors are not neatly separated into levels ('multi-level governance'), but functionally overlap with several initiatives taken to create some level of coherence – although this might actually be increased a lot more still (van Asselt and Zelli, 2018). On this basis one can conclude that the concept of polycentric governance is very apt as a descriptive device. In the next section, we examine how capable polycentric theory is of *explaining* the emergence and functioning of the system using the propositions outlined above.

5.5 Explaining Climate Change Governance

5.5.1 Local Action

Ostrom originally proposed that many actors would address climate change to reap so-called 'co-benefits' such as improved human health and better local air quality. In general, non-state actors are making a rational calculation to act against climate change. They are not waiting to be told what to do (Ostrom, 2010b: 6). It is important to note that Proposition 1 does not necessarily hold that all actors have the capacity or the motivation to act locally. The uneven geographies of transnational climate action have already been noted above. In these and other sub-domains, action may only occur when a particular type of actor is present – variously referred to as a policy entrepreneur, a leader, or an orchestrator (Boasson, 2018; Abbott, 2018). But if a relatively small number of actors play such a disproportionate role, then perhaps the system as a whole is possibly not as robust as the Ostroms claimed. Moreover, local action has generated many new forms of governance, but it has not yet triggered a significant, economy-wide process of deep decarbonisation. Perhaps polycentric governance is better thought of as a means to encourage experimentation within a particular trajectory of climate governance, rather than providing a step change.

5.5.2 Mutual Adjustment

Consequential linkages can in principle form between many different units and domains. The linkage that has attracted most scholarly attention is that connecting the international and transnational sub-domains. The general argument here is that transnational climate governance emerges in the 'shadows' of the UNFCCC process (Bulkeley et al., 2012: 693), giving substance to issues that have only been partially determined by international negotiators. Yet, the international sub-domain has, as noted above, adjusted to these developments, suggesting that a *polycentric system* of sorts has emerged, which goes beyond fragmentation. With hindsight, the Copenhagen COP was a 'critical juncture' in the development of two-way interlinkages (Hale, 2016: 15). Another significant axis of mutual adjustment is that connecting the international and national domains, namely that national actors use the negotiation of international agreements as a window of opportunity to push for stronger commitments at a national level (Setzer and Nachmany, 2018).

5.5.3 Experimentation

If 'experimentation' is defined loosely to refer to tinkering with new governing devices, then climate governance is arguably awash with experiments. From cities to private companies, to nation states and even within the UNFCCC, experimentation has been widely cited as both an enabler of and a motivation for the explosive growth in climate governance. Emissions trading is probably the most emblematic of this trend, having moved from being a rather theoretical idea in applied economics, to a small scale deviation from 'normal' government regulations in the USA and certain other jurisdictions, to a solution endorsed by international organisations the world over, including the UNFCCC (see Voß and Schroth, 2018; Biedenkopf and Wettestad, 2018).

However, if an experiment is defined narrowly as a process of investigation under controlled conditions, then the extent of experimentation is probably considerably less than the Ostroms expected. Moreover, has experimentation produced innovations in governance, as they claimed? Much depends on how narrowly or broadly one defines 'innovation'. If it is taken to mean the development of radically different policy and governance approaches (i.e. entirely new to the world), then the fruits of all the experimental activity have not been that spectacular because attention grabbing new forms of governance have not come from it, at least thus far (Jordan et al. 2018a, 2018b). And is experimentation generating societal learning? In monocentric systems, higher authorities manage and legitimise learning activities. But when governance is more polycentric, it becomes harder to work out who is doing what, let alone evaluate their activities and learn universally applicable lessons. In fact, different units may well adopt approaches to evaluation that actually conflict with and/or fail to share their findings with neighbours (remember that polycentric governance thinking is strongly focused on local democracy and acting in ways that fit with local preferences).

5.5.4 Trust

As the landscape of climate governance becomes more congested with governance initiatives, researchers have uncovered evidence of collective self-organisation born of trust, but actually also of conflicting priorities and approaches where trust is minimal or absent. In principle, many different types of interaction are possible and should be carefully documented: initiatives and policies could complement one another without actually interacting; but they could also merge, compete and conflict with one another, or some may actively replace other types (Jordan et al., 2015). For example, in the sub-domain of city-level initiatives, competition for financial investment has emerged between networks of members – here strong trust *within* networks goes hand in hand with low levels of trust *between* networks. In relation to carbon finance, banks and NGOs compete with one another to shape its flow, 'creating problems of duplication and turf wars over who funds what' (Bulkeley and Newell, 2010: 106) and thus low levels of trust. Meanwhile, in relation to adaptation, if measures are not taken in a planned and coordinated fashion, they may not be sufficiently 'synchronised' (e.g. a flood defence system that ends at a political border between two administrative units). In short, the relationship between initiatives could be a conditional one (e.g. complementary in some conditions, but substitutive in others) and thus be built on trust per se (Andonova et al., 2017).

5.5.5 Overarching Rules

Finally, the UNFCCC is a key source of significant rules, norms, and values. It is arguably the 'centre of gravity' of the whole system (Hickmann, 2017: 17). It certainly satisfies one of Ostrom's conditions for a rule to be deemed 'overarching' – that is it defines the broad goals of climate governance. It has also demonstrated its own agency, through advancing sectoral engagements with the fashion and sports sectors (see above). However, as the Trump administration in the USA demonstrated in 2017, any party that wishes to withdraw from all or parts of it is quite at liberty to do so. In other words, the rules may be 'overarching', but their enforceability is limited. Similarly no actor (or sector) can be compelled to enrol in a UN-led sectoral engagement activity such as Sports for Climate Action.

Nevertheless, there most certainly are enforceable rules at other levels and in other subdomains. Two prominent examples are Norway's Climate Change Settlement and the Climate Change Act in the UK. Although these rules are not universally overarching, they may have longer-term potency, for example in facilitating the subsequent development of more specific and binding laws in certain jurisdictions and/or governing particular sub-issues (Setzer and Nachmany, 2018). Interest groups and even businesses have, in turn, sought to enforce these laws by initiating legal actions in national and sub-national courts. For example, in the 2019 Urgenda case the supreme court in the Netherlands ruled that the Dutch government was under a binding legal obligation to prevent dangerous climate change by reducing its emissions in line with its human rights obligations. The case was brought by a charity – Urgenda – on behalf of 886 Dutch citizens. And in 2021, a lower court in the same country decreed that the international oil company Shell (partly Dutch at that time) was obliged to reduce its worldwide emissions in line with the Paris agreement too. Each of these, and similar, verdicts are expected to have (and are already having) an impact on court decisions elsewhere.

5.6 Conclusion and Future Research

Polycentric governance thinking offers a popular and distinct framework for understanding and actively reshaping environmental issues. It is certainly different to run of the mill international relations approaches, for which the main point of reference remains international actors and international level processes. It does share some similarities with variants of multi-level governance theory and political federalism (McGinnis and Ostrom, 2011: 15). But unlike them, it is more directly concerned with the role of non-governmental units and/or situations in which jurisdictions overlap. It has most in common with theories of networked governance (Jordan and Schout, 2006), with which it shares a concern with how and why centralised and decentralised forms of coordination emerge and find some coexistence (McGinnis and Ostrom, 2011: 15).

There is a rich and expanding literature on polycentric governance covering a range of empirical contexts. One of the main advantages of polycentric thinking is that it challenges a number of well-known simplifications in environmental governance, including the idea that 'global problems' require 'global solutions'. Elinor Ostrom's personal contribution lay not so much in establishing polycentric governance as a new theoretical perspective, but in sensing that polycentric thinking could be applied to a wide range of different contexts and problem areas. Her core message was unashamedly positive: she suggested that these activities, although initially small in size and few in number, would become 'cumulatively additive' over time (Ostrom, 2010a: 551, 555) and were 'likely to expand in the future' (Ostrom, 2010a: 555). By and large, in the realm of climate change governance, history has proven her right.

Polycentric governance is arguably best thought of as a meso-level concept around which other concepts and theories can be brought into a more productive dialogue with one another (see also Galaz et al., 2012: 22). As a *descriptive* device, we have shown that it is a useful means to account for the rapidly changing contours of governance particularly in a relatively young and dynamic area of governing such as climate change. In the past, climate governance has been examined from the standpoint of single levels and domains. Polycentric theorists seek to offer a more holistic perspective which furnishes a more synoptic appreciation of all the landscape's component parts and, even more crucially, the interactions between them. We think that the five propositions outlined above provide a good basis for a shared programme of *explanatory* work on climate governance. However, structural issues, such as the exercise of political power, legitimacy, and accountability, are not yet fully accounted for. Finally, Elinor Ostrom promoted academic research that was doubly engaged – it addressed real-world problems *and* understood the real-world complexity that governors confront on a daily basis. We think that the five propositions provide a rich source of practical policy *prescriptions* that can be applied to steer systems of governance. For example, the Local Action proposition suggests that governors should act locally where possible (acknowledging the principle of subsidiarity) and build on local motivations where possible (e.g. by employing a 'co-benefits' framing) (for other examples, see Jordan et al., 2018b: 21).

Where should the literature head next? One important priority is to understand better the role that governments and thus public policy should play in polycentric governance. The Ostroms have often been misread as being completely fixated with local action, when polycentric theory has in fact always been concerned with the dynamic *balance* between monocentric and polycentric forces. At present, a rather binary view of the state risks taking hold in climate governance scholarship. One line of argument is that the state, even if not completely hollowed out by austerity and captured by neoliberal forces, is not capable of – and does not have the legitimacy to – governing complex issues such as climate change alone.

Another line of argument is that although pure monocentricity may be a non-starter, the state nonetheless holds important cards such as the power to regulate and tax. Polycentric theory is well placed to work across this binary conception. We know for example that the structure of national systems (including the courts) exerts a passive effect through affecting the political opportunity structures encountered by sub-national and non-state actors (Roger et al., 2017). States also facilitate the diffusion of governance initiatives by funding learning exercises such as policy evaluations. Similarly, it is important to understand how state structures affect how new ideas (e.g. emission trading) circulate and become transplanted in national policy systems.

Promising attempts have been made to pay more attention to the role of political power in the functioning of polycentric governance systems. For example, Morrison et al. (2019) distinguish between: the *power to design* institutions – often resting with national governments and sometimes international institutions; *pragmatic power* – often to do with the interpretation and implementation of policies; and *framing power* – which shapes the way problems are discussed and perceived in the first place and which is often influenced by NGOs and the media. They show that power – in each of these three forms – does affect the functioning of polycentric governance regimes, but much more work could be done on this. It would for instance be interesting to study how power affects the realisation of the five propositions. To give one example: mutual adjustment between actors often takes place on an unequal footing, because some actors have greater levels of design and framing power than others, and they can use this power to foreclose certain discussions which they do not see as being in their interest.

Another priority is: How long does polycentric governance take to emerge and how and why does it change over time? Critics of the UNFCCC process contend that it has taken far too long – more than three decades – to emerge and affect the trajectory of global emissions, yet a fully functioning polycentric governance system may take at least as long to emerge. Still, scientists argue that global emissions should be brought to zero in the next 30 years – which corresponds to a massive acceleration in the performance of governance. Which raises another tricky issue: How long do new forms of non-state governance actually last? Polycentric thinking reminds us that bottom-up governance is a perilous activity, vulnerable to lapses in funding and state support. Experience suggests that many initiatives are actually rather ephemeral and quietly 'sink' (Benson et al., 2012), often after states withdraw their support. Around 40 per cent of the public–private partnerships adopted at the 2002 World Summit on Sustainable Development have already suffered this fate (Hale, 2016: 18). If survival is such a challenge, it may explain why many forms of bottom-up governance set relatively unambitious targets and incorporate weak monitoring systems (Jordan et al. 2018). It may also explain why the higher degree of polycentricity endorsed by the Paris Agreement has yet to evolve into a fully functional *polycentric system*.

Finally, do polycentric systems make a material contribution to the resolution of environmental problems? Green (2014) helpfully reminds us that for many of the newer forms of governance, 'process' contributions (sharing knowledge, enhancing awareness, etc.) were a significant initial motivation, not the rapid reduction of emissions. The increasingly urgent issue of how to achieve emission reductions and other substantive effects will eventually have to be addressed, as will the (non) delivery of broader, system-wide functions such as facilitating equity, justice, legitimacy, and accountability. One possibility is that greater polycentricity is providing new opportunities to address these issues that have not been satisfactorily delivered by monocentric governance. Another equally plausible alternative is that polycentric governance will struggle to address systemic challenges, such as the age-old North–South divide in environmental politics. The relatively limited participation of countries from the Global South in the design and running of many city networks and international cooperative networks tells its own, far from unique, story about the spatial reach of polycentric climate governance.

Box 5.1 Chapter Summary

- Polycentric governance thinking offers a distinct framework for understanding and actively reshaping environmental issues.
- Polycentric thinking challenges a number of well-known simplifications in environmental governance, including the idea that 'global problems' require 'global solutions'.
- Polycentric thinking is a useful way to account for the rapidly changing contours of governance, particularly in a relatively young and dynamic area such as climate change.
- The state, even if not completely hollowed out by austerity and captured by neoliberal forces, is not capable of – and does not have the legitimacy to – governing complex issues such as climate change alone.

Notes

1 For Marks and Hooghe (2014), Type I multi-governance corresponds to a situation in which authority is consciously dispersed to a limited number of non-overlapping jurisdictions and a limited number of levels. Thus it shares many similarities with certain types of federal political systems.
2 This section and the one after it draw on material published in Jordan et al. (2018a, 2018b).
3 https://unfccc.int/climate-action/sectoral-engagement/sports-for-climate-action
4 https://unfccc.int/climate-action/sectoral-engagement/fashion-for-global-climate-action

References

Abbott, K. (2018). Orchestration. In: Jordan, A., D. Huitema, H. van Asselt, J. Forster (ed.). *Governing Climate Change: Polycentricity in Action*? Cambridge: Cambridge University Press, pp 188–209.

Aligica, P. and Tarko, V. (2012). Polycentricity: from Polanyi to Ostrom, and beyond. *Governance*, 25(2), 237–262.

Andonova, L., Hale, T. and Roger, C. (2017). National policy and transnational governance of climate change. *International Studies Quarterly*, 61(2), 253–268.

Benson, D., Jordan, A., Smith, L. S. and H. Cook (2012). Collaborative environmental governance: are watershed partnerships swimming or are they sinking? *Land Use Policy* 30, 748–757.

Berardo, R. and M. Lubell (2019). The ecology of games as a theory of polycentricity: recent advances and future challenges, *Policy Studies Journal*, 47, 6–26.

Biedenkopf, K. and J. Wettestad (2018). Harnessing the Market. In: Jordan, A., D. Huitema, H. van Asselt, J. Forster (ed.). *Governing Climate Change: Polycentricity in Action*? Cambridge: Cambridge University Press, pp 231–247.

Biermann, F., Pattberg, P. and van Asselt, H. et al. (2009). The fragmentation of global governance architectures: a framework for analysis. *Global Environmental Politics*, **9**(4), 14–40.

Boasson, E. (2018). Entrepreneurship. In: Jordan, A., D. Huitema, H. van Asselt, J. Forster (ed.). *Governing Climate Change: Polycentricity in Action*? Cambridge: Cambridge University Press, pp 117–134.

Bulkeley, H., Andonova, L. and Bäckstrand, K. et al. (2012). Governing climate change transnationally: assessing the evidence from a database of sixty initiatives. *Environment & Planning C*, **30**(4), 591–612.

Bulkeley, H., Andonova, L. and Betsill, M. et al. (2014). *Transnational Climate Change Governance*. Cambridge: Cambridge University Press.

Bulkeley, H. and Newell, P. (2010). *Governing Climate Change*. Routledge: London.

Cole, D. (2015). Advantages of a polycentric approach to climate change policy. *Nature Climate Change*, **5**(2), 114–118.

Dorsch, M. and Flaschland, C. (2017). A polycentric approach to global climate governance. *Global Environmental Change*, **17**(2), 45–64.

Falkner, R. (2016). The Paris Agreement and the new logic of international climate politics. *International Affairs*, **95**(2), 1–28.

Galaz, V., Crona, B. and Osterblom, H. et al. (2012). Polycentric systems and interacting planetary boundaries: emerging governance of climate change–ocean acidification–marine biodiversity. *Ecological Economics*, **81**, 21–32.

Green, J. (2014). *Rethinking Private Authority: Agents and Entrepreneurs in Global Environmental Governance*. Princeton: Princeton University Press.

Hale, T. (2016). 'All hands on deck': the Paris Agreement and non-state climate action. *Global Environmental Politics*, **16**(3), 12–21.

Hickmann, T. (2017). The reconfiguration of authority in global climate governance. *International Studies Review*, **19(**3), 430–451.

Hoffmann, M. (2011). *Climate Governance at the Crossroads: Experimenting with a Global Response after Kyoto*. Oxford: Oxford University Press.

Jordan, A., Huitema, D. and Hildén, M. et al. (2015). Emergence of polycentric climate governance and its future prospects. *Nature Climate Change*, **5**(11), 977–982.

Jordan, A., D. Huitema, H. van Asselt, J. Forster (2018a). Governing climate change: the promise and limits of polycentric governance. In: Jordan, A., D. Huitema, H. van Asselt, J. Forster (ed.). *Governing Climate Change: Polycentricity in Action*? Cambridge: Cambridge University Press, pp 357–383.

Jordan, A., D. Huitema, H. van Asselt, J. Forster (ed.) (2018b). *Governing Climate Change. Polycentricity in Action?* Cambridge: Cambridge University Press.

Jordan, A. and Schout, A. (2006). *The Coordination of the European Union*. Oxford: Oxford University Press.

Jordan, A.J. (2008). The governance of sustainable development: taking stock and looking forwards. *Environment and Planning C*, 26, 17–33.

Keohane, R. and Oppenheimer, M. (2016). Paris: towards the climate dead end through pledge and review? *Politics and Governance*, **4**(3), 142–151.

Kooiman, J. (1993). *Governing and Governance*. London: Sage.

Lemos M, Agrawal A, (2006). Environmental governance. *Annual Review of Environmental Resources* 31, 297–325.

Marks, G. and L. Hooghe (2014). Contrasting visions of multi-level governance. In: I. Bache and M. Flinders (eds) *Multi-level Governance*. Oxford University Press,

McGinnis, M. and Ostrom, O. (2011). Reflection on Vincent Ostrom, public administration and polycentricity. *Public Administration Review*, **72**(1), 15–25.

Morrison, T.H., W.N. Adger and K. Brown et al. (2019). The black box of power in polycentric environmental governance. *Global Environmental Change*, **57**, 101934.

Ostrom, E. (2009). *A Polycentric Approach for Coping with Climate Change*. World Bank Policy Research Working Paper Series

Ostrom, E. (2010a). Polycentric systems for coping with collective action and global environmental change. *Global Environmental Change*, **20**(4), 550–557.

Ostrom, E. (2010b). A long polycentric journey. *Annual Review of Political Science*, **13**, 1–23.

Ostrom, V. (1999). Polycentricity – Part 1. In M. McGinnis (ed) *Polycentricity and Local Public Economies*. Ann Arbor: University of Michigan Press, pp 52–74.

Ostrom, V., C. Tiebout and R. Warren (1961). The Organisation of government in metropolitan areas. *American Political Science Review*, **55**(4), 831–842.

Polanyi, M. (1951). *The Logic of Liberty*. Chicago: University of Chicago Press.

Roger, C., Hale, T. and Andonova, L. (2017). The comparative politics of transnational climate governance. *International Studies*, **43**(1), 1–25.

Rosenau J, (1992). Governance, order and change in world politics', in *Governance without Government* (Eds.) J Rosenau, E-O Czempiel, Cambridge University Press, Cambridge, pp 1–29.

Setzer, J. and M. Nachmany (2018). National governance. In: Jordan, A., D. Huitema, H. van Asselt, J. Forster (ed.). *Governing Climate Change. Polycentricity in Action?* Cambridge: Cambridge University Press, pp 47–62.

Skelcher, C. (2005). Jurisdictional integrity, polycentrism, and the design of democratic governance. *Governance*, **18**, 89–110.

Sovacool, B., M. Tan-Mullins, D. Ockwell, P. Newell (2017). Political economy, poverty, and polycentrism in the global environment facility's least developed countries fund (LDCF) for climate change adaptation. *Third World Quarterly*, **38**, 1249–1271.

Tarko, V. (2017). *Elinor Ostrom: An Intellectual Journey*. London and New York: Rowman and Littlefield.

Thiel, A., Blomquist, W.A. and Garrick, D.E. (eds.) (2019). *Governing Complexity: Analyzing and Applying Polycentricity*. Cambridge: Cambridge University Press.

Van Asselt, H., & Zelli, F. (2018). International Governance: Polycentric Governing by and beyond the UNFCCC. In: A. Jordan, D. Huitema, H. Van Asselt, & J. Forster (Eds.), *Governing Climate Change: Polycentricity in Action*? (pp. 29–46). Cambridge: Cambridge University Press.

Voß, J-P. and F. Schroth (2018). Experimentation. The politics of innovation and learning in polycentric governance. In: Jordan, A. et al. (Eds.) (2018). *Governing climate change. Polycentricity in action?* Cambridge: Cambridge University Press, pp. 99–116.

6
ECOLOGICAL MODERNIZATION AND BEYOND

Martin Jänicke and Helge Jörgens

6.1 Introduction

Until recently, the environmental policies of developed and, to a lesser extent, developing countries have been relatively successful (see Chapter 7). The political strategy underlying the substantial advances, which marked the period from the early 1980s to the mid-2000s, especially in areas such as pollution control, energy efficiency, resource management, and the transition to renewable energies, has been described as "ecological modernization" (Jänicke, 1985; Hajer, 1995; Gouldson and Murphy, 1998; Jänicke, 2009; Mol et al., 2009b). For several decades, the greatest strength of the ecological modernization program was its compatibility with economic growth (Cohen, 2013), which turned it into the industrialized countries' primary answer to the "Limits to Growth" (Meadows et al., 1972). Policies under the ecological modernization paradigm managed to reconcile economic growth with environmental preservation by systematically making products, processes, and services more eco-efficient. Instead of stifling economic growth, ecological modernization triggered an impressive *growth of the limits* that lasted for decades.

Today ecological modernization is a well-established policy strategy that intends to preserve or restore environmental quality by resource-efficient innovation. Moreover, ecological modernization has emerged as a "powerful political discourse, in which economic growth, environmental protection and energy security are mutually reinforcing" (Machin, 2019, p. 208). This process of eco-innovation has led to rapid and global technological change driven by pioneers such as Germany, Denmark, Sweden, California, and China. It has become a real global industrial revolution (Rifkin, 2011).

However, despite these successes, in recent years – and particularly in light of the threat of global warming and the international consensus to keep the global temperature rise below 2 °C (preferably 1.5 °C) – the gap between environmental goals and actual improvements has begun to widen (UN Environment, 2019a; IPCC, 2018; OECD, 2019; European Environment Agency, 2019). In the face of an "increasing scale, global reach and speed of change in [many] drivers of environmental change" (UN Environment, 2019b, p. 7), some of the former strengths of the ecological modernization approach to environmental policymaking may have turned into weaknesses. Its reliance on relative, not absolute, reductions in emissions and resource consumption is increasingly being compensated by absolute growth (for energy efficiency, see

 DOI: 10.4324/9781003043843-7

European Environment Agency, 2019, p. 154). At the same time, its strong dependence on the availability of feasible technological solutions (Fisher and Freudenburg, 2001, p. 702; Mol and Jänicke, 2009, p. 20) reduces its effectiveness in areas such as climate change, biodiversity loss, and land use where technological innovation is either not available, still in an experimental stage of development, too expensive, or strongly conflicts with entrenched polluter interests.

This recognition that today's best practice in environmental policymaking is insufficient for solving the most persistent and potentially disruptive environmental challenges – above all climate change and loss of biodiversity – has triggered a debate about the future of ecological modernization and of environmental policy as a whole.

In this chapter we explore the contradictory nature of ecological modernization. On the one hand this modernization is characterized by an extremely high dynamic. It has become highly successful, for example regarding the use of renewable energy in the power sector, waste recycling, or eco-efficient water supply. In addition, several economic co-benefits, such as additional employment, reduced production costs, or the stimulation of technological innovation, have helped increase the political feasibility of policies under the ecological modernization paradigm. There are on the other hand significant weaknesses: the environmental effectiveness of ecological modernization is often only selective and restricted to market-based solutions. Its high political feasibility can lead to a preference for "easy" policies. Ecological modernization is also not an effective alternative where it only adds clean(er) technology to an existing "dirty" branch structure. There are also problems of equity regarding the attribution of costs or the global distribution of benefits. The strengths of ecological modernization are as remarkable as its weaknesses. This will be exemplified by the case of Germany.

We aim to contribute to a differentiated evaluation of ecological modernization as an environmental policy strategy. We argue that its political feasibility, and consequently its success, has been high with regard to policies aiming to increase the eco-efficiency of products, production processes, and services, though it is much lower if the aim is the ecological transformation of entire sectors of the economy (e.g. phasing out fossil fuels). After a presentation and evaluation of the concept of ecological modernization, we try to identify some characteristics of environmental policies which move beyond its current conceptualization. In particular, we argue that a better understanding and a clear definition of environmental objectives *beyond* "traditional" ecological modernization is necessary. A policy of structural change of industries and infrastructures is the necessary supplement to ecological modernization on the road to sustainable development. Strict goal-oriented approaches and increasing the political feasibility of second-generation ecological modernization policies through capacity building and innovative modes of governance also appear to be important if the full potential of ecological modernization is to be realized. Regarding the "planetary boundaries" of resource-intensive growth, there is another imperative that goes beyond the logic of ecological modernization. It is the old idea of sufficiency (see Chapter 25), which so far does not have the same success story as the modernization approach. Sufficiency as a political strategy could be strengthened if it had a more selective focus on the consumption of only those products and services that are characterized by a high resource- and energy-intensity and were supported by adequate policy instruments.

6.2 The Concept of Ecological Modernization

Ecological modernization refers to the innovation and diffusion of technologies that have a positive impact on the environment and the efficiency of resource use. This process has so far been essentially market based (whether it could transcend the logic of markets could be a relevant topic for the future). Furthermore, ecological modernization is the intended transition from a highly polluting techno-structure to an *ex ante* more environmentally friendly technology base of the economy. So far the process of ecological modernization has been primarily policy-driven. Therefore, in the field of environmental policy analysis, the term "ecological modernization" is used to describe environmental policies that move beyond traditional pollution control towards an innovation-oriented "green industrial policy" (Walz, 2015; Altenburg and Assmann, 2017). The expected result of ecological modernization policies is a decoupling of economic growth and environmental degradation. Today there are several synonyms or similar concepts such as "green development", "green innovation", "green growth", "Green New Deal", or transition towards a "green economy" (OECD, 2011; UNEP (United Nations Environment Programme) 2011).

The term "ecological modernization" has its origins in the German environmental policy debate of the 1980s. Later on it turned into a political formula of the red-green German government (1998–2005) with the aim of connecting environmental policy with the fields of innovation and employment policy. It was first used by one of the authors of this chapter in a programmatic parliamentary speech in January 1982 (Abgeordnetenhaus von Berlin, 1982, pp. 756–757) and later on in a special issue of the journal *NATUR* (Jänicke, 1983; 1985). The term then was adopted by the so-called "Berlin School of Environmental Policy Research" (Simonis, 1988; Zimmermann et al., 1990; Prittwitz, 1993; Foljanty-Jost, 1995; Mez and Weidner, 1997; Weidner, 2002; Jänicke and Jacob, 2006) and later by other authors in the German public environmental policy debate (e.g. Hauff and Müller, 1985).

The basic idea of ecological modernization as a distinct approach to environmental policy-making is to use the inherent pressure for innovation in competitive market economies to transform the resource- and environment-intensive mode of industrialism. Its basic assumption is that protecting the environment can be compatible with economic growth. The main approach behind ecological modernization policies, thus, is to increase the eco-efficiency of production processes, products, and services by means of technological innovation. This includes, but is not limited to, reducing the input of materials and energy per produced unit, reducing pollution and emissions from products, production processes, and services, and recycling products and materials after their use. Thus, as a political strategy, ecological modernization comprises policies that promote technological solutions to increase the eco-efficiency of products, production processes, and services. A similar idea is that of "ecologizing the economy" which had been developed roughly at the same time (Huber, 1982).

Since the early 1990s, the concept of ecological modernization has been firmly established in the environmental policy debate (Spaargaren and Mol, 1992; Weale, 1992; Hajer, 1995; Young, 2000; Mol, 2001; Mol et al., 2009a; Andersen and Massa, 2000). While authors such as Mol and Spaargaren discussed ecological modernization more broadly in the context of the sociological theory of modernity, the German discourse had a stronger policy orientation and focus on policy advice. Jänicke and Jacob stressed for instance the multiple interactions between policy and technology development. This was illustrated in particular by the example of policy-induced *lead markets* for clean technologies (Jänicke and Jacob, 2006).

Today, policies under the ecological modernization paradigm constitute the dominant environmental policy approach in most industrialized countries. One of the main reasons for their broad acceptance and diffusion is their high political feasibility. Following May (2005, pp. 129–130) we define political feasibility as the political risk decision-makers face when supporting or opposing (a policy's) adoption. Several distinct characteristics of ecological modernization policies explain why their political feasibility is high compared to other types of policy. First, their basic compatibility with the prevailing growth-oriented economic model of Western industrial societies leads to a fundamental acceptance of these measures. Second, the economic system's inherent dynamic of continuous improvement of processes and products is used while only the criteria for improvement need to be changed to include ecological ones. Third, ecological modernization measures have a high potential for win–win solutions (environmental protection, competitiveness, job effects, etc.), which improves their political acceptance and enforceability (see Section 6.3.1). Fourth, ecological modernization measures are directed at a relatively small number of collective actors (companies, associations), which makes it more likely that the agreed measures will be implemented.

6.3 Ecological Modernization as a Global Green Industrial Revolution

The growing attention to ecological modernization and the idea of greening the economy resulted, among other factors, from the study of success stories of market-based environmental innovations (Jänicke and Weidner, 1995; Jänicke, 1996). These innovations are characterized by *high speed* and their *global dimension* of change. Today, most countries have a rapidly growing "green" *clean-tech sector*, that is, an industry sector producing "environmental technology and resource efficient products, processes and services" (Bundesministerium für Umwelt et al., 2018, p. 90). The companies in this sector provide the innovative technologies that underlie strategies of ecological modernization in the entire economy.

A study by the Roland Berger think tank estimates the global market volume of the main segments of the green clean-tech industry to have been €4.628 billion in 2020 (Bundesministerium für Umwelt et al., 2021). The authors expect a market volume of €9.383 billion by 2030, with an annual growth of 7.3%. The growth rate has increased in recent years. Table 6.1 shows the clean-tech segments and their annual growth rates.

The speed of change in the renewable energy sector is particularly high. In past decades, many countries experienced a surprising and largely unexpected acceleration of the diffusion of wind and solar power (Jänicke, 2012a). Generally, a dynamic of positive political feedback can be observed, which often led to a continuous increase in ambition. For example, Germany started in 2000 with a target of 20% of power from renewable energy to be achieved by 2020. This goal was controversial at the time and considered unrealistic by many. But due to a rapid diffusion of renewable energy technology, the target was raised to 30% nine years later and another year later it was raised to 35%. The share of electricity from renewable energy sources of 45.4% actually achieved in 2020 even exceeded these ambitious targets (Umweltbundesamt, 2022). This race between targets and achievement can also be observed in other countries. The most interesting example is China (see Figure 6.1) which increased its target for photovoltaic capacity six times between 2010 and 2018 and nevertheless surpassed its goals in the target year 2020.

Table 6.1 Annual growth rate of the main sub-groups of the global clean technology industry, 2020–2030 (forecast %)

Sustainable agriculture and forestry	11.3
Sustainable mobility	8.7
Green energy supply	8.5
Material efficiency	8.4
Energy efficiency	6.3
Waste management and recycling	5.9
Sustainable water management	4.2
Total clean technology	**7.3**

Source: Bundesministerium für Umwelt et al. (2021).

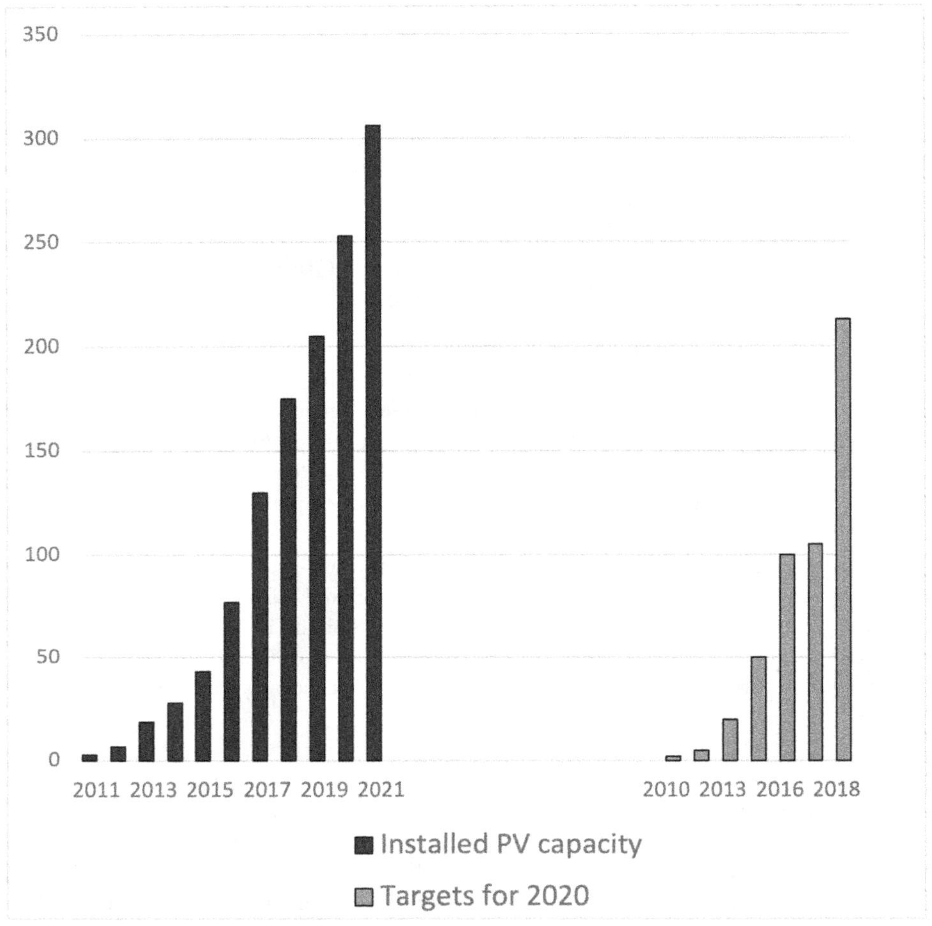

Figure 6.1 Installed PV capacity 2011–2021 in China + targets for 2020 (GW).

The diffusion of climate-related *targets and policies* is similarly rapid and is also global in scope. In 2017, a total of 157 countries worldwide had adopted national targets for energy efficiency (REN21, 2018) and, in 2019, 172 countries had introduced national renewable energy targets (REN21, 2020). It is remarkable that these diffusion processes are completely voluntary and mainly based on "lesson-drawing" from pioneer countries (Jänicke and Wurzel, 2019). The rapid catching-up process in developing countries is another interesting aspect of this rapid global change (REN21, 2018).

6.3.1 Co-Benefits of Ecological Modernization

How can this rapid worldwide political and technological change be explained? One explanation is of course the influence of global policies such as environmental and climate policy, supported by the policy formulation process of sustainable development since the UN summit in Rio de Janeiro in 1992. This was an interactive process of policy learning in the global system of multi-level governance which has developed from that time on (Zürn, 2012; Ostrom, 2010).

Peer-to-peer learning about ecological modernization as a policy strategy began as a process between national governments, as lesson-drawing from pioneers. Meanwhile this "horizontal" learning process can be observed also at the level of provinces/states as well as cities or even villages (Jänicke, 2017b).

A second explanation of the global dynamics of ecological modernization is the observation that policies to increase the eco-efficiency of products, production processes, and services often bring about relevant co-benefits beyond the specific environmental goals of these policies. Co-benefits were first discussed in the context of climate mitigation by Mayrhofer and Gupta (2016). But the concept can be extended to ecological modernization and the green economy in general. Early on, co-benefits became a "no-regret" argument according to which many climate policy measures can be legitimized by their positive economic side-effects alone (Adler, 2000). Over time, in addition to positive side-effects, *multiple* benefits were increasingly addressed. In 2014, the International Energy Agency (IEA) published a list of 15 potential co-benefits which may result from increased energy efficiency (IEA, 2014). The Fifth Assessment Report issued by the Intergovernmental Panel on Climate Change (IPCC, 2014) arrived at 18 potential economic, ecological, and social co-benefits of climate change mitigation (see Figure 6.2).

The global co-benefits of the transition towards renewable energy have repeatedly been reported by the International Renewable Energy Agency (IRENA, 2017). For 2050 it expects 100 million jobs in the transformed energy sector, an increase of 72% from 2017 (IRENA 2017, 2020, p. 100). The World Bank has calculated the health and energy benefits from climate mitigation in China, India, USA, EU, Brazil, and Mexico for 2030 to be $1.23 trillion (World Bank, 2014).

Reduced material flows	**Economic co-benefits**	**Environmental co-benefits**	**Social co-benefits**
Mining ⇩	Lower costs of:	Reduced:	Improved:
Basic industries ⇩	- Materials - Energy - Water - Transport - Land use - Pollution control	- Emissions - Dissipative losses - Waste - Loss of living space - Loss of species and functions	- Health - Employment - Taxes - Rural development - …
Manufacturing ⇩	Competitiveness		
Retail trade ⇩	Innovation		
Final consumption ⇩	New materials		
Waste management	Income		
	Balance of trade		

Figure 6.2 Potential economic, ecological, and social co-benefits of resource efficiency along the supply chain.

Source: Jänicke (2017a).

It is not easy to explain why the "multiple benefit approach" has been restricted to climate protection. Ecological modernization may have even more co-benefits as it has a broader focus which goes beyond energy efficiency to include all kinds of eco-efficiency. Like policies to reduce the carbon intensity of the economy, ecological modernization polices are characterized by the double advantage of increasing resource-efficiency and lowering environmental damage which often go hand in hand with a stimulation of technological innovation, the reduction of production costs through more efficient resources use, the possibility of entering new markets, and the creation of additional jobs. A study by Altenburg and Assmann (2017) describes the potential co-benefits of a green industrial policy also for developing countries, ranging from health benefits and energy security to the opening of new markets, increased net employment, and higher productivity. Co-benefits therefore are an important argument for ecological modernization in general and a major reason for the comparatively high political feasibility of policies under the ecological modernization paradigm. Together with environmental and resource-related concerns they are an important explanation for the Green Industrial Revolution.

This can be exemplified by the flow of material resources along the value chain. At each stage of the production process there is a broad variety of resource uses, the reduction of which can lead to economic, ecological, and social co-benefits (Figure 6.2).

6.4 The Case of Germany

The OECD has characterized Germany "as a highly innovative country", drawing "the maximum benefits and opportunities of globalization to address environmental problems while boosting its environmental industry sector" (OECD, 2007, p. 43). This may be a good description of the German environmental modernization strategy, which from the beginning had a strong orientation towards economic co-benefits. The country is a dominant player in the global clean technology market. According to the Global Innovation Index it belongs to the leading top ten countries of innovation (Bundesministerium für Umwelt et al., 2021). There are other countries with a similar eco-innovation approach to environmental policy such as Sweden, Denmark, China, Japan, or US states like California. Germany is an interesting case because its approach is explicitly based on the concept – and term – of ecological modernization.

Compared with other industrialized countries, Germany has not only been a forerunner of ecological modernization policies but has also introduced the term "ecological modernization" into the language of environmental policymaking. As Weale (1992, p. 79) argues, "there were elements in the ideological and institutional traditions of German public policy that made certain elements of ecological modernisation both a legitimising device of public policy developments and a potential source of policy principles". When the red-green government, a coalition of Social Democrats and Greens, came into power in 1998, it explicitly placed green industrial policy in Germany under the heading of ecological modernization. The second coalition agreement of this government, adopted in 2002, explicitly mentions the potential co-benefits of ecological modernization, such as the "integration of labor and environment", based on "increased eco-efficiency, lower production costs and improved competitiveness" (SPD und Bündnis 90/Die Grünen, 2002).

Since 2005, the German government has been publishing a regular report on "GreenTech in Germany". The report is a regular statistical review of the GreenTech industry (see Table 6.1). This is the hard core of the larger "green economy" of the country, which also includes such activities as organic farming, eco-tourism, and green finance. The GDP share of the turnover of the GreenTech sector grew steadily from 8% (2007) to 15% (2016) (Bundesministerium für Umwelt et al., 2018). The expected annual growth rate between 2020 and 2025 is 9.9% (Bundesministerium für Umwelt et al., 2021). The expected growth of employment during 2020–2025 is 6.8%. No other industrial sector has similar growth in Germany, neither the successful car industry nor the strong mechanical engineering sector. The term "ecological modernization" continues to be central to the German government's communication about its green industrial policy: "environmental technology and resource efficiency promote the green transformation in all branches of industry, injecting powerful stimulus for ecological modernization" (Bundesministerium für Umwelt et al., 2018, p. 13).

The estimated employment effect of the German GreenTech sector is 1.5 million jobs (Bundesministerium für Umwelt et al., 2018). The expected growth rate up until 2025 is 6.8% (Bundesministerium für Umwelt et al., 2018). Employment in the broader environmental sector in Germany was calculated in a different study at 2.8 million in 2017 (Umweltbundesamt, 2020).

Not only the employment but also the productivity of the German industry seems to have benefited from the policy-driven ecological modernization of the economy. This is true at least for the chemical industry, which, according to a study by the German Chemicals Industry Association (Verband der Chemischen Industrie, VCI), has derived more than one additional benefit from ecological modernization. With sales growth of 41.2% between 2000 and 2013, it has reduced energy consumption by 12.6% in absolute terms, water consumption by 20.8%, and waste generation by 61.9% (Verband der Chemischen Industrie, 2015). This, of course, also means a reduction in costs, which seems to have contributed to the export successes of this sector. The German company BASF claims to have systematically increased its energy efficiency and optimized its production processes in recent decades. This has led to a 50% reduction in emissions since 1990, although production has doubled in the same period (BASF, 2019). This improved resource-efficiency in the German chemical industry also seems to have reduced the danger of carbon leakage, that is, the relocation of production to countries with less stringent emissions regulations.

An indicator of the acceptability and feasibility of resource efficient eco-innovation may be the occurrence of illegal pollution in Germany, which increased until 1998 and was steadily reduced in the following years. The main issue was illegal waste disposal, which gradually became obsolete, not only due to better waste collection and regulation. An important reason was the increased value of recycled waste. The recycling quota of total waste in Germany was 69% in 2017 (Statistisches Bundesamt, 2019).

6.5 Strengths and Weaknesses: Evaluation of Ecological Modernization as a Global Environmental Policy Strategy

Irrespective of the undisputed environmental successes that ecological modernization policies and the idea of greening the economy have stimulated, a differentiated assessment is necessary. Using generally accepted criteria of policy evaluation (e.g. Wollmann, 2007; IPCC, 2014, p. 1156) – *effectiveness, economic efficiency, distributional equity, and political feasibility* – the following general assessment can be made. Ecological modernization has achieved a high speed of change and a global scale of diffusion because of its conformity with market mechanisms and economic growth. Its *political feasibility is high.* One reason is the high *economic efficiency* of the approach which aims at resource saving and often leads to lower production costs, including lower investment in unproductive end-of-pipe technology. This is part of the potential economic *co-benefits* of ecological modernization, from increased competitiveness to innovation and employment. The realization of co-benefits can trigger the dynamics of positive policy feedback (Pierson, 1993). In essence, then, ecological modernization is an approach based more on interests and less on normative standards (van Schaik and Schunz, 2012), is more voluntary than legally binding, and is less of a burden than an opportunity. The strength of ecological modernization and its global dynamic as an industrial revolution are economic efficiency and political and technological feasibility.

There are however also weaknesses, which have been criticized in recent times. This critique refers mainly to the evaluation criteria of *effectiveness* and *equity*. The effectiveness (the degree of achievement of the intended goal) has been disputed by several authors (York and Rosa, 2003; Ewing, 2017). And indeed, many environmental problems have not been solved or they have even worsened. Soil pollution, waste prevention, loss of species, and particularly climate change are examples of these largely unsolved environmental problems.

Another weakness of ecological modernization is growth effects, for example the renewed increase in environmental pollution after a temporary improvement (Jänicke, 1985). Some authors emphasize the rebound effect, where increased resource efficiency may be neutralized by higher resource consumption due to financial savings (Gillingham et al., 2014). The alternative to both problems could be a focus on *radical* innovations (photovoltaics, energy-plus buildings, electric vehicles), where the positive environmental effect cannot easily be neutralized by higher consumption. Other options are dynamic targets and standards, or a clear cap for absolute emissions or levels of pollution.

A significant problem is the selectivity of ecological modernization measures. As an innovation-based approach, it has so far been essentially restricted to technical and marketable solutions. However, not every environmental problem has a technical solution (e.g. the loss of biodiversity). And focusing on technology can lead to an underestimation of non-technical solutions such as behavioral changes. In many cases, there is also no market, or the prevailing market rationale may undermine the policy goal of improving the environment (Machin, 2019). Often a strict policy intervention is necessary to prevent environmental deterioration. The loss of productive soil is an example. Another problem is the one-sided focus on hard technologies. The "greening of the economy" must, of course, include first and foremost truly green and nature-based solutions and not all of them can be market-based. In many parts of the world – especially in poor rural villages – green and nature-based development has become a great opportunity. This is also green innovation and modernization, but it depends largely on public investment.

There is still another weakness of ecological modernization which has been criticized in recent times: the dimension of *equity*, or the fairness of this process, for instance regarding its global impacts (e.g. Bonds and Downey, 2012; Ewing, 2017). Often, measures of ecological modernization do not take into account the polluter-pays principle. A German study on the *Energiewende* shows overwhelming support of 90% for the transition towards renewable energy, but 67% of the respondents say the costs of this policy will be borne by "ordinary people" (Setton et al., 2017). This constitutes another equity problem: the relocation of "dirty industries" from industrialized countries to less advanced countries. The final production in developed countries may be relatively "clean"; however, the early stages of the production line are often highly polluting and located in developing countries. Consumption of imported goods in the rich world can be generally cleaner than the production in poorer countries.

6.5.1 Insufficient Structural Change

Insufficient structural change is one of the main explanations of the deficits of ecological modernization. Phasing-out "dirty" sectors or infrastructures is a difficult task. Its feasibility has been particularly low so far. Its aim is not only technological change but the change of basic interest constellations (Jänicke, 1985). It can rely neither on relevant co-benefits nor on market rationality. Although there is a common belief that innovation leads to "creative destruction" of the former technology (Schumpeter, 1942), this is not always true: "green" innovation can coexist with the "brown" sector which remains more or less untouched. In other words, ecological modernization is often only additive where it should be substitutive.

This can be illustrated by the example of Germany. The country has so far been the world's leading player in the lignite sector. Despite remarkable green innovation in the energy sector with a rapid increase in renewables, the share of lignite coal remained stable through 2018. Another example is the German chemical industry. Thanks to green innovations, it has become very resource-efficient and has thus reduced its main emissions. However, the impact of its *products*, for example on biodiversity, is often seen as extremely negative. The use of pesticides in German agriculture has increased, although the agricultural area is decreasing. The rural biodiversity indicator in 2013 was one-third lower than in 1975 (Statistisches Bundesamt, 2017).

In other countries, too, ecological modernity and traditional "dirty" industries coexist: organic farming and industrialized agriculture, fuel-efficient cars and an oil industry that continues to grow. Ecological modernization therefore needs structural change in order to become sustainable, namely the gradual phase-out of polluting industries. Ecological restructuring must be seen as a necessary second step of ecological modernization (Jänicke, 1985). We refer to this second step as *ecological modernization 2.0* (see Section 6.7). Policies to steer and accelerate ecological restructuring processes are characterized by significantly lower political feasibility compared to traditional ecological modernization measures. The main reason for this is that the economic losers of the ecological transformation are clearly identifiable and will tend to oppose any policy aimed at accelerating this process.

Recently, however, there have been attempts to overcome the weaknesses of ecological modernization, including the difficult task of *eco-restructuring*. As for the energy core of the "Green Industrial Revolution," phasing out coal has become the standard in some countries such as Canada or the United Kingdom. German lignite can again be used as an example after a phase-out policy was adopted in January 2019. This change in policy was brought about by a special commission on a broad stakeholder basis. Generous compensation was provided for the losses of both employees and companies. The EU launched a general process of ecological restructuring in 2019 (now conceived as "just transition") involving massive compensation payments. US President Biden launched a similar new approach in 2020.

6.6 Ecological Modernization as Political Modernization

This reminds us that ecological modernization and the idea of greening the economy have always had a broader political dimension beyond its main goal of protecting the environment. Innovation of institutions, policy, politics, and governance has been the necessary condition of this process. As Caldwell stated already in 1974, the "holistic character of environmental and ecological concepts has influenced many public environmental policies and has required a greater degree of agency cooperation and coordination than has traditionally characterized the administration of government" (Caldwell, 1974, p. 727). Consequently, environmental policy integration, that is, the incorporation of "environmental concerns into the decision-making procedures of non-environmental policies" (Jacob and Volkery, 2004, p. 291), has been an important pillar of environmental policymaking since the late 1980s (see Chapter 9). Mechanisms and institutions of environmental policy integration have supported new environmental and climate policy approaches in other policy domains and sectors of the economy. In the field of renewable energy, the innovative policy instrument of feed-in tariffs as a support scheme for electricity from renewable sources has been a strong driver of green electricity, including its global diffusion (Busch and Jörgens, 2012). An interesting policy innovation was the Japanese "Top Runner Program" regarding energy efficiency of products, followed by the (less ambitious) European Eco-Design directive. Another example of innovative policy instruments developed to accelerate the ecological modernization of the economy is the creation of national lead markets supporting the technical learning process which enables a global diffusion (Jacob et al., 2005; Beise and Rennings, 2003) Environmental innovations have also been supported by new modes of participation and networking. Goal-oriented approaches have steadily increased their influence since the 1990s (Kanie and Biermann, 2017). The above-mentioned phenomenon of a race between targets and achievements has led to a new reflexive governance mechanism of stimulating climate policy ambition in the Paris Agreement (Article 3). Emission trading was another important innovation in climate policy. The policy of phasing out coal is one of the more recent examples; its rate of international diffusion is low but increasing.

New institutional arrangements and governance mechanisms supporting ecological modernization could also be observed in the *international* policy arena. The development of a global system of multi-level governance after the UN Summit in Rio de Janeiro (1992) was an important framework condition for horizontal and vertical "green" policy learning. Multi-level governance has become a motor of innovation and diffusion by lesson-drawing particularly in climate policy (Zürn, 2012; Jänicke, 2017b). A strategic global policy innovation – relevant also for ecological modernization – has been the introduction of the 17 Sustainable Development Goals. They have created a globally accepted long-term development program and new standards for national public policy. This has also become a system of synergies, where the co-benefits of ecological modernization can find reinforced attention.

Such green governance innovations show that the ecological modernization of politics and governance goes beyond the logic of markets. It has its own inherent logic based on necessities and outcomes. The concept of ecological modernization should therefore not be limited to market-based solutions. The phasing-out of "dirty" branches which cannot be transformed as such is the most critical aspect of structural change as a condition of successful ecological modernization. Changing infrastructures (e.g. the transportation system) or consumption patterns are other types of structural change that have so far been characterized by low political enforceability. However, the political feasibility of these approaches can be increased through policy learning and capacity building (Jänicke and Weidner, 1997). This can be particularly the case when the pressure for change becomes more and more dramatic, as will probably be the case with the impacts of climate change.

6.7 Beyond Ecological Modernization

The recognition that environmental policies under the ecological modernization paradigm so far have been insufficient for solving some of the most persistent and potentially disruptive environmental challenges – above all climate change and loss of biodiversity – has renewed the debate about alternative policy paradigms. We distinguish two major types of policy that move beyond the present approach of ecological modernization (see also Table 6.2).

First, *ecological modernization 2.0* type policy programs aim to move from marginal increases in the eco-efficiency of products, processes, and services, towards a comprehensive restructuring of public and private infrastructures, an "upgrading of the entire production system to environmental and resource-saving processes and products" (Jänicke, 2012b, p. 14), including the imperative of just transition. This type of policy would aim for a sustainable "green economy". While being non-incremental and based on an absolute rather than a relative decoupling of

Table 6.2 Policy paradigms in environmental and climate governance

Policy paradigm	*Interpretive framework*	*Policy objectives*	*Policy instruments*	*Institutional requirements*
Ecological modernization	Compatibility of environmental protection and economic growth	Increase eco-efficiency through technological innovation	Eco taxes, incentives for green/eco-efficient investment, top-runner program, etc.	Institutional mechanisms for environmental policy integration
Ecological modernization 2.0	Absolute decoupling of growth from emissions/resource consumption, climate engineering	Green transformation of branches and infrastructures, "green solutions"	Long-term investment programs, consensual transition management, etc.	Green new-deal type long-term strategy
Sufficiency	Current lifestyles and economic system are unsustainable, limited capacity of the Earth system	Change lifestyles and consumption patterns; impose limits on polluting economic activities	Individual carbon allowances, bans on disposable plastic bags, dietary rules, etc.	Modification of national accounts to include non-material aspects like well-being

economic growth from environmental pollution and resource consumption, these programs are – at least in theory – compatible with continued, albeit dematerialized, economic growth (Ferguson, 2015; Jackson and Victor, 2019). Policies under this paradigm include large-scale investments into national energy, transport, or building infrastructures (for a detailed review of policy instruments in the building sector, see Lucon et al., 2014, pp. 715–722; Ürge-Vorsatz et al., 2007) but also compensatory mechanisms. Typically they also comprise attempts to systematically integrate environmental indicators into systems of national accounts (United Nations Statistical Commission, 2017).

Another type of environmental policy innovation, which can also be subsumed under the label of "ecological modernization 2.0" and might become relevant in climate policy, consists of large-scale technological interventions to deliberately modify the Earth's climate system. These interventions are discussed under the terms *geoengineering* or *climate engineering* (The Royal Society, 2009; Burns and Nicholson, 2017; Jinnah et al., 2019, see also Chapter 27). In contrast to most environmental policy approaches, climate engineering does not attempt to reduce emissions, but aims to break the causal chain between emissions and potentially harmful environmental change (Dessler and Parson, 2005, p. 90). Policies under the geoengineering paradigm include attempts to reduce the amount of sunlight absorbed by the Earth's atmosphere (solar radiation management) (Jinnah et al., 2019) and programs to remove CO_2 from the atmosphere and store it safely, ranging from ocean iron fertilization (Abate, 2013) and large-scale afforestation to bioenergy with carbon capture and storage (BECCS) (Michaelson, 2013; Geden et al., 2019). What distinguishes these approaches from classical ecological modernization policies is that they are based on high-risk technologies with potentially catastrophic consequences. The higher risk associated with them reduces their public acceptance and political feasibility (Chapter 27).

Second, the apparent limits of efficiency-based policies have brought *sufficiency-based policies* back into the debate (Princen, 2003; Büchs and Koch, 2017; García et al., 2017; Spangenberg and Lorek, 2019; Burger et al., 2020, see also Chapter 25). What they have in common is the idea that the planet's environmental capacities are limited and that policy constraints on consumption and production are needed (Gaffney and Rockström, 2021). Building on Herman Daly's (1972) concept of a steady-state economy, the main thrust of these proposals is to slow down or even halt economic growth. Policies under this paradigm either attempt to change individual lifestyles by "limiting the consumption of environmentally unsound goods and services" (Spengler, 2016, p. 922; see also Gilg et al., 2005; Jackson and Smith, 2018) or to try to define and impose limits on ecologically harmful economic activities (Alexander, 2012; Koch, 2020). While scholars agree that sufficiency "needs to be supported by ... policy frameworks" (Lorek and Spangenberg, 2019, p. 26), we still lack empirical assessments of the extent to which policies based on the sufficiency principle have found their way into national environment and climate policy mixes. This principle is older than the concept of ecological modernization. However, it has not had the same success story so far. This may be partly due to the often-unclear connection with the GDP growth debate, partly due to a lack of practical policy instruments. Without fully entering the growth debate, we would limit ourselves to arguing that zero growth of polluting industries is not enough if absolute reduction or phase-out is required. However, investing in "green" alternatives means nothing other than economic growth. Even when combined with phase-out processes, this will lead to at least moderate overall GDP growth. Therefore, a differentiated approach to growth is needed.

Following Heyen et al. (2013, pp. 14–17), we can distinguish four types of instruments for sufficiency policy: regulatory instruments (e.g. product bans, car bans in city centers, speed limits, regulating planned obsolescence), economic instruments (e.g. kerosene taxes, property taxes, personal carbon allowances), planning and infrastructure provisions (e.g. priority lanes for car-sharing, vegetarian offers in public canteens), and information-based instruments (e.g. public campaigns to eat less meat or to lower room temperatures in winter) (see also Maitre-Ekern and Dalhammar (2016) for the regulation of planned obsolescence; Parag et al. (2011) for personal carbon allowances).

6.8 Conclusion and Future Research

Ecological modernization and the transition to a greener economy are taking place with unexpected speed and the underlying policies are spreading worldwide. By now, ecological modernization has reached the dimensions of a global green industrial revolution. It is strong where radical technology-based improvements take place – for example the transition to renewable energy, energy-plus buildings, electric vehicles, drought-resistant plants, water recovery techniques, or certain bio-based products. As a rule, these transformation processes are supported and accelerated by environmental policy measures. Ecological modernization has often succeeded in combining ecological necessities with the pressure to innovate which is inherent in competitive market economies. Policies under the ecological modernization paradigm, thus, are driven by both the pressures of the environmental and climate crisis and the co-benefits of resource-efficient innovation. Their global diffusion has been the result of processes of policy learning which, in turn, have been strongly supported by the polycentric framework of multi-level governance (Chapter 5) as a multi-impulse system of interactive learning.

However, the environmental effectiveness of ecological modernization and the underlying policies is increasingly reaching its limits and the distributive justice and fairness of policies under this paradigm are controversial. The environmental improvements achieved through ecological modernization have often been canceled out by growth and rebound effects. "Green" sectors often coexist with "brown" industries (especially the fossil fuel sector) because some sectors have successfully resisted structural change. Success is often selective: environmental problems such as loss of biodiversity typically cannot be tackled by marketable technologies. The positive side-effects of eco-innovation in rich countries often go hand in hand with increased environmental damage in the countries where the resource-intensive input into the value chain is produced. Distributive justice is also violated when the polluter-pays principle is not respected. Ecological modernization policies are attractive because of their high acceptance and feasibility. However, the option of easy measures can undermine the political will to adopt more difficult measures that need to overcome strong opposition. Moreover, the dominance of market rationality in the ecological modernization discourse can weaken the political rationality of those policies that are ecologically necessary but less compatible with economic growth. Overall, ecological modernization has demonstrated that market dynamics can be successfully put at the service of environmental policy. However, strong and stable political leadership is needed to realize this potential. Left to their own devices, markets lack the ultimate responsibility for effective and equitable solutions.

Thus, while ecological modernization is a paradise of feasibility and opportunities, it is not a general solution to all environmental problems. Market-based green innovations are no substitute for a strict problem-based and goal-oriented environmental policy. Their inherent logic is different. This leads to the wider political dimension of ecological modernization. It has been always important but now it has become crucial. Political modernization has been the main driver of ecological modernization. This process is essential as far as environmental problems with a lower feasibility are concerned. Feasibility can be increased by policy learning and capacity building. There has been a push for green political modernization in the EU from 2019 and in the USA from 2020. This process of policy learning and capacity building must not only go on, but it should also be intensified. In addition, the traditional idea of sufficiency could be revitalized as a supporting guideline. It could be strengthened if it is (1) focused on physical production and consumption (avoiding the general growth debate) and (2) supported by specific policy instruments.

Future research could focus primarily on four topics. First, we need systematic and comparative analyses of the limits of ecological modernization as an environmental policy strategy and how these limits can be extended. To what extent can environmental policies be designed to not only increase the relative eco-efficiency of products, production processes, and services, but also to bring about absolute improvements despite continued economic growth? What strategies are needed to transform or replace traditional polluting sectors of the economy, such as the fossil fuel sector? How can these necessary transformations be accelerated politically?

Second, ecological modernization as a policy strategy has been extensively studied in highly industrialized countries, especially in the European Union and OECD member states. But we still lack research on the potential, feasibility, and effectiveness of this type of environmental policy in developing countries. How can successful measures to increase the eco-efficiency of products, production processes, and services in industrialized countries be extended to those countries where much of the resource extraction and manufacturing occurs? A mapping of best practices of ecological modernization policies in countries of the Global South could fill this gap and contribute to a better understanding of the factors that facilitate or hinder the adoption and implementation of these policies. Successful technology transfer is an important precondition for a broader adoption and implementation of policies under the ecological modernization paradigm in developing countries. Research could focus on the opportunities and restrictions of technology transfer under the United Nations Framework Convention on Climate Change (UNFCCC) to explore the potential for accelerating the global diffusion of ecological modernization policies.

Third, the weaknesses of the traditional ecological modernization approach suggest that paradigmatic policy change in the environmental and climate policy domain might already be underway. Consequently, there is a need for the development of forward-looking typologies of environmental policy paradigms and empirical assessments of actual environmental policy change in industrialized and developing countries. What types of environmental policies and measures have been adopted in past years? Are policies that belong to the ecological modernization paradigm still the dominant type? What other types of environmental policies are emerging? Do we witness a shift towards ecological modernization 2.0 or towards sufficiency-based environmental policies?

Finally, more systematic research on the feasibility of different types of environmental policy is needed. Low political feasibility, rather than lack of "political will", has been the major obstacle to faster and more effective environmental and climate policies (Skodvin et al., 2010; Gilligan and Vandenbergh, 2014; Aakre, 2016; Jewell and Cherp, 2020). But political feasibility is still a relatively vague and under-researched concept. In order to improve its usefulness for environmental policy research, it must be operationalized and empirically applied. Revisiting earlier research on capacity-building for environmental policy (Jänicke and Weidner, 1997; OECD, 1995) could help explore the extent to which the lower political feasibility of transformative environmental policies can be compensated through an increase of national and international political capacities.

Box 6.1 Chapter Summary

- Ecological modernization is the policy-driven innovation and diffusion of resource efficient clean technologies. Policies under the ecological modernization paradigm attempt to reconcile economic growth with environmental preservation by systematically making products, processes, and services more eco-efficient.
- Ecological modernization as a political strategy has been at the core of the substantial environmental policy successes from the early 1980s to the mid-2000s, especially in areas such as pollution control, energy efficiency, resource management, and the transition to renewable energies.
- Ecological modernization policies are characterized by a high degree of political feasibility, which results from their general compatibility with markets and economic growth and several economic and social co-benefits.
- Despite these successes, environmental pressures have increased in recent years and the gap between environmental goals and actual improvements has widened. Ecological modernization as the dominant environmental policy strategy seems to be reaching its limits.
- Its environmental improvements so far have been typically selective and mainly restricted to market-based solutions. Often it coexists with the older "dirty" structures (such as the fossil fuel sector) instead of fully replacing or transforming them. There are also problems of equity regarding the attribution of costs or the global distribution of benefits.
- To overcome the weaknesses of ecological modernization policies it is necessary to proceed to structural solutions regarding economic sectors, infrastructures, land use, and even lifestyles. We propose the term *ecological modernization 2.0* to describe this advanced strategy. To compensate for the lower political feasibility of ecological modernization 2.0 policies, political capacities must be strengthened.

References

Aakre, Stine (2016): The Political Feasibility of Potent Enforcement in a Post-Kyoto Climate Agreement. *International Environmental Agreements* 16 (1), pp. 145–159.

Abate, Randall S. (2013): Ocean Iron Fertilization; Science, Law, and Uncertainty. In William C. G. Burns, Andrew L. Strauss (Eds.): *Climate Change Geoengineering: Philosophical Perspectives, Legal Issues, and Governance Frameworks*. Cambridge: Cambridge University Press, pp. 221–251.

Abgeordnetenhaus, von Berlin (1982): 9. Wahlperiode. Plenarprotokoll 9/14. 14. Sitzung. Berlin, Freitag, 22. Januar 1982. Berlin: Abgeordnetenhaus.

Adler, Jonathan (2000): *Greenhouse Policy without Regrets. A Free Market Approach to the Uncertain Risks of Climate Change*. Washington, D.C.: The Competitive Enterprise Institute.

Alexander, Samuel (2012): Planned Economic Contraction: The Emerging Case for Degrowth. In *Environmental Politics* 21 (3), pp. 349–368.

Altenburg, Tilman; Assmann, Claudia (2017): *Green Industrial Policy. Concept, Policies, Country Experiences*. Geneva, Bonn: UN Environment; German Development Institute.

Andersen, Mikael Skou; Massa, Ilmo (2000): Ecological Modernization - Origins, Dilemmas and Future Directions. *Journal of Environmental Policy and Planning* 2 (4), pp. 337–345.

BASF (2019): Innovationen für eine Klimaschonende Chemieproduktion. Press Release, 10.01.2019.

Beise, Marian; Rennings, Klaus (2003): Lead Markets of Environmental Innovations: A Framework for Innovation and Environmental Economics. ZEW Discussion Paper No. 03-01. Mannheim: ZEW.

Bonds, Eric; Downey, Liam (2012): "Green" Technology and Ecologically Unequal Exchange: The Environmental and Social Consequences of Ecological Modernization in the World-System. *Journal of World-Systems Research* 18 (2), pp. 167–186.

Büchs, Milena; Koch, Max (2017): *Postgrowth and Wellbeing: Challenges to Sustainable Welfare*. Cham: Palgrave Macmillan.

Bundesministerium, Für Umwelt Naturschutz; Nukleare Sicherheit (2018): *GreenTech made in Germany 2018 – Umwelttechnik-Atlas für Deutschland*. Berlin: BMU.

Bundesministerium, Für Umwelt Naturschutz; Nukleare Sicherheit (2021): *GreenTech Made in Germany 2021. Umwelttechnik-Atlas für Deutschland*. Berlin: BMU.

Burger, Paul; Sohre, Annika; Schubert, Iljana (2020): Governance for Sufficiency: A New Approach to a Contested Field. In Philippe Hamman (Ed.): *Sustainability Governance and Hierarchy*. London: Routledge, pp. 157–177.

Burns, Wil; Nicholson, Simon (2017): Bioenergy and Carbon Capture with Storage (BECCS): The Prospects and Challenges of an Emerging Climate Policy Response. *Journal of Environmental Studies and Sciences* 7 (4), pp. 527–534.

Busch, Per-Olof Jörgens Helge (2012): Europeanization through Diffusion? Renewable Energy Policies and Alternative Sources for European Convergence. In Francesc Morata, Israel Solorio Sandoval (Eds.): *European Energy Policy: An Environmental Approach*. Cheltenham: Edward Elgar, pp. 66–84.

Caldwell, Lynton K. (1974): Environmental Policy as a Catalyst of Institutional Change. *American Behavioral Scientist* 17 (5), pp. 711–730. DOI: 10.1177/000276427401700506.

Cohen, Maurie J. (2013): Sustainable Development and Ecological Modernisation: National Capacity for Rigorous Environmental Reform. In Denis Requier-Desjardins, Clive L. Spash, Jan van der Straaten (Eds.): *Environmental Policy and Societal Aims*. Dordrecht: Springer, pp. 103–128.

Daly, Herman E. (1972): In Defense of a Steady-State Economy. *American Journal of Agricultural Economics* 54 (5), pp. 945–954.

Dessler, Andrew E.; Parson, Edward A. (2005): *The Science and Politics of Global Climate Change: A Guide to the Debate*. Cambridge: Cambridge University Press.

European Environment Agency (2019): *The European Environment - State and Outlook 2020: Knowledge for Transition to a Sustainable Europe*. Copenhagen: EEA.

Ewing, Jeffrey A. (2017): Hollow Ecology: Ecological Modernization Theory and the Death of Nature. *Journal of World-Systems Research* 23 (1), pp. 126–155.

Ferguson, Peter (2015): The Green Economy Agenda: Business as Usual or Transformational Discourse? *Environmental Politics* 24 (1), pp. 17–37.

Fisher, Dana R.; Freudenburg, William R. (2001): Ecological Modernization and Its Critics: Assessing the Past and Looking Toward the Future. *Society and Natural Resources* 14, pp. 701–709.

Foljanty-Jost, Gesine (Ed.) (1995): *Ökonomie und Ökologie in Japan. Politik zwischen Wachstum und Umweltschutz*. Opladen: Leske + Budrich.

Gaffney, Owen; Rockström, Johan (2021): *Breaking Boundaries: The Science of Our Planet. With assistance of Greta Thunberg*. New York: DK.

García, Ernest; Martínez-Iglesias, Mercedes; Kirby, Peadar (Eds.) (2017): *Transitioning to a Post-Carbon Society: Degrowth, Austerity and Wellbeing*. London: Palgrave Macmillan.

Geden, Oliver; Peters, Glen P.; Scott, Vivian (2019): Targeting Carbon Dioxide Removal in the European Union. *Climate Policy* 19 (4), pp. 487–494.

Gilg, Andrew; Barr, Stewart; Ford, Nicholas (2005): Green Consumption or Sustainable Lifestyles? Identifying the Sustainable Consumer. *Futures* 37 (6), pp. 481–504.

Gilligan, Jonathan M.; Vandenbergh, Michael P. (2014): Accounting for Political Feasibility in Climate Instrument Choice. *Virginia Environmental Law Journal* 32 (1), pp. 1–26.

Gillingham, Kenneth; Rapson, David; Wagner, Gernot (2014): The Rebound Effect and Energy Efficiency Policy. Discussion Paper RFF DP 14-39. Washington, D.C.: Resources for the Future.

Gouldson, Andrew; Murphy, Joseph E. (1998): *Regulatory Realities: The Implementation and Impact of Industrial Environmental Regulation*. London: Earthscan.

Hajer, Maarten A. (1995): *The Politics of Environmental Discourse: Ecological Modernization, and the Policy Process*. Oxford: Clarendon Press.

Hauff, Volker; Müller, Michael (Eds.) (1985): *Umweltpolitik am Scheideweg. Die Industriegesellschaft zwischen Selbstzerstörung und Aussteigermentalität*. München: Beck.

Heyen, Dirk A.; Fischer, Corinna; Barth, Regine; Brunn, Christoph; Grießhammer, Rainer; Keimeyer, Friedhelm; Wolff, Franziska (2013): When Less is More. Sufficiency – Need and Options for Policy Action. Oeko-Institute's Working Paper 3/2013.

Huber, Joseph (1982): *Die verlorene Unschuld der Ökologie: Neue Technologien und superindustrielle Entwicklung*. Frankfurt am Main: Fischer.

IEA (International Energy Agency) (2014): *Capturing the Multiple Benefits of Energy Efficiency*. Paris: OECD.

IPCC (Intergovernmental Panel on Climate Change) (2014): Climate change 2014. *Mitigation of Climate Change: Working Group III Contribution to the Fifth Assessment Report of the Intergovernmental Panel on Climate Change*. New York: Cambridge University Press.

IPCC (Intergovernmental Panel on Climate Change) (2018): Global Warming of 1.5°C. An IPCC special report on the Impacts of Global Warming of 1.5 °C Above Pre-Industrial Levels and Related Global Greenhouse Gas Emission Pathways, in the Context of Strengthening the Global Response to the Threat of Climate Change, *Sustainable Development, and Efforts to Eradicate Poverty. Summary for Policymakers*. Geneva: IPCC.

IRENA - International Renewable Energy Agency (2017): *Renewable Energy Benefits: Understanding the Socio-Economics*. Abu Dhabi: IRENA.

IRENA - International Renewable Energy Agency (2020): *Global Renewables Outlook: Energy Transformation 2050*. Abu Dhabi: IRENA.

Jackson, Tim; Smith, Carmen (2018): Towards Sustainable Lifestyles: Understanding the Policy Challenge. In Alan Lewis (Ed.): *The Cambridge Handbook of Psychology and Economic Behaviour*. 2nd ed. Cambridge: Cambridge University Press, pp. 481–515.

Jackson, Tim; Victor, Peter A. (2019): Unraveling the Claims For (and Against) Green Growth. *Science* 366 (6468), pp. 950–951.

Jacob, Klaus; Beise, Marian; Blazejczak, Jürgen; Edler, Dietmar; Haum, Rüdiger; Jänicke, Martin et al. (2005): *Lead Markets for Environmental Innovations*. Heidelberg: Physica-Verlag.

Jacob, Klaus; Volkery, Axel (2004): Institutions and Instruments for Government Self-Regulation: Environmental Policy Integration in a Cross-Country Perspective. *Journal of Comparative Policy Analysis: Research and Practice* 6 (3), pp. 291–309.

Jänicke, Martin (1983): Beschäftigungspolitik. *Natur* (4).

Jänicke, Martin (1985): Preventive Environmental Policy as Ecological Modernisation and Structural Policy. Discussion Paper 85/2. Berlin: Social Science Research Center (WZB).

Jänicke, Martin (Ed.) (1996): *Umweltpolitik der Industrieländer. Entwicklung – Bilanz – Erfolgsbedingungen*. Berlin: Edition Sigma.

Jänicke, Martin (2009): On Ecological and Political Modernization. In Arthur P. J. Mol, David A. Sonnenfeld, Gert Spaargaren (Eds.): *The Ecological Modernisation Reader: Environmental Reform in Theory and Practice*. London: Routledge, pp. 28–41.

Jänicke, Martin (2012a): Dynamic Governance of Clean-Energy Markets: How Technical Innovation Could Accelerate Climate Policies. *Journal of Cleaner Production* 22 (1), pp. 50–59.

Jänicke, Martin (2012b): "Green Growth": From a Growing Eco-Industry to Economic Sustainability. *Energy Policy* 48, pp. 13–21.

Jänicke, Martin (2017a): Ecological Modernization as Global Industrial Revolution. *Journal of Environmental Policy and Administration* 25 (S), pp. 1–32.

Jänicke, Martin (2017b): The Multi-level System of Global Climate Governance – The Model and its Current State. *Environmental Policy and Governance* 27 (2), pp. 108–121.

Jänicke, Martin; Jacob, Klaus (Eds.) (2006): *Environmental Governance in Global Perspective. New Approaches to Ecological Modernisation.* Berlin: Freie Universität Berlin.

Jänicke, Martin; Weidner, Helmut (Eds.) (1995): *Successful Environmental Policy: A Critical Evaluation of 24 Cases.* Berlin: Edition Sigma.

Jänicke, Martin; Weidner, Helmut (Eds.) (1997): *National Environmental Policies. A Comparative Study of Capacity-Building. With assistance of Helge Jörgens.* Berlin: Springer.

Jänicke, Martin; Wurzel, Rüdiger K.W. (2019): Leadership and Lesson-Drawing in the European Union's Multilevel Climate Governance System. *Environmental Politics* 28 (1), pp. 22–42.

Jewell, Jessica; Cherp, Aleh (2020): On the Political Feasibility of Climate Change Mitigation Pathways: Is it too Late to Keep Warming Below 1.5°C? *WIREs Climate Change* 11 (1), p. 19.

Jinnah, Sikina; Nicholson, Simon; Morrow, David R.; Dove, Zachary; Wapner, Paul; Valdivia, Walter et al. (2019): Governing Climate Engineering: A Proposal for Immediate Governance of Solar Radiation Management. *Sustainability* 11 (14), p. 3954.

Kanie, Norichika; Biermann, Frank (Eds.) (2017): *Governing Through Goals: Sustainable Development Goals As Governance Innovation.* Cambridge, Mass.: MIT Press.

Koch, Max (2020): The State in the Transformation to a Sustainable Postgrowth Economy. *Environmental Politics* 29 (1), pp. 115–133.

Lorek, Sylvia; Spangenberg, Joachim H. (2019): Identification of Promising Instruments and Instrument Mixes to Promote Energy Sufficiency. *EUFORIE – European Futures for Energy Efficiency.* Deliverable 5.5 [Available online: https://ec.europa.eu/research/participants/documents/downloadPublic?documentIds=080166e5c39c2b51&appId=PPGMS].

Lucon, Oswaldo; Ürge-Vorsatz, Diana; A. Zain, Ahmed; Akbari, Hashem; Bertoldi, Paolo; Cabeza, Luisa F. et al. (2014): Buildings. In Ottmar Edenhofer, Ramón Pichs-Madruga, Youba Sokona, Jan C. Minx, Ellie Farahani, Susanne Kadner et al. (Eds.): *Climate Change 2014: Mitigation of Climate Change. Working Group III Contribution to the Fifth Assessment Report of the Intergovernmental Panel on Climate Change.* New York: Cambridge University Press, pp. 671–738.

Machin, Amanda (2019): Changing the Story? The Discourse of Ecological Modernisation in the European Union. *Environmental Politics* 28 (2), pp. 208–227.

Maitre-Ekern, Eléonore; Dalhammar, Carl (2016): Regulating Planned Obsolescence: A Review of Legal Approaches to Increase Product Durability and Reparability in Europe. *RECIEL* 25 (3), pp. 378–394.

May, Peter J. (2005): Policy Maps and Political Feasibility. In Iris Geva-May (Ed.): *Thinking Like a Policy Analyst: Policy Analysis as a Clinical Profession. Policy analysis as a clinical profession.* New York: Palgrave Macmillan, pp. 127–151.

Mayrhofer, Jan P.; Gupta, Joyeeta (2016): The Science and Politics of Co-benefits in Climate Policy. *Environmental Science & Policy* 57, pp. 22–30.

Meadows, Donella; Meadows, Dennis; Randers, Jørgen; Behrens, William W. (1972): *The Limits to Growth.* New York: Universe Books.

Mez, Lutz; Weidner, Helmut (Eds.) (1997): *Umweltpolitik und Staatsversagen. Perspektiven und Grenzen der Umweltpolitikanalyse. Festschrift für Martin Jänicke zum 60. Geburtstag.* Berlin: Edition Sigma.

Michaelson, Jay (2013): Geoengineering and Climate Management: From Marginality to Inevitability. In William C. G. Burns, Andrew L. Strauss (Eds.): *Climate Change Geoengineering: Philosophical Perspectives, Legal Issues, and Governance Frameworks.* Cambridge: Cambridge University Press, pp. 81–114.

Mol, Arthur P. J. (2001): *Globalization and Environmental Reform: The Ecological Modernization of the Global Economy.* Cambridge, Mass.: MIT Press.

Mol, Arthur P. J.; Jänicke, Martin (2009): The Origins and Theoretical Foundations of Ecological Modernisation Theory. In Arthur P. J. Mol, David A. Sonnenfeld, Gert Spaargaren (Eds.): *The Ecological Modernisation Reader: Environmental Reform in Theory and Practice.* London: Routledge, pp. 17–27.

Mol, Arthur P. J.; Sonnenfeld, David A.; Spaargaren, Gert (Eds.) (2009a): *The Ecological Modernisation Reader: Environmental Reform in Theory and Practice.* London: Routledge.

Mol, Arthur P. J.; Spaargaren, Gert; Sonnenfeld, David A. (2009b): Ecological Modernisation: Three Decades of Policy, Practice and Theoretical Reflection. In Arthur P. J. Mol, David A. Sonnenfeld, Gert Spaargaren (Eds.): *The Ecological Modernisation Reader: Environmental Reform in Theory and Practice.* London: Routledge, pp. 3–14.

OECD (1995): *Developing Environmental Capacity. A Framework for Donor Envolvement.* Paris: OECD.
OECD (2007): *Environmental Innovation and Global Markets, ENV/EPOC/ GSP(2007)2/REVI.* Paris: OECD.
OECD (2011): *Towards Green Growth.* Paris: OECD.
OECD (2019): *Environment at a Glance. OECD Indicators.* Paris: OECD.
Ostrom, Elinor (2010): Beyond Markets and States: Polycentric Governance of Complex Economic Systems. *American Economic Review* 100 (3), pp. 641–672.
Parag, Yael; Capstick, Stuart; Poortinga, Wouter (2011): Policy Attribute Framing: A Comparison Between Three Policy Instruments for Personal Emissions Reduction. *Journal of Policy Analysis and Management* 30 (4), pp. 889–905.
Pierson, Paul (1993): When Effect Becomes Cause: Policy Feedback and Political Change. *World Politics* 45 (4), pp. 595–628.
Princen, Thomas (2003): Principles for Sustainability: From Cooperation and Efficiency to Sufficiency. *Global Environmental Politics* 3 (1), pp. 33–50.
REN21 (2018): Renewables 2018: Global Status Report. Paris: REN21.
REN21 (2020): Renewables 2020: Global Status Report. Paris: REN21.
Rifkin, Jeremy (2011): *The Third Industrial Revolution: How Lateral Power is Transforming Energy, the Economy, and the World.* Basingstoke: Palgrave Macmillan.
Schumpeter, Joseph A. (1942): *Capitalism, Socialism and Democracy.* New York: Harper & Brothers.
Setton, Daniela; Matuschke, Ira; Renn, Ortwin (2017): Social Sustainability Barometer for the German Energiewende 2017. *2017: Core Statements and Summary of the Key Findings.* Potsdam: IASS.
Simonis, Udo Ernst (Ed.) (1988): *Präventive Umweltpolitik.* Frankfurt/Main: Campus.
Skodvin, Tora; Gullberg, Anne Therese; Aakre, Stine (2010): Target-Group Influence and Political Feasibility: The Case of Climate Policy Design in Europe. *Journal of European Public Policy* 17 (6), pp. 854–873.
Spaargaren, Gert; Mol, Arthur P.J. (1992): Sociology, Environment, and Modernity: Ecological Modernization as a Theory of Social Change. *Society & Natural Resources* 5 (4), pp. 323–344.
Spangenberg, Joachim H.; Lorek, Sylvia (2019): Sufficiency and Consumer Behaviour: From Theory to Policy. *Energy Policy* 129, pp. 1070–1079.
SPD und Bündnis 90/Die Grünen (2002): *Koalitionsvertrag 2002-2006: Erneuerung – Gerechtigkeit – Nachhaltigkeit.* Berlin.
Spengler, Laura (2016): Two Types of 'Enough': Sufficiency as Minimum and Maximum. *Environmental Politics* 25 (5), pp. 921–940.
Statistisches Bundesamt (2017): *Nachhaltige Entwicklung in Deutschland – Indikatorenbericht 2016.* Wiesbaden: Statistisches Bundesamt.
Statistisches Bundesamt (2019): *Statistisches Jahrbuch Deutschland.* Wiesbaden: Statistisches Bundesamt.
The Royal Society (2009): *Geoengineering the Climate: Science, Governance and Uncertainty.* London: The Royal Society.
Umweltbundesamt (2020): *Beschäftigung im Umweltschutz: Entwicklung und gesamtwirtschaftliche Bedeutung.* Dessau: Umweltbundesamt.
Umweltbundesamt (2022): *Erneuerbare Energien in Deutschland. Daten zur Entwicklung im Jahr 2021.* Dessau: UBA.
UN Environment (2019a): *Global Environment Outlook – GEO-6: Healthy Planet, Healthy People.* Cambridge: Cambridge University Press.
UN Environment (2019b): *Global Environment Outlook GEO-6. Summary for Policymakers.* Cambridge: Cambridge University Press.
UNEP (United Nations Environment Programme) (2011): *Towards a Green Economy: Pathways to Sustainable Development and Poverty Eradication.* Geneva: UNEP.
United Nations Statistical Commission (2018): Global Assessment of Environmental-Economic Accounting and Supporting Statistics 2017. *Background Document. Statistical Commission, Forty-ninth Session*, 6–9 March 2018. New York: StatCom.
Ürge-Vorsatz, D.; Koeppel, Sonja; Mirasgedis, Sebastian (2007): Appraisal of Policy Instruments for Reducing Buildings' CO_2 Emissions. *Building Research & Information* 35 (4), pp. 458–477.
van Schaik, Louise; Schunz, Simon (2012): Explaining EU Activism and Impact in Global Climate Politics: Is the Union a Norm- or Interest-Driven Actor? *Journal of Common Market Studies* 50 (1), pp. 169–186. DOI: 10.1111/j.1468-5965.2011.02214.x.
Verband der Chemischen Industrie (2015): *Chemie3 – Fortschrittsbericht 2015.* Frankfurt a.M: VCI.

von Prittwitz, Volker (Ed.) (1993): *Umweltpolitik als Modernisierungsprozeß. Politikwissenschaftliche Umweltforschung und -lehre in der Bundesrepublik Deutschland.* Opladen: Leske + Budrich.

Walz, Rainer (2015): Green Industrial Policy in Europe. *Intereconomics* 50 (3), pp. 145–152.

Weale, Albert (1992): *The New Politics of Pollution*. Manchester: Manchester University Press.

Weidner, Helmut (2002): Capacity Building for Ecological Modernization: Lessons From Cross-National Research. *American Behavioral Scientist* 45 (9), pp. 1340–1368.

Wollmann, Hellmut (2007): Policy Evaluation and Evaluation Research. In Frank Fischer, Gerald J. Miller, Mara S. Sidney (Eds.): *Handbook of Public Policy Analysis: Theory, Politics, and Methods*. Boca Raton: CRC Press, pp. 393–402.

World Bank (2014): *Building Competitive Green Industries: The Climate and Clean Technology Opportunity for Developing Countries*. Washington, D.C.: World Bank.

York, Richard; Rosa, Eugene A. (2003): Key Challenges to Ecological Modernization Theory: Institutional Efficacy, Case Study Evidence, Units of Analysis, and the Pace of Eco-Efficiency. *Organization & Environment* 16 (3), pp. 272–288.

Young, Stephen C. (2000): *The Emergence of Ecological Modernisation: Integrating the Environment and the Economy?* London: Routledge.

Zimmermann, Klaus W.; Hartje, Volkmar J.; Ryll, Andreas (Eds.) (1990): *Ökologische Modernisierung der Produktion. Strukturen und Trends*. Berlin: Edition Sigma.

Zürn, Michael (2012): Global Governance as Multi-Level Governance. In David Levi-Faur (Ed.): *The Oxford Handbook of Governance*. Oxford: Oxford University Press, pp. 730–744.

PART II

Determinants of Environmental Policy Performance

7

DETERMINANTS OF PERFORMANCE IN NATIONAL ENVIRONMENTAL POLICIES

Detlef Jahn and Lisa Klagges

7.1 Introduction

The performance of national environmental policies encompasses many aspects and as a consequence various factors explain the variety of national performance (Fiorino, 2011). Comparative research on environmental policies and their impacts has a research tradition of more than three decades. Following a first generation of more descriptive studies, a second generation of research was more analytical (Andersen, 2000). However, studies with a comparative theoretical framework only began in the 1990s. These studies often examined a large number of countries in order to draw solid empirical conclusions. Therefore, we focus here on a macro-comparative analysis of highly industrialized democracies.[1] These countries are at the centre of attention because they have a long history of pollution and a comparable and open institutional environment for combating environmental degradation. Many aspects need to be considered when looking at democracies, such as domestic institutions, policy processes and actions, as well as international aspects including international treaties, globalization, and supranational organizations. A particular challenge is to filter the effects of political factors in relation to structural, economic, and geographical influences.

We will address these aspects in turn and proceed in six steps. First, we provide an overview of political science studies on environmental policy performance (Section 7.2). Based on this assessment, we outline different conceptions of environmental policy performance in Section 7.3. In Section 7.4, we turn to the socio-economic and structural determinants of environmental performance. Section 7.5 takes a specific focus on the role of domestic political factors, while the role of international factors is addressed in Section 7.6. In Section 7.7 we draw general conclusions from past and current research for future work on comparative research on environmental policy and performance.

7.2 Political Studies on Environmental Performance

International comparative studies initially focused on case studies (Enloe, 1975; Lundqvist, 1980; Vogel, 1986). In the 1990s, the first comprehensive macro-comparative studies were conducted (Andersen, 2000). A pioneering study was the analysis of 32 countries by Martin

DOI: 10.4324/9781003043843-9

Jänicke (1992). He showed that the material, institutional, and socio-cultural capacities of a country are more relevant for the outcome of environmental policy than the choice of policy instruments. Jänicke also developed the highly influential concept of ecological modernization in order to show how environmental challenges could be addressed. Consensus building remained an important explanatory variable in subsequent macro-comparative statistical studies. Crepaz (1995), Jahn (1998), and Scruggs (1999) analyzed the effects of corporatism and showed that it has a strong positive impact on environmental degradation. Poloni-Staudinger (2008) also showed that a country's consensus capacity is important, using Lijphart's concept of consensual democracies. Scruggs (2003) was the first to conduct a book-length comprehensive macro-comparative study of 17 OECD countries. He analyzed step by step which factors can affect environmental performance and confirmed the influence of corporatist arrangements. Surprisingly, most of these studies did not consider effects of political parties. Only Jahn (1998) examined the party effect and showed that the strength of the social democratic party has a positive effect on environmental performance. However, when social democrats are in power, this effect disappears. Even more important than party politics is the politicization of environmental issues by social movements and left-liberal parties. Still, much of the results depend on the way environmental performance is conceptualized and operationalized.

7.3 Conceptualizing Environmental Performance

Environmental performance is an impact variable defined in relation to other countries, a comparison over time, or a set standard (Jahn, 2018, pp. 91–97). In this way, it is possible to assess whether the environmental indicator is lower or higher in a given country or at a given time, or whether it meets the standard. Environmental performance must correspond to indicators that can be influenced by political action. A volcanic eruption can change the state of the environment, but cannot be used as an indicator of political action because an eruption cannot be changed by political actors (due to a lack of accountability). In some cases, authors use output variables (policy actions such as laws and regulations) as performance indicators. However, high political activity may not translate into a change in environmental conditions (Limberg et al., 2020).

Some studies use single indicators such as sulfur dioxide emissions, water pollution, or waste generation. Others use comprehensive indices that include several indicators. The problem is that, on the one hand, single indicators do not cover the whole spectrum of environmental degradation and, on the other hand, though composite indices summarize some important information, it is difficult to detect trends and identify causal relationships since the indicators included in such an index show overlapping or opposing developments.[2] One way out of this dilemma is to identify the dimensionality of environmental issues and arrive at some relevant dimensions (Poloni-Staudinger, 2008). Recently Jahn (2018) elaborated this concept using a first time-series study of 21 OECD countries on environmental performance. He finds three dimensions (general pollution, water pollution, and mundane environmental policies), but adds environmental performance indices that focus on the most important environmental problems in a country, on the one hand, and on the most successfully combated environmental problems, on the other. This comprehensive study shows that air pollution and water pollution have been successfully tackled over the past three decades whereas waste generation, carbon dioxide emissions, and nuclear waste are currently the most important environmental problems of highly industrialized societies.

The general definition of environmental performance can be refined by considering specific situations. For instance, one can use water withdrawal as an environmental indicator and observe that the lower the rate of water extraction, the better the environmental performance. However, water withdrawal has a very different meaning for countries with water scarcity, such as the Mediterranean countries, and countries with abundant water resources, such as Canada or the Nordic countries. To capture this aspect, country specific environmental performance indicators and contextualized comparative analyses are recommended (Jahn, 2018, pp. 143–146).

Environmental performance can also take a more specific focus. Some indicators provide information on how environmental measures are designed and whether they fundamentally change or only slightly change environmental conditions (Jahn, 2014; Poloni-Staudinger, 2008; Neumayer, 2003b; Klagges, 2019).

One issue that is highly neglected in the analysis of national environmental performance is the extent to which wealthy societies export their pollution to other countries, whether in the form of waste exports or the externalizing of polluting production processes.

7.4 Socio-Economic and Structural Determinants of Environmental Performance

Environmental performance depends on many factors. Many of these factors are not political or are not directly influenced by political factors. In particular, when examining the relationship between policy decisions and their environmental impacts, it is important to consider these factors in the empirical analysis.

First, a country's environmental impact depends on its resources. A country like Sweden, which has no oil, coal, or gas resources, must meet its energy needs differently than Poland or Germany, which have rich coal resources. It is conceivable, for example, that Sweden would cover a large part of its energy needs through nuclear power and Poland through coal. Other counties are in the position of using renewable energy such as hydro-power, for example Austria, Norway, and Switzerland. Although national energy resources have a major influence on a country's energy mix, this mix also depends on political decisions. This applies in particular to fossil fuel consumption and the share of renewable energies.

The production structure of a country is another factor that has a major impact on emissions (Jänicke et al., 1989). A country that bases its economy on heavy (dirty) industry has different environmental impacts than a service society. The transition of a society from an industrial to a service society may have positive environmental impacts. However, a growing service sector does not necessarily imply an improvement in the environment. Increasing leisure activities such as mass tourism leads to increasing air and car travel and unsustainable tourism practices.

The relationship between economic prosperity and the environment has been controversial. The obvious expectation that increasing production leads to more pollution has been questioned, and it has been postulated that growth in wealthy societies leads to environmental improvements. The trend that growth causes more pollution in poorer societies and less pollution in richer societies is represented by an inverted U-curve or the environmental Kuznets curve (EKC) (Grossman & Krueger, 1991; Stern, 2018). In this context, some scholars assume that ecological modernization – that is, the replacement of old polluting products and production processes with new, more environmentally friendly products and technologies – leads to an overall improvement in the environment. The advantage of this view is that it reconciles the economic growth necessary for industrial societies with the environment. This strategy – along with the sustainable development program[3] – has been the prevailing belief of many key policy actors, such as governments and international organizations (see Chapter 12). Recent evidence shows that environmental improvements have been much smaller than expected and that the EKC concept is not as valid as assumed (Stern, 2018). Empirical studies addressing the economic crisis or the coronavirus pandemic show that the link between economic growth and environmental degradation remains intact (Jahn, 2018; Helm, 2020; Rume & Islam, 2020). However, when looking at empirical results, it is important to note that such results are highly dependent on the environmental indicators used and whether economic growth is considered in addition to GDP per capita (Shafik, 1994; Stern, 2018). Finally, it may be that environmental improvements in some areas are offset by increasing pollution in others, or that we see rebound effects, so that the relationship between economic growth and environmental performance is more of an N-shaped curve (Jänicke, 2007). For instance, the introduction of electric cars on a large scale may reduce CO_2 emissions substantially, but create new environmental challenges such as the use of natural resources for the production of batteries and the disposal of old batteries. In addition, electric cars are not a solution to the harmful effects of private transport, since, for example, the absolute number of cars produced or problems such as rubber abrasion remains unaffected. This trend becomes clear when we look at wealth levels and environmental performance in general, water pollution, and country-specific pollution. Only water pollution shows an inverted U-curve, while general environmental performance and country-specific pollution show an N-shaped trend.

Other determinants are geographical aspects of a country. Spatially large countries have particular problems with long distances for the transport of goods and people, which causes high emissions. On the other hand, they also have the possibility of hiding environmentally harmful products or polluting production facilities in remote areas. In countries with a high population density, the presence of environmental problems becomes more evident. This, in turn, shapes the public's attitude toward environmental issues. As mentioned earlier, countries also have country-specific environmental problems. Finally, climate has been shown to have an impact on environmental performance. A country like New Zealand with a moderate climate has a lower heating demand than Canada or Finland and a lower cooling demand than, for example, the southern states of the USA. In this context, Jahn (2013) found a significant impact of annual climate conditions on some air emissions. In what follows, we structure the impact of political factors by looking alternately at domestic and international politics.

7.5 Domestic Politics and Environmental Performance

In political science, there is no specific approach that focuses on the link between policy and environmental performance. Most studies tend to use general concepts or those from other areas of comparative politics. The pioneering studies have focused on institutional variables. Corporatism, in particular, has had great explanatory power (Crepaz, 1995; Jahn, 1998; Scruggs, 1999). Although corporatist arrangements were created to address macroeconomic problems, Scruggs (2003, p. 160) claims that "corporatist institutions contribute positively and strongly to environmental policy performance". This argument is consistent with studies that use Lijphart's pattern of democracy as an explanatory factor. Consensus democracies appear to have better environmental performance than majority democracies (Poloni-Staudinger, 2008; Lijphart, 2012; Ozymy & Rey, 2013). Wälti (2004) shows that multilevel structures such as federalism promote environmental performance. The problem with these institutional analyses is that institutions change slowly, making it difficult to detect changes in environmental performance over time. For example, the USA was a leader in environmental protection in the 1960s and early 1970s, but is now one of the laggards. This change has occurred even though the US institutional structure is unchanged. It is not surprising, therefore, that studies examining environmental change over time conclude that corporatism has no effect on

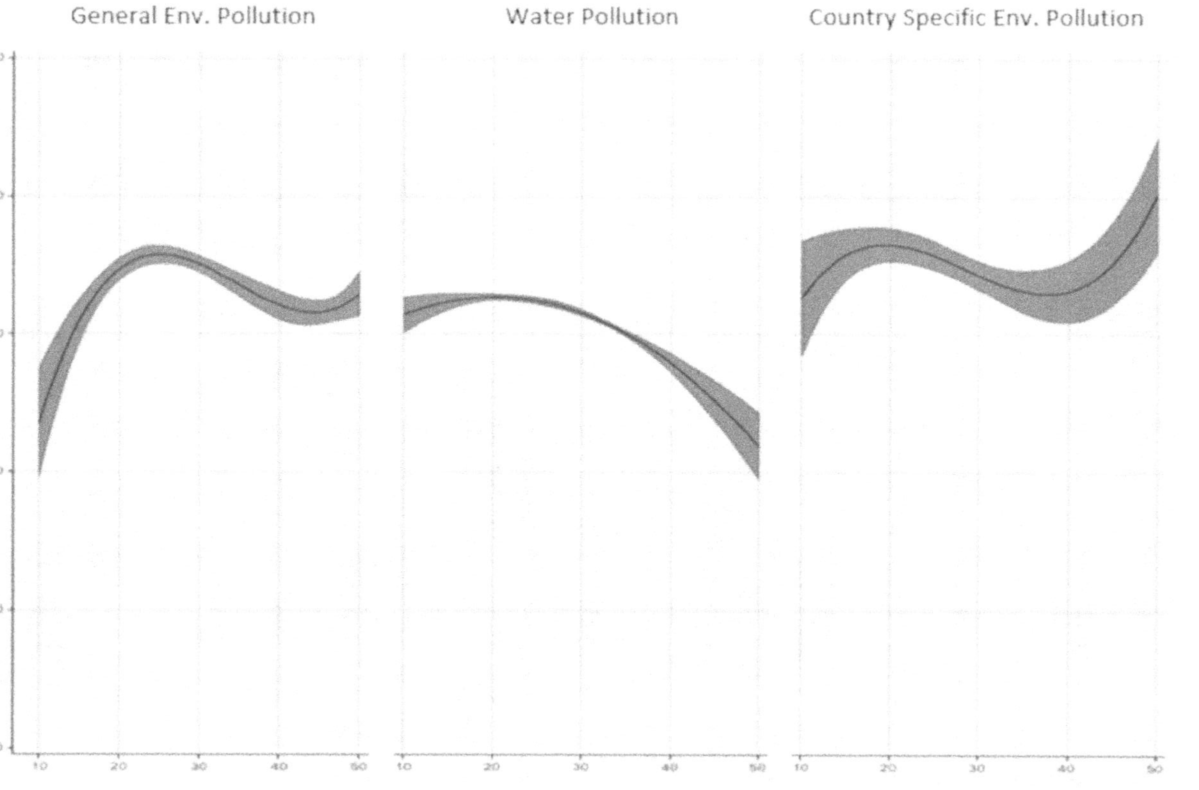

Figure 7.1 Economic wealth and environmental performance, 1980–2007.

Notes: Various atmospheric emission indicators, municipal and radioactive waste, and water abstraction form the index of General Environmental Pollution. The index of Water Pollution contains the pollution of rivers and lakes as well as fertilizer consumption. For country-specific environmental pollution, the three most relevant environmental issues in the 1980s were identified for each country and summarized in an indicator, which thus represents the national handling of the most pressing environmental problems (Jahn, 2018, pp. 143–146). All indicators were scaled to a range of values from 0 to 100, with higher values indicating higher environmental exposure.

Source: Jahn (2018, p. 162).

environmental performance (Neumayer, 2003a). However, using a time-variant corporatism index, Jahn (2018) showed that the influence of corporatism has changed over time. In the early years until the mid-1990s, corporatist arrangements were a kind of "economic growth coalition" (Schnaiberg, 1980; Offe, 1985). Over time, however, some actors in corporatist arrangements changed their attitudes toward environmental protection, and as this view took hold, consensus-based decision-making led corporatism to become a driving force for environmental improvements (Jahn, 2018). However, such a time-varying effect implies that corporatism may play a different role in the future, or that the period of de-corporatism may have specific effects on environmental performance in some countries. It would be a major advance if the institutional aspect of corporatism could be combined with the programmatic positions of the political actors involved.

Neumayer's (2003a) study not only questioned the effects of corporatism, but also showed that the strength of green parties is an important aspect in explaining environmental performance. Looking at the strength of green parties in elections or in parliament is a poor indicator because the causal relationship is unclear. Are green parties strong because pollution is low in a country, or is pollution low because of the success of green parties? A better way to examine the impact of green parties is to look at their role in governments. With this focus, partisan theory comes into the spotlight. Many studies have shown that left-wing parties have a positive impact on environmental performance. This is particularly the case when they have to compete with green parties (Carter, 2013; see also Jensen & Spoon, 2011; Abou-Chadi, 2014; Spoon et al., 2014; Fagerholm, 2016; Farstad, 2018) and when unions are receptive to environmental issues (Jahn, 1993a). However, recent studies on climate emissions show that the impact of left-wing parties is much weaker than that of green parties and that the impact of populist parties in government is harmful for the environment (Jahn, 2021, 2022; Buzogány & Mohamad-Klotzbach, 2021; Böhmelt, 2021). Less in focus is the finding in these studies that non-religious centrist parties have a negative impact on performance. More research is needed here to explore the impact of parties on environmental performance.

Another question is how to classify political parties in terms of the environmental dimension. The party family approach can only say with certainty that green parties give high priority to environmental issues. But what are the environmental positions of the other party families? There is no conclusive answer to this question. The alternative is to use expert judgments that incorporate the environmental dimension (Benoit & Laver, 2006) or party manifesto data (Volkens et al., 2020). However, most studies take only the one environmental issue of the manifesto coding (coding item: pro-environment; per 501), while others focus on specific issues such as climate change and recode manifestos (Farstad, 2018). There are very few attempts to construct an ideological dimension for environmental studies, similar to a left–right index (see Box 7.1).

The causal mechanisms that drive policymakers toward stronger environmental positions are difficult to determine in macro-comparative studies. EKC suggests that a larger proportion of the population is environmentally conscious in richer societies than in poorer societies. The reason is that new (post-materialist) values emerge as wealth increases. However, there is mixed evidence that public opinion about the environment has a causal influence on environmental performance. While Scruggs (2003, chapter 4) finds no robust results, other studies observe a positive relationship (Weaver, 2008; Shum, 2009; Andersen et al., 2017). Conversely, there is a strong influence of radical environmental movements and a high politicization of production on performance (Kitschelt, 1988; Jahn, 2018). The need to translate values into politics is shown by studies examining the success of green parties (Müller-Rommel, 1998; Grant & Tilley, 2019).

Box 7.1 Measuring Green-Growth Positions in Party Research

To date, the Party Manifesto Project is the only source of longitudinal data for more than 1,000 parties in over 50 countries from 1945 until today. For party research, the left–right indices derived from these data are important tools for comparative policy analysis. For environmental performance analysis, Jahn (2018, pp. 43–49) developed a green-growth index from party platforms, drawing on green political theory (Eckersley, 1992; Goodin, 1992; Dobson, 1995). The core of the green-growth dimension is formed by five statements included in the coding of party manifestos that reflect key concepts of green theory. The results show that these statements vary in radicalness. An anti-growth statement is much stronger than the more moderate pro-environment statement. The result presented is the first dimension of a multidimensional scaling analysis. The figure shows how strong individual statements of the Party Manifesto codings (percentage of overall statements) are situated on the green-growth dimension.

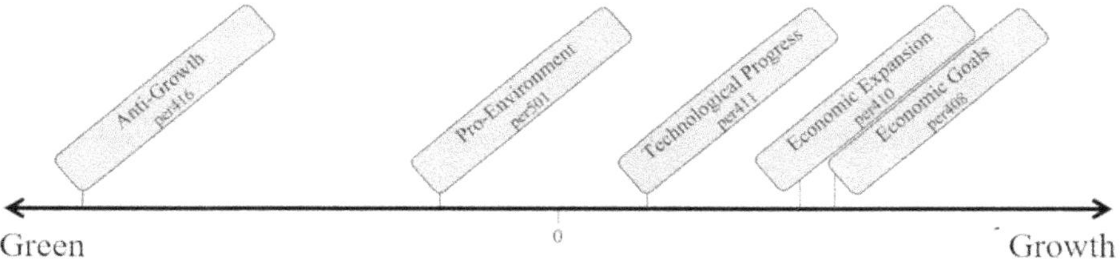

Figure 7.2 Statements of the green-growth dimension.

Source: Own illustration based on Jahn (2018, p. 48).

The green-growth index building on these core issues can be used for: government positions; veto players such as second chambers, coalitions, and EU positions; as well as ideological misfits between the EU and its member states (Jahn, 2018).

However, this success is not only dependent on values in a society, but also on party politics in a political system (Jahn, 1993b; Spoon et al., 2014; Abou-Chadi, 2014; Schulze, 2021).

Far less research has been devoted to the factors that impede environmental improvement. As noted earlier, there are many studies analyzing the impact of green and left parties on environmental performance, but there are far fewer studies looking at the other camp: institutions and actors that hinder environmental performance improvement. As noted earlier, non-religious centrist parties correlate with poor performance, and even when some industries reveal their preferences for environmental issues, there is no systematic study that analyzes the impact of industry, employer associations, and military actions on environmental performance. The aspect most often considered is the influence of veto players (Madden, 2014). The veto player approach became popular in comparative politics around the turn of the millennium (Tsebelis, 2002; Jahn, 2010). Tsebelis defines veto players as actors who are involved in the policy-making process. If they are plentiful and diverse in their opinions, there is little reason to believe in policy change. Because environmental improvement needs to change the status quo to become more effective, the consequence of veto players is that they prevent action that

Box 7.2 Causal Mechanisms in Macro-Comparative Research

Causality is a controversial topic in comparative research. In this context, some consider the macro-comparative approach to be a "crazy method" or one that is inconsistent with the ontology of modern comparative research (Kittel, 2006; Hall, 2003). Others suggest a more productive way of capturing causality in a macro-comparative framework (Franzese, 2009). Macro-comparative research works with broad concepts and inference based on significant correlations. Of course, correlations are not synonymous with causality, and the causal mechanism is usually implicit or theorized in these studies. Not exclusively, but probably more than in other research methods, causal mechanisms are difficult to determine. The simple assumption that structural conditions drive public opinion, which in turn drives political parties to shift preferences, which again leads to certain policies that ultimately affect performance, is a long chain of causal mechanisms. Each link can (and has been) challenged. Consider just the link between preferences, policies, and impacts. Most macro-comparative studies analyze preferences and outcomes or preferences and output. The few studies that focus on policies and impacts conclude that this link is far from perfect (Knill et al., 2012). And even if we look at government preferences (which are often a compromise of different parties or factions) and outcomes, the causal mechanism does not necessarily lead through policy. Governments have other ways to influence impacts. For example, through their personnel policies (heads of environmentally important institutions, judges, selection of expert groups and reports, etc.), through favorable opportunities, (re)allocation of resources, or subsidization of industries. Thus, a concrete causal path may be less relevant than assumed, and parties in government may distribute power through unobservable channels. Research on populist parties (in government) provides us with illustrative examples of how such preference dispersion may occur (Pappas, 2019; Escartin, 2020). So even if macro-comparative environmental studies cannot establish causal relationships with certainty, they can rule out causal relationships when there is no significant correlation. And that is more than most other methods can do, which rely on subjective interpretation and selection bias.

will improve performance. This effect has already been noted in the case of green taxes and climate change policy (Ward & Cao, 2012; Madden, 2014).

Research on institutions and political actors suggests that policy outcomes are determined by an interaction of both. With the turn of macro-comparative analysis to interactive models, there are a large number of studies that analyze various interactive relationships. For instance, Lim and Duit (2018) find strong partisan effects when considering different welfare state regimes, Chang et al. (2018) show that left-wing partisan effects are only relevant for countries with low levels of pollution, and Jahn and Wälti (2007) demonstrate that federal structures interact with the degree of corporatism.

Recent studies have focused on institutional and bureaucratic capacities in the environmental field, as well as the diversity of instruments in order to explain the effectiveness of environmental policies (Limberg et al., 2020; Fernández-i-Marín et al., 2021). These studies allow us to follow the causal mechanisms between institutions, policies, and outcomes. Another approach focuses on the parliamentary sphere and develops an agenda-setting power model that combines the preferences of governments and veto players on the one hand and models the time lag of the policy process on the other (Jahn, 2018). Both approaches promote the development of macro-comparative environmental policy and go beyond the correlation of single variables or arbitrarily chosen interactions of two variables. The path taken by these studies is a fruitful one for making macro-comparative studies an indispensable tool for analyzing the determinants of environmental performance.

7.6 International Politics and Environmental Performance

Like most other policy impacts, domestic environmental performance cannot be explained solely by domestic factors (Garrett, 1998; Drezner, 2007; Hays, 2009; Solingen, 2009). States interact with other states, and this interaction influences domestic politics and policy decisions. Analyzing domestic and international factors with respect to environmental policy, Knill et al. (2014, p. 74) conclude that external factors such as membership in international organizations are the "main determinant" for environmental policy adoption. International factors are influential in many ways, from the just mentioned membership in international or supranational organizations, to participation in multilateral or bilateral environmental agreements, to the effects of more diffuse processes such as globalization and international interdependence.

International multilateral environmental agreements and treaties are well researched in the field of international relations.[4] In particular, there are numerous case studies on individual agreements and treaties that describe and analyze the origins, formation, content, and context of environmental treaties. Although some studies explicitly focus on the impact of international environmental regimes (IERs) (Young, 1999, 2014; Brandi et al. 2019; Kim et al. 2017), there are very few that systematically examine their impact comparatively by analyzing the response of individual countries (Bernauer, 1995; Victor et al., 1998; Weiss & Jacobson, 1998; Helm & Sprinz, 2000; Steinberg & VanDeveer, 2012; Biesenbender & Tosun, 2014). In recent years, scholars have moved away from case study research and have collected data for a larger number of environmental regimes and countries. In particular, the Columbia Center for International Earth Science Information Network (CIESIN, 2007), Breitmeier et al. (2006) and Mitchell (2008, 2010) have collected such data. The main conclusion of these studies is that international environmental agreements matter. Breitmeier et al. (2006: 188) emphasize the fact that international environmental agreements based on consensus or unanimity as a decision rule are most effective. About three-quarters of consensus-based international environmental agreements have produced improvements in regard to the problem. The comparable figure for international environmental agreements with unanimity rule is slightly more than 50 percent. Even when noncompliance does not lead to the imposition of sanctions, participation in international environmental agreements can lead to environmental improvements because countries expect national benefits if they comply with international standards. Others contend that consensus and unanimity rules within the non-hierarchical structures characteristic of international environmental agreements lead to the implementation of policies desired by the least ambitious member of the group, resulting in a preference for the status quo (Downs et al., 1996; Miles et al., 2002).

There is also a gap between the adaptation to international treaties (ratification) and their accommodation at the national level. Biesenbender and Tosun (2014) show that instrument adaptation and accommodation are distinct processes that follow different decision-making logics. Finally, there is an implementation gap that may lead to IERs not having the impact intended by the initiators. These findings show that the causal chain includes international agreement, adaptation of individual countries, domestic accommodation, and finally implementation. Although the causal chain is long, there is evidence that international commitments and treaty involvement are an important factor leading to improvements in domestic environmental performance (Jahn, 2018, pp. 247–263).

Another major field of research is the impact of the EU on its member states. Environmental policy is one of the most important activities of the EU. In no other area is it as active as in environmental regulations and directives (Hix & Høyland, 2011, p. 205). In addition, there are several environmental action programs and on the international stage the EU represents its member states. This leads some to conclude that the EU has created a system of "environmental governance" in Europe and that it enforces stricter environmental norms and standards in its member states (Weale et al., 2000; Knill & Liefferink, 2007). Recent studies show that the EU's initiative varies over time. In the period between the Maastricht Treaty in the early 1990s and the onset of the economic crisis in 2008, the EU's positive influence was strongest (Jahn, 2018; see also Lenschow et al., 2020). The economic crisis was a turning point and led to policy dismantling (Steinebach & Knill, 2017; Burns & Tobin, 2020; Knill et al., 2020, Knill & Liefferink, 2021). Moreover, comparative studies show that the impact of the EU on its member states depends on member states' willingness to implement EU policies. Member states differ fundamentally, and there are some countries that lead in their environmental activities while others lag behind. The reason for this is that some countries have more ambitious domestic policies than the EU sets out. But there are also countries that do not implement EU initiatives to the same extent as others (Börzel & Buzogány, 2019). Although it appears that the implementation gap has narrowed, at the same time the demands of EU environmental policy on member states is less stringent. In a comparative perspective that includes EU member states and non-EU countries, the conclusion is that the EU's influence on national environmental policy "is not as dominant as one would expect" (Sommerer et al., 2008, p. 180). This conclusion is confirmed for environmental performance with an innovative approach to conceptualizing the EU's environmental policy position and ideological misfits with its member states (Jahn, 2018). The influence of the EU can be found in water pollution, mundane environmental policies, and even country-specific environmental performance, but there is no significant influence in the important issues of (nuclear) waste, carbon dioxide, and soil degradation.

Finally, the impact of globalization and diffusion on environmental performance is a contentious issue. Some see a race to the bottom. That is, increased international competition and trade are leading to environmental degradation as national environmental regulations prevent economic efficiency. There is also a relocation of polluting industries abroad, so that wealthy industrialized democracies export dirty industries and even their wastes to less developed countries with little environmental regulation ("pollution heavens"). But there is also the other view that international interaction leads to a race to the top. The causal mechanism is that environmentally friendly industries set a standard that other industries cannot disregard (Konisky, 2007). Prakash and Potoski (2006) show that international companies that adhere to environmental standards (ISO 14001) have a positive effect on the same standards in target countries. The same is true for the introduction of environmental tax reforms that spill over from the market leader to other countries (Ward & Cao, 2012).

7.7 Conclusion and Future Research

We can conclude, in line with Robert Franzese's (2009) insight from quantitative comparative research, that "almost everything causes almost everything else" and "the effect of almost everything depends on almost everything else". This is true, but we can still draw some conclusions from the research findings of the more than three decades of macro-quantitative studies on environmental performance. First, institutions matter, but they only reinforce political positions mobilized by political actors. These positions, in turn, depend on many factors, and economic crises or other issues such as the coronavirus pandemic can eclipse environmental issues. Government positions, veto players, corporatism, and the mobilization of environmentally conscious citizens are key factors that determine political influence on environmental performance. The hope that technological progress and ecological modernization will solve environmental problems has proven to be an illusion. Even if some pollutants have been reduced by technological progress, other – often new – problems have emerged. Much evidence suggests that the relationship between economic growth and environmental degradation does not take the form of an inverted U, but rather follows an N-shaped curve.

In previous and current studies, too much emphasis has been placed on what factors can improve environmental performance. Too little attention has been paid to factors that hinder environmental improvement or even promote pollution. Concepts such as sustainable development or ecological modernization have helped to keep environmental concerns from being a marginal issue in society to the realm of mainstream politics. However, these concepts have also given the impression that economic growth and environmental concerns could be reconciled, which is controversial (see above regarding the EKC).

Another major challenge for comparative environmental research in analyzing the determinants of environmental performance is the lack of reliable data and relevant concepts for many countries and over time. Enloe (1975, p. 321) stated in her seminal study: "What can be measured can be turned into a theme, a goal, a performance criterion." This is true both for the dependent variable of environmental performance and for key independent variables such as environmental policy measures, ideologies, and institutions. There has been great progress in all of these areas, but most of these achievements depend on individual researchers or research teams, and the data are not always as freely available as desired.

Box 7.3 Relevant Data and Concepts for Macro-Comparative Analysis

Data for the environmental performance are collected primarily from the OECD and EuroStat, although the two datasets are often difficult to combine. The UN and World Bank have only very basic data for a global survey of countries. The challenge with these data is that time series often have breaks and some data series are abandoned. Thus, even when there are elaborate and appropriate indices for policy analysis, it is often difficult to obtain long time series for a large number of countries.

For domestic determinants in environmental analysis, the University of Greifswald data provide concepts and time series for established OECD countries and beyond (Jahn et al., 2022). There are concepts for government positions, veto players, corporatism, heating and cooling needs, environmental institutionalization, EU positions, ideological misfits between the EU and its member states, and so on.

For environmental policies, studies often use data from the International Environmental Agency (IEA) or expert judgments (Knill et al., 2012). The former is difficult to use because it collects agreements without paying attention to the effectiveness and scope of the measures. The ECOLEX database (FAO et al., 2021), which collects environmental laws, also has this problem.

For international environmental agreements the IEA database by Ronald Mitchell is a reliable source collecting bilateral and multilateral agreements (Mitchell et al., 2020).

Macro-comparative environmental studies have followed the general trend in macro-comparative studies of shifting policy analysis from the consideration of individual key factors to the study of interactions. However, there are countless interactions, and studies have not yet developed a unified concept of which interactions are truly relevant. A first attempt is the agenda-setting power model, which combines agenda setters and veto players and models the political process for macro-comparative analysis. Another is the more policy-oriented approach, which combines features of policy instruments and bureaucratic capacity to explain environmental performance (Limberg et al., 2020; Fernández-i-Marín et al., 2021). This means that both approaches focus on factors that promote environmental improvements as well as those that hinder them.

A further challenge is to model and to analyze the interconnectedness between domestic and international politics. It is surprising that relatively few attempts have been made to combine international with national environmental studies, and when they have, they focus mainly on single links in the causal chain (Bernauer et al., 2010; Cao & Prakash, 2012; Ward & Cao, 2012; Schulze, 2014). The likely reason is that the causal chain is long and obscure. This implies that the process takes a considerable amount of time. A relatively simple example: When government representatives meet for an international agreement such as the Paris Agreement, 196 parties must find a way to come to an agreement. Such an agreement is the result of lengthy negotiations leading up to the meeting. Once agreement is reached, national governments must find ways to take action to meet the agreed-upon goals. The ratification process is, by and large, quite similar to the domestic legislative process (Barrett, 2003; Schulze, 2014). This means that we need to model the multi-level game in quantitative analysis. But this is challenging. In the case of democratic states, the government that made the agreement need not be the same one that has to implement the measures. What we need is a model for multilevel process analysis. While it is clear that because of this complexity there are few studies that combine all of these aspects, it is equally clear that the interplay of all of these aspects is essential for understanding the effectiveness of human activities in reducing environmental degradation. In this respect, we need an integrated theory which is suitable for macro-comparative research. But also in this regard, we need to follow Franzese's advice (2009, p. 67): "context matters: so model it". If macro-comparative studies heed this advice, it will be a very efficient method that will help to reach comprehensive and important representative conclusions.

Notes

1 There are also some macro-comparative studies that reach beyond highly industrialized democracies (Ward, 2008; Bättig & Bernauer, 2009; Escher & Walter-Rogg, 2020). These studies mainly focus on whether democracies are more efficient in combating environmental degradation than authoritarian societies. Most of these studies show that democracies are more effective in their environmental performance.

2 There are some other environmental performance indices that have been developed by international think tanks and attract a lot of attention in the media, but they do not meet rigorous scientific requirements such as reproducibility or transparency. For example, the Environmental Performance Index of the Yale Center of Environmental Law and Policy (Wendling et al., 2020) or the Ecological Footprint Index (Wackernagel & Rees, 1996; Rees, 2018).

3 Sustainable development is a concept of international politics and was brought into focus by the UN World Commission on Environment and Development report (1987). The major message was that present generations should not determine the fate of future generations and that the development of the northern hemisphere should not be at the expense of the southern hemisphere. However, this concept has often been misused to legitimize economic growth (Whitehead, 2018).

4 For overviews see, for instance, Biermann and Pattberg (2012), Haas et al. (1993), Keohane and Levy (1996), Luterbacher and Sprinz (2001), Mitchell (1994, 2009, 2010), Young (1997), and Oberthür and Gehring (2006).

References

Abou-Chadi, T. (2014). Niche party success and mainstream party policy shifts – how green and radical right parties differ in their impact. *British Journal of Political Science, 46*(2), 417–436. https://doi.org/10.1017/S0007123414000155

Andersen, B., Böhmelt, T., & Ward, H. (2017). Public opinion and environmental policy output: A cross-national analysis of energy policies in Europe. *Environmental Research Letters, 12*(11). https://doi.org/10.1088/1748-9326/aa8f80

Andersen, M. S. (2000). Ecological modernisation capacity: Finding patterns in the mosaic of case sudies. In S. C. Young (Ed.), *The Emergence of Ecological Modernisation: Integrating the Environment and the Economy?* (pp. 107–132). Routledge.

Barrett, S. (2003). *Environment and Statecraft: The Strategy of Environmental Treatymaking*. Oxford University Press.

Bättig, M. B., & Bernauer, T. (2009). National institutions and global public goods: Are democracies more cooperative in climate change policy? *International Organization, 63*(2), 281–308. https://doi.org/10.1017/S0020818309090092

Benoit, K., & Laver, M. (2006). *Party Policy in Modern Democracies*. Routledge.

Bernauer, T. (1995). The effect of international environmental institutions: how we might learn more. *International Organization, 49*(2), 351–377. https://doi.org/10.1017/S0020818300028423

Bernauer, T., Kalbhenn, A., Koubi, V., & Spilker, G. (2010). A comparison of international and domestic sources of global governance dynamics. *British Journal of Political Science, 40*(3), 509–538. https://doi.org/10.1017/S0007123410000098

Biermann, F., & Pattberg, P. H. (Eds.). (2012). *Earth System Governance: A Core Research Project of the International Human Dimensions Programme on Global Environmental Change Global Environmental Governance Reconsidered*. MIT Press.

Biesenbender, S., & Tosun, J. (2014). Domestic politics and the diffusion of international policy innovations: How does accommodation happen? *Global Environmental Change, 29*, 424–433. https://doi.org/10.1016/j.gloenvcha.2014.04.001

Böhmelt, T. (2021). Populism and environmental performance. *Global Environmental Politics, 21*(3), 97–123. https://doi.org/10.1162/glep_a_00606

Börzel, T. A., & Buzogány, A. (2019). Compliance with EU environmental law. The iceberg is melting. *Environmental Politics, 28*(2), 315–341. https://doi.org/10.1080/09644016.2019.1549772

Brandi, C., Blümer, D., Morin, J. (2019). When do international treaties matter for domestic environmental legislation. *Global Environmental Politics, 19*(4), 14–44. https://doi.org/10.1162/glep_a_00524

Breitmeier, H., Young, O. R., & Zürn, M. (2006). *Analyzing International Environmental Regimes: From Case Study to Database. Global Environmental Accord*. MIT Press.

Burns, C., & Tobin, P. (2020). Crisis, climate change and comitology: Policy dismantling via the backdoor? *Journal of Common Market Studies, 58*(3), 527–544. https://doi.org/10.1111/jcms.12996

Buzogány, A., & Mohamad-Klotzbach, C. (2021): Populism and nature – the nature of populism: New perspectives on the relationship between populism, climate change, and nature protection. *Zeitschrift für Vergleichende Politikwissenschaft, 15*(2), 155–164. https://doi.org/10.1162/glep_a_00606

Cao, X., & Prakash, A. (2012). Trade competition and environmental regulations: Domestic political constraints and issue visibility. *The Journal of Politics, 74*(1), 66–82. https://doi.org/10.1017/S0022381611001228

Carter, N. (2013). Greening the mainstream: Party politics and the environment. *Environmental Politics, 22*(1), 73–94. https://doi.org/10.1080/09644016.2013.755391

Chang, C.-P., Wen, J., Dong, M., & Hao, Y. (2018). Does government ideology affect environmental pollutions? New evidence from instrumental variable quantile regression estimations. *Energy Policy, 113*, 386–400. https://doi.org/10.1016/j.enpol.2017.11.021

CIESIN (Ed.). (2007). *Environmental Treaties and Resource Indicators (entri)*. CIESIN. http://sedac.ciesin.columbia.edu/data/collection/entri/entri-service

Crepaz, M. M. L. (1995). Explaining national variations in air pollution levels: Political institutions and their impact on environmental policy making. *Environmental Politics, 4*(3), 391–414. https://doi.org/10.1080/09644019508414213

Dobson, A. (1995). *Green Political Thought* (2. ed.). Routledge.

Downs, G. W., Rocke, D. M., & Barsoom, P. N. (1996). Is the good news about compliance good news about cooperation? *International Organization, 50*(3), 379–406. https://doi.org/10.1017/S0020818300033427

Drezner, D. W. (2007). *All Politics is Global: Explaining International Regulatory Regimes*. Princeton University Press. https://doi.org/10.2307/j.ctt7st6p

Eckersley, R. (1992). *Environmentalism and Political Theory: Toward an Ecocentric Approach*. State University of New York Press.

Enloe, C. H. (1975). *The Politics of Pollution in a Comparative Perspective: Ecology and Power in Four Nations*. McKay Company.

Escartin, A. R. (2020). Populist challenges to EU foreign policy in the southern neighborhood: An informal and illiberal europeanisation? *Journal of European Public Policy*, *27*(8), 1195–1214. https://doi.org/10.1080/13501763.2020.1712459

Escher, R., & Walter-Rogg, M. (2020). *Environmental Performance in Democracies and Autocracies*. Palgrave.

Fagerholm, A. (2016). Social democratic parties and the rise of ecologism: A comparative analysis of Western Europe. *Comparative European Politics*, *14*, 547–571. https://doi.org/10.1057/cep.2014.34

FAO, ICUN, & UNEP. (2021). *Ecolex*. https://www.ecolex.org/

Farstad, F. M. (2018). What explains variation in parties' climate change salience? *Party Politics*, *24*(6), 698–707. https://doi.org/10.1177/1354068817693473

Fernández-i-Marín, X., Knill, C., & Steinebach, Y. (2021). Studying policy design quality in comparative perspective. *American Political Science Review*, *115*(3), 931–947. https://doi.org/10.1017/S0003055421000186

Fiorino, D. J. (2011). Explaining national environmental performance: Approaches, evidence, and implications. *Policy Sciences*, *44*(4), 367–389. https://doi.org/10.1007/s11077-011-9140-8

Franzese, R. J. (2009). Multicausality, context-conditionality, and endogeneity. In C. Boix & S. C. Stokes (Eds.), *Oxford Handbook of Comparative Politics* (pp. 27–72). Oxford University Press.

Garrett, G. (1998). *Partisan Politics in the Global Economy*. Cambridge University Press.

Goodin, R. E. (1992). *Green Political Theory*. Polity Press.

Grant, Z. P., & Tilley, J. (2019). Fertile soil: Explaining variation in the success of green parties. *West European Politics*, *42*(3), 495–516. https://doi.org/10.1080/01402382.2018.1521673

Grossman, G. M., & Krueger, A. B. (1991). *Environmental impacts of a North American free trade agreement*. NBER Working Papers 3914.

Haas, P. M., Keohane, R. O., & Levy, M. A. (Eds.). (1993). *Global Environmental Accords Series. Institutions for the Earth: Sources of Effective International Environmental Protection*. MIT Press.

Hall, P. A. (2003). Aligning ontology and methodology in comparative politics. In J. Mahoney & D. Rueschemeyer (Eds.), *Comparative Historical Analysis in the Social Sciences* (pp. 373–404). Cambridge University Press.

Hays, J. C. (2009). *Globalization and the New Politics of Embedded Liberalism*. Oxford University Press.

Helm, C., & Sprinz, D. F. (2000). Measuring the effectiveness of international environmental regimes. *Journal of Conflict Resolution*, *44*(5), 630–652. https://doi.org/10.1177/0022002700044005004

Helm, D. (2020). The environmental impacts of the coronavirus. *Environmental and Resource Economics*, *76*(1), 1–18. https://doi.org/10.1007/s10640-020-00426-z

Hix, S., & Høyland, B. (2011): *The Political System of the European Union* Palgrave Macmillan.

Jahn, D. (1993a). *New Politics in Trade Unions: Applying Organizational Theory to the Ecological Discourse on Nuclear Energy in Sweden and Germany*. Dartmouth.

Jahn, D. (1993b). The rise and decline of new politics and the greens in Sweden and Germany. *European Journal of Political Research*, *24*(3), 177–194. https://doi.org/10.1111/j.1475-6765.1993.tb00375.x

Jahn, D. (1998). Environmental performance and policy regimes: Explaining variations in 18 OECD-countries. *Policy Sciences*, *31*(2), 107–131. https://doi.org/10.1023/A:1004385005999

Jahn, D. (2010). The veto player approach in macro-comparative politics: Concepts and measurement. In T. König, M. Debus, & G. Tsebelis (Eds.), *Reform Processes and Policy Change: Veto Players and Decision-Making in Modern Democracies* (pp. 43–68). Springer.

Jahn, D. (2013). The impact of climate on atmospheric emissions: constructing an index of heating degrees for 21 OECD countries from 1960-2005. *Weather, Climate, and Society*, *5*(2), 97–111. https://doi.org/10.1175/WCAS-D-11-00050.1

Jahn, D. (2014). The three worlds of environmental politics. In A. Duit (Ed.), *State and Environment. The Comparative Study of Environmental Governance* (pp. 81–109). MIT Press.

Jahn, D. (2018). *The Politics of Environmental Performance: Institutions and Preferences in Industrialized Democracies*. Cambridge University Press.

Jahn, D. (2021). Quick and dirty: How populist parties in government affect greenhouse gas emissions in EU member states. *Journal of European Public Policy*, *28*(7), 980–997. https://doi.org/10.1080/13501763.2021.1918215

Jahn, D. (2022). Party families and greenhouse gas emissions: A new perspective on an old concept. *Zeitschrift für Vergleichende Politikwissenschaft, 15*(4), 477–496. https://doi.org/10.1007/s12286-021-00504-6

Jahn, D., Düpont, N., Baltz, E., Andorff-Woller, M., Klagges, L. & Suda, S. (2022): Parties, Institutions & Preferences: PIP Collection [Version 2020-04]. Chair of Comparative Politics, University of Greifswald. https://doi.org/10.7910/DVN/KRXPH4

Jahn, D., & Wälti, S. (2007). Umweltpolitik und Föderalismus: Zur Klärung eines ambivalenten Zusammenhangs. In K. Jacob, F. Biermann, P.-O. Busch, & P. H. Feindt (Eds.), *Politische Vierteljahresschrift Sonderhefte: Vol. 39. Politik und Umwelt* (pp. 262–282). VS Verlag für Sozialwissenschaften.

Jänicke, M. (1992). Conditions for environmental policy success: An international comparison. *The Environmentalist, 12*(1), 47–58.

Jänicke, M. (2007). Ecological modernisation: New perspectives. In M. Jänicke & K. Jacob (Eds.), *Environmental Governance in Global Perspective: New Approaches to Ecological Modernisation* (2nd ed., pp. 9–29). Free University of Berlin.

Jänicke, M., Mönch, H., Ranneberg, T., & Simonis, U. E. (1989). Structural change and environmental impact. *Intereconomics, 24*(1), 24–35. https://doi.org/10.1007/BF02928545

Jensen, C. B., & Spoon, J.-J. (2011). Testing the 'party matters' thesis: Explaining progress towards Kyoto protocol targets. *Political Studies, 59*(1), 99–115. https://doi.org/10.1111/j.1467-9248.2010.00852.x

Keohane, R. O., & Levy, M. A. (1996). *Institutions for Environmental Aid: Pitfalls and Promise. Global environmental accords.* MIT Press.

Kim, Y., Tanaka, K., & Matsuoka, S. (2017). Institutional mechanisms and the consequences of international environmental agreements. *Global Environmental Politics, 17*(1), 77–98. https://doi.org/10.1162/GLEP_a_00391

Kitschelt, H. (1988). Left-libertarian parties: Explaining innovation in competitive party systems. *World Politics, 40*(2), 194–234. https://doi.org/10.2307/2010362

Kittel, B. (2006). A crazy methodology? On the limits of macro-quantitative social science research. *International Sociology, 5*(21), 647–677. https://doi.org/10.1177/0268580906067835

Klagges, L. (2019). *Ökologische Performanz von Mehrheits- und Konsensdemokratien. Eine empirische Analyse der OECD-Länder 2000 bis 2015.* https://doi.org/10.25358/openscience-2367

Knill, C., & Liefferink, D. (2007). *Environmental Politics in the European Union: Policy-making, Implementation and Patterns of Multi-level Governance.* Manchester University Press.

Knill, C., & Liefferink, D. (2021): The establishment of EU environmental policy. In A. Jordan & V. Gravey (Eds.), *Environmental Policy in the EU* (pp. 13–32). Routledge.

Knill, C., Schulze, K., & Tosun, J. (2012). Regulatory policy outputs and impacts: exploring a complex relationship. *Regulation & Governance, 6*(4), 427–444. https://doi.org/10.1111/j.1748-5991.2012.01150.x

Knill, C., Shikano, S., & Tosun, J. (2014). Explaining environmental policy adoption: A comparative analysis of policy development in 24 OECD countries. In A. Duit (Ed.), *State and Environment. The Comparative Study of Environmental Governance* (pp. 53–80). MIT Press.

Knill, C., Steinebach, Y., & Fernández-i-Marín, X. (2020). Hypocrisy as a crisis response? Assessing changes in talk, decisions, and actions of the European commission in EU environmental policy. *Public Administration, 98*(2), 363–377. https://doi.org/10.1111/padm.12542

Konisky, D. M. (2007). Regulatory competition and environmental enforcement: Is there a race to the bottom? *American Journal of Political Science, 51*(4), 853–872. https://doi.org/10.1111/j.1540-5907.2007.00285.x

Lenschow, A., Burns, C., & Zito, A. (2020). Dismantling, disintegration or continuing stealthy integration in European Union environmental policy? *Public Administration, 98*(2), 340–348. https://doi.org/10.1111/padm.12661

Lijphart, A. (2012). *Patterns of Democracy: Government Forms and Performance in Thirty-Six Countries.* Yale University Press.

Lim, S., & Duit, A. (2018). Partisan politics, welfare states, and environmental policy outputs in the OECD countries, 1975-2005. *Regulation & Governance, 12*, 220–237. https://doi.org/10.1111/rego.12138

Limberg, J., Steinebach, Y., Bayerlein, L., & Knill, C. (2020). The more the better? Rule growth and policy impact from a macro perspective. *European Journal of Political Research, 60(2)*, 438–454. https://doi.org/10.1111/rego.12297

Lundqvist, L. (1980). *The Hare and Tortoise: Clean Air Policies in the United States and Sweden.* University of Michigan Press.

Luterbacher, U., & Sprinz, D. F. (Eds.). (2001). *Global Environmental Accord. International Relations and Global Climate Change.* MIT Press.

Madden, N. J. (2014). Green means stop: Veto players and their impact on climate-change policy outputs. *Environmental Politics, 23*(4), 570–589. https://doi.org/10.1080/09644016.2014.884301

Miles, E. L., Andresen, S., Carlin, E. M., Skjærseth, J. B., Underdal, A., & Wettestad, J. (2002). *Environmental Regime Effectiveness: Confronting Theory with Evidence. Global environmental accord*. MIT Press.
Mitchell, R. B. (1994). *Intentional Oil Pollution at Sea: Environmental Policy and Treaty Compliance. Global environmental accords series*. MIT Press.
Mitchell, R. B. (Ed.). (2008). *International Environmental Politics*. Sage.
Mitchell, R. B. (2009). The influence of international institutions: Institutional design, compliance, effectiveness, and endogeneity. In H. V. Milner & A. Moravcsik (Eds.), *Power, Interdependence, and Nonstate Actors in World Politics* (pp. 66–83). Princeton University Press.
Mitchell, R. B. (2010). *International Politics and the Environment*. Sage.
Mitchell, R. B., Andonova, L. B., Axelrod, M., Balsiger, J., Bernauer, T., Green, J. F., Hollway, J., Kim, R. E., & Morin, J.-F. (2020). What we know (and could know) about international environmental agreements. *Global Environmental Politics*, *20*(1), 103–121. https://doi.org/10.1162/glep_a_00544
Müller-Rommel, F. (1998). Explaining the electoral success of green parties: A cross-national analysis. *Environmental Politics*, 7(4), 145–154. https://doi.org/10.1080/09644019808414428
Neumayer, E. (2003a). Are left-wing party strength and corporatism good for the environment? Evidence from panel analysis of air pollution in OECD countries. *Ecological Economics*, *45*(2), 203–220. https://doi.org/10.1016/S0921-8009(03)00012-0
Neumayer, E. (2003b). *Weak versus Strong Sustanability*. Edward Elgar.
Oberthür, S., & Gehring, T. (2006). *Institutional Interaction in Global Environmental Governance: Synergy and Conflict among International and EU Policies. Global environmental accord*. MIT Press.
Offe, C. (1985). New social movements: challenging the boundaries of institutional politics. *Social Research*, *52*(4), 817–868. https://doi.org/10.1007/978-3-658-22261-1_12
Ozymy, J., & Rey, D. (2013). Wild spaces or polluted places: Contentious policies, consensus institutions, and environmental performance in industrialized democracies. *Global Environmental Politics*, *13*(4), 81–100. https://doi.org/10.1162/GLEP_a_00199
Pappas, T. S. (2019). Populists in power. *Journal of Democracy*, *30*(2), 70–84. https://doi.org/10.1353/jod.2019.0026
Poloni-Staudinger, L. M. (2008). Are consensus democracies more environmentally effective? *Environmental Politics*, *17*(3), 410–430. https://doi.org/10.1080/09644010802055634
Prakash, A., & Potoski, M. (2006). Racing to the bottom? Trade, environmental governance, and iso 14001. *American Journal of Political Science*, *50*(2), 350–364. https://doi.org/10.1111/j.1540-5907.2006.00188.x
Rees, W. (2018). Ecological footprint. In N. Castree, M. Hulme, & J. D. Proctor (Eds.), *Companion to Environmental Studies* (pp. 43–48). Routledge.
Rume, T., & Islam, S. M. D.-U. (2020). Environmental effects of Covid-19 pandemic and potential strategies of sustainability. *Heliyon*, *6*(9). https://doi.org/10.1016/j.heliyon.2020.e04965
Schnaiberg, A. (1980). *The Environment: From Surplus to Scarcity*. Oxford University Press.
Schulze, K. (2014). Do parties matter for international environmental cooperation? An analysis of environmental treaty participation by advanced industrialised democracies. *Environmental Politics*, *23*(1), 115–139. https://doi.org/10.1080/09644016.2012.740938
Schulze, K. (2021). Policy characteristics, electoral cycles, and the partisan politics of climate change. *Global Environmental Politics*, *21*(2), 44–72. https://doi.org/10.1162/glep_a_00593
Scruggs, L. (1999). Institutions and environmental performance in seventeen western democracies. *British Journal of Political Science*, *29*(1), 1–31. https://doi.org/10.1017/S0007123499000010
Scruggs, L. (2003). *Sustaining Abundance: Environmental Performance in Industrial Democracies*. Cambridge University Press.
Shafik, N. (1994). Economic Development and Environmental Quality. An Econometric Analysis. *Oxford Economic Papers*, *46*(0), 757–773.
Shum, R. Y. (2009). Can attitudes predict outcomes? Public opinion, democratic institutions and environmental policy. *Environmental Policy and Governance*, *19*(5), 281–295. https://doi.org/10.1002/eet.518
Solingen, E. (2009). The global context of comparative politics. In M. I. Lichbach & A. S. Zuckerman (Eds.), *Comparative Politics: Rationality, Culture, and Structure* (2nd ed., pp. 220–259). Cambridge University Press.
Sommerer, T., Holzinger, K., & Knill, C. (2008). The pair approach: What causes convergence of environmental policies? In K. Holzinger, C. Knill, & B. Arts (Eds.), *Environmental Policy Convergence in Europe: The Impact of International Institutions and Trade* (pp. 144–195). Cambridge University Press.
Spoon, J.-J., Hobolt, S. B., & de Vries, C. (2014). Going green: Explaining issue competition on the environment. *European Journal of Political Research*, *53*(2), 363–380. https://doi.org/10.1111/1475-6765.12032

Steinberg, P. F., & VanDeveer, S. D. (Eds.). (2012). *Comparative Environmental Politics: Theory, Practice, and Prospects*. MIT Press.

Steinebach, Y., & Knill, C. (2017). Still an entrepreneur? The changing role of the European commission in eu environmental policy-making. *Journal of European Public Policy, 24*(3), 429–446. https://doi.org/10.1080/13501763.2016.1149207

Stern, D. L. (2018). The environmental kuznets curve. In N. Castree, M. Hulme, & J. D. Proctor (Eds.), *Companion to Environmental Studies* (pp. 49–54). Routledge.

Tsebelis, G. (2002). *Veto Players: How Political Institutions Work*. Princeton University Press.

Victor, D. G., Raustiala, K., & Skolnikoff, E. B. (Eds.). (1998). *Global Environmental Accord. The Implementation and Effectiveness of International Environmental Commitments: Theory and Practice*. MIT Press.

Vogel, D. (1986). *National Styles of Regulation: Environmental Policy in Great Britain and the United States*. Cornell University Press.

Volkens, A., Burst, T., Krause, W., Lehmann, P., Matthieß, T., Merz, N., Regel, S., Weßels, B., Zehnter, L., & Wissenschaftszentrum Berlin Für Sozialforschung. (2020). *Manifesto Project Dataset*. https://doi.org/10.25522/MANIFESTO.MPDS.2020B

Wackernagel, M., & Rees, W. E. (1996). *Our Ecological Footprint. Reducing Human Impact on the Earth*. New Society Publishers.

Wälti, S. (2004). How multilevel structures affect environmental policy in industrialized countries. *European Journal of Political Research, 43*(4), 599–634. https://doi.org/10.1111/j.1475-6765.2004.00167.x

Ward, H. (2008). Liberal democracy and sustainability. *Environmental Politics, 17*(3), 386–409. https://doi.org/10.1080/09644010802055626

Ward, H., & Cao, X. (2012). Domestic and international influences on green taxation. *Comparative Political Studies, 45*(9), 1075–1103. https://doi.org/10.1177/0010414011434007

Weale, A., Pridham, G., Cini, M., Konstadakopulos, D., Porter, M., & Flynn, B. (2000). *Environmental Governance in Europe: An Ever Closer Ecological Union?* Oxford University Press.

Weaver, A. A. (2008). Does protest behavior mediate the effects of public opinion on national environmental policies? A simple question and a complex answer. *International Journal of Sociology, 38*(3), 108–125. https://doi.org/10.2753/IJS0020-7659380305

Weiss, E. B., & Jacobson, H. K. (Eds.). (1998). *Global Environmental Accord: Strategies for Sustainability and Institutional Innovation. Engaging Countries: Strengthening Compliance with International Environmental Accords*. MIT Press.

Wendling, Z., Emerson, J. W., de Sherbinin, A., & Esty, D. C. (2020). *Environmental Performance Index 2020*. Yale Center for Environmental Law and Policy.

Whitehead, M. (2018). Sustainable development. In N. Castree, M. Hulme, & J. D. Proctor (Eds.), *Companion to Environmental Studies* (pp. 110–114). Routledge.

World Commission on Environment and Development. (1987). *Our Common Future [Brundtland-Report]*. Oxford University Press.

Young, O. R. (Ed.). (1997). *Global environmental accords series. Global Governance: Drawing Insights from the Environmental Experience*. MIT Press.

Young, O. R. (Ed.). (1999). *Global environmental accord. The Effectiveness of International Environmental Regimes: Causal Connections and Behavioral Mechanisms*. MIT Press.

Young, O. R. (2014). The effectiveness of international environmental regimes: Existing knowledge, cutting-edge themes, and research strategies. In M. Betsill, K. Hochstetler, & D. Stevis (Eds.), *Advances in International Environmental Politics* (pp. 273–299). Palgrave Macmillan.

8

BUREAUCRACY AND ENVIRONMENTAL POLICY

Christoph Knill and Yves Steinebach

8.1 Introduction

Bureaucracy and administration are often seen as an obstacle and barrier to effective environmental and climate protection. They are deemed too inflexible, undemocratic, and too much devoted to rules and red tape to allow for timely and innovative policy actions. Yet, bureaucracies are clearly underrated with regard to their importance for environmental matters (Biesbroek et al., 2018). In fact, it is *bureaucrats* who draft environmental policies, convert abstract political goals into concrete policy measures, and forge compromises between official rules and reality on the ground.

In this chapter, we examine the role of bureaucracies in environmental policy. We focus on the role of bureaucracies in policy-making and implementation. We show that effective bureaucracies are crucial for the proper functioning of environmental policies throughout the different policy stages. Independent and well-equipped administrations not only produce better-designed environmental policies but are also faster and more effective in applying them. We conclude by showing that, despite their importance, environmental administrations are under 'siege' in modern democracies as their organizational capacities do *not* keep pace with the strong expansion of implementation tasks.

8.2 Bureaucracy and Environmental Policy-Making

The effectiveness of environmental policies presumes high-quality policy design (Peters et al., 2018). This means that (1) policies correctly identify the underlying causes of the respective problem (Pressman & Wildavsky, 1984) and (2) that policy-makers pick the 'right' policy instruments that actually work for a given policy context and adequately take account of both the problem and the target group characteristics (Weaver, 2014). Moreover, environmental policies must be (3) ambitious enough to make a real-world difference (Steinebach, 2021). This latter aspect implies that, for instance, air pollutant limits are set low enough to stimulate emission reductions or that green taxes are high enough to push citizens to permanently change their consumption habits (Green, 2021).

To be able to propose and draft such well-designed and ambitious policies, the bureaucracies involved in policy formulation – usually ministerial or agency bureaucracies – need "[the]

DOI: 10.4324/9781003043843-10

competencies and capabilities necessary to perform [their] policy functions" (Wu, et al., 2015). As discussed in the following, competencies imply that bureaucracies are able to make policy decisions *independent* of imminent political influences. Capabilities, in turn, are a function of the analytical capacities of the people working in the administration and their ability to overcome both inner- and outer-institutional boundaries (Hsu, 2015; Davies et al., 2000).

8.2.1 Bureaucratic Autonomy

While *no* bureaucracy in a democratic system has the authority to act without political oversight or public scrutiny, there are a wide variety of ways in which its mandate might be issued, and politicians can interfere with the administration's day-to-day operations (Fukuyama, 2013). The degree to which environmental bureaucracies and the professionals working within them can draft policies *independently of imminent political influences* has been identified as a major determinant of the quality and effectiveness of environmental policies.

Studying different hotspots of mining in Peru, Orihuela et al. (2021) show that environmental administrations without a substantial level of bureaucratic autonomy are only "paper institutions", that is, "organizations that do not fulfil their legally-mandated functions" (p. 10; see also Boullier et al., 2019). Similarly, Potoski and Woods (2000) find that "hardwiring administrative procedures" (p. 203) increases the influence of politicians over the US state-level environmental protection agencies and that this increased influence goes hand in hand with overall less ambitious environmental standards (see also Potoski, 2002).

Wellstead and Biesbroek (2022) qualify this argument by arguing that public servants involved in environmental and climate policy-making "need to be shielded from the involvement of political actors (elected and appointed officials), but … [in other contexts] may be subordinate with regard to the larger climate change goals" (p. 2). This "sweet spot" (p. 1) between external influences and bureaucratic autonomy, in turn, depends on a country's administrative tradition and the extent to which societal actors are used to accepting top-down rules and orders (Biesbroek et al., 2018). Moreover, overly autonomous bureaucracies might tend to 'encapsulate' themselves from the outside, so that new and potentially controversial policy ideas find it more difficult to enter the administration (ibid.). In consequence, bureaucratic autonomy and openness must be carefully balanced (Bach, 2022). Making a similar argument, Fukuyama (2013) argues that an appropriate degree of bureaucratic autonomy does *not* mean that bureaucrats should be entirely isolated from external influences but that administrations "need to be shielded from certain influences … but also subordinate to the society with regard to larger goals" (p. 358).

Meckling and Nahm (2018) also take on the issue of bureaucratic autonomy in environmental and climate policy-making but approach it from a *procedural* perspective. More precisely, the authors argue that it makes a difference whether broader policy goals are first set by the legislators and then, in a second step, passed onto autonomous environmental bureaucracies to adopt concrete policy measures ("bureaucratic policy design"); or whether it is the bureaucracy that determines the broader goals and the legislator that specifies the measures ("legislative policy design"). Comparing the cases of California and Germany, the authors find that the legislative policy design dominant in Germany prevents the attainment of emissions reduction goals, as politicians have a greater incentive to respond to vested interests (see also Meckling & Nahm, 2022). This renders the policy-making process more vulnerable to 'regulatory capture', that is, a situation in which public representatives are more beholden to the target groups they are supposed to regulate than to the public interest (Dal Bó, 2006).

Despite the importance of the concept of bureaucratic autonomy in the literature, scholars have found it difficult to agree on a common conceptualization and measurement. For instance, while some studies have focused on the extent to which the bureaucratic leadership needs to confirm with the responsible minister, others have looked at the agencies' funding structures (for a broader discussion, see e.g. Verhoest et al., 2004). In empirical terms, the most common approach is to rely on expert surveys where perceptions of experts on a given bureaucracy are recorded (Fukuyama, 2013). An alternative approach has been put forward by Jordana et al. (2018; see also Jordana et al. 2011). The authors assess the autonomy of regulatory agencies based on their level of (1) managerial autonomy, (2) political independence, (3) public accountability, and (4) regulatory capabilities. While the level of managerial autonomy captures the extent to which political leaders can directly intervene in the bureaucracy's organizational structure and budgetary planning, political independence assesses how easy (or difficult) it is for the ruling majority to exchange the bureaucracy's leadership. Public accountability, in turn, reflects the activities agencies need to perform to justify their policy decisions and judgments, such as mandatory public reporting and participatory and other transparency requirements. Regulatory capabilities indicate the level of formal regulatory power (capacity to unilaterally set regulation, sanctions, etc.) delegated to an agency. The respective data is collected through the *formal* rules laid down in administrative laws and decrees.

Figure 8.1 plots the data on the four autonomy 'dimensions' for regulatory agencies in environmental matters for a global sample of 47 countries. These are *all* countries worldwide with an environmental protection agency operative in 2010. The figure reveals several aspects. First, it shows that countries substantially differ with regard to the extent to which autonomy is granted to the environmental bureaucracy. Second, it reveals that the different dimensions are somewhat related but also tend to diverge. In Malta, for instance, the Environment and Resources Authority has quite some independence with regard to managerial autonomy, political independence, and public accountability, but possesses relatively few substantial policy competencies. Third, the figure shows that the agencies' regulatory capabilities are typically less advanced than the other three autonomy dimensions.

Overall, however, it needs to be mentioned that the bureaucratic autonomy in environmental matters is *generally* lower than in other policy areas assessed by Jordana et al. (2018).[1] For instance, while the mean managerial autonomy in all other policy sectors (competition, electricity, food security, etc.) is 0.21, it is –0.33 in environmental policy. With regard to political independence, this difference is even more pronounced with a mean of 0.45 in all other areas and –0.49 in the environment.

8.2.2 Analytical Skills and Effective Exchange

Another insight from the literature is that it is not only the formal autonomy that matters but also whether the bureaucracy can leverage its 'freedom'. This is due to two aspects. First, *bureaucracies need analytical skills* as they have to identify and select the best policy instrument (mix) to solve a specific environmental problem (Hsu, 2015). Given that "policy design is invariably about the future and how to get there" (Bali et al., 2019: 3), bureaucrats need the analytical and anticipatory capabilities to be able to go through the scientific evidence available and to assess the policy option available based on logic and cogitation (Adam et al., 2019).

Second, bureaucrats must be able to *overcome institutional boundaries* to come up with well-designed environmental policies. Many environmental problems, including climate change, are multifaceted and cut across different policy domains. At the same time, the responsibility

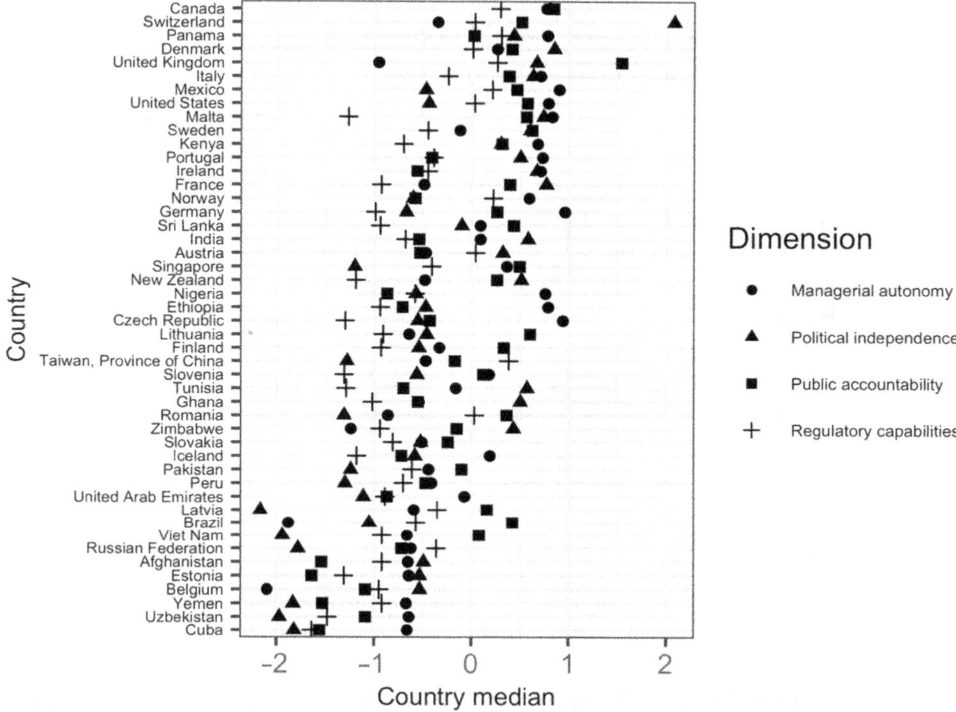

Figure 8.1 Bureaucratic autonomy of environmental protection agencies in 47 countries.

to develop policy solutions rests with fragmented and compartmentalized bureaucratic 'silos' (Den Uyl & Russel, 2018). Consequently, effective environmental policy-making requires that bureaucrats effectively coordinate their actions *horizontally across different administrative units* (Duffy & Cook, 2019). In the absence of this capacity, other parts of the bureaucracy might tend to produce redundant or – even worse – contradictory policy measures (Howlett & Ramesh, 2014). Moreover, effective policy-making hinges on the administration's capacity to ensure an exchange between the policy-formulating bureaucracies at the 'top' and the implementing authorities at the 'bottom'. While policies are produced centrally, the local implementers experience what works in practice and what does not (Matland, 1995). *Vertical feedback channels and loops* can thus help to analyse why specific policies have succeeded or failed in achieving their goals and, based on this, to learn how (future) policy designs could be improved and optimized (Knill et al., 2021; see also Chapter 9 in this book).

There are multiple empirical studies demonstrating the importance of bureaucratic capacities for environmental policy-making. Colenbrander and Bavinck (2017), for instance, show for the case of coastal management in Cape Town that poor levels of inner-bureaucratic coordination have led to diverging and inconsistent policy goals "manifesting in a deteriorating coastal environment and mal-adaptive impacts" (p. 48). Desveaux et al. (1994) provide a similar finding studying the federal health, environment, and energy bureaucracies in Canada. They show that the ability to come up with innovative policy solutions needs not only generous budgets and expertise but also the capacity "to transcend professional compartmentalization and routine" (p. 493).

Box 8.1 Average Instrument Diversity: A Novel Measure of Environmental Policy Design Quality

Environmental policy instruments and instrument combinations can be either more or less the same across all policy targets or vary from one policy target to the other. In the former case, governments tend to pick 'off-the-rack' solutions. In the latter case, governments can be assumed to generally opt for more 'tailor-made' interventions. Fernández-i-Marín et al. (2021) propose and apply the concept of average instrument diversity (AID) to assess the extent to which policy-makers tend toward either of these options. The AID index indicates the probability that two policy instruments randomly drawn from a policy portfolio are of a *different* kind – with a higher index value indicating a more diversely composed policy portfolio and a lower index value indicating a more uniform one. The AID is found to be (1) systematically related to higher levels of environmental performance and to be determined by (2) a country's institutional constraints and (3) the bureaucratic capacities available.

While all these studies identify bureaucratic capacity as a crucial determinant of environmental policy success, they typically focus on single cases or countries. A notable exemption in this context is the work by Fernández-i-Marín et al. (2021). The authors show for a sample of 21 advanced democracies that countries with greater bureaucratic capacities tend to produce better-designed environmental policies. More precisely, countries with greater analytical capabilities within the bureaucracy and well-functioning intra-bureaucratic coordination structures tend to apply a more diverse set of different policy instruments and instrument combinations. A greater "average instrument diversity" (p. 931), in turn, is associated with higher levels of environmental performance, that is, less air and water pollution, waste production, and excessive fertilizer use.

8.3 Bureaucracy and Environmental Policy Implementation

A central issue in environmental policy is ensuring that the (ambitious) policy measures taken 'on paper' are also translated into concrete policies 'in action'. The policy implementation process usually involves two steps (for a discussion, see Chapter 10). The first one is the step of policy transposition. Policy transposition is typically discussed in the context of EU policies, where national governments must integrate supranational policies into domestic legislation (Steunenberg, 2006; Zhelyazkova, 2013). In fact, however, it can refer to *any* multi-level structure (international agreements, federal states, etc.) in which governments have to transpose 'higher-level' policies into 'lower-level' policy frameworks. The second step is the practical application (execution and enforcement) of environmental policies. Depending on the exact policy in question, this can involve a range of different activities, such as the carrying out of on-the-spot inspections, the granting of environmental permits, or the collection of green taxes.

To implement and apply environmental policies in a timely, correct, and effective manner, bureaucracies require various capabilities. As discussed in more detail in the following, they need sufficient *staff and organizational resources* allocated for implementation. Moreover, they need well-designed *implementation structures* as well as some *discretionary power.*

8.3.1 *Staff and Organizational Resources*

A central research question in EU compliance studies is to examine what determines at which speed and correctness EU laws are transposed in the national context (Treib, 2014). A central insight from this research strand is that countries with well-resourced bureaucracies, that is, administrations with more "staff ... designated to support the implementation of policy instruments", "informational resources, involved to support policy implementation", and "financial resources, allocated to the implementation of policy goals" (Bondarouk and Mastenbroek, 2018: 20), will find it easier to make EU policies work at the national level.

Steinebach and Limberg (2022) highlight that these insights are not restricted to the EU context but that the existence of sufficient bureaucratic resources is equally important in the context of the international climate agreements (see also Lederer & Höhne, 2021; Hickmann et al., 2017). Transposing international climate policies into the domestic context requires the creation of national legal frameworks and the establishment of public authorities in charge of execution and enforcement. The central challenge of international climate agreements is that they often involve the transfer of 'new' types of policies (e.g. carbon trading schemes) to developing countries with relatively low administrative resources. Analysing the policy implementation of the Clean Development Mechanism (CDM),[2] the authors find that the lack of national administrative resources constitutes a major obstacle to the successful implementation of international schemes. Countries with lower levels of administrative capacities do not only take longer to set up the necessary domestic environmental institutions (so-called 'designated national authorities') but also attract far fewer CDM projects over time. Simply put, CDM projects are foreign investments in carbon-saving projects in developing countries that can be traded to industrialized countries to offset their emissions (for a critical assessment of CDMs see Schneider, 2011). The authors argue that international support programmes that "train bureaucrats, develop a country's sustainable criteria, and help to develop the institutional legal framework to establish a DNA" (Steinebach & Limberg, 2022: 1160) might compensate for countries' lacking administrative capacities.

Staff and organizational resources, however, are not only relevant in the context of policy transposition. Even more so, they matter for environmental policy execution, enforcement, and delivery (Povitkina & Bolkvadze, 2019). Shimshack and Ward (2005) show that the threat of credible policy enforcement creates massive spill-over effects that do *not* only affect and change the behaviour of the inspected plant but also the overall violation rate (see also Telle, 2009). Bureaucracies that lack the organizational capacities to uphold this threat are thus "an unlikely vehicle for effective policy implementation" (Ringquist, 1993: 1026).

A modification of this general argument is that it is not (only) the mere number of resources possessed by an authority that has an impact on the proper application of environmental policies but that it ultimately also depends on the type of policy requiring implementation (Knill and Lenschow, 1998). The environmental administrations of EU member states, for instance, massively struggled to implement the EU Water Framework Directive (WFD) – not because they had not been staffed adequately but because the expertise and skills available did *not* match the requirements set out by the directive (Newig & Koontz, 2014). The WFD was adopted to succeed and replace the traditional regulatory practices relying on so-called 'end-of-pipe' control. Under this old approach, specific parameters of water quality and pollution were monitored at the point of discharge. Thus, the main task water authorities had to perform was permitting and controlling polluting facilities. The regulatory approach put forward by the WFD, in turn, required the management of the entire ecological system. Unlike other environmental regulations that usually prescribe specific targets, the WFD specified that all river basins and water bodies must be in a 'good' ecological state (Voulvoulis et al., 2017). In other words: it was no longer enough that administrations possessed sufficient staff to inspect and authorize industrial plants; they now also required the expertise to 'manage' entire eco-systems, an "implementation function for which they were initially neither trained nor staffed" (Adam et al., 2019: 100).

Limberg et al. (2021) further qualify that the need for staff and resources in environmental policy implementation is not 'static' but requires constant adaption and expansion. The authors show that the 'right' level of administrative resources *cannot* be assessed in general as it is determined by the totality of policy responsibilities an environmental bureaucracy is in charge of. In other words: a government that wants to improve environmental quality effectively needs to constantly ensure that the bureaucracy's capacities keep up with the rise in environmental regulations.

8.3.2 Bureaucratic Structures and Coordination Needs

The capacity to successfully implement environmental policy does not only depend on the capacities within the bureaucracy. It also makes a difference whether administrators have to coordinate their actions with *other* public or private sector actors involved in policy implementation (Pressman & Wildavsky, 1984). The need for coordination increases with the number of authorities and divisions involved in the process of policy implementation. Intra- and inter-organizational bureaucratic boundaries, in turn, might hamper the free flow of information. Moreover, they can lead to conflicts between the organizations. As highlighted by Paavola (2016), "the essence of many coordination problems is in fact a conflict. When several ways of conducting matters exist, and one of them has to be chosen, this choice typically entails differential costs and benefits to the actors involved" (p. 144). In consequence, the literature suggests that having clear responsibilities and a rather low number of administrative actors involved reduces the risks of potential maladministration and bureaucratic failure (Hepburn, 2010).

Brinkerhoffs (1996), for instance, shows for Madagascar's 'Environmental Action Plan' that "the multiplicity of hierarchies involved …, those internal to the implementing agencies as well as the interagency ones, makes the shaping of consistent action on everyone's part extremely difficult" (p. 1505). In consequence, the author recommends the reduction of tight administrative interdependencies through "less frequent formal reporting or supervision, more operational autonomy once contracts and work plans are approved, [and] more reliance on informal collaborative arrangements" (p. 1506).

Steinebach (2022) mapped the characteristics of all bureaucratic structures in charge of enforcing clean air policies in 14 OECD countries over a period of 25 years (1990 to 2014) for a total of about 400 environmental policies. Figure 8.2 shows at the country level the share of environmental policies being executed and enforced (1) by a central/federal authority; (2) in cooperation with private sector actors; and (3) by multiple environmental authorities. About 50 percent of the policies are implemented by a central/federal authority. About the same share of policies is implemented in cooperation with private sector actors.[3] In 20 percent of the cases, multiple environmental authorities enforce the *same* environmental policy.

The majority of clean air policies are thus put into practice through implementation structures in which a single central authority cooperates with private sector actors during policy implementation. Especially the strong involvement of private sector actors supports the notion of shift from 'government' to 'governance', that is, the observation that environmental policy implementation is increasingly done through organizational networks encompassing both private and public actors. At the same time, however, the data also shows that in most cases a *single* public authority is responsible for environmental policy implementation. This

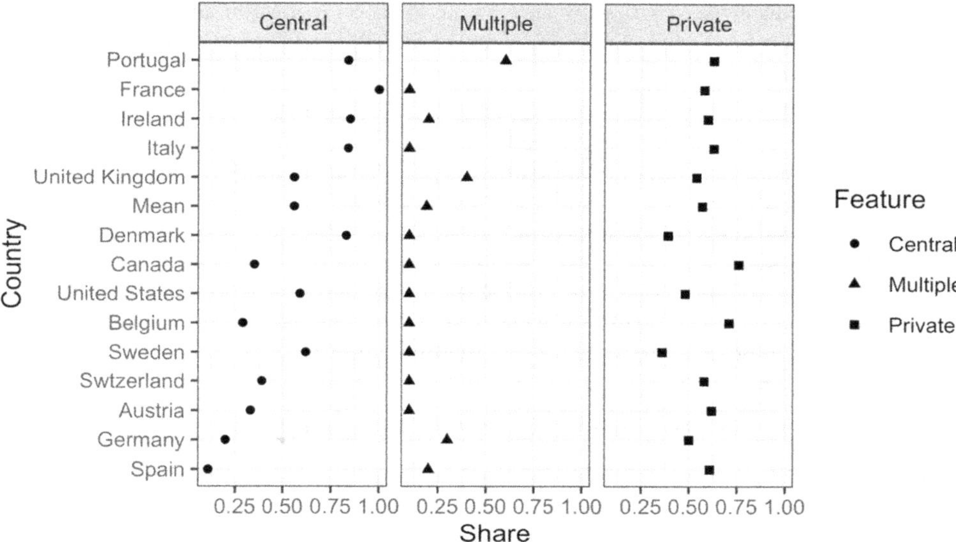

Figure 8.2 Bureaucratic structures in charge of enforcing clean air policies.

observation provides some support to Kettl's (1990) argument that the scholarship on policy implementation has slightly exaggerated the role of inter-organizational policy implementation as most policy programmes are ultimately handled through rather simplistic and hierarchical bureaucratic arrangements.

8.3.3 *Administrative Tradition and Discretionary Power*

Environmental bureaucracies are part of a broader institutional setup and administrative traditions.[4] These traditions, in turn, determine the way bureaucracies function (Peters, 2021; Peters and Painter, 2010). Amongst other aspects, they affect how tightly rules must be followed and how much discretion is left to the administrators in deciding in which way and how strict regulations should be applied. This flexibility in administering public policies seems to be generally higher in so-called managerial systems than in legalistic systems (Christensen et al., 2011). Managerial systems refer to administrations where administrators are expected to run policy programmes as efficiently and smoothly as possible. In legalistic systems, by contrast, the public sector is considered to be rule-following with administrators being primarily responsible for ensuring compliance with prevailing laws, rules, and regulations (Kickert, 2005). The bureaucratic approach towards environmental policy implementation in Great Britain – a country that is often said to be the managerialist archetype – is described as "one in which agencies have great discretion to make 'deals' with regulated firms" (Schmidt, 2006, p. 137; Vogel, 1986). Similar patterns can be found in the USA. Here US enforcement authorities possess substantial leeway "to decide what level of emission controls will be required … and when emission controls must be achieved" (Rosenbaum, 2017, p. 207). Moreover, they are allowed to grant "some delay [to the regulated firms] in achieving emissions controls that are otherwise required under law" (ibid.). The exact opposite is said about the German *Rechtsstaat* ('constitutional state') tradition. Here, Versluis (2003) highlights that German administrators follow "a strict enforcement approach where inspectors do not give regulated second chances but punish immediately ... [and] show an extreme focus on legislation and inspect with the primary aim to check whether companies comply with all details" (p. 36).

Steinebach (2022b) shows that a country's administrative tradition and, based on this, the bureaucracies' discretionary power has a *direct* impact on the effectiveness of environmental regulation. More precisely, the author finds that the effect of regulations on air pollutant emissions is stronger in countries with a managerial administrative tradition. These differences, however, are not the same for all kinds of regulatory changes. While both managerial and legalistic bureaucracies manage to implement minor policy reforms, it is primarily managerial systems that can effectively handle more ambitious regulations. A potential explanation for this finding is that administrators with more discretionary power will find it generally easier to implement environmental regulation – especially when these regulations are ambitious and thus might cause resistance and opposition on the side of the target group.

8.4 Environmental Administrations under Siege

In the previous sections, we have seen that well-equipped and well-staffed bureaucracies are necessary for effective environmental policy. Consequently, governments should constantly ensure that their bureaucracies have what they need to make environmental policies work (bureaucratic autonomy, well-thought implementation arrangements, plenty of resources, etc.). In reality, however, we can observe an increasing mismatch between the policies up for implementation and the administrative capacities available.

In the ongoing ACCUPOL[5] research project, we have conducted about 50 semi-structured interviews with members of the environmental authorities in Denmark, Germany, Ireland, Italy, and Portugal. The interviewees from the five countries were responsible for a large variety of implementation-related activities, ranging from the supervision of subordinate entities to the granting of permits and the monitoring and inspections of industrial plants or water basins in environmental policy implementation. In these interviews, we asked several questions about the bureaucrats' workload, the key reasons for and sources of their workload, and their ways and strategies to cope with it. In these interviews, we essentially made the following three observations.

The first insight is that, across all countries, the environmental administrators reported that their *workload has constantly increased over time* because of additional implementation duties associated with new environmental policies. They indicated, for instance, that "we can definitely talk about an immense explosion of tasks", that the "constant production of policies does not only happen with regard to legislative changes but also in administrative orders", and that the "work is not only becoming more ... [but that it] is also becoming more and more complex". Moreover, the administrators indicated that, in most instances, the administrations had *not* been compensated for the increased workload along with the additional administrative capacities. They state that the resources "did not grow in proportion to the requirements or regulations that were being published" and that the implementation tasks rain "down on [them] from above".

The second insight is that, despite the general trend of an increasing gap between the policies up for implementation and the administrative capacities available, there are also *pronounced variations across the countries*. The Italian bureaucrats, for instance, stated that there is a "big disconnect between the political and the technical [implementation] side" and that "politicians [simply] go their own way". In Denmark, by contrast, we found overall less frustration and dissatisfaction. Here, the administrators stated that "the politicians, the government, the parliament have been quite aware that new policies have costs. And they are aware that the financial opportunities must be increased – so we may achieve the goals that they put on our organization". While the observed variation can easily be explained by the differences in general state capacities among the countries (Hanson & Sigman, 2021), it might also suggest that the institutionalized ties between the policy-making and policy-implementing level (see Chapter 9 of this Handbook) make an important difference for the effectiveness of environmental policies.

The third and final takeaway from the interviews is that when being confronted with overload, that is, more policies and administrative burdens that can effectively be handled by the bureaucracy, administrations *start developing coping practices and strategies* with negative repercussions for the functioning of environmental policies. One interview partner noted, for instance, that

> "it's practically the responsibility of the decision-makers on the spot as to how specifically they go about their task. It's like the duvet that's too short – it's sometimes by the nose, sometimes by the toes – but it's never enough to cover everything".

Another interviewee confirmed this claim stating that "the overload is of such an extent that we often have to prioritize our actions – and some effectiveness is definitely lost when we are forced to prioritize tasks".

This finding is well in line with insights gained by Kaplaner and Steinebach (2022) on the implementation of the Industrial Emission Directive (IED) in the German state of Baden-Württemberg.[6] The EU IED prescribed the use of so-called 'best available techniques' (BAT) in large industrial installations and set mandatory requirements on when and how on-the-spot visits and inspections should be carried out. Analysing more than 2,000 inspection reports, Kaplaner and Steinebach (2022) find that environmental administrators (1) tend to inspect many industrial sites later than they are required to by law and (2) that they tend to *systematically* prioritize closer over more distant, and (3) less risky over more risky plants. This might indicate that even in well-advanced democracies such as Germany, administrative capacities are under siege and on the edge of becoming overstressed and overstrained. A recent survey performed on behalf of the German Federal Environmental Agency (*Umweltbundesamt*) with state-level environmental inspectors makes a similar observation (Ziekow et al., 2018). Here, 60 percent of the surveyed environmental administrators answered that they have no longer sufficient time and resources to properly carry out their implementation tasks.

8.5 Conclusion and Future Research

The central aim of this chapter has been to offer a systematic perspective on the role of bureaucracies in environmental policy. We have learned that, contrary to popular opinion, bureaucracy is not only an obstacle but also a chance for effective environmental policy. Independent and sufficiently equipped administrations are able to formulate and come up with well-designed and ambitious environmental policies. Likewise, bureaucracies that possess sufficient staff and organizational resources, are operative in appropriate implementation structures, and possess some discretionary power are more effective in implementing and applying environmental policies. Despite the insights gained in the literature on bureaucracy and environmental policy, however, some important questions remained unanswered and are worth exploring in future research.

First, environmental policies and bureaucracies are typically seen as a reflection of the same underlying phenomena: the commitment and willingness of the government to address environmental declaration and climate change, though it is also well possible that there is a 'sequence' involved (Dubash, 2021). In other words, the institutionalization of environmental and climate issues in the state bureaucracy (the creation of environmental ministries, agencies, committees, etc.) might boost and accelerate later policy actions. Future research might thus examine whether and, if so, how the formal organization of the environmental policy development process affects later policy outputs.

Second, we need more and better data on the capacities of the environmental administration in charge of implementation. As highlighted by Moynihan (2022), "administrative capacity may be too broad a unit of analysis, but we should talk about it anyway". Current *comparative* studies typically use aggregated indices such as the World Bank's Worldwide Governance Indicators. In reality, however, administrative capacities are very context-specific and depend on the exact tasks an authority pursues (Williams, 2021). In consequence, more efforts are needed to assess which capacities different policy types require and whether the administrations in charge of implementation possess these capacities.

Lastly, research on environmental policy has more or less exclusively focused on the dismantling of environmental *policies* – with the central insight that policies are easier to make than to terminate. A frightening recognition of the years of the Trump Administration is, however, that tackling the administration, that is to "strip agencies' budgets and resources used to fund science, experts and research" and "to purge academic scientists from these agencies' scientific advisory boards" (Bomberg 2021, p. 631), might be a more effective strategy to undermine environmental protection. As highlighted by Pierson (2007), this strategy might not be limited to environmental regulation but a more general strategy of the political right to undermine "the nagging durability of government activism" (p. 35). Future research might thus examine how often and how successful pro-growth and less environmentally friendly governments use attacks on the administration as a distinct form of policy dismantling (for a first approach see Milhorance, 2022 and also Adam et al., 2007).

Box 8.2 Chapter Summary

- Bureaucracies are underrated with regard to their importance for environmental matters.
- Effective bureaucracies are crucial for the proper functioning of environmental policies throughout the different policy stages. Independent and well-equipped administrations *not* only produce better-designed environmental policies, they are also faster and more effective in applying them.
- The ability of bureaucracies to produce and implement environmental policies depends on different factors, such as the analytical capacities of the people working in the administration, their capability to overcome inner and outerinstitutional boundaries, and their independence (autonomy) from direct political intervention.
- Despite their importance, environmental administrations are under 'siege' in modern democracies as their organizational capacities do *not* keep pace with the massive expansion of implementation tasks.

Notes

1 Overall, Jordana et al. (2018) assesses 17 policy sectors. These are central bank, financial services, telecommunications, securities and exchange, electricity, insurance, nuclear safety, competition, gas, pharmaceuticals, pensions, food safety, postal services, water, environment, health services, and work safety.

2 The CDM was the world's first international market-based mechanism to reduce greenhouse gas emissions, established under the Kyoto Protocol. Under the CDM, industrialized countries (Annex B countries) can fund carbon projects in developing countries (Non-Annex B countries) to offset their emissions at home. Given that emission reductions in developing countries can be realized more cost-effectively, developed countries can achieve their reduction targets at a much lower price.

3 Here, private sector involvement refers to the practical application of environmental policies through or with the help of private actors. This explicitly excludes other forms of cooperation with the public sector such as public hearings or consultations. The respective execution and enforcement activities by private actors include, amongst others, vehicle safety and emissions inspections, the collection of toll charges, or the verification of industrial plants' emission reports.

4 Administrative traditions can be conceived as a part of the broader concept of administrative culture. Generally speaking, administrative culture is studied at three different levels: (1) the micro-level, including the values, roles, and behaviours of individual members of the administration, as well as the attitudes of the general public towards administrations; (2) the macro-level of administrative traditions; and (3) the meso-level of administrative styles, understood as the standard operating procedures of administrative behaviour and decision-making (Bayerlein et al., 2020).

5 "Unlimited Growth? A Comparative Analysis of Causes and Consequences of Policy Accumulation", funded by the European Research Council.

6 Baden-Württemberg is among the wealthiest states in Germany and, since 2011, has been led by a green government that has put a strong emphasis on environmental concerns (Hörisch and Wurster, 2019). It thus presents somewhat of an 'unlikely' case for the occurrence of large-scale implementation deficits. Moreover, German public administration is underpinned by a strong legalistic tradition that stresses strict adherence to rules and processes (Bach et al., 2017). It thus presents an overall rather unlikely case for large implementation deficits and delays.

References

Adam, C., Bauer, M.W., Knill, C., & Studinger, P. (2007). The termination of public organizations: Theoretical perspectives to revitalize a promising research area. *Public Organization Review* 7: 221–236.

Adam, C., Hurka, S., Knill, C., & Steinebach, Y. (2019). *Policy Accumulation and the Democratic Responsiveness Trap*. Cambridge University Press.

Bach, T. (2022). *Bureaucracies and Policy Ideas*. Oxford Research Encyclopedia of Politics. https://oxfordre.com/politics/view/10.1093/acrefore/9780190228637.001.0001/acrefore-9780190228637-e-1421

Bayerlein, L., Knill, C., & Steinebach, Y. (2020). *A Matter of Style? Organizational Agency in Global Public Policy*. Cambridge: Cambridge University Press.

Bó, E. Dal (2006). "Regulatory capture: A review." *Oxford Review of Economic Policy*, 22(2): 203–225.

Bomberg, E. (2021). "The environmental legacy of President Trump." *Policy Studies*, 42(5–6): 628–645.

Bondarouk, E., & Mastenbroek, E. (2018). "Reconsidering EU compliance: Implementation performance in the field of environmental policy". *Environmental Policy & Governance*, 28: 15– 27.

Boullier, H., Demortain, D., & Zeeman, M. (2019). "Inventing prediction for regulation: The development of (Quantitative) structure-activity relationships for the assessment of chemicals at the US environmental protection agency". *Science & Technology Studies* 32(4): 137–157.

Brinkerhoff, D. W. (1996). "Coordination issues in policy implementation networks: An illustration from Madagascar's Environmental Action Plan". *World Development*, 24(9): 1497–1507.

Christensen, R. K., Goerdel, H. T., & Nicholson-Crotty, S. (2011). Management, law, and the pursuit of the public good in public administration. *Journal of Public Administration Research and Theory*, 21(Suppl. 1), i125–i140.

Colenbrander, D., & Bavinck, M. (2017). "Exploring the role of bureaucracy in the production of coastal risks, City of Cape Town, South Africa." *Ocean & Coastal Management*: 35–50.

Davies, H. T. O., Nutley, S. M., & Smith, P. C. (Eds.) (2000). *What Works? Evidence-based Policy and Practice in Public Services*. Policy Press.

Den Uyl, R. M., & Russel, D. J. (2018). "Climate adaptation in fragmented governance settings: the consequences of reform in public administration". *Environmental Politics*, 27(2): 341–361.

Desveaux, J., Lindquist, E., & Toner, G. (1994). "Organizing for policy innovation in public bureaucracy: AIDS, energy and environmental policy in Canada." *Canadian Journal of Political Science*, 27(3): 493–528.

Dubash, N. (2021). Varieties of climate governance: the emergence and functioning of climate institutions. *Environmental Politics* 30 (1): 1–25

Duffy, R. J., & Cook, J. (2019). "Overcoming bureaucratic silos? Environmental policy integration in the Obama administration." *Environmental Politics*, 28(7): 1192–1213.

Fernandez-i-Marin, X., Knill, C., & Steinebach, Y. (2021). "Studying policy design quality in comparative perspective." *American Political Science Review*, 115(3): 931–947.

Fukuyama, F. (2013). "What is governance?" *Governance*, 26: 347–368.

Green, J. G. (2021). "Does carbon pricing reduce emissions? A review of ex-post analyses." *Environmental Research Letters*, 16(4): 1–17.

Hanson, J. K., & Sigman, R. (2021). "Leviathan's latent dimensions: Measuring state capacity for comparative political research." *The Journal of Politics* 83(4): 1495–1510.

Hickmann, T., Fuhr, H., Höhne, C., Lederer, M., & Stehle, F. (2017). "Carbon governance arrangements and the nation-state: the reconfiguration of public authority in developing countries." *Public Administration and Development*, 37(5): 331–343.

Howlett, M., & Ramesh, M. (2014). "The two orders of governance failure: Design mismatches and policy capacity issues in modern governance." *Policy and Society*, 33(4): 317–327.

Hsu, Angel (2015). "Measuring policy analytical capacity for the environment: A case for engaging new actors." *Policy and Society*, 34(3–4): 197–208.

Jordana, J., Fernández-i-Marín, X., & Bianculli, A.C. (2018). "Agency proliferation and the globalization of the regulatory state: Introducing a data set on the institutional features of regulatory agencies." *Regulation & Governance*, 12: 524–540.

Jordana, J., Levi-Faur, D., & Fernández-i-Marín, X. (2011). "The global diffusion of regulatory agencies: Channels of transfer and stages of diffusion." *Comparative Political Studies*, 44(10): 1343–1369.

Kaplaner, C. & Steinebach, Y. (2022). "Coping Practices and the Spatial Dimension of Agency Design." *Unpublished manuscript.*

Kettl, D. F. (1990). "The perils—and prospects—of public administration". *Public Administration Review*, 50(4): 411–419.

Kickert, W. J. M. (2005). "Distinctiveness in the study of public Management in Europe: A historical-institutional analysis of France, Germany and Italy". *Public Management Review*, 7(4): 537–563.

Knill, C., & Lenschow, A. (1998). "Coping with Europe: the impact of British and German administrations on the implementation of EU environmental policy." *Journal of European Public Policy*, 5(4): 595–614.

Lederer, M., & Höhne, C. (2021). "Max Weber in the tropics: How global climate politics facilitates the bureaucratization of forestry in Indonesia." *Regulation & Governance*, 15: 133–151.

Meckling, J., & Nahm, J. (2018). "The power of process: State capacity and climate policy." *Governance* 2018(31): 741–757.

Meckling, J., & Nahm, J. (2022). "Strategic state capacity: How states counter opposition to climate policy". *Comparative Political Studies*, 55(3): 493–523.

Milhorance, C. (2022). "Policy dismantling and democratic regression in Brazil under Bolsonaro: Coalition politics, ideas, and underlying discourses". *Review of Policy Research*, 39: 752– 770.

Moynihan, Donald P. (2022). "Why is American administrative capacity in decline?" In *Can We Still Govern?*, https://donmoynihan.substack.com/p/why-is-american-administrative-capacity

Newig, J., & Koontz, T. M. (2014). "Multi-level governance, policy implementation and participation: The EU's mandated participatory planning approach to implementing environmental policy". *Journal of European Public Policy*, 21(2): 248–267.

Orihuela, J. C., Mendieta, A., Pérez, C., & Ramírez, T. (2021). "From paper institutions to bureaucratic autonomy: Institutional change as a resource curse remedy." *World Development*, 143: 105463.

Painter, M., & Peters, B. G. (2010). *Tradition and public administration*. Palgrave Macmillan.

Peters, B. G. (2021). *Administrative traditions. Understanding the roots of contemporary administrative behavior.* Oxford University Press.

Pierson, P. (2007). "The rise and reconfiguration of activist government" In: Pierson, P. & Skopcol, T. (Eds). *The Transformation of American Politics: Activist Government and the Rise of Conservatism.* Oxford: Oxford University Press.

Potoski, M. (2002). "Designing bureaucratic responsiveness: Administrative procedures and agency choice in state environmental policy." *State Politics & Policy Quarterly*, 2(1): 1–23.

Povitkina, M., and Bolkvadze, K. (2019). "Fresh pipes with dirty water: How quality of government shapes the provision of public goods in democracies." *European Journal of Political Research*, 58: 1191–1212.

Pressman, J. L., & Wildavsky, A. (1984). *Implementation: How Great Expectations in Washington are Dashed in Oakland; Or, Why it's Amazing that Federal Programs Work at All, This Being a Saga of the Economic Development Administration as Told by Two Sympathetic Observers Who Seek to Build Morals on a Foundation.* Berkeley: University of California Press.

Ringquist, E. J. (1993). "Does regulation matter?: Evaluating the effects of state air pollution control programs." *The Journal of Politics* 55(4): 1022–1045.

Rosenbaum, W. A. (2017). *Environmental Policies and Politics*. New York: SAGE.

Schneider, L. R. (2011). "Perverse incentives under the CDM: An evaluation of HFC-23 destruction project." *Climate Policy*, 11(2): 851–864.

Shimshack, J. P., & Ward, M. B. (2005). Regulator reputation, enforcement, and environmental compliance, *Journal of Environmental Economics and Management*, 50(3): 519–540.

Steinebach, Y. (2022a). "Instrument choice, implementation structures, and the effectiveness of environmental policies: A cross-national analysis." *Regulation & Governance*, 16: 225–242.

Steinebach, Y. (2022b). "Administrative traditions and the effectiveness of regulation", *Journal of European Public Policy*, online first: 1–20.

Steinebach, Y., & Limberg, J. (2022). "Implementing market mechanisms in the Paris era: the importance of bureaucratic capacity building for international climate policy". *Journal of European Public Policy*, 29(7): 1153–1168.

Telle, K. (2009). "The threat of regulatory environmental inspection: impact on plant performance". *Journal of Regulatory Economics* 35: 154–178.

Treib, O. (2014). Implementing and complying with EU governance outputs, *Living Reviews in European Governance* 9: 1–47.

Verhoest, K., Peters, B.G., Bouckaert, G., and Verschuere, B. (2004). "The study of organizational autonomy: a conceptual review". Public Administration & Development, 24: 101–118.

Versluis, E. (2003). *Enforcement Matters: Enforcement and Compliance of European Directives in Four Member States*. Eburon.

Vogel, D. (1986). *National Styles of Regulation: Environmental Policy in Great Britain and the United States.* New York: Cornell University Press.

Voulvoulis, N., Arpon, K. D., & Giakoumis, T (2017). "The EU water framework directive: From great expectations to problems with implementation." *Science of the Total Environment* 575: 358–366.

Weaver, R. K. (2014). "Compliance regimes and behavioral change." *Governance*, 27: 243–265.

Wellstead, A., & Biesbroek, R. (2022). "Finding the sweet spot in climate policy: balancing stakeholder engagement with bureaucratic autonomy". *Current Opinion in Environmental Sustainability*, 54: 1–8.

Williams, M. J. (2021). "Beyond state capacity: Bureaucratic performance, policy implementation and reform." *Journal of Institutional Economics* 17(2): 339–357.

Wu, X., Ramesh, M., & Howlett, M. (2015). "Policy capacity: A conceptual framework for understanding policy competencies and capabilities". *Policy and Society*, 34(3–4): 165–171.

Ziekow, J., Bauer, C., Steffens, C., and Willwacher, H. (2018). Deutsches forschungsinstitut für öffentliche verwaltung. Dialogue with experts on the EU legislative act on environmental inspections – Exchange on possible changes in the implementation of EU environmental law. *Umweltbundesamt*. Available at: https://www.umweltbundesamt.de/sites/default/files/medien/1410/publikationen/2018-03-01_texte_21-2018_umweltinspektionen.pdf

9

ANALYTICAL PERSPECTIVES ON ENVIRONMENTAL POLICY INTEGRATION

Christina Steinbacher

9.1 Introduction

Natural disasters, extreme weather events, and smog warnings are becoming increasingly common; marine pollution is getting worse; global warming and melting glaciers are picking up speed. As problem pressure rises, so does societal and political awareness of the need for environmental protection, climate action, and sustainability. Despite many complaints that too little is being done to protect the environment, growing environmental awareness has led to governmental action: environmental ministries have become standard in modern democracies (Aklin and Urpelainen, 2014) and national environmental policy portfolios tremendously increased in recent decades (Busch and Jörgens, 2005; Adam *et al.*, 2019; Limberg *et al.*, 2021).

Yet, doubts about the sufficiency and effectiveness of existing measures to combat environmental degradation are unequivocal. The chances of environmental policy-making succeeding and effectively conserving and protecting the environment are put at risk by several factors. These factors can be summarized as three cascading challenges. The first challenge is the *complexity challenge*. This implies that policy effectiveness is undermined by insufficient implementation capacities (Adam *et al.*, 2019; Limberg *et al.*, 2021), along with inconsistencies, redundancies, and negative interactions between different policies (Grabosky, 1995; Gunningham and Sinclair, 2017; Howlett and Rayner, 2018). Besides being situated within increasingly complex policy contexts, environmental policy per se shows high levels of complexity (see, e.g., Adelle and Russel, 2013; Briassoulis, 2004). 'Good' intentions may never be (fully) realized, or they may even be reversed on the ground.

Especially for environmental policies, the problem of increasing complexity is further reinforced by a second challenge, the *cross-sectoral challenge*. Environmental issues cut across multiple sectors (Peters, 2015). For environmental policies to be successful in protecting soil, water, air, and nature, other policy sectors – such as industry, agriculture, transport, or energy – must include environmental policy objectives in their agendas. Yet, this sector-spanning quality of environmental policy clashes with traditional sectoral ministries' tendencies to reinforce patterns of 'siloization' or 'departmentalism' (Lægreid and Rykkja, 2015; Pollitt, 2003).

Finally, the third challenge environmental policy is confronted with can be referred to as the *political power challenge* (Hertin and Berkhout, 2003). Environmental interests and policy goals compete with a multitude of other interests. Many of these interests, such as economic growth,

 DOI: 10.4324/9781003043843-11

are better organized and persistently backed by robust stakeholder groups. The strongest rivals are those policy fields upon which the success of environmental policies relies. Environmental ministries usually have a weaker 'voice' than other ministries (e.g. the finance ministry or the ministry of the economy). That is why environmental policies depend more on pronounced political commitment than other policy fields (Burns *et al.*, 2019).

The observation that environmental policy-making comes with challenges that need to be met through coordination, harmonization, or integration efforts is not new and dates back to the very emergence of the environmental policy domain in the 1970s (Caldwell, 1974; Jacob and Volkery, 2004; Lenschow, 2002; Tosun and Peters, 2018; Weale, 1992). Since then, environmental policy integration (EPI) has been commonly presented as a potential 'antidote' to the various problems of environmental policy-making.

However, despite the expected positive effects of EPI, researchers have not yet been able to fully demonstrate that EPI initiatives increase environmental policy performance. More than three decades of intensive research has produced mixed and sometimes controversial results. In part, this is because EPI research itself does not live up to its name. Research approaches tend to diverge, incorporating new areas and issues instead of converging and systematically building upon what is already there. Qualitative single-case or small-N studies dominate EPI research. Their insights, however, are neither being followed up nor sustained by larger-scale comparisons or quantitative investigations. Problems of assessing the actual implementation and effects of EPI measures – the question of what works, where, and why – are widely pointed out (see, e.g., Runhaar *et al.*, 2020), but they are far from being resolved. Despite the high plausibility of the integration idea, this outcome predicament may profoundly shake the motivation for and relevance of EPI research.

Against this background, in this chapter, I argue that current shortcomings may be alleviated by a more novel analytical distinction of different perspectives on EPI. More specifically, a differentiation is made between horizontal and vertical integration on the one hand and two different objects of integration on the other. That is, whether the focus of integration is procedural (relating to processes of policy-making) or substantial (relating to policy outputs to be integrated).

In this regard, *horizontal* EPI primarily intends to meet the *cross-sectoral challenge* through the integration of policies or processes *between* pre-existing sectoral responsibilities (e.g., between an agricultural ministry and the ministry for the environment). The less-researched *vertical* dimension, in contrast, may present answers to the *complexity challenge* through an integration of policies or processes *within* existing sectoral frames, along the 'up and down' axis within politico-administrative 'silos' of responsibility (Lafferty and Hovden, 2003, p. 13). For example, within the realm of environmental policy-making, such a 'silo' may extend from an environmental ministry at the 'top', over an environmental agency and its regional subdivisions in the 'middle', and down to local implementation authorities at the 'bottom'.

Horizontal EPI focuses primarily on *policy formulation*, that is, on defining and setting the goals and thrusts of environmental policy-making. Therefore, horizontal *procedural* EPI focuses on integrating the phases of environmental agenda setting, policy formulation, and policy adoption between the policy sectors relevant for achieving the environmental goal in question. Consequently, when it is about the integration of the *policy substance*, the public policies as formulated take centre stage as the object of horizontal integration. *Vertical* EPI, on the other hand, exhibits a stronger *policy implementation* focus. Vertical *procedural* integration is about integrating the interface between implementation processes – that is, implementation, evaluation, and policy maintenance – and political and ministerial decision-making. Therefore, the vertical integration of *policy substance* is oriented toward instruments rather than goals and concentrates on the outcomes of public policies. Put simply, horizontal integration is about managing policy targets. Its aim is to overcome goal conflicts and the distance between sectors so that environmental policy objectives may be adopted across the relevant policy sectors. Vertical integration, in contrast, is about managing the policies' operation, confronting policy complexity, instrumental conflicts, and the availability of implementation capacities.

This chapter starts by explaining the evolution and core principles of environmental policy integration to provide the basis for the subsequent systematization of the different analytical perspectives on EPI. These perspectives refer, first, to the horizontal integration of environmental policy-making processes; second, to the horizontal integration of environmental policies; third, to the vertical integration of environmental policy-making processes; and, finally, the vertical integration of environmental policies. After discussing the relationship of these different perspectives, the chapter concludes with an appeal for greater systematization within EPI research.

9.2 What Constitutes Environmental Policy Integration?

In 1987, the Brundtland Report of the UN's World Commission on Environment and Development (WCED) strenuously expressed the need to tackle the challenges with which environmental policy-making is confronted. The idea that had been put forward was to interweave and harmonize environmental policy goals with those of challenger policy fields by giving *principled priority* to the concept of sustainable development (WCED, 1987). From the 1990s onwards and with the European Union and the international level as its epicentre, this general 'strategy' for achieving sustainable development and environmental protection flourished (Lenschow, 2002; Runhaar *et al.*, 2020) and started to attract more and more scholarly attention (Persson *et al.*, 2018; Trein *et al.*, 2019, p. 336). Under the name of EPI, it nowadays marks a specific branch within policy integration research (Knill *et al.*, 2020) for which it provides fertile ground.

In line with its genesis to tackle environmental degradation, EPI has primarily been perceived as the substantive integration of environmental policy goals and concerns into other concurrent policy fields (see, e.g., Lafferty and Hovden, 2003), such as transport, industry, or agriculture. In this context, Lafferty and Hovden (2003) already distinguish between horizontal environmental policy integration (HEPI) and its vertical counterpart (VEPI): HEPI refers to the breadth of central policy integration (strategies) *across* different policy sectors, whereas VEPI is concerned with its depth *within* individual sectors. Hence, substantive integration, sometimes called normative or strong integration (Jordan and Lenschow, 2010), primarily focuses on policy outputs. This means that it concentrates on the extent to which environmental concerns have prevailed in the political decision-making scramble and were merged into other sectors' policy portfolios under the maxim of environmental policy prioritization. Accordingly, the performance of substantive EPI is typically evaluated against "the relative weight of environmental objectives in sectoral policies, ranging from avoiding conflicts ('coordination') and striving for synergies ('harmonisation') up to favouring environmental objectives ('prioritization')" (Persson and Runhaar, 2018, p. 141).

Among these different interpretations of EPI, the idea of giving principled priority to environmental policies over other, traditional sectoral policies (Casado-Asensio and Steurer, 2014; Jordan and Lenschow, 2010; Lafferty and Hovden, 2003) deserves special attention. Rather than being perceived as a contextual factor related to political will or commitment and facilitating environmental policy integration, for some scholars prioritization has become the goal of integration itself. In essence, principled priority implies that environmental protection must be the fundamental premise for any policy decision: every policy proposal must be assessed for its environmental consequences, and, in the case of conflict, preference will be given to environmental matters. Strictly following this principled priority EPI approach, environmental degradation cannot be swept aside, the preservation of nature is made a superordinate societal objective (Lafferty and Hovden, 2003, p. 9). In practice, this means that principled priority would orient the set-up and powers of environmental ministries towards those of the treasury. In their 2021 federal election campaign, the German Green Party, for example, proposed to set up a new climate protection ministry with a general veto power over all legislative projects that would violate climate goals (see Hansen and Carrel, 2021). However, this prime example of principled priority was not implemented by the new German government coalition despite the Greens' electoral success and subsequent government participation.

But as the above example suggests, the normative and exclusively output-oriented conception of EPI is often at odds with the reality of democratic policy-making (see, e.g., Jordan and Lenschow, 2008). When the bar for integration is set as high as principled priority envisions, the analytical net is too wide to catch 'smaller fish'. This way, it is inevitable that empirical analyses of environmental integration will primarily reveal deficits (Persson and Runhaar, 2018) and fail to identify the various nuances of EPI. Consequently, the more moderate perspectives within EPI research have prevailed over time, understanding EPI as a basic principle or strategy for balancing environmental and other sectoral interests (Lenschow, 2002, p. 6; Liberatore, 1997) in an effective way (Lundqvist, 2004). Even some of the strongest proponents of the idea of principled priority have backed down, acknowledging that EPI might be better understood as insurance to minimize unacceptable environmental risks or impacts of other sectoral policies (see, e.g., Knudsen and Lafferty, 2016). Finally, it was also these 'tamer' perspectives on EPI that ceased to focus exclusively on policy outputs, paving the way for more process-oriented EPI approaches rooted in policy analysis (Biesbroek, 2021; Candel and Biesbroek, 2016; Jordan and Lenschow, 2010; Runhaar *et al.*, 2020).

Looking at EPI as a process opens up different dimensions and mechanisms of integration. Emphasis is placed on the sensitivity of EPI to the context and conditions in which integrated policy-making is to take place. This way, EPI is no longer treated as a black-and-white concept in which integration is either achieved or not achieved (Adelle and Russel, 2013). Candel and Biesbroek's (2016) procedural conceptualization of (general) policy integration as a specific process of policy and institutional change, for example, allows us to acknowledge the multifaceted nature of EPI and serves as a popular reference point for recent research on it. When considering EPI as a process, a distinction can be made between horizontal and vertical integration (Biesbroek, 2021). In this context, the horizontal dimension still refers to the cross-sectoral aspects of EPI, albeit to its policy-procedural rather than the policy-output-related elements. The vertical dimension, on the other hand, is linked to the multilevel component of policy-making, including both the formulation and implementation phases. This means that vertical *procedural* integration concentrates on the extent to which EPI passes through different levels of government and is reflected in implementation (Howlett and del Río, 2015; Howlett *et al.*, 2017; Knill *et al.*, 2020, 2021; Lenschow, 2002).

Undoubtedly, such process-oriented frameworks have directed scholarly focus toward a more comprehensive picture of EPI. However, to date, the specificity of integrating either *public policies in terms of substance* or *policy-making processes* has been diluted both empirically and analytically. Furthermore, most research conducted so far has primarily concentrated on the horizontal aspects of EPI. Its vertical dimension, in contrast, remains understudied. Consequently, we do not yet know what difference it makes for EPI to target processes rather than policy outputs, how these two objects relate to each other, and what role their horizontal and vertical dimensions play.

9.3 Towards an Integrative Systematization: Four Analytical Perspectives on EPI

Policy integration scholars have already acknowledged the concurrence of the integration of policy processes – i.e., the structures, procedures, and arrangements behind policy-making – on the one hand, and the integration of the policy 'substance' – i.e., the concrete policy goals and instruments which constitute the output of policy-making – on the other (Tosun and Lang, 2017). Reforms aiming at these types of integration were also referred to as 'administrative coordination reforms' instead of 'policy integration reforms' (Trein and Maggetti, 2020; Trein *et al.*, 2019). In empirical terms, policy integration reforms describe "legislative changes that connect or combine existing laws, changes in the mix of policy instruments, or new political strategies that embody future visions or plans that link various policy sectors", whereas administrative coordination refers to "reforms changing the relationship between different organizational units that elaborate and implement policies in the sense of improving their interaction and cooperation" (Trein *et al.*, 2021, pp. 5f.).

Nonetheless, this intuitive understanding of policy integration and procedural integration ignores the distinctiveness of the vertical and horizontal components of both concepts. Successful EPI is not only about the most frequently mentioned cross-sectoral, horizontal integration of environmental goals into other policy domains (HEPI), but also about their (vertical) integration (and implementation) down individual sectoral 'silos' (VEPI). Without a clear dimensional distinction, it remains unclear whether or to what extent a policy or policy-making system can be said to be integrated or disintegrated, why this is the case, and what effect is produced. To organize these different facets of EPI into a more coherent system, the framework presented here distinguishes not only between the integration of policy-making processes, on the one hand, and the integration of policy 'substance' (i.e., public policies at different stages of the policy cycle), on the other, but also between the two dimensions of integration (Knill *et al.*, 2020). In this context, horizontality refers to the integration *between* policy sectors, whereas the vertical dimension is concerned with integration *within* a given policy sector. The result of the proposed systematization are four different analytical perspectives on environmental policy integration. These are illustrated in Figure 9.1 and discussed in greater detail in the following sections.

9.3.1 The Horizontal Dimension: Governing Cross-Cutting Policy Goals

Following a problem-led approach, horizontal integration is primarily intended to address the cross-sectoral challenge. Tackling environmental problems requires the involvement of a variety of other policy sectors. Hence, in relation to the policy substance (or output) to be integrated, horizontal EPI strives first and foremost at the integration of environmental policy goals into the portfolios of other sectors that are relevant for these goals' achievement. It follows that, in procedural terms, the crucial policy stages for achieving horizontal integration are agenda-setting, policy formulation and adoption, but not yet policy implementation, that is, putting policies into effect on the 'ground'. Consequently, two different perspectives on EPI can be distinguished within the system of horizontality. First, the integration of the policy-making process in its decision-making stage, and second, the integration of policy output, that is, the extent to which other relevant sectors commit to and adopt environmental policy goals.

The *first analytical perspective on EPI* designates the *horizontal integration of policy-making processes* systematically linking the sectors (and actors) relevant (Christensen and Lægreid, 2007; Pollitt,

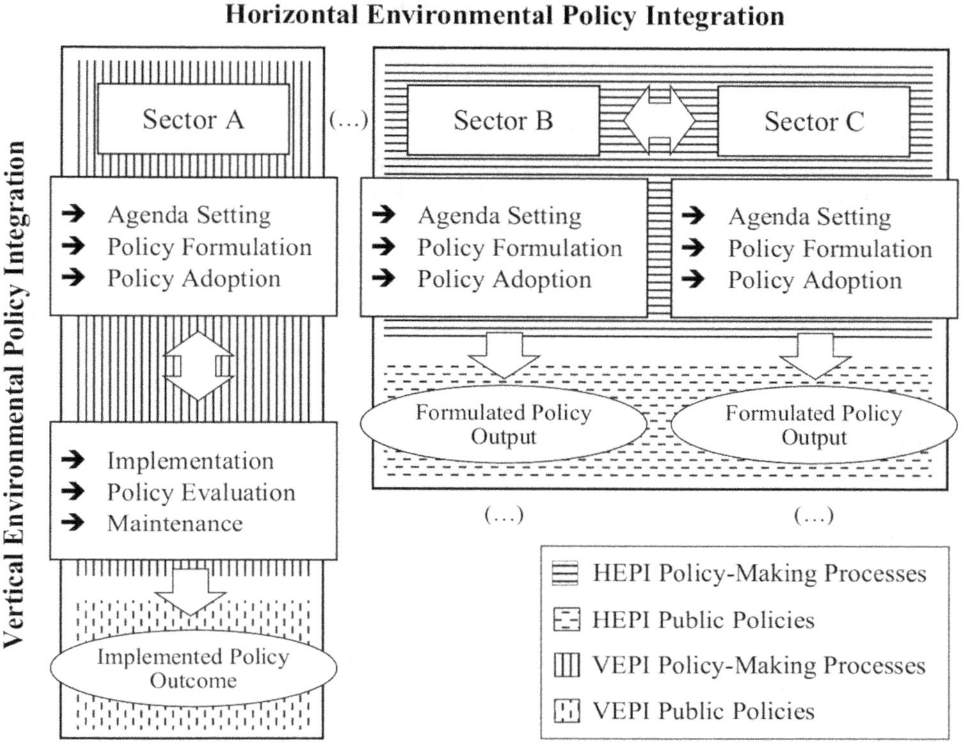

Figure 9.1 Illustration of the four different analytical perspectives on EPI systematized according to their spatial dimension and their objects.

2003) for solving an environmental issue. Following the logic of horizontal management, horizontal process integration enables superior policy outputs relative to those that can be achieved in sectoral isolation (Bakvis and Juillet, 2004; Ling, 2002; Peters, 2015). There are many ways in which different extents of *horizontal procedural integration* can be attained. The different approaches are best distinguished by their intensity of intervention: they may proceed, for example, through voluntary cooperation or sustained contact (e.g., exchange of civil servants), through 'emulsion' (e.g., environmental departments within environmentally relevant sectors), through procedures requiring involvement or participation of the environmental sector (e.g., joint commissions), or through hierarchical command and control instruments (e.g., an environmental authority with the power to issue instructions towards other policy sectors). Of course, there may also be combinations of different approaches.

Generally, it is assumed that cooperative models of policy-making are needed to integrate policy-making processes in a horizontal way (Hertin and Berkhout, 2003). Hence, reform strategies promoting horizontal process integration usually include the introduction of shared budgeting and a high degree of transparency regarding information, tasks, authority, and performance (e.g., Halligan *et al.*, 2011; Peters, 2015). Moreover, the development of an integrative team culture that is built upon trust, embeddedness, and joint missions facilitates horizontal procedural integration (Bardach, 1998; Halligan *et al.*, 2011; Ling, 2002; O'Flynn *et al.*, 2011). Within EPI literature, it has been shown that "supportive organizational structures" as well as horizontal "managerial and procedural instruments are the most effective strategies for realizing environmental integration" (Runhaar *et al.*, 2020, p. 197).

The *second analytical perspective on EPI* refers to the *horizontal integration of public policies (as formulated public policy outputs)*, that is, the horizontal integration of environmental policy outputs across sectors as they have formally been adopted. Within the EPI literature, this resembles most closely the substantive approaches within EPI research (Jordan and Lenschow, 2010; Lafferty and Hovden, 2003; Persson and Runhaar, 2018). Hence, this second perspective focuses on the extent to which environmental policy goals have been formally integrated into other (relevant) policy domains. There are two complementary ways in which the degree of horizontal integration of public policies is evaluated: first, it is assessed by the extent of sectoral inclusion, that is, the ratio of sectors adopting environmental policy objectives to those whose participation would be required to achieve them. Second, horizontal public policy integration is measured by the weight given to environmental objectives when they are adopted. In this respect, a distinction is made in ascending order between 'coordination', 'harmonization', and 'prioritization' (Lafferty and Hovden, 2003; Persson *et al.* 2018; Storbjörk and Isaksson, 2014). Since the term 'prioritization' implies strong normative claims, this latter aspect, that is, the weight given to environmental policy goals by non-environmental sectors, could alternatively be assessed in terms of consistency, coherence, and congruence (Howlett and Rayner, 2018) between the 'extra-sectoral' environmental policies and the 'normal' sectoral objectives. In fact, this second perspective means that what some scholars have previously called 'vertical' integration is here classified as horizontal. Specifically, this concerns the understanding of Lafferty and Hovden (2003). For them, 'vertical' integration refers to the extent to which the "decision premise[s]" and "steering domain[s]" of other sectors are subjected to "greening" (2003, p. 12f.). In the classification offered here, however, this understanding counts as *horizontal integration of public policies (as formulated public policy outputs)*.

Furthermore, and although this systematization attempts to keep the fragmentation in EPI research in check, the *second analytical perspective on EPI* subsumes two different approaches in academic practice. While both strands do consider the substance of a particular public policy as the object of integration, the nature of this 'substance' differs. One strand focuses on a 'meta-policy' (Jann and Wegrich, 2019), in the sense of superordinate strategic integration policies. These policies may be process-directed, such as large-scale policy campaigns to push sectoral agenda setting (e.g., Nunan *et al.*, 2012), or policy-directed, such as the introduction of obligatory environmental impact assessments with respect to other sectors' policies (see, e.g., 'policy integration reforms' in Trein *et al.*, 2021). The second branch, in contrast, focusses on the results of (integrated) policy-making processes. In doing so, this interpretation of policies' *horizontal integration* acknowledges the organizational and procedural components of horizontal EPI (see, e.g., Persson *et al.*, 2016; Runhaar *et al.*, 2020). It introduces a sequential logic into EPI research, in the sense that the integration of policy-making processes precedes the integration of public policies. Consequently, it is this second branch that is of particular theoretical importance and is valuable for empirical assessments of EPI success or failure.

9.3.2 The Vertical Dimension: Governing Complexity in Implementation

While the horizontal dimension of EPI intends to address the *cross-sectoral challenge*, it is the less explored vertical dimension (Knill *et al.*, 2020) that is supposed to address the *complexity challenge*. Environmental policies exhibit a high degree of complexity associated with advancing technological standards, trade-off solutions, and increasing regulatory density (Adam *et al.*, 2019). However, potential inconsistencies, negative externalities, and the resulting administrative burdens only fully materialize when environmental policies are implemented. What is more, the *cross-sectoral challenge* described above not only impacts environmental policy formulation. It also amplifies the *complexity challenge* that environmental policy implementation is facing: putting legislation into effect on the 'ground' is compounded by the fact that the different interests and claims of a multitude of actors come into play, jeopardizing the state's ability to act on environmental matters. In consequence, *vertical environmental policy integration* exhibits a stronger focus on policy implementation and policy instruments, as opposed to the concentration of horizontal EPI on policy formulation and policy goals.

Against this background, the *third analytical perspective on EPI* designates the *vertical integration of policy-making processes*. Vertical *procedural* integration is about integrating the interface between implementation processes, that is, implementation, evaluation and policy maintenance, and political and ministerial decision-making (Bolleyer and Börzel, 2010; Howlett *et al.*, 2017; Jänicke and Jörgens, 2006; Knill *et al.*, 2021; Urwin and Jordan, 2008). Coordinating these two stages of the policy cycle and their respective actors matters in two ways: first, policy implementation serves as an important source of expertise and feedback with regard to the 'implementability' and quality of policies in different contexts. Second, it ensures that the policy formulation level sufficiently engages with and is committed to effective implementation. To evaluate the extent of procedural vertical integration within policy sectors, Knill *et al.* (2021) have developed a corresponding indicator scheme which captures the costs and responsibilities assumed by the policy formulation level for policy implementation, as well as the ability and opportunities of the implementation level to feed its input into policy-making.

Analogously to procedural horizontal integration, one can think of several strategic initiatives for fostering vertical procedural EPI within different sectoral settings: green budgeting, for example, may serve not only as a horizontal integrative tool, but also as a vertical information and accountability generator to achieve political commitment to environmental policy implementation *within* non-environmental policy sectors (Bova, 2021; Runhaar, 2016). The same holds for sectoral supervisory obligations in terms of reporting and monitoring (Lafferty, 2002) or for strategic environmental assessments (Runhaar *et al.*, 2013). Consultative schemes, forums, policy appraisal, and other communicative tools may furthermore serve as facilitating structures for acquiring implementation expertise and feedback (Lafferty, 2002; Jordan and Lenschow, 2008; Runhaar, 2016).

The final and *fourth analytical perspective on EPI* refers to the *vertical integration of public policies (as implemented public policy outcomes)*, that is, the operational effects environmental public policies exhibit when implemented *within* a specific sector. In contrast to its horizontal counterpart, this vertical integration of policy 'substance' is oriented toward instruments rather than goals. In this vein, a fully vertically integrated policy designates a policy that is fully implementable, that is, whose realization is not obstructed by internal contradictions or negative interactions across governmental levels within a certain sector and whose design matches the sectoral implementation realities (Howlett and Del Río, 2015; Howlett *et al.*, 2017). This definition of vertical integration seems at odds with some common terminology used in EPI literature (e.g., with Lafferty and Hovden's (2003) notion of vertical integration). However, it is much in line with the origins of EPI research. As early as 1980, Underdal, a pioneer of EPI, developed the criterion of vertical consistency to assess the extent to which "implementary measures conform to more general guidelines and to policy goals" while permeating "all policy levels and all government agencies involved in its execution" (Underdal, 1980, p. 162). In doing so, Underdal emphasized the role that policy implementation and policy instruments play in integrating environmental policy. Over time, however, this concept of verticality became mixed with horizontal integration, as scholars focused primarily on policy formulation and environmental goals, rather than on their implementation.

Although EPI literature remains somewhat opaque with respect to this *fourth analytical perspective*, the literature does identify some direct measures to promote policies' vertical integration: policy experiments and pilots, for example, are a promising way to identify potential inconsistencies, negative interactions, and preconditions for effectively putting a policy into practice (Nair and Howlett, 2016). It is also argued that the creation of checklists improves the vertical consistency and coherence of policies and reduces the risk of implementation failure by responding to challenges that have been identified *a priori* (Persson, 2004; OECD, 2002). Another option is to establish project management and intervention teams for environmental policy implementation in non-environmental sectors. In Denmark, for example, so-called 'travel teams' support and refine policy implementation (Ravn-Christensen and Iversen, 2013).

9.3.3 Discussion: The Relationship Between the Four Analytical Perspectives on EPI

The four analytical perspectives on EPI are related to each other in both reciprocal and sequential ways. First, procedural integration prepares the groundwork for substantive integration. If environmental concerns are represented within institutionalized policy-making processes – whether they are about policy formulation (horizontal) or implementation (vertical) – it can be reasonably assumed that the chances for integrated policy outputs are generally higher than in a setting without any procedural integration. Hence, measures to achieve procedural integration constitute a form of *indirect integration* that takes place on the sectoral, not the policy, level, and that alters the chance of integrated policy outputs. In contrast, initiatives aiming at policy output integration present a form of *direct integration*. Operating at the policy level, single integrative initiatives may or may not influence the integration of the procedural set-up, but no systematic link can be established. Auld *et al.* (2014) find a similar relationship between integrative environmental measures directed at planning and processes (*indirect integration*) and measures directed at an explicit environmental performance goal (*direct integration*) at the company level. Hence, in terms of long-term and sustainable environmental policy integration (cf. Biesbroek, 2021), emphasis should be put on the integration of policy-making processes.

Second, regarding the relationship between horizontality and verticality, it can be stated that vertical EPI represents downstream integration in relation to its horizontal dimension. In contrast to the cross-sectoral idea of horizontality, vertical integration takes place *within* the sectoral 'silo' into which – as targeted by horizontal integration – an environmental policy objective is to be integrated. Vertical integration, thus, corresponds to the idea of intra-sectoral integration (Briassoulis, 2004; Persson, 2007) and depends on the decisions made in horizontal integration. In this way, the distinction between horizontal and vertical integration balances the unresolved debate about the timing of integration (for a discussion, see Runhaar, 2016). To summarize: a logical sequencing of EPI analysis and practice would start with horizontal procedural integration and end with vertical substantial integration (see Figure 9.2). Errors and failures diagnosed within substantial integration could indicate problems with respect to procedural integration, whereas vertical implementation issues could well be associated with horizontal EPI negligence during policy formulation.

Finally, and despite its consecutive character, the relationship between horizontality and verticality is reciprocal: while horizontal integration is directed at remedying the fact that

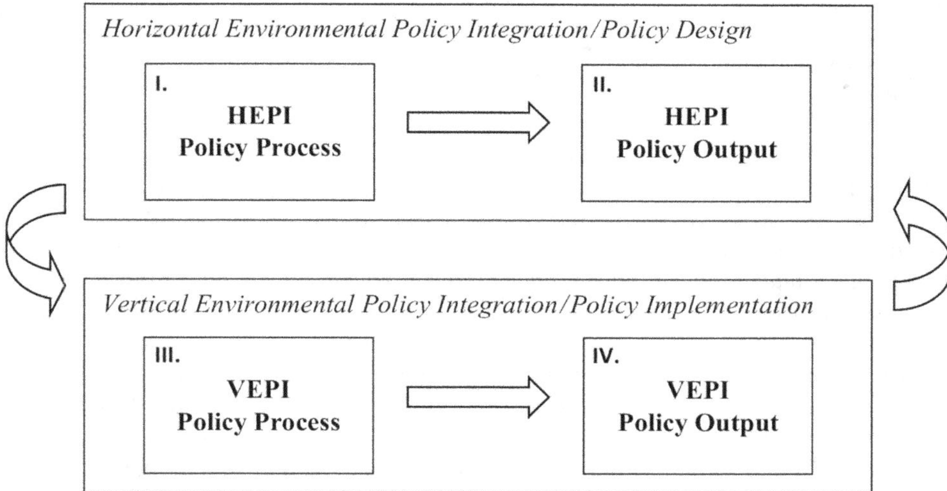

Figure 9.2 Theorizing the relationships between the four analytical perspectives on environmental policy integration.

environmental policy-making clashes with traditional policy sectors (*cross-sectoral challenge*), vertical integration aims at mitigating the complexity associated with the implementation of these environmental policies within these sectors, and at providing the implementation capacity to do so (*complexity challenge*). Hence, horizontal integration determines parts of the context of vertical integration, but vertical integration serves as an important tool to inform the horizontal integration of environmental goals and co-determines the effect of integrative initiatives on the ground. This way it is suggested that paying attention to verticality within EPI research might put an end to the puzzle surrounding the implementability of integrative measures. Furthermore, it may serve as a new driver for policy learning approaches associated with environmental policy integration (Nilsson and Eckerberg, 2007; Runhaar *et al.*, 2020; Storbjörk and Isaksson, 2014).

9.4 Conclusion and Future Research

Today, the functioning of environmental policy is widely recognized as more important than ever. To make it work, environmental policy integration is key. The four analytical perspectives on EPI attempt to analytically separate its different lenses. It has been argued that the integration of policy-making processes (co-)determines and influences the integration of policy-making output: the set-up and design of policy-making processes condition how integrated policies can be effectively formulated and implemented (Domorenok *et al.*, 2021). Second, as conceptualized in this chapter, horizontal integration and vertical integration go hand in hand (as also demonstrated by Runhaar *et al.*, 2020): horizontal integration initially determines the context of vertical integration, which in turn informs and influences the horizontal integration of goals and instruments. Furthermore, vertical integration co-determines the effect of integrative initiatives on the ground and strengthens the availability of sufficient implementation capacity. Yet, both dimensions aim to address different challenges that environmental policy-making faces. While horizontality strives to remedy the *cross-sectoral challenge*, verticality focuses on the *complexity challenge* and on reducing the implementation gap diagnosed by EPI research. As Candel suggests (2017), EPI initiatives should not be evaluated based on overall environmental performance but rather on intermediate outcomes offered by this kind of problem-oriented approach.

The political power challenge remains outside the four analytical perspectives on EPI. The idea of strict *prioritization* of environmental policy goals has always been argued to provide the matching David to this Goliath. Prioritization is frequently perceived as another component of EPI and even became a 'degree' of measuring integration (cf. Persson and Runhaar, 2018). Yet, strict prioritization is at odds with democratic policy-making and even follows a different logic than integration. To meet the *political power challenge*, a certain degree of 'external' prioritization in terms of awareness and commitment may constitute a necessary condition to kick-start EPI and to put environmental policies in a position to catch up with challenger policy fields. Having said this, prioritization (or political commitment) does constitute a contextual factor influencing the likelihood of EPI success or failure, but not an EPI component. Moreover, some examples indicate that EPI can persist even if prioritization fades (Persson *et al.*, 2016). The reasons for this are institutionalized integrative practices as captured here by the presented systematization: institutionalized integration can make prioritization obsolete if just enough synergies have been achieved and exploited.

Box 9.1 Chapter Summary

- Research on EPI suffers from a lack of systematization. Therefore, this chapter has offered four novel analytical perspectives on EPI, distinguishing between (1) horizontal and vertical EPI and (2) policy substance versus policy process orientations.
- By following a problem-led approach and paying attention to policy design and implementation stages, the proposed systematization facilitates future theory building and testing.
- It has been argued that the integration of policy-making processes (co-)determines and influences 'substantial' EPI: processes condition how integrated environmental policies can be effectively formulated and implemented.
- Horizontal EPI strives to remedy the cross-sectoral challenge of environmental policy-making, while vertical EPI seeks to manage complexity and to reduce the present EPI implementation gap.
- The reader has been provided with a wide range of empirical examples and graphic illustrations to clarify the argument.

References

Adam, C., Hurka, S., Knill, C., and Steinebach, Y. (2019) *Policy Accumulation and the Democratic Responsiveness Trap*. Cambridge: Cambridge University Press.

Adelle, C., and Russel, D. (2013) 'Climate policy integration: a case of déjà vu?', *Environmental Policy and Governance*, 23(1), pp. 1–12.

Aklin, M., and Urpelainen, J. (2014) 'The global spread of environmental ministries: Domestic–international interactions', *International Studies Quarterly*, 58(4), pp. 764–780.

Auld, G., Mallett, A., Burlica, B., Nolan-Poupart, F. (2014) 'Evaluating the effects of policy innovations: lessons from a systematic review of policies promoting low carbon technology', *Global Environmental Change* 29, pp. 444–458.

Bakvis, H., and Juillet, L. (2004) 'The strategic management of horizontal issues: Lessons in interdepartmental coordination in the Canadian government', paper for 20th Anniversary of International Political Science Association Research Committee on the Structure and Organization of Government, *Conference on Smart Practices Toward Innovation in Public Management*, University of British Columbia, Vancouver, 15–17 June.

Bardach, E. (1998) *Getting Agencies to Work Together: The Practice and Theory of Managerial Craftsmanship*. Washington, DC: Brookings Institution Press.

Biesbroek, R. (2021) 'Policy integration and climate change adaptation', *Current Opinion in Environmental Sustainability*, 52, pp. 75–81.

Bolleyer, N., and Börzel, T. A. (2010) 'Non-hierarchical policy coordination in multilevel systems', *European Political Science Review*, 2(2), pp. 157–185.

Bova, E. (2021) 'Green budgeting practices in the EU: A first review', *European Economy Discussion Papers 140.*

Briassoulis, H. (2004) 'The institutional complexity of environmental policy and planning problems: The example of Mediterranean desertification' *Journal of Environmental Planning and Management*, 47(1), pp. 115–135.

Burns, C., Tobin, P., and Sewerin, S. (2019) 'European environmental policy at a time of crisis: Benign neglect, or a leader losing pace?', in C. Burns, P. Tobin, and S. Sewerin (Eds.) *The Impact of the Economic Crisis on European Environmental Policy*. Oxford: Oxford University Press, pp. 199–217.

Busch, P. O., and Jörgens, H. (2005) 'International patterns of environmental policy change and convergence', *European Environment*, 15(2), pp. 80–101.

Caldwell, L. K. (1974) 'Environmental policy as a catalyst of institutional change', *American Behavioral Scientist*, 17(5), pp. 711–730.

Candel, J. J. (2017) 'Holy Grail or inflated expectations? The success and failure of integrated policy strategies', *Policy Studies*, 38(6), pp. 519–552.

Candel, J. J., and Biesbroek, R. (2016) 'Toward a processual understanding of policy integration', *Policy Sciences*, 49(3), pp. 211–231.

Casado-Asensio, J., and Steurer, R. (2014) 'Integrated strategies on sustainable development, climate change mitigation and adaptation in Western Europe: Communication rather than coordination', *Journal of Public Policy*, 34(3), pp. 437–473.

Christensen, T., and Lægreid, P. (2007) 'The whole-of-government approach to public sector reform', *Public Administration Review*, 67(6), pp. 1059–1066.

Domorenok, E., Graziano, P., and Polverari, L. (2021) 'Introduction: policy integration and institutional capacity: Theoretical, conceptual and empirical challenges', *Policy and Society*, 40 (1), pp. 1–18.

Grabosky, P. N. (1995) 'Regulation by reward: On the use of incentives as regulatory instruments', *Law and Policy*, 17(3), pp. 257–282.

Gunningham, N., and Sinclair, D. (2017) 'Smart regulation', in P. Drahos (ed.) *Regulatory Theory: Foundations and Applications*. Acton, ACT: ANU Press, pp. 133–148.

Halligan, J., Buick, F., and O'Flynn, J. (2011) 'Experiments with joined-up, horizontal and whole-of-government in Anglophone countries', in A. Massey (Ed.) *International Handbook on Civil Service Systems*. Cheltenham, UK: Edward Elgar Publishing, pp. 74–99.

Hansen, H., and Carrel, P. (2021) 'Germany's Greens pitch climate ministry to boost faltering campaign', *Reuters*, 03 August. Available at: https://www.reuters.com/world/europe/germanys-greens-pitch-climate-ministry-boost-faltering-campaign-2021-08-03/ (Accessed: 01 November 2022).

Hertin, J., and Berkhout, F. (2003) 'Analysing institutional strategies for environmental policy integration: the case of EU enterprise policy', *Journal of Environmental Policy and Planning*, 5(1), pp. 39–56.

Howlett, M., and del Río, P. (2015) 'The parameters of policy portfolios: Verticality and horizontality in design spaces and their consequences for policy mix formulation', *Environment and Planning C: Government and Policy*, 33(5), pp. 1233–1245.

Howlett, M., and Rayner, J. (2018) 'Coherence, congruence and consistency in policy mixes', in M. Howlett and I. Mukherjee (Eds.) *Routledge Handbook of Policy Design*. New York: Routledge, pp. 389–403.

Howlett, M., Vince, J., and del Río, P. (2017) 'Policy integration and multi-level governance: dealing with the vertical dimension of policy mix designs', *Politics and Governance*, 5(2), pp. 69–78.

Jacob, K., and Volkery, A. (2004) 'Institutions and instruments for government self-regulation: Environmental policy integration in a cross-country perspective', *Journal of Comparative Policy Analysis: Research and Practice* 6 (3), pp. 291–309.

Jänicke, M., and Jörgens, H. (2006) 'New approaches to environmental governance', in M. Jänicke and K. Jacob (Eds.) *Environmental Governance in Global Perspective. New Approaches to Ecological and Political Modernisation*. Berlin: Freie Universität Berlin, pp. 167–209.

Jann, W., and Wegrich, K. (2019) 'Generalists and specialists in executive politics: Why ambitious meta-policies so often fail', *Public Administration*, 97(4), pp. 845–860.

Jordan, A., Lenschow, A. (eds.) (2008) *Innovation in Environmental Policy? Integrating the Environment for Sustainability*. Cheltenham: Edward Elgar Publishing.

Jordan, A., and Lenschow, A. (2010) 'Environmental policy integration: A state of the art review', *Environmental Policy and Governance*, 20(3), pp. 147–158.

Knill, C., Steinbacher, C., and Steinebach, Y. (2020) 'Policy integration: Challenges for public administration', in Peters, B.G. & Thynne, I. (Eds.) *Oxford Research Encyclopedia of Politics*. Oxford: Oxford University Press.

Knill, C., Steinbacher, C., and Steinebach, Y. (2021) 'Balancing trade-offs between policy responsiveness and effectiveness: The impact of vertical policy-process integration on policy accumulation', *Public Administration Review*, 81(1), pp. 157–160.

Knudsen, J. K., and Lafferty, W. M. (2016) 'Environmental policy integration: The importance of balance and trade-offs', in D. Fisher (Ed.) *Research Handbook on Fundamental Concepts in Environmental Law*. Cheltenham: Edward Elgar Publishing, pp. 337–368.

Lægreid, P., and Rykkja, L. H. (2015) 'Organizing for "wicked problems" - analyzing coordination arrangements in two policy areas: Internal security and the welfare administration', *International Journal of Public Sector Management*, 28(6), pp. 475–493.

Lafferty, W. (2002) *Adapting government practice to the goals of sustainable development. Improving Governance for Sustainable Development*. OECD Seminar 22–23 November 2001, Paris: OECD.

Lafferty, W., and Hovden, E. (2003) 'Environmental policy integration: towards an analytical framework', *Environmental politics*, 12(3), pp. 1–22.

Lenschow, A. (Ed.) (2002) *Environmental Policy Integration: Greening Sectoral Policies in Europe*. London: Earthscan.

Liberatore, A. (1997) 'The integration of sustainable development objectives into EU policymaking', in D. Richardson (Ed.) *The Politics of Sustainable Development: Theory, Policy and Practice within the European Union*. London, UK: Routledge, pp. 107–126.

Limberg, J., Steinebach, Y., Bayerlein, L., and Knill, C. (2021) 'The more the better? Rule growth and policy impact from a macro perspective', *European Journal of Political Research*, 60(2), pp. 438–454.

Ling, T. (2002) 'Delivering joined-up government in the UK: dimensions, issues and problems', *Public Administration*, 80(4), pp. 615–642.

Lundqvist, L. (2004) 'Integrating Swedish water resource management: a multi-level governance trilemma', *Local Environment*, 9(5), pp. 413–424.

Nair, S., and Howlett, M. (2016) 'Meaning and power in the design and development of policy experiments', *Futures*, 76, pp. 67–74.

Nilsson, M., and Eckerberg, K. (Eds.) (2007) *Environmental Policy Integration in Practice Shaping Institutions for Learning*. Sterling, VA: Earthscan.

Nunan, F., Campbell, A., and Foster, E. (2012) 'Environmental mainstreaming: the organisational challenges of policy integration', *Public Administration and Development*, 32(3), pp. 262–277.

OECD (2002) *Improving Policy Coherence and Integration for Sustainable Development: A Checklist*. Paris: OECD.

O'Flynn, J., Buick, F., Blackman, D., and Halligan, J. (2011) 'You win some, you lose some: experiments with joined-up government', *International Journal of Public Administration*, 34(4), pp. 244–254.

Persson, Å. (2004) 'Environmental Policy Integration: An Introduction', *PINTS–Policy Integration for Sustainability Background Paper*. Stockholm: Stockholm Environment Institute.

Persson, Å. (2007) 'Different perspectives on EPI', in M. Nilsson and K. Eckerberg (Eds.) *Environmental Policy Integration in Practice: Shaping Institutions for Learning*. Sterling, VA: Earthscan, pp. 25–47.

Persson, Å., Eckerberg, K., and Nilsson, M. (2016) 'Institutionalization or wither away? Twenty-five years of environmental policy integration under shifting governance models in Sweden', *Environment and Planning C: Government and Policy*, 34(3), pp. 478–495.

Persson, Å., and Runhaar, H. (2018) 'Conclusion: Drawing lessons for environmental policy integration and prospects for future research', *Environmental Science and Policy*, 85, pp. 141–145.

Persson, Å., Runhaar, H., Karlsson-Vinkhuyzen, S., Mullally, G., Russel, D., and Widmer, A. (2018) 'Editorial: Environmental policy integration: Taking stock of policy practice in different contexts', *Environmental Science and Policy*, 85, pp. 113–115.

Peters, B. G. (2015) *Pursuing Horizontal Management*. Lawrence, KS: University Press of Kansas.
Pollitt, C. (2003) 'Joined-up government: a survey', *Political Studies Review*, 1(1), pp. 34–49.
Ravn-Christensen, C, and Iversen, S. D. (2013) *Implementering af Klimatilpasningsplanen – Hvem, Hvad, Hvordan?* Afgangsprojekt, Land Management, Aalborg Universitet 2013. Available at: https://projekter.aau.dk/projekter/files/77462536/Implementering_af_klimatilpasningsplanen.pdf (Accessed: 01 November 2022).
Runhaar, H. (2016) 'Tools for integrating environmental objectives into policy and practice: What works where?', *Environmental Impact Assessment Review*, 59, pp. 1–9.
Runhaar, H., van Laerhoven, F., Driessen, P., and Arts, J. (2013) 'Environmental assessment in the Netherlands: effectively governing environmental protection? A discourse analysis', *Environmental Impact Assessment Review*, 39, pp. 13–25.
Runhaar, H., Wilk, B., Driessen, P., Dunphy, N., Persson, Å., Meadowcroft, J., and Mullally, G. (2020) 'Policy integration', in F. Biermann and R. Kim (Eds.) *Architectures of Earth System Governance: Institutional Complexity and Structural Transformation*. Cambridge: Cambridge University Press, pp. 183–206.
Storbjörk, S., and Isaksson, K. (2014) 'Learning is our Achilles heel: Conditions for long-term environmental policy integration in Swedish regional development programming', *Journal of Environmental Planning and Management*, 57 (7), pp. 1023–1042.
Tosun, J., and Lang, A. (2017) 'Policy integration: Mapping the different concepts', *Policy Studies*, 38(6), pp. 553–570.
Tosun, J., and Peters, B. G. (2018) 'Intergovernmental organizations' normative commitments to policy integration: The dominance of environmental goals', *Environmental Science and Policy*, 82, pp. 90–99.
Trein, P., and Maggetti, M. (2020) 'Patterns of policy integration and administrative coordination reforms: A comparative empirical analysis', *Public Administration Review*, 80(2), pp. 198–208.
Trein, P., Maggetti, M., and Meyer, I. (2021) 'Necessary conditions for policy integration and administrative coordination reforms: an exploratory analysis', *Journal of European Public Policy*, 28(9), pp. 1410–1431.
Trein, P., Meyer, I., and Maggetti, M. (2019) 'The integration and coordination of public policies: A systematic comparative review', *Journal of Comparative Policy Analysis: Research and Practice*, 21(4), pp. 332–349.
Underdal, A. (1980) 'Integrated marine policy: What? why? how?', *Marine Policy*, 4(3), pp. 159–169.
Urwin, K., and Jordan, A. (2008) 'Does public policy support or undermine climate change adaptation? Exploring policy interplay across different scales of governance', *Global Environmental Change*, 18(1), pp. 180–191.
WCED – World Commission on Environment and Development (1987) *Our Common Future*. Oxford: Oxford University Press.
Weale, A. (1992) *The New Politics of Pollution*. Manchester: Manchester University Press.

10

ENVIRONMENTAL POLICY IMPLEMENTATION

Jale Tosun and Simon Schaub

10.1 Introduction

In academia, 'policy implementation' refers to how policies are put into practice by administrative and other actors and to what extent they impact the behaviour of their target populations (Lasswell, 1956). After a period of stagnation, a 'crisis of implementation' (Mazmanian and Sabatier, 1983) in the United States in the 1980s drew scholarly attention to policy implementation and resulted in several influential early works on the subject. Diverging from the pervasive notion of implementation as a self-runner once policies had been adopted, these scholars showed that implementation, in fact, was mostly difficult (Mazmanian and Sabatier, 1983) and suggested incorporating it in the design phase of policies (Pressman and Wildavsky, 1984b). They further developed theories and conceptual frameworks to capture the influence of organisational structures and influential political variables (Montjoy and O'Toole, 1979; O'Toole and Montjoy, 1984; Sabatier and Mazmanian, 1980) and discussed conceptually the role of implementation in policymaking (Hjern, 1982; Linder and Peters, 1987).

In Europe, scholarly interest in policy implementation has been on the rise since the 1990s due to two developments (Tosun and Treib, 2018). First, scholars analysing policymaking processes in the European Union (EU) began to address the question of how EU policies are implemented by the individual member states. In the EU context, this is an important topic since the advantages of the Common Market can only materialise when policies become harmonised to some extent. However, when policies are harmonised on paper but not put into practice, this reduces the effectiveness of the Common Market. Second, the emergence of the governance approach within public administration and policy studies renewed interest in policy implementation.

Both perspectives have been used to study the implementation of environmental policy. Regarding the first, the work by Knill and Lenschow, for example, has stimulated a prolific research agenda on how national administrative traditions impact the implementation of EU environmental policy (Knill and Lenschow, 1998) and whether and to what degree 'new' policy instruments and public participation could reduce the implementation gap in this environmental policy (Knill and Lenschow, 2000). The governance perspective drew attention to the influence of non-governmental actors, institutions, and policy networks on policymaking, including the implementation of policies (Hill and Hupe, 2002). This incorporates literature

DOI: 10.4324/9781003043843-12

dealing with the question of how participatory governance affects the implementation of environmental policy (Jager *et al.*, 2020; Newig and Fritsch, 2009). It is worth noting that, in the pertinent literature, environmental policy is often mentioned in one breath with environmental governance, which implies that modern environmental policy can hardly be considered without the governance side.

Interestingly, the default expectation concerning the implementation of environmental policy is that it is prone to deficits. This expectation appears substantiated upon inspection of the literature. While it is less surprising to observe that environmental policy implementation has flaws in developing countries (e.g. Olmstead and Zheng, 2021), even countries considered as 'environmental leaders' (Liefferink and Wurzel, 2017), such as Denmark or Sweden, can experience difficulties in implementing environmental policy (e.g. Graversgaard *et al.*, 2021; Zingraff-Hamed *et al.*, 2020). Generally speaking, the two main causes of problems with the implementation of environmental policy are the lack of capacity and will (Holzinger and Knoepfel, 2000). The focus on implementation deficits is a consequence of the dominant theoretical perspective that regards implementation as a process in which the original stipulations of a policy should be executed with as little deviation as possible (see most prominently Pressman and Wildavsky, 1984a). It also results from the insight that poorly implemented environmental policies have numerous side effects. These include environmental and social costs, economic costs born of the creation of an unequal playing field for economic actors, and political costs stemming from the credibility loss of government bodies.

In this chapter, we take a broader perspective on environmental policy implementation. Not only do we cover insights provided by the existing research literature on environmental policy outcomes, but we elaborate in detail on the empirical characteristics of the implementation process. The chapter unfolds as follows. First, we present a conceptual model, from which we derive the components of the subsequent sections. Then, we shed light on the role of policy design, implementation structure, agency decision-making, and target group behaviour for environmental policy implementation. After this, we discuss the relationship between policy implementation and policy impacts, before summarising the main insights and offering some concluding remarks.

10.2 Key Concepts

Policy implementation is a process that requires different types of actors to use different types of actions so that existing policies become effective and achieve the intended policy impacts (Tosun 2012, 2018). Central for implementation are those actors who are targeted by a given policy measure. In other words, policy implementation is, in essence, the process of bringing about changes in the behaviour of a target group, which can comprise individual or collective actors. To induce behavioural change, governments can use nodality (or information), authority, treasure, or organisational resources (Hood and Margetts, 2007). Originally, environmental policy was prone to using authority, but over time there has been an increased use of 'new' policy instruments based on treasure (also known as 'economic' or 'market-based' instruments) and nodality (Knill and Lenschow, 2000; Pacheco-Vega, 2020; Tews *et al.*, 2003). Other sources contend that market-based management is commonly adopted on emissions, while top-down-oriented management, such as legislation and prohibition, tend to be applied in the context of toxicity and other health-related problems (Sevä and Jagers, 2013).

The question of which policy instruments should be used in the implementation process is paramount since they have differing consequences for the implementation structure. Specifically, the choice of policy instrument determines which organisations become responsible

Table 10.1 Components of policy implementation

	Influencing factors
Policy design	Agenda-setting dynamics, political support, societal support, public communication
Implementation structure	Extensive implementation structure: impact on horizontal (e.g., cross-sectoral) and vertical coordination Concentrated implementation structure: insufficient capacity
Agency decision-making	Interconnectedness and cohesion between agencies and target groups, agencies' constraints, nature of the policy target, negotiation vs. legalism
Target group behaviour (policy outcome)	Degree of organisation, homogeneity, networks, capacity (unity of positions and threat)
Policy impact	Policy outputs and outcomes

for providing a service or infrastructure or for monitoring and enforcing compliance with the stipulations of an environmental policy. From this perspective, the roots of implementation problems can be found at the stage of policy formulation (Winter, 2012). Therefore, Vancoppenolle *et al.* (2015), building on Winter (2012), recommend assessing policy implementation by focusing on the following components and their interrelations: (i) policy design, (ii) implementation structure, (iii) agency decision-making, (iv) target group behaviour, and (v) policy impact.

Policy design refers to the features we have already touched upon (i.e., policy instruments), whereas implementation structure denotes the public and private organisations in charge of implementing a policy. Implementing organisations act according to decisions taken within them, which connects to the third aspect: agency decision-making. Target group behaviour is the fourth aspect. It feeds into the more general assessment of how a policy is put into practice and whether the policy leads to the intended policy outcome (i.e., change in the target groups' behaviour), which may then contribute to achieving the desired policy impact (i.e., overall environmental or societal objective), such as the reduction of environmental pollution or climate change mitigation[1] (Knill and Tosun, 2020; Limberg *et al.*, 2021).

In the next sections, we address each of these aspects in turn to give an overview of the key insights of the state of research on environmental policy implementation. See Table 10.1 for an overview.

A factor Vancoppenolle *et al.* (2015) do not identify as a separate variable is government capacity, which studies on environmental policy implementation in developing countries in particular have stressed as a critical variable (e.g., Heichel *et al.*, 2014). We will pay attention to government capacity but as part of the implementation structure, thereby keeping the conceptual model proposed by the authors intact. Indeed, what makes the conceptual model by Vancoppenolle *et al.* (2015) compelling is that it is likely to facilitate a balanced discussion on the different facets of environmental policy implementation.

10.3 Policy Design

The first dimension of the integrated implementation model put forth by Vancoppenolle *et al.* (2015) concerns policy design, which includes the choice of policy objectives and (the mix of) policy instruments. The policy design is the result of policy formulation and can be influenced by how a policy was placed on the political agenda and what kind of political support it received. According to the authors, the very result of these influences can affect the implementation of a policy. In other words, Vancoppenolle *et al.* (2015) consider policy design to comprise the politics of the chosen policy instruments.

One study that provides insights into the relationship between the politics of an environmental policy measure and its implementation is provided by Tosun (2018). It investigates the market positioning of fuel mixtures containing up to 10% ethanol (E10), which were introduced to produce various positive effects on the environment. The E10 fuel was envisaged to reduce fossil fuels and carbon dioxide emissions. In Germany, the uptake of the fuel mixtures was significantly below expectation. One of the reasons for the poor implementation of the policy was that it lacked broad political and societal support and, while the competent federal minister launched the new fuel mixture together with the president of the German automobile club, the communication strategy was poor and created mistrust among car drivers as to whether it was a desirable alternative to the already existing fuels. In other European countries, the uptake of fuel mixtures was better. In Sweden, for instance, the consumption of fuel mixtures rose until 2012, but there, too, it has declined since. In addition to concerns about harmful effects on car engines, the debate on whether biofuel production contributes to food crises further reduced public support in Sweden (Andersson *et al.*, 2020).

Turning to the question of how the choice of environmental policy instrument affects policy implementation, Steinebach (2019) shows for a sample of advanced democracies that authority-based policy instruments (regulations) have an impact on air pollutant emissions, but information and treasure (market-based) instruments do not yield the same effects. A complementary perspective is offered by Daugbjerg and Sønderskov (2012), who stress that environmental policy comes in packages comprising of both supply-side and demand-side policy instruments. The analysis of green markets in Denmark, Sweden, the United Kingdom, and the United States demonstrates that cross-country variation in policy implementation is explained by differences in the mix of policy instruments. A different perspective is offered by Nilsson *et al.* (2012), who draw attention to the fact that policy mixes can be coherent at the level of policy objectives and instruments but conflict at the level of implementation. The authors illustrate this by offering case studies on the implementation of policies on biodiversity, habitats, resource efficiency, and water.

Another cause of problems in the implementation of environmental policy, especially in developing or transition countries, is the common practice of copying policy instruments from international organisations or developed countries. Mexico, for example, in the context of negotiating a free trade area, adopted a substantial share of its standards for regulating industry discharges into water from the United States, even though it did not have the same implementation capacity, which resulted in poor implementation (Heichel *et al.*, 2014). There exist several examples that corroborate the finding that a copy-and-paste practice in formulating environmental policy causes implementation problems.

10.4 Implementation Structure

Implementation structure denotes the number and types of organisations involved in the implementation process. It is difficult to identify an ideal structure for implementing environmental policy. On the one hand, a concentrated implementation structure (i.e., similar and a small number of organisations) can reduce transaction costs, but the organisations involved in it could easily suffer from insufficient capacity since no redundancies exist (Vancoppenolle *et al.*, 2015). On the other hand, an extensive implementation structure (i.e., diverse and a large number of organisations) can result in agency problems while providing a more robust structure that is less likely to suffer from insufficient capacity (Vancoppenolle *et al.*, 2015). As a rule, the implementation structure should be as extensive as necessary and as concentrated as possible. However, as we will discuss below, if we expect organisations to learn from their own past experiences of implementing environmental policy, or from the experience of others, more capacity and therefore a more extensive implementation structure appears favourable.

It is often the case that two or more ministries or agencies are responsible for implementing a policy. This is especially true with cross-sectoral policies, such as environmental policy, where the various organisations in charge of policy implementation need to coordinate their activities in what is known as horizontal coordination (Peters, 2015). For example, in Germany, both the Federal Ministry of the Environment, Nature Conservation and Nuclear Safety and the Federal Ministry of Food and Agriculture are responsible for developing policies that target the safeguarding and promotion of biodiversity. The latter comes into play since the use of plant protection (pesticides and herbicides) in agriculture is one of the main reasons for the loss of insect biodiversity. The same division of competences applies to the ministries at the level of the German states, which are responsible for implementing the federal law on biodiversity. The different ministries do not only have to invest in coordinating their positions but are also likely to support a policy design that favours either environmental interests over agricultural ones or the other way around. Such constellations result in the use of vague stipulations of policies (Knill and Lehmkuhl, 1999), which put implementing actors in the difficult position of having to find out and negotiate what those vague formulations mean within a given context (Beunen and Duineveld, 2010). Another example concerns the implementation of the EU Water Framework Directive, where staff members of national and regional water agencies in charge of implementation have stated that lacking intersectoral coordination, especially between watershed management and agricultural administrations, has been one of the main obstacles (Zingraff-Hamed *et al.*, 2020).

The need for horizontal coordination when implementing policies is not limited to the coordination of ministries or agencies representing different sectoral interests. In multi-level polities, horizontal coordination refers to constellations that involve units located at the same level of a governmental system. For example, biophysical planning requires horizontal coordination between riparian territorial jurisdictions located at the same level of a political system (Newig and Koontz, 2014). Furthermore, horizontal coordination can refer to ministries and agencies in charge of environmental policy implementation located at the same level of a political system.

In multi-level polities, vertical coordination may also be needed, as organisations located at different levels of government are responsible for policy implementation (Knill *et al.* 2020). A case in point is the river basin planning for the river Ems, which is constituted by the two German Federal states of North Rhine-Westphalia and Lower Saxony as well as the Netherlands (Newig and Koontz, 2014). A vast literature has investigated the implementation of EU environmental policy by the member states (Knill and Lehmkuhl, 1999; Knill and Lenschow, 1998; 2000). An overarching finding of this rich and insightful literature is that some member states have faced few problems in implementing EU environmental policies, while others have been plagued by poor implementation (Knill and Liefferink, 2013). When assessing the implementation performance of individual EU member states, it is important to decide whether one wants to concentrate on the cumulative number of implementation deficits experienced in only one year or a few. Most studies opt for the first approach, which is noteworthy since some of the reported implementation deficits have existed for several years or even decades.

While an extensive overview of this literature would go beyond the scope of this chapter, it is worth alluding to the most recent literature to assess the impact of economic crisis on the

Box 10.1 Citizen Science

Citizen science comprises practices employed by individuals to generate knowledge. The basic premise behind citizen science is the belief that the knowledge of citizens, which is embedded in a specific social context and closely related to specific living conditions, may constitute a supplement to expert knowledge. One of the areas in which citizens are engaged in generating knowledge is the environment. Since the 1970s, citizens have launched numerous science projects to collect data on air pollution by using sensors (Wróblewski *et al.*, 2021). Because sensors for monitoring have become more affordable over the years and air pollution is directly connected to the quality of life, citizen science projects have expanded considerably over the last few years. While air quality has been the main focus of citizen science, individuals also participate in the monitoring of other environmental issues, including water pollution and noise pollution. Further, citizen science, in the form of counting birds or plants, is regarded as a promising format for generating knowledge on biodiversity (Peter *et al.*, 2021). The richer the knowledge base on environmental issues, the better the possibility for connecting environmental outcomes to the adoption and implementation of environmental policies. And this gives non-governmental organisations and other civic organisations further justification for demanding more ambitious environmental action.

implementation of environmental policy (Melidis and Russel, 2020). A particularly innovative twist to the topic of how EU member states implement environmental policy is provided by studies that examine how digitalisation and technical advances have improved the institutional capacity of the European Commission as well as the national and subnational implementing actors, such as towns and cities (Bürgin, 2020). A related factor for explaining the changes in the implementation of environmental policy is the improvement of data on environmental policy outcomes as produced, *inter alia*, by lay people within the frame of citizen science (see Box 10.1).

To improve the implementation of EU environmental law, the European Union Network for the Implementation and Enforcement of Environmental Law (IMPEL) was founded in 1992. Initially an informal network, it transformed into an international non-profit association under Belgian law in 2008. IMPEL aims to increase the capacity of the environmental authorities of the EU, acceding and candidate countries of the EU, European Economic Area and European Free Trade Area countries, and potential candidates for joining the European Community. To this end, IMPEL facilitates an exchange of information and experiences on implementation (usually best practice cases) and organises peer reviews among the participating environmental authorities as well as offering additional support for capacity building (Jordan and Tosun, 2012).

10.5 Agency Decision-Making

Agency decision-making refers to the process of making legal stipulations more concrete and therefore implementable. This requires the members of one or several competent organisations to decide on these concretisations and to work out a procedure for the delivery of the policies. The implementation of environmental policy is executed in various forms and by various actors. Most studies tend to focus upon the relationship between regulators, regulations, and firms and companies (Sevä and Jagers, 2013).

Bressers and O'Toole (1998) have already suggested that decision-making in agencies and the design of policies, specifically the choice of policy instruments, are interrelated and have implications for implementation. They postulate that the design of policies differs depending on the relationship (i.e., interconnectedness and cohesion) between governmental agencies and the target group – that is, the actors to be targeted by the specific policy. In the case of water policy in Germany, the United Kingdom, the Netherlands, and the United States, the authors showed how policy instruments changed from subsidies and information towards regulatory instruments alongside decreasing levels of interconnectedness and cohesion (Bressers and O'Toole, 1998).

Building on this and other earlier works (e.g., Bruijn and Heuvelhof, 1997; Linder and Peters, 1989), Howlett (2004) suggests that the combination of constraints on government agencies (with regard to resources and legitimacy) and the nature of the policy target (which relates to exchange or policy actors) lead to the development of 'implementation styles' comprising typical combinations of substantive and procedural policy instruments that are employed to achieve a given set of policy goals. Based on these two dimensions, he distinguishes four ideal-typical 'implementation styles'.

Institutionalised voluntarism is the implementation style of choice when the policy targets constitute a large group and state constraints are high. It corresponds to an exhortation-based manipulation of market actors and the institutionalisation of networks. *Regulatory corporatism* is chosen when state constraints are high, but the policy targets constitute a small group. This implementation style regulates market actors and manipulates their interest-articulation system by means of financial incentives. It corresponds to a 'corporatist' model of economic planning in industrial policymaking. The third implementation style, *directed subsidisation*, is expected to be chosen when governmental actors face low constraints and interact with large groups of policy targets. Directed subsidisation refers to the extensive use of financial instruments to steer the behaviour of market actors, coupled with the use of authority, to recognize network actors in order of preference. The fourth implementation style, *public provision with oversight*, corresponds to a mobilisation model, in which governmental organisations use resources to provide goods and services to small groups of policy targets. At the same time, the government seeks to manipulate actor networks through information. This implementation style is chosen when the number of policy addressees is small and so the constraints on the state are low.

Lesnikowski *et al.* (2021) apply Howlett's typology of implementation styles to investigate how local governments implement climate change adaption measures. Their analysis reveals that of these four implementation types, directed subsidisation constitutes a much smaller share. However, the authors acknowledge that their findings could depend on the specific selection of measures they examined. More broadly speaking, there are environmental policy sectors in which directed subsidisation is applied more frequently. As Tosun and Treib (2018) argue, the regional governments in Germany have promoted the transition to a bio-based economy (also known as the bioeconomy; for an overview, see, e.g., Wydra *et al.*, 2021) by subsidising the industries concerned as well as investing heavily in research activities that could potentially support this process.

Implementation styles in environmental policy can also be split into legalism and negotiation (Howlett, 2000). Both styles differ essentially in respect of the relationship between government agencies and the target group. For example, environmental policy implementation in the United States until 1990 took the form of 'adversarial legalism', as environmental laws empowered citizens to enforce implementation through litigation. This implementation style is characterised by procedural instruments that facilitate citizen suits, judicial activism, and action-forcing statutes (Howlett, 2000). In contrast, the implementation style of negotiation is characterised by initiatives that grant agencies the ability to incorporate industry and industrial associations into the process of policy implementation. This encourages 'voluntary' approaches and consultations with stakeholders, improving implementation by enhancing the legitimacy of the policies adopted. The use of procedural instruments – i.e., negotiations with industry on standards and enforcement, multi-stakeholder consultations, and environmental treaties – can be viewed as 'collaborative governance' between the state and the target group (Howlett, 2000).

Current policymaking in German water protection policy constitutes an example of the implementation style of negotiation. With the aim of mitigating the entry of micropollutants into surface waters, the Federal Ministry of the Environment initiated a multi-stakeholder consultation over a period of more than three years. Overall, the intention was to encourage voluntary action and legitimise precautionary approaches before adopting new environmental policies (Schaub and Tosun, 2021).

China, as an authoritarian state, has experimented with different types of implementation (Jia and Chen, 2019). Performance evaluation has been central to the implementation of environmental policy targets in China. Since the 2000s, they have been included in the country's Five-Year Plans and implemented via the target responsibility system. Since 2006, Chinese central planners have granted veto power to the majority of binding environmental targets within the target responsibility system. Local government has the authority at each administrative level to decide how to allocate targets to subordinate governments as well as state-owned enterprises and public institutions (Kostka and Goron, 2021).

Campaign style is another implementation type with which China has experimented. This involves the mobilisation of administrative resources under political sponsorship to achieve salient policy targets and has been developed and applied in many countries (Jia and Chen, 2019). The Communist Party of China has adopted this political method usually to rectify shortcomings when regular governance policies fail because the defined tasks are incompatible with other policy goals (Zhao *et al.*, 2020).

As we have seen, the focus of most pertinent research is on top-down implementation (Tosun and Treib, 2018). However, it should be noted that environmental policy implementation can be executed in various forms and by various actors. A complementary form is bottom-up implementation, which brings street-level bureaucrats to the fore. Sevä and Jagers (2013) explain that in the field of environmental policy, street-level bureaucrats differ from traditional bureaucrats in that the daily, face-to-face encounters are with clients or stakeholders. According to the authors, what makes 'environmental bureaucrats at the frontline' comparable to street-level bureaucrats is that they are rather autonomous and have some level of discretionary power. Against this background, research has shown that the enforcement styles of environmental bureaucrats affect the compliance of stakeholders with rules and regulations (May and Winter, 2000; Shimshack, 2014). Studies have also demonstrated that the norms and values of street-level bureaucrats impact their policy implementation style (Sevä, 2013).

10.6 Target Group Behaviour

Target group behaviour denotes the role that policy addressees play in the implementation process – their actions as well as their needs. Target groups are key in policy implementation. Their (change in) behaviour is essential for achieving the overall policy objective. Therefore, it is crucial for policy implementation to be aware of their characteristics and to anticipate their behaviour when designing policies. Since at least the formulation of the Dutch Target Group Policy in

1989, there has been the assumption that the transition to sustainability is too large a goal to be achieved without collaboration between the state and target groups. Furthermore, this has been linked to the notion that it is important to approach collaboration differently depending on differences in the structures of the target groups (Hofman and Schrama, 2005).

In their analysis of the Dutch Target Group Policy, Hofman and Schrama (2005) propose six features that governments can choose when addressing target groups. First is the degree of freedom of choice granted to the target groups. Target groups may not be able to attain certain requirements in the short term, such as developing an innovative solution to reduce emissions. Instead, it might be more promising for governments to propose long-term goals and leave the path to achieving these targets open to negotiations. Second, governments have the option to vary the degree of collaboration with the target group, from simple bilateral negotiations to interactive stakeholder consultations. The third feature is the level of stringency imposed on the target group, and the fourth is the timeframe for complying with legal requirements. The latter needs a good balance since overly strict short-term requirements may result in suboptimal outcomes – such as businesses adopting existent but ineffective solutions just to ensure compliance, instead of investing in the development of more effective solutions – while lax long-term requirements may lead to postponement of action. The fifth feature is the choice of substantive policy instruments, and the sixth is the selection of the appropriate addressees of policy (Hofman and Schrama, 2005). The last feature is especially relevant when incorporating differences in the characteristics of target groups.

Hofman and Schrama (2005) point to at least three ways in which target groups may differ and offer means of addressing them. First, target groups differ in how well they are organised and how easily they can be reached. For instance, instead of targeting consumers directly, it may be more effective to address them indirectly through well-organised industrial branches. Second, a target group might be heterogeneous, making it necessary to differentiate within the group, for instance, between 'frontrunners' and 'laggards'. Third, some target groups might be tightly organised into networks and thus it may be more effective to address the network instead of individual businesses.

Empirically, Hofman and Schrama (2005) show how Dutch authorities incorporated differences between target groups when formulating the Target Group Policy. First, the authorities formulated long-term emissions reduction targets for the whole industry and then, second, prioritised the branches responsible for the majority of environmental pollution. Third, they successfully negotiated agreements with the trade associations representing these branches. In a fourth step, individual companies of the branches were informed and persuaded to join the agreement with the help of an independent agency. This agreement was then adapted at the level of individual companies. Whereas the former was confronted with uniform implementation plans, companies in the heterogeneous sector had to develop their own implementation plans. Finally, the companies had to incorporate these plans into their environmental practices and adjust their environmental licences.

Overall, the authors conclude that both government and industry were pleased with the outcome. In addition, they state that the negotiated agreement was more successful in respect of well-organised target groups (Hofman and Schrama, 2005).

Another part of the literature suggests that differences in the capacity of target groups to influence policymaking can impact policy implementation (Jänicke, 1992, 1997; Skodvin *et al.*, 2010). It takes the perspective of target groups and how they vary in their ability to enforce their interests, usually to the disadvantage of environmental protection. Skodvin *et al.* (2010) show in the case of EU climate policy that target groups are able to reduce the room for manoeuvre in respect of politically feasible policy design when they have control over resources which decision-makers depend on. For instance, target groups can possess the expert knowledge necessary for optimising policies as well as harnessing the public support that decision-makers need in order to raise their chances of re-election. The capacity of target groups is dependent on at least two factors: the unity of their positions and the effectiveness of their threats (Skodvin *et al.*, 2010).

Target groups can comprise different types of units. They can be individuals and companies in a given country or the country itself. Regarding the latter, countries are target groups themselves when they sign international environmental agreements and thereby declare that they are willing to comply with their respective stipulations. Since no global authority exists for enforcing compliance with international environmental agreements, there is no alternative mechanism for bringing about changes in the environment-related decisions taken by state governments, which are known as monitoring, reporting, and verification systems (see Box 10.2).

Turning to individuals, these can be citizens in the broadest sense or segments of society – for instance, 'car drivers' or 'houseowners'. The selection of the type of environmental policy determines the likelihood or speed at which individuals will change their environment-related behaviour. Some instruments provide positive incentives such as subsidies, and others negative incentives, such as financial sanctions. The choice of the policy instrument depends on the nature of the environmental problem that a given policy addresses. The most drastic instrument is a ban on products or activities. Despite their hierarchical nature, citizens tend to prefer these to market-based instruments, such as environmental taxes (Rhodes *et al.*, 2017; Tosun *et al.*, 2020).

Apart from the chosen policy instrument, research has shown that citizens are more likely to comply with environmental policies they regard as legitimate (Jagers *et al.*, 2012). In this context, Jager *et al.* (2020) show how participatory governance arrangements increase the willingness of individuals to comply with environmental policy. The authors stress that the

Box 10.2 Monitoring, Reporting, and Verification

The Enhanced Transparency Framework, adopted under the Paris Agreement in 2015, builds on the monitoring (sometimes also termed 'measurement'), reporting, and verification system of the United Nations Framework Convention on Climate Change (UNFCCC). Together, monitoring, reporting, and verification represent key principles for checking the implementation of the national pledges to curb carbon dioxide emissions, for which the UNFCCC has developed detailed guidance. Monitoring, reporting, and verification describe all measures which countries take to collect data on emissions and mitigation actions. Monitoring denotes the direct measurement or estimated calculations of emissions and emissions reductions following guidance and protocols. Reporting refers to the documentation for informing all interested parties on issues such as methodology or data. Verification indicates the specific procedures or expert reviews used to verify the quality of the data and estimations. Some transnational city networks also have such surveillance systems in place to assess whether and to what extent their members implement local-level climate policies. For example, the transnational city network on climate governance, the Covenant of Mayors for Climate & Energy, uses a surveillance regime that aligns with the logic of monitoring, reporting, and verification (De Francesco *et al.*, 2020).

convergence of stakeholder perspectives, including social aspects of conflict resolution, trust building, mutual gains, and the building of shared norms, is particularly important for facilitating compliance with environmental policy.

Environmental policy comprises a wide range of sectors. In particular, policies concentrating on emissions (into air or water) and environmental quality tend to target businesses and industries. In the case of promoting biofuels, for instance, policy implementation crucially depends on the behaviour of the fuel producers and suppliers as well as on consumers' buying choices. The behaviour of fuel stations was one of several reasons why the introduction of E10 in Germany faced implementation challenges. Instead of offering E10 in addition to other fuel mixes, numerous fuel stations simply replaced the other fuel mixes with E10, resulting in mistrust among car drivers (Tosun, 2018). Another case in point is the implementation of the bioeconomy, which a growing number of countries have embraced as a strategy for achieving sustainable development. In a nutshell, the establishment of a bioeconomy involves replacing inputs based on fossil fuels and non-renewable materials with renewable ones. The implementation of the bioeconomy depends on the provision of policy incentives as well as on the behaviour of business and industry (Bezama *et al.*, 2019).

Generally, it is difficult to connect the various steps in the implementation process to policy outcome. However, there exist some settings in which this is methodologically feasible. Particularly insightful are studies on how environmental policies are implemented in China, which has experimented with different strategies and thus makes for a good case in point. Of the existing research, we would like to mention the study by Shen *et al.* (2020), for two reasons: first, it uses an alternative measurement of policy outcomes, as it focuses on environmental innovations; second, it examines the effect of different types of environmental policy instruments on innovations. In this regard, the study reveals that regulation has a positive effect on innovations relating to end-of-pipe treatment and green products, whereas pollution charges matter for promoting innovation in green processes.

Another literature that is instructive for developing a better understanding of how policy implementation relates to policy outcomes investigates the implementation of sustainability. This concept lends itself particularly well to such an analysis since there exist quantifiable means of assessing the degree to which it was delivered. The rich body of research contends that environmental sustainability has not been delivered. Analysing this literature, Howes *et al.* (2017) allude to various reasons as to why it has been difficult to achieve environmental sustainability, including a conflict between the objectives of environmental policies and those of economic policies, a lack of incentives to implement environmental policies, and a failure to communicate objectives to key stakeholders.

10.7 Policy Impact

The last component of policy implementation is the policy impact – the overall objective of the policy. This is determined by the previous components and serves as a benchmark for how successfully a policy has been implemented. The literature on environmental policy includes several studies that provide measurements for policy impact, which is either measured using data for individual indicators, such as (changes in) water quality, or composite indicators for several types of environmental impact (for an overview, see e.g. Tosun and Schnepf, 2020). Nevertheless, changes in the policy impact may not necessarily be the result of changes in policy outcome. An important research gap that requires filling is the causal link between policy outcome and policy impact.

10.8 Conclusion and Future Research

Since the 1970s, environmental policy has developed into a policy sector in its own right (Knill and Liefferink, 2013). In addition, environmental concerns have been considered as important enough not to be realised through environmental policy alone but also by mainstreaming environmental concerns into other sectoral policies (Tosun *et al.*, 2019). While the adoption of environmental policy is a necessary step to reduce or prevent environmental harm, it equally requires that the policies adopted are implemented. The implementation of environmental policy, in turn, comprises various actions by various actors, and can take many different forms. In this chapter, we have given an overview of the state of research on environmental policy implementation by concentrating on (i) policy design, (ii) implementation structure, (iii) agency decision-making, (iv) target group behaviour, and (v) policy impacts.

By following this structure, we have been able to identify topics on which ample research exists as well as areas in need of further investigation. Concerning the latter, we consider three aspects particularly promising for advancing the literature. The first could be addressed by systematically connecting the degree of political (dis)agreement when a given environmental policy instrument has been adopted to the behaviour of the target groups and the broader policy impacts. The second aspect relates to the limited attention street-level bureaucrats have received in the literature on environmental policy implementation, which is surprising given that it could benefit from the insights provided by the existent literature on regulatory inspections. Third, although we were able to identify exciting work on how participatory arrangements affect target group behaviour and policy impacts, thereby challenging the classic take on implementation as a top-down process, there is room to develop this specific research perspective further and to connect it in a systematic fashion to other subdisciplines in political science, such as political sociology or political psychology.

Overall, despite the extensive literature on environmental policy implementation and the valuable insights it has provided, there remain knowledge gaps that future research should consider reducing.

Note

1 The term 'policy outcome' has been used inconsistently in the pertinent literature. To clarify, some authors have termed the overall policy objective as 'policy outcome' instead of 'policy impact' (e.g., Adam *et al.*, (2018); Tosun and Schnepf, 2020). We differentiate between 'policy outcome' as target group behaviour and 'policy impact' as the overall objective in the sense of addressing the policy problem that triggered the policy process.

References

Adam, C., Steinebach, Y. and Knill, C. (2018) 'Neglected challenges to evidence-based policy-making: the problem of policy accumulation', *Policy Sciences* 51(3): 269–90, doi:10.1007/s11077-018-9318-4.

Andersson, L., Ek, K., Kastensson, Å. and Wårell, L. (2020) 'Transition towards sustainable transportation– What determines fuel choice?', *Transport Policy* 90: 31–38.

Beunen, R. and Duineveld, M. (2010) 'Divergence and convergence in policy meanings of European environmental policies: the case of the birds and habitats directives', *International Planning Studies* 15(4): 321–33.

Bezama, A., Ingrao, C., O'Keeffe, S. and Thrän, D. (2019) 'Resources, collaborators, and neighbors: The three-pronged challenge in the implementation of bioeconomy regions', *Sustainability* 11 (24): 7235.

Bressers, H.T.A. and O'Toole, L.J. (1998) 'The selection of policy instruments: A network-based perspective', *Journal of Public Policy* 18(3): 213–39, doi:10.1017/S0143814X98000117.

Bürgin, A. (2020) 'Compliance with European Union environmental law: An analysis of digitalization effects on institutional capacities', *Environmental Policy and Governance* 30(1): 46–56.

Daugbjerg, C. and Sønderskov, K.M. (2012) 'Environmental policy performance revisited: Designing effective policies for green markets', *Political Studies* 60(2): 399–418, doi:10.1111/j.1467-9248.2011.00910.x.

de Bruijn, J.A. and ten Heuvelhof, E.F. (1997) 'Instruments for network management', in W.J.M. Kickert, E.-H. Klijn and J.F.M. Koppenjan (eds), *Managing Complex Networks: Strategies for the Public Sector*, London: Sage publications, pp. 119–36.

De Francesco, F., Leopold, L. and Tosun, J. (2020) 'Distinguishing policy surveillance from policy tracking: transnational municipal networks in climate and energy governance', *Journal of Environmental Policy & Planning* 22(6): 857–69.

Graversgaard, M. et al. (2021) 'Policies for wetlands implementation in Denmark and Sweden–historical lessons and emerging issues', *Land Use Policy* 101: 105206.

Heichel, S., Pape, J. and Tosun, J. (2014) 'Regulation of industrial discharges into surface water', in H. Jörgens, A. Lenschow and D. Liefferink (eds), *Understanding Environmental Policy Convergence: The Power of Words, Rules and Money*, Cambridge: Cambridge University Press, pp. 64–103.

Hill, M. and Hupe, P. (2002) *Implementing Public Policy: Governance in Theory and in Practice*, London: Sage Publications.

Hjern, B. (1982) 'Implementation research — The link gone missing', *Journal of Public Policy* 2(3): 301–08, doi:10.1017/S0143814X00001975.

Hofman, P. and Schrama, G. (2005) 'Dutch target group policy', in Bruijn, Theo J. N. M. and V. Norberg-Bohm (eds), *Industrial transformation: Environmental policy innovation in the United States and Europe*, Cambridge, MA: MIT Press, pp. 39–63.

Holzinger, K. and Knoepfel, P. (2000) *Environmental Policy in a European Union of Variable Geometry? The Challenge of the Next Enlargement*, Basel, Genf, München: Helbing und Lichtenhahn.

Hood, C.C. and Margetts, H.Z. (2007) *The Tools of Government in the Digital Age*, Macmillan International Higher Education.

Howes, M. et al. (2017) 'Environmental sustainability: A case of policy implementation failure?', *Sustainability* 9(2): 165, doi:10.3390/su9020165.

Howlett, M. (2000) 'Beyond legalism? Policy ideas, implementation styles and emulation-based convergence in Canadian and US environmental policy', *Journal of Public Policy* 20(3): 305–29, doi:10.1017/S0143814X00000866.

Howlett, M. (2004) 'Beyond good and evil in policy implementation: Instrument mixes, implementation styles, and second generation theories of policy instrument choice', *Policy and Society* 23(2): 1–17.

Jager, N.W., Newig, J., Challies, E. and Kochskämper, E. (2020) 'Pathways to implementation: Evidence on how participation in environmental governance impacts on environmental outcomes', *Journal of Public Administration Research and Theory* 30(3): 383–99.

Jagers, S.C., Berlin, D. and Jentoft, S. (2012) 'Why comply? Attitudes towards harvest regulations among Swedish fishers', *Marine Policy* 36(5): 969–76.

Jänicke, M. (1992) 'Conditions for environmental policy success: An international comparison', *The Environmentalist* 12(1): 47–58, doi:10.1007/BF01267594.

Jänicke, M. (1997) 'The political system's capacity for environmental policy', in M. Jänicke, H. Jörgens and H. Weidner (eds), *National Environmental Policies*, Berlin, Heidelberg: Springer Berlin Heidelberg, pp. 1–24.

Jia, K. and Chen, S. (2019) 'Could campaign-style enforcement improve environmental performance? Evidence from China's central environmental protection inspection', *Journal of Environmental Management* 245: 282–90, doi:10.1016/j.jenvman.2019.05.114.

Jordan, A. and Tosun, J. (2012) 'Policy implementation', in A. Jordan and C. Adelle (eds), *Environmental Policy in the EU*, Routledge.

Knill, C. and Lehmkuhl, D. (1999) 'How Europe Matters. Different Mechanisms of Europeanization', *European Integration online Papers (EIoP)* 3.

Knill, C. and Lenschow, A. (1998) 'Coping with Europe: The impact of British and German administrations on the implementation of EU environmental policy', *Journal of European Public Policy* 5(4): 595–614.

Knill, C. and Lenschow, A. (2000) *Implementing EU Environmental Policy: New Directions and Old Problems*, Manchester University Press.

Knill, C. and Liefferink, D. (2013) *Environmental Politics in the European Union: Policy-Making, Implementation and Patterns of Multi-Level Governance*, Manchester University Press.

Knill, C., Steinbacher, C. and Steinebach, Y. (2020) 'Sustaining statehood: A comparative analysis of vertical policy-process integration in Denmark and Italy', *Public Administration*.

Knill, C. and Tosun, J. (2020) *Public Policy: A New Introduction*, London: Red Globe Press.

Kostka, G. and Goron, C. (2021) 'From targets to inspections: the issue of fairness in China's environmental policy implementation', *Environmental Politics* 30(4): 513–37, doi:10.1080/09644016.2020.1802201.

Lasswell, H.D. (1956) *The Decision Process: Seven Categories of Functional Analysis*, University Park: University of Maryland Press.

Lesnikowski, A., Biesbroek, R., Ford, J.D. and Berrang-Ford, L. (2021) 'Policy implementation styles and local governments: the case of climate change adaptation', *Environmental Politics* 30(5): 753–90, doi:10.1080/09644016.2020.1814045.

Liefferink, D. and Wurzel, R.K.W. (2017) 'Environmental leaders and pioneers: agents of change?', *Journal of European Public Policy* 24(7): 951–68, doi:10.1080/13501763.2016.1161657.

Limberg, J., Steinebach, Y., Bayerlein, L. and Knill, C. (2021) 'The more the better? Rule growth and policy impact from a macro perspective', *European Journal of Political Research* 60(2): 438–54, doi:10.1111/1475-6765.12406.

Linder, S.H. and Peters, B.G. (1987) 'A design perspective on policy implementation: The fallacies of misplaced prescription', *Review of Policy Research* 6(3): 459–75, doi:10.1111/j.1541-1338.1987.tb00761.x.

Linder, S.H. and Peters, B.G. (1989) 'Instruments of government: Perceptions and contexts', *Journal of Public Policy* 9(1): 35–58, doi:10.1017/S0143814X00007960.

May, P. and Winter, S. (2000) 'Reconsidering styles of regulatory enforcement: Patterns in Danish agro-environmental inspection', *Law & Policy* 22(2): 143–73.

Mazmanian, D. and Sabatier, P. (1983) *Implementation and Public Policy*, Glenview: Scott, Foresman.

Melidis, M. and Russel, D.J. (2020) 'Environmental policy implementation during the economic crisis: an analysis of European member state 'leader-laggard'dynamics', *Journal of Environmental Policy & Planning* 22(2): 198–210.

Montjoy, R.S. and O'Toole, L.J. (1979) 'Toward a theory of policy implementation: An organizational perspective', *Public Administration Review* 39(5): 465, doi:10.2307/3109921.

Newig, J. and Fritsch, O. (2009) 'Environmental governance: participatory, multi-level–and effective?', *Environmental Policy and Governance* 19(3): 197–214.

Newig, J. and Koontz, T.M. (2014) 'Multi-level governance, policy implementation and participation: the EU's mandated participatory planning approach to implementing environmental policy', *Journal of European Public Policy* 21(2): 248–67.

Nilsson, M. et al. (2012) 'Understanding policy coherence: Analytical framework and examples of sector-environment policy interactions in the EU', *Environmental Policy and Governance* 22(6): 395–423, doi:10.1002/eet.1589.

Olmstead, S. and Zheng, J. (2021) 'Water pollution control in developing countries: Policy instruments and empirical evidence', *Review of Environmental Economics and Policy* 15(2): 261–80.

O'Toole, L.J. and Montjoy, R.S. (1984) 'interorganizational policy implementation: A theoretical perspective', *Public Administration Review* 44(6): 491, doi:10.2307/3110411.

Pacheco-Vega, R. (2020) 'Environmental regulation, governance, and policy instruments, 20 years after the stick, carrot, and sermon typology', *Journal of Environmental Policy & Planning* 22(5): 620–35.

Peter, M., Diekötter, T., Höffler, T. and Kremer, K. (2021) 'Biodiversity citizen science: Outcomes for the participating citizens', *People and Nature* 3(2): 294–311.

Peters, B.G. (2015) *Pursuing Horizontal Management*, JSTOR.

Pressman, J.L. and Wildavsky, A. (1984a) *Implementation: How Great Expectations in Washington are Dashed in Oakland; Or, Why it's Amazing that Federal Programs Work at All, This being a Saga of the Economic Development Administration as Told by Two Sympathetic Observers Who Seek to Build Morals on a Foundation*, University of California Press.

Pressman, J.L. and Wildavsky, A.B. (1984b) *Implementation: How Great Expectations in Washington are Dashed in Oakland*, Berkeley: University of California Press.

Rhodes, E., Axsen, J. and Jaccard, M. (2017) 'Exploring citizen support for different types of climate policy', *Ecological Economics* 137: 56–69, doi:10.1016/j.ecolecon.2017.02.027.

Sabatier, P. and Mazmanian, D. (1980) 'The implementation of public policy: A framework of analysis', *Policy Studies Journal* 8(4): 538–60, doi:10.1111/j.1541-0072.1980.tb01266.x.

Schaub, S. and Tosun, J. (2021) 'Politikgestaltung im Dialog? Umweltgruppen und ihre Mitwirkung bei der Regulierung von Spurenstoffen in Gewässern', *ZPol Zeitschrift für Politikwissenschaft* 31(2): 291–325, doi:10.1007/s41358-021-00278-z.

Sevä, M. (2013) 'A comparative case study of fish stocking between Sweden and Finland: explaining differences in decision making at the street level', *Marine Policy* 38: 287–92.

Sevä, M. and Jagers, S.C. (2013) 'Inspecting environmental management from within: the role of street-level bureaucrats in environmental policy implementation', *Journal of Environmental Management* 128: 1060–70, doi:10.1016/j.jenvman.2013.06.038.

Shen, C., Li, S., Wang, X. and Liao, Z. (2020) 'The effect of environmental policy tools on regional green innovation: Evidence from China', *Journal of Cleaner Production* 254(1): 120122, doi:10.1016/j.jclepro.2020.120122.

Shimshack, J.P. (2014) 'The economics of environmental monitoring and enforcement', *Annual Review of Resource Economics* 6(1): 339–60.

Skodvin, T., Gullberg, A.T. and Aakre, S. (2010) 'Target-group influence and political feasibility: the case of climate policy design in Europe', *Journal of European Public Policy* 17(6): 854–73, doi:10.1080/13501763.2010.486991.

Steinebach, Y. (2019) 'Instrument choice, implementation structures, and the effectiveness of environmental policies: A cross-national analysis', *Regulation & Governance* 01(03): 209, doi:10.1111/rego.12297.

Tews, K., Busch, P.-O. and Jörgens, H. (2003) 'The diffusion of new environmental policy instruments 1', *European Journal of Political Research* 42(4): 569–600.

Tosun, J. (2012) 'Environmental monitoring and enforcement in Europe: A review of empirical research', *Environmental Policy and Governance* 22(6): 437–48.

Tosun, J. (2018) 'The behaviour of suppliers and consumers in mandated markets: the introduction of the ethanol–petrol blend E10 in Germany', *Journal of Environmental Policy & Planning* 20(1): 1–15.

Tosun, J., De Francesco, F. and Peters, B.G. (2019) 'From environmental policy concepts to practicable tools: Knowledge creation and delegation in multilevel systems', *Public Administration* 97(2): 399–412.

Tosun, J., Schaub, S. and Fleig, A. (2020) 'What determines regulatory preferences? Insights from micropollutants in surface waters', *Environmental Science & Policy* 106: 136–44, doi:10.1016/j.envsci.2020.02.001.

Tosun, J. and Schnepf, J. (2020) 'Measuring change in comparative policy analysis: concepts and empirical approaches', *Handbook of Research Methods and Applications in Comparative Policy Analysis*, Edward Elgar Publishing.

Tosun, J. and Treib, O. (2018) 'Linking policy design and implementation styles', *Routledge Handbook of Policy Design*, Routledge, pp. 316–30.

Vancoppenolle, D., Sætren, H. and Hupe, P. (2015) 'The politics of policy design and implementation: A comparative study of two Belgian service voucher programs', *Journal of Comparative Policy Analysis: Research and Practice* 17(2): 157–73.

Winter, S.C. (2012) 'Implementation perspectives: Status and reconsideration', *Handbook of Public Administration* The SAGE: 265–78.

Wróblewski, M., Suchomska, J. and Tamborska, K. (2021) 'Citizens or consumers? Air quality sensor users and their involvement in sensor.community. results from qualitative case study', *Sustainability* 13(20): 11406, doi:10.3390/su132011406.

Wydra, S. et al. (2021) 'Transition to the bioeconomy–Analysis and scenarios for selected niches', *Journal of Cleaner Production* 294: 126092.

Zhao, Y., Zhang, X. and Wang, Y. (2020) 'Evaluating the effects of campaign-style environmental governance: Evidence from environmental protection interview in China', *Environmental Science and Pollution Research* 27: 28333–47.

Zingraff-Hamed, A. et al. (2020) 'Perception of bottlenecks in the implementation of the European water framework directive', *Water Alternatives* 13(3): 458–83.

11
ENVIRONMENTAL POLICY EVALUATION

Jonas J. Schoenefeld

11.1 Introduction

Impactful environmental policies are in high demand. Pressing environmental issues, including air and water pollution but of course also climate change, have generated a need for effective government action in reaching environmental aims such as clean air and water or a stable climate. But, unfortunately, not every government policy is effective. Newspapers and political commentaries tend to be littered with scathing critiques of ineffective policies, wasted public money, and insufficient government action. For example, the European Emissions trading system has been subject to significant critique on account of limited coverage, the low carbon prices it has generated in the past, and instances of fraud.[1]

But how can one distinguish between "what works" and what does not (Sanderson, 2002)? Policy evaluation or the "careful retrospective assessment of the merit, worth and value of administration, output and outcome of government interventions, which is intended to play a role in future, practical action situations" (Vedung, 1997, 3) is a practice that has developed in order to provide such assessments. This definition draws on earlier thinking on policy evaluation by Scriven (1991), who understands evaluation as an activity that assesses "the merit or worth or value of [public policy]; or the product of that process" (Scriven, 1991, 139). Focusing on environmental policy, Crabbe and Leroy (2012) in turn define evaluation as "a scientific analysis of a certain policy area, the policies of which are assessed for certain criteria, and on the basis of which recommendations are formulated" (p. 1). The latter definition broadens the remit of evaluation because a scientific analysis could, for example, also assess the design and adequacy of a policy. This chapter takes the latter definition as a point of departure, but also recognises the point made by earlier scholars that evaluation is about assessing a policy against specific values. As practitioners and scholars began to discuss different evaluation criteria and approaches, they quickly came to recognise that evaluation not only generates methodological and technical questions, but is also an inevitably political exercise, meaning that it is inextricably linked with interests in, and sometimes conflicts over, resources (Weiss, 1993; Bovens et al., 2006). For example, a negative evaluation may lead to the termination of a public policy, which may run counter to the interests of policy-makers who originally promoted it. As a result of these

DOI: 10.4324/9781003043843-13

insights, the simple question of "What works?" becomes a lot more complicated, for example "What works for whom, under what circumstances, with what consequences?".

These different aspects of evaluation are especially relevant in the field of environmental policy, where intense societal and political debates have typically been the context in which policies have emerged and developed (see for example Hulme, 2009). Learning more about *de facto* evaluation use in political contexts, (environmental) evaluation scholars have gone about adjusting their approaches and their methodologies to better suit the needs of environmental policy-makers and refine their practices to increase the relevance of their work (e.g., Patton, 2008). While environmental policy is not the first field where evaluation appeared – its origins date back to the assessments of the social programmes that emerged in the USA and in Europe in the middle of the 20th century (Knaap and Tschangho, 1998) – it has become a feature in demand in modern-day environmental policy-making. Many policy-makers now see evaluation as an important element of their work (e.g., European Environment Agency, 2016; United States Environmental Protection Agency, 2020) and politicians have often highlighted a perceived need for evaluation and science-based policy-making. In fact, in some cases, evaluation has also become a legally mandated rule-making feature (Bussmann, 2005). Likewise, the supply side of evaluation has also developed, as for example dedicated professional associations, journals, and consultant industries have emerged to offer evaluation services and platforms for exchange and to consolidate a growing professional field, which now also boasts specific university degree programmes and a range of professional opportunities.

Many different strands of evaluation have developed over time. One important distinction concerns the temporal orientation of evaluation (Crabbe and Leroy, 2012). *Ex ante* evaluation (also known as impact assessment) is a practice that assesses the future (i.e., projected) impact of public policies. Impact assessment has been subject to considerable scholarly debate and has become an important and increasingly mandatory element of environmental policy-making (Paul, 2020; Turnpenny et al., 2009). Furthermore, Crabbe and Leroy (2012) add *ex nunc* evaluation, that is, the evaluation of ongoing public policies. The authors highlight that, strictly speaking, most evaluation tends to be of this type, given that policies are rarely fully terminated (see Adam et al., 2018). Finally, *ex post* evaluation is a retrospective assessment of public policy. It is this retrospective evaluation perspective on which this chapter focuses.

Another important aspect of the debate on evaluation concerns evaluation criteria. In the early days, evaluation first focused on policy effectiveness in relation to the original policy targets, as the headline criterion. Over time, however, scholars recognised that evaluation requires a broader remit of criteria to adequately capture policy outcomes, such as for example effects on equity and justice, or on the efficiency with which policy aims are being achieved. Furthermore, policies may generate intended and unintended consequences that require attention, and their effects may spill over to other policy areas or governance levels. These debates have given rise to a broader argument on the need for multi-criteria evaluation (e.g., Hanberger, 2013; Konidari and Mavrakis, 2007). Taken together, evaluation has sought to respond to the growing complexity in the policy environment by increasing the sophistication of its own approaches.

Against this background, this chapter assesses the current state of the field without claiming to be comprehensive. To do so, it first focuses on prominent evaluation approaches, methodological debates, and *de facto* environmental policy evaluation practices in Section 11.2 before turning to the multiple rationales for evaluation in Section 11.3. Section 11.4 offers an in-depth discussion of the cutting edge of environmental policy evaluation and especially future directions for scholars and practice. Section 11.5 concludes with reflections on the current and future role of policy evaluation in governing the environment in the coming decades.

11.2 Environmental Evaluation Approaches and Practices

11.2.1 Insights into Current Environmental Evaluation Activities

Policy evaluation in general and environmental policy evaluation in particular have been a growth area in recent decades and have rapidly become institutionalized. Many countries and regions such as the EU have seen the creation of national and regional evaluation associations,[2] including the American Evaluation Association,[3] the Japan Evaluation Society,[4] and the European Evaluation Society.[5] These associations/societies are networks of evaluators, practitioners, government officials, academics, and others with an interest in policy evaluation. They organise regular conferences as platforms for exchange and some have sections or working groups that deal with dedicated policy sectors, such as the environment. For example, the German Evaluation Society[6] has an ad hoc working group on the environment.[7] International networks focusing on the environment have also been created, such as the Environmental Evaluators Network (EEN)[8] or the associated European Environmental Evaluators Network (EEEN). These networks also organise regular conferences such as the EEEN Forum.[9] Another sign of institutionalisation has been the emergence of dedicated journals as platforms for exchange on evaluation (Jacob et al., 2015), including the *American Journal of Evaluation*, its European counterpart, *Evaluation*, or the *Evaluation Journal of Australasia*. These journals have carried individual pieces on environmental evaluation and some journals, such as *New Directions for Evaluation*, have also published dedicated special issues, such as one on "Environmental Program and Policy Evaluation: Addressing Methodological Challenges" (Birnbaum and Mickwitz, 2009; see also Stephenson et al., 2019; Straßheim and Schwab, 2020). Furthermore, publications such as the *International Atlas of Evaluation* (Furubo et al., 2002), the *Encyclopedia of Evaluation* (Mathison, 2004), or *The SAGE Handbook of Evaluation* (Shaw et al., 2006) bring together a range of evaluation-related topics (see also van Voorst et al., 2023).

The growth of evaluation activities has also become evident through catalogues of evaluation studies (for a concrete example of evaluations in the context of the German Climate Change Adaptation Strategy, see Box 11.1). While more comprehensive, library-like systems do not (yet) exist in the area of evaluation, scholars and practitioners have begun to collect, categorise, and analyse evaluations in certain sub-systems. For example, Mastenbroek et al. (2016) found 2016 *ex post* legislative evaluations that the European Commission (the EU's executive body) produced between 2000 and 2014. In the area of the environment, Huitema et al. (2011) found 259 climate policy evaluations in the EU between 1998 and 2007. Likewise, Schoenefeld (2023) identified 618 climate policy evaluations at the EU level, in the United Kingdom, and Germany published between 1997 and 2014, while Sandin et al. (2019) analysed 33 evaluations focusing on energy efficiency policy in Sweden, and Fujiwara et al. (2019) identified 236 climate policy evaluations at the EU level and in six of its Member States. Practitioners have also begun to bring together evaluations; for instance, the European Environment Agency has recently published the "EEA catalogue of European environment and climate policy evaluations", which contains nearly 600 evaluations and is also available as a searchable online database (Ramboll, 2020).[10]

Most analyses of existing evaluation studies in the area of the environment and climate change do however highlight shortcomings, suggesting that existing evaluations tend to be restrictive in their foci, criteria, or methods and that they often do not capture the full thrust of policy effects (see Mickwitz, 2021; Schoenefeld, 2023; Fujiwara et al., 2019; Sandin et al., 2019). For instance, Uyl and Driessen (2015) argue that restricted evaluation criteria can only unpack parts of the necessary factors for sustainable development; drawing on various literatures, they thus propose a new framework that assesses "equity, democracy, legitimacy, the

Box 11.1 Evaluating Adaptation Policy

The severe floods in Europe and Asia as well as heat waves in northern countries and large-scale forest fires in the Western United States and Australia between 2019 and 2021 visibly demonstrated the urgent need to adapt to a warming climate. To pick one example, adaptation to climate change had already been on the agenda of German policy-makers for some time. In December 2008, the German Federal Government agreed on the German Adaptation Strategy.[11] The strategy set out to create an action plan, which includes steps to engage in dialogue with a range of societal actors, increase the (scientific) knowledge base on adaptation, and put forward concrete adaptation policies and measures. Notably, it contains a concrete requirement to regularly monitor and evaluate its implementation (i.e., an evaluation clause).

The evaluation of the strategy has various components. In late 2015 the federal government published a progress report and presented a second, revised adaptation action plan. In late 2020 a second progress report and a third revised action plan for adaptation followed. In addition, there were two evaluations, one by the German Federal Environment Agency (UBA) in 2015[12] and another evaluation in 2019.[13] Even though the most recent evaluation was commissioned by a governmental agency, the UBA stresses the independence of their external evaluation consultants, who conducted the evaluation.[14] The evaluation builds on a methodology that had previously been developed with external stakeholders and an inter-ministerial working group. It draws on program theory, which generated five central evaluation questions. It uses multiple methods (document analysis, interviews, a survey, and indicator analysis). While the participating institutions generally saw the creation of central documents as part of the adaptation strategy as positive, challenges emerged when involving non-organised citizens, who were harder to reach in participatory processes. The evaluation identified a perceived need for more concrete targets, but a new adaptation strategy was not viewed as necessary. Gaps were for example identified at the municipal level, which requires more resources from the federal and the state level to adapt. The evaluation then presents a range of recommendations for the future implementation of the strategy. It recommends, for example, increasing personnel resources for adaptation at the federal level. Taken together, the evaluation(s) of the German Adaptation Strategy contain(s) many of the elements and approaches that have been recommended by environmental evaluation scholars.

handling of scale issues and the handling of uncertainty issues" in order to better assess policy making in the Dutch fen landscape, which comprises peat lowlands with extensive usage for grazing and other activities (Uyl and Driessen, 2015, 186). Taken together, while the evaluation of environment and climate change policy has clearly become an established feature, ample room remains for future improvements and research.

11.2.2 Key Evaluation Approaches

A challenging feature of environmental policy-making is that neither the physical-ecological systems, nor the human interactions within them, are fully understood (Mickwitz, 2003). This means that policy evaluation constantly runs the risk of being incomplete by missing potentially important factors or aspects in the assessment. Responding to this challenge, Mickwitz (2003, 2021) proposes expanding the long-standing evaluation focus on the criterion of effectiveness with a view to policy goals to also assess side effects, to work with intervention theories (i.e., a set of axioms about the particular functioning of an environmental policy), to use multiple evaluation criteria (including, e.g., effectiveness, relevance, transparency, and legitimacy), to apply triangulation (i.e., multiple methods, data sources, analysis, and theories), and to widen participation in evaluation to numerous actors.

In combination, a broader reach of evaluation increases the probability of detecting relevant policy effects and outcomes that may otherwise fall through the cracks of more restricted evaluation approaches. However, it should also be recognised that such demands of evaluation and evaluators produce challenges, not only in terms of resources (i.e., financing wide-ranging evaluations), but also in bringing together highly skilled, interdisciplinary evaluation teams that are willing and able to conduct environmental policy evaluation to a high standard. Evaluations with greater complexity in turn increase the demand on evaluation users in terms of understanding and acting upon evaluation findings. This is especially the case because scholars have detected increasing trends towards "policy accumulation", meaning that new policies are added without removing older ones, increasing the overall stock (Adam et al., 2018). Policy accumulation adds another layer of complexity and challenge for evaluators, who must tease apart the multifarious interactions between different policy instruments in order to attribute them to concrete outcomes. To do so, Adam et al. (2018) recommend methodological triangulation, as well as a focus on causal mechanisms rather than just outputs and outcomes.

But even when executed to a very high standard, single policy evaluation (studies) may not suffice to build comprehensive knowledge on policy interventions. In the mid-2000s, evaluation scholars started arguing that the field of evaluation should pivot "from studies to streams" of evaluative knowledge (Rist and Stame, 2006). These arguments formed part of a broader debate, which addresses evaluation systems, including evaluation institutions and evaluation cultures that started to emerge as evaluation began to spread across countries and policy fields (Furubo et al., 2002; Jacob et al., 2015; Stern, 2009; Toulemonde, 2000). There are clear signs that evaluation systems and cultures have also emerged in the field of environmental policy.

11.3 Why Evaluate Environmental Policy?

The rationale for engaging with environmental policy evaluation ultimately returns to the question of evaluation use. Early in the debate, Weiss (1979) identified "many meanings of research utilization", or policy evaluation use. This chapter focuses on three prominent motivations for evaluating environmental policy-making, namely to improve learning, increase accountability, and to gain political advantages (Schoenefeld and Jordan, 2019). Different goals of evaluation may be complementary, but also potentially (at least in part) antagonistic (Schoenefeld and Jordan, 2019).

11.3.1 Evaluation for Learning

Many consider learning, or the "updating of beliefs about public policy" (Dunlop and Radaelli, 2018, 53), as the holy grail of evaluation use. While scholars have long argued that learning tends to be a complex and multi-faceted process (Bennett and Howlett, 1992; Dunlop and Radaelli, 2013), evaluation debates frequently assume that learning is a rational process where insights from evaluations feed directly into policy-making and thus drive improvement. It is a persistent and attractive belief that has seen countless repetitions by zealous politicians and policy-makers seeking to justify their evaluation activities and aiming to demonstrate that they work in a rationally informed way. Contrary to such overblown expectations, a more realistic perspective on the role of evaluation in environmental policy-making is that it is one (potentially important) input in to the policy process, but by no means the *only* one. Numerous other aspects, such as political interests and competition (see below), policy legacies and path dependence, or preferred solutions by policy-makers, also play a role in policy decisions and may even take precedence over evaluation findings. As Weiss (1993, 94–95) put it nearly three decades ago, "the programs with which the evaluator deals are not neutral, antiseptic, laboratory-type entities. They emerged from the rough and tumble of political support, opposition, and bargaining". These elements may still be present and impact on the evaluation itself and on the corresponding use of evaluation knowledge.

The view of evaluation as one input into the environmental policy-making process comes closer to the "enlightenment" function of evaluation that Weiss (1979) proposed and which others have subsequently developed in public policy contexts (Sabatier, 1988). Evaluators have recognised this challenge and begun to adjust their practices and approaches to increase the opportunities for learning through evaluation. One of the most prominent is that of "utilization-focused evaluation", which argues that evaluation should focus on "intended use by intended users" (Patton, 2008, 37). Doing so involves, for example, working with the intended users to generate buy-in to evaluate and address the potential fears of it, collaborating with participants to formulate evaluation questions and recognising that governments may also suffer from information overload (Patton, 2008). Likewise, practitioners have begun to better institutionalise evaluative practices in their work, for example by adding evaluation clauses (requirements) in legislation (Bussmann, 2005). The European Union (EU) has for instance implemented an "evaluate first" principle in its Better Regulation Guidelines (European Commission, 2017), meaning that policy revisions require a previous evaluation of existing policies; similarly, new policy proposals must first be subject to a published impact assessment (*ex ante* evaluation), before they can be adopted. Given that the EU is a relevant player in environmental policy-making, both internally and internationally (see Jordan and Gravey, 2021), this demonstrates how environmental policy evaluation has been institutionalised in that particular context.

Empirical evidence on the learning function of evaluation in environmental policy-making remains scarce, but there are some notable exceptions. For example, Borrás and Højlund (2015) found that environmental policy-learning took place at the policy officer level in the European Commission, while Hildén (2011) detected a range of different types of learning, such as rhetorical and political learning, in the case of climate policy in Finland. In a similar vein, Uitto (2014) presents two evaluations from the nexus between international development and the environment, arguing that insights from evaluation informed discussions and decision-making within the UN Development Programme. In sum, some evaluation-based policy learning has been detected, but it tends to be the exception rather than the norm.

11.3.2 Evaluation for Accountability

Accountability is one of the key systemic attributes of modern governance in democratic systems, but it may also play a role in other governance systems. It may be understood as a "broader concept, in which accountability is seen as a personal or organisational virtue, and the narrower concept, in which accountability is defined as a social relation or mechanism" (Bovens, 2010, 948). As Bovens (2010) explains, accountability as a virtue relates to, for example, notions of "good governance", whereas a mechanism may relate to the relationship between two actors, such as a government and a parliament, where the former accounts for its actions to the latter. Evaluation may play a role with a view to both types of accountability as a perceived part of "good governance" and as a concrete mechanism that enables accountability (Stame, 2006).

This has also been highlighted in the case of environmental policy (Mickwitz, 2021, 242). Uitto (2014, 54) argues that there is a relationship between learning and accountability, because "learning from the past and improving future performance is key to achieving greater accountability". This idea chimes with Bundi (2018), who found that Swiss parliamentarians often used evaluation to control implementation or obtain information about a policy rather than using the findings for learning. However, in a study of the European Parliament, Zwaan et al. (2016) found that evaluations were used more for agenda-setting than holding the Commission to account. Very few empirical studies exist that explicitly deal with evaluation's role in enabling accountability for environmental policy (Schoenefeld and Jordan, 2019).

11.3.3 Evaluation and Politics

Aside from learning and good governance/accountability, policy evaluation – including in the area of the environment – is an inherently political act and may also be used for strategic political purposes. The term "evaluation" contains the concept of value, and societal values – that is, what we deem important and worthy – tend to be contested and require some level of agreement for an evaluation (Fischer, 2006). Evaluation inevitably prioritises some values over others, adding to its political dimension, given that some actors may have an interest in certain values but not others.

The political dimension of evaluation also emerges in a more instrumental way when political or societal actors use evaluation for the strategic purposes of governing and steering. For example, evaluations may be conducted to delay a process or avoid a decision, or its main aim may be to support a preconceived direction of policy-making or empower the views of certain actors over others (Schoenefeld and Jordan, 2019). Evaluation may also be done to emulate what other governments do in order to appear like a rational administration which bases its ideas on "science". These political purposes of evaluation may have little to do with the learning and accountability functions detailed above, and they are typically characterised as somewhat problematic (Bovens et al., 2006; Parkhurst, 2017). Yet, from an environmental policy perspective, this need not be so. If, for example, an evaluation delays an environmentally harmful project (such as President Obama's approach to delay the construction of the Keystone XL pipeline from Canada to Nebraska),[15] it may be viewed as a positive outcome from the perspective of protecting the environment and avoiding harm. See Box 11.2.

Box 11.2 Evaluation and the Paris Agreement

Environmental policy evaluation has been a growing practice around the world. The often-expressed hope is that evaluation will lead to better policies and thus a healthier, cleaner, and more stable environment and climate by providing knowledge that enables continuous policy development. Expressing this view, Articles 13 and 14 of the 2015 Paris Agreement for example prescribe that countries must track and evaluate their efforts to mitigate climate change, and a "global stock take" brings together this information. A specific focus on monitoring and evaluation is also contained in the specific adaptation provisions in Article 7 of the Agreement, which requires "all Parties to engage in assessments of impacts and vulnerability, the adoption of national adaptation plans, the determination of nationally prioritized actions, and the implementation of monitoring and evaluation of these actions" (Lesnikowski et al., 2017, 827). The underlying assumption of the Paris Agreement is thus that evaluative practices will foster sufficient collective action by generating transparency and related peer pressure to reach the overall aim of the Agreement of limiting global warming to no more than 2.0 degrees and ideally 1.5 degrees Celsius. This chapter however demonstrates that policy evaluation is more than a checking and transparency device and includes a range of features that may contribute to such aims, but also detract from them. Some of the political motivations for evaluation (Schoenefeld and Jordan, 2019) may for example generate very different consequences from those envisaged by the Paris Agreement.

11.4 Future Directions for Environmental Policy Evaluation

11.4.1 Evaluation Patterns and Knowledge Defragmentation

Learning and improvement requires knowledge, especially when uncertainty runs high as is the case in many environmental policy settings. Policy evaluation has often been praised as a source of precisely that knowledge, but extant debates suggest that only cumulative insights may generate the streams of knowledge (Rist and Stame, 2006) and indeed the "enlightenment" that prominent evaluation scholars have described as a realistic influence of evaluation on policy-making processes (Weiss, 1979; Sabatier, 1988). Doing so requires an infrastructure to systematically collect and ultimately analyse environmental policy evaluations for their joint insights (but also potential differences in their findings). The idea is that evaluations better connect to earlier arguments and contribute to an ever-growing web of evaluative policy knowledge. Cumulative and especially international evaluation databases do not (yet) exist, but are a key requisite for the advancement of environmental policy knowledge.

11.4.2 Evaluation and Policy-Making

Once the overall patterns of the existing environmental policy evaluation landscape become clearer, two questions become central. The first question concerns how to better integrate evaluation knowledge into policy-making. While the heuristic of a simple policy cycle (from agenda-setting and policy-making through to implementation and evaluation) may hold some value in explaining policy-making to novices, decades-long research and modelling of policy processes have highlighted the severe limits of the heuristic (Sabatier and Weible, 2014). Alternative conceptualisations such as the multiple stream or the garbage-can models depict messier and not at all stage-like policy processes (see e.g. Kingdon and Stano, 1984). Thinking seriously about such models generates demanding and by-and-large unanswered questions on policy evaluation – namely what role it currently plays and could in future play in these "messier" policy environments. One approach is to see evaluation as one important ingredient of public and expert debate on environmental policies, and also a platform to engage in exchanges about different criteria that underwrite assessments of policy success and failure (Mickwitz, 2021). By pondering such questions, scholars and practitioners may over time offer new insights about the role of evaluative knowledge in environmental policy-making.

11.4.3 The Governance of Evaluation

When considering the contribution of evaluation to policy-making systems, the governance of evaluation itself comes into focus. Schoenefeld and Jordan (2017) have highlighted with reference to climate policy that there are in principle multiple options to govern evaluation, ranging from hierarchical ones (i.e., a single or very few evaluation actors, such as a central evaluation agency) to more polycentric, bottom-up evaluation where multiple actors and institutions conduct and support evaluations (see also Schoenefeld, 2023). At the same time, evaluators (and potentially their funders) could be either state-driven (e.g., governments, parliaments, or courts) or they could be society-driven (e.g., non-governmental organisations and business associations). Each model comes with a set of unique advantages and disadvantages (e.g., governmental evaluation may increase the use of evaluation findings, but also potentially lack independence, while non-governmental evaluation may have greater independence, but may also lack crucial access to policy and policy-makers). Such governance questions have been little addressed in the field of (environmental policy) evaluation (Schoenefeld and Jordan, 2017) and merit greater attention if the practice is to evolve. Likewise, evaluation policy within single organisations or institutions (Stern, 2009; Al Hudib and Cousins, 2019) has drawn interest and is especially important for big environmental policy actors such as the European Commission or national ministries (Eckhard and Jankauskas, 2019), which now see themselves confronted with mounting societal demands for fast and effective environment and climate policy on an unprecedented scale (Schoenefeld, 2021; Schoenefeld et al., 2021).

11.4.4 Linking Different Evaluation Practices

The fourth challenge involves better linking of different evaluative practices, such as impact assessment, policy monitoring, and policy evaluation. Focusing on the different temporal aspects of evaluation (e.g., impact assessment) and on different general approach (monitoring means regularly keeping track of policies against defined indicators), bringing together different evaluative elements, may help to gain a better understanding of policy development, but also of the methodological aspects. For example, a combination of impact assessments and *ex post* evaluations could help assess the quality of policy predictions and potentially the assumptions that drive them. By the same token, impact assessment can serve as an anchor for *ex post* policy evaluation and as an appreciation of the circumstances under which the policy first emerged. Particularly in the EU, there has been a growing debate on environmental policy monitoring (Tosun, 2012; Bürgin, 2020). The EU's efforts to monitor its climate policies have been subject to multiple inquiries (Hyvarinen, 1999; Hildén et al., 2014; Schoenefeld and Jordan, 2020; Schoenefeld et al., 2019; Schoenefeld et al., 2023; Schoenefeld et al., 2021), but their integration with the considerable climate policy evaluation activities discussed above (Schoenefeld, 2023; Huitema et al., 2011; Fujiwara et al., 2019; Ramboll, 2020) has not yet been addressed sufficiently and is an area for future work by practitioners and scholars alike (see also Rossi et al., 2018). A notable exception is the European Environment Agency (EEA)'s work on combining different data sources in the evaluation of energy efficiency policy, which shows that doing so can address knowledge gaps (European Environment Agency, 2018).

11.4.5 Evaluation Standards and Guidelines

The fifth aspect for future work concerns the formulation of environmental policy evaluation standards or guidelines. While individual evaluations will always remain unique given specific policy contexts and circumstances (see Rog, 2012), a common standards-supported baseline (perhaps formulated as guidelines rather than immovable standards) may help to ensure some minimum level of evaluation quality (Stake and Schwandt, 2006; Dahler-Larsen, 2019) and a level of comparability and complementarity that may facilitate the emergence of the cumulative knowledge discussed above. Evaluative standards have been subject to debate in evaluation communities for some time and some institutions have published evaluation standards, such as the European Commission (2007). This has also been a debate at the international level where "standards are seen as having several benefits, including reassuring citizens that they will not be harmed by evaluations, reassuring customers that they will receive a quality service and encouraging consistency and good practice within the evaluation community" (Stern, 2006, 309–311). Yet, there are also drawbacks, as the creation of an international evaluation market may lose focus of specific local conditions and circumstances (Stern, 2006) – aspects of high relevance for environmental policy.

11.4.6 Evaluating Global Sustainability

One source of fresh ideas for environmental policy evaluation with a global perspective emerges from the new initiative of "blue marble evaluation" for a sustainability transition, a global systems evaluation perspective that aims to transcend sectoral boundaries and think more holistically to assess local–global interactions (Patton, 2019; Patton, 2020). Specifically, Patton (2020) proposes six key criteria for evaluation, namely (1) transformation fidelity (i.e., are the policy initiatives enough to generate a transformation); (2) complex systems framing (the focus should be on systems); (3) eco-efficiency full-cost accounting (including economic, social, and environmental costs); (4) adaptive sustainability (assess resilience in socio-ecological systems); (5) diversity/equity/inclusion (how the transformation expresses these values); (6) interconnectedness (how interconnections between people, networks, and institutions drive the transformation and how to strengthen them). In a similar vein, Mickwitz et al. (2021) also argue that existing evaluation frameworks and practices are insufficient to support large-scale sustainability transitions. They offer an even broader framework to analyse the extent to which existing evaluations already possess helpful characteristics and where the gaps are. To do so, their framework centres on (1) the context of evaluation; (2) the evaluation focus; (3) evaluation design, methods, and data; (4) evaluation criteria; and (5) how the evaluations encourage use (Mickwitz et al., 2021). These emerging approaches offer new perspectives to reform evaluation standards and, indeed, evaluation practice more generally with a view to supporting the much-needed transition to sustainability.

11.5 Conclusion and Future Research

What has become clear over time is that environmental policy generates particular challenges for evaluation on account of the complexity of socio-environmental systems and the potentially multiple and unforeseen effects that policy interventions may generate. Many existing environment and climate policy evaluations have been shown to be too limited and too unsophisticated to meet the demands of generating comprehensive insights. Furthermore, recently developing debates showcase the need for deeper reform in policy evaluation in order to support and drive (global) sustainability transitions in order to protect the planet's basic life support for humanity and many other forms of life (Mickwitz et al., 2021; Patton, 2019; Patton 2020).

This chapter has demonstrated that different evaluation actors may pursue different ends, even when it comes to the same evaluation. For example, while a government may wish to learn through an evaluation, parliamentarians may use it for accountability. By the same token, an interest group could use the evaluation in order to influence the public debate or set the political agenda. It is therefore important to recognise that the "many uses" that Weiss (1979) identified may emerge in the context of a single environmental policy evaluation and in many cases across them. Such uses have a bearing on the role of evaluation in policy-making. Environmental policy evaluation has come some way since it first emerged in the middle of the 20th century, but the climb will remain steep and challenging for the foreseeable future.

Acknowledgements

The author is grateful to Andrew Jordan for comments on an earlier version of this chapter and to the editors of this volume for their constructive feedback and support throughout the writing process. The preparation of this chapter was supported by the Fritz Thyssen Foundation, Grant/Award Number: 10.19.1.024PO.

Notes

1 See, for example
https://www.euronews.com/2021/07/16/why-is-the-eu-s-new-emissions-trading-system-so-controversial
https://www.theguardian.com/environment/2011/apr/28/overhaul-europe-carbon-trading-scheme
https://www.nytimes.com/2013/02/21/business/energy-environment/21iht-green21.html?searchResultPosition=7
2 For a list of national evaluation associations, see the website of the International Organization for Cooperation in Evaluation: https://ioce.net/national-organizations
3 https://www.eval.org/
4 http://evaluationjp.org/english/index.html
5 https://europeanevaluation.org/
6 https://www.degeval.org/en/home/
7 https://www.degeval.org/en/working-groups/
8 http://www.environmentalevaluators.net/
9 https://www.syke.fi/en-US/Current/Events/EEEN2020_European_Environmental_Evaluato(55530)
10 http://poleval-catalogue.apps.eea.europa.eu
11 https://www.bmu.de/fileadmin/bmu-import/files/pdfs/allgemein/application/pdf/das_gesamt_bf.pdf
12 https://www.umweltbundesamt.de/sites/default/files/medien/378/publikationen/climate_change_13_2015_evaluierung_der_das.pdf
13 https://www.umweltbundesamt.de/sites/default/files/medien/1410/publikationen/politikanalyse_zur_evaluation_der_deutschen_anpassungsstrategie_an_den_klimawandel_das_-_evaluationsbericht.pdf
14 https://www.umweltbundesamt.de/publikationen/politikanalyse-zur-evaluation-der-deutschen
15 *The New York Times* reported the political reasons for the delay, moving the issue until after the presidential election in 2012: https://www.nytimes.com/2011/11/11/us/politics/administration-to-delay-pipeline-decision-past-12-election.html. The official justification was that the State Department did not have sufficient information in order to make a decision on the pipeline and its potential route: https://ecommons.cornell.edu/xmlui/bitstream/handle/1813/78018/CRS_Keystone_XL_Pipeline_Project_Key_issues_1213.pdf?sequence=1

References

Adam, Christian, Steinebach, Yves, Knill, Christoph (2018). Neglected challenges to evidence-based policy-making: The problem of policy accumulation. *Policy Sciences* 51 (3), 269–290.

Al Hudib, Hind, Cousins, J. Bradley (2019). Understanding evaluation policy and organizational capacity for evaluation: An interview study. *American Journal of Evaluation* 43 (2), 234–254.

Bennett, Colin J., Howlett, Michael (1992). The lessons of learning: Reconciling theories of policy learning and policy change. *Policy Sciences* 25 (3), 275–294.

Birnbaum, Matthew, Mickwitz, Per (2009). *Environmental program and policy evaluation: Addressing methodological challenges: new directions for evalution*, Number 122. Jossey-Bass.

Borrás, Susana, Højlund, Steven (2015). Evaluation and policy learning: The learners' perspective. *European Journal of Political Research* 54 (1), 99–120.

Bovens, Mark (2010). Two concepts of accountability: Accountability as a virtue and as a mechanism. *West European Politics* 33 (5), 946–967.

Bovens, Mark, Hart, Paul't, Kuipers, Sanneke (2006). The politics of policy evaluation. In: M. Moran, M. Rein, R. E. Goodin (Eds.). *The Oxford Handbook of Public Policy*. Oxford, Oxford University Press, 319–335.

Bundi, Pirmin (2018). Parliamentarians' strategies for policy evaluations. *Evaluation and Program Planning* 69, 130–138.

Bürgin, Alexander (2020). Modernization of environmental reporting as a tool to improve the European commission's regulatory monitoring capacity. *Journal of Common Market Studies* 59 (2), 354–370.

Bussmann, Werner (2005). Typen und Terminologie von Evaluationsklauseln. *LeGes: Gesetzgebung & Evaluation* 16 (1), 97–102.

Crabbe, Ann, Leroy, Pieter (2012). *The Handbook of Environmental Policy Evaluation*. Routledge.

Dahler-Larsen, Peter (2019). *Quality: From Plato to Performance*. Springer.

Dunlop, Claire A., Radaelli, Claudio M. (2013). Systematising policy learning: From monolith to dimensions. *Political Studies* 61 (3), 599–619.

Dunlop, Claire A., Radaelli, Claudio M. (2018). Does policy learning meet the standards of an analytical framework of the policy process? *Policy Studies Journal* 46, S48–S68.

Eckhard, Steffen, Jankauskas, Vytautas (2019). The politics of evaluation in international organizations: A comparative study of stakeholder influence potential. *Evaluation* 25 (1), 62–79.

European Commission (2007). Responding to strategic needs: Reinforcing the use of evaluation. Available online at https://ec.europa.eu/transparency/documents-register/api/files/SEC(2007)213_0/de00000000727670?rendition=false (accessed 7/28/2021).

European Commission (2017). *Better Regulation Guidelines. European Union*. Brussels. SWD (2017) 350. Available online at https://ec.europa.eu/info/sites/default/files/better-regulation-guidelines.pdf (accessed 7/27/2021).

European Environment Agency (2016). Environment and Climate Policy Evaluation. *European Union*. Luxembourg. 18/2016. Available online at https://www.eea.europa.eu/publications/environment-and-climate-policy-evaluation (accessed 7/27/2021).

European Environment Agency (2018). Using Member States' Information on Policies and Measures to Support Policymaking: Energy Efficiency in Buildings. *European Union*. Available online at https://www.eea.europa.eu/publications/using-member-states-information-on (accessed 7/29/2021).

Fischer, Frank (2006). *Evaluating Public Policy*. Mason, Cengage Learning.

Fujiwara, Noriko, van Asselt, Harro, Böβner, Stefan, Voigt, Sebastian, Spyridaki, Niki-Artemis, Flamos, Alexandros, Alberola, Emilie, Williges, Keith, Türk, Andreas, Donkelaar, Michael Ten (2019). The practice of climate change policy evaluations in the European Union and its member states: results from a meta-analysis. *Sustainable Earth* 2 (1), 1–16.

Furubo, Jan-Eric, Rist, Ray C., Sandahl, Rolf (2002). *International Atlas of Evaluation*. Transaction Publishers.

Hanberger, Anders (2013). Framework for exploring the interplay of governance and evaluation. *Scandinavian Journal of Public Administration* 16 (3), 9–27.

Hildén, Mikael (2011). The evolution of climate policies–the role of learning and evaluations. *Journal of Cleaner Production* 19 (16), 1798–1811.

Hildén, Mikael, Jordan, Andrew J., Rayner, Tim (2014). Climate policy innovation: Developing an evaluation perspective. *Environmental Politics* 23 (5), 884–905.

Huitema, Dave, Jordan, Andrew, Massey, Eric, Rayner, Tim, van Asselt, Harro, Haug, Constanze, Hildingsson, Roger, Monni, Suvi, Stripple, Johannes (2011). The evaluation of climate policy: Theory and emerging practice in Europe. *Policy Sciences* 44 (2), 179–198.

Hulme, Mike (2009). *Why We Disagree About Climate Change: Understanding Controversy, Inaction and Opportunity*. Cambridge University Press.

Hyvarinen, Joy (1999). The European community's monitoring mechanism for CO2 and other greenhouse gases: The Kyoto Protocol and other recent developments. *Review of European, Comparative & International Environmental Law*, 8 (2), 191–197.

Jacob, Steve, Speer, Sandra, Furubo, Jan-Eric (2015). The institutionalization of evaluation matters: Updating the International Atlas of Evaluation 10 years later. *Evaluation* 21 (1), 6–31.

Jordan, Andrew J., Gravey, Viviane (Eds.) (2021). *Environmental Policy in the EU. Actors, Institutions and Processes*. 4th ed. Rourtledge.

Kingdon, John W., Stano, Eric (1984). *Agendas, Alternatives, and Public Policies*. Little, Brown Boston.

Knaap, Gerrit J., Tschangho, J. K. (Eds.) (1998). *Environmental Program Evaluation: A Primer*. University of Illinois Press.

Konidari, Popi, Mavrakis, Dimitrios (2007). A multi-criteria evaluation method for climate change mitigation policy instruments. *Energy Policy* 35 (12), 6235–6257.

Lesnikowski, Alexandra, Ford, James, Biesbroek, Robbert, Berrang-Ford, Lea, Maillet, Michelle, Araos, Malcolm, Austin, Stephanie E. (2017). What does the Paris Agreement mean for adaptation? *Climate Policy* 17 (7), 825–831.

Mastenbroek, Ellen, van Voorst, Stijn, Meuwese, Anne (2016). Closing the regulatory cycle? A meta evaluation of ex-post legislative evaluations by the European Commission. *Journal of European Public Policy* 23 (9), 1329–1348.

Mathison, Sandra (2004). *Encyclopedia of Evaluation*. Sage Publications.

Mickwitz, Per (2003). A framework for evaluating environmental policy instruments: context and key concepts. *Evaluation* 9 (4), 415–436.

Mickwitz, Per (2021). Policy evaluation. In: Andrew J. Jordan, Viviane Gravey (Eds.). *Environmental Policy in the EU. Actors, Institutions and Processes*. 4th ed. Oxon, UK, Rourtledge, 241–258.

Mickwitz, Per, Neij, Lena, Johansson, Maria, Benner, Mats, Sandin, Sofie (2021). A theory-based approach to evaluations intended to inform transitions toward sustainability. *Evaluation* 23 (3), 281–306, https://doi.org/10.1177/1356389021997855

Parkhurst, Justin (2017). *The Politics of Evidence: From Evidence-Based Policy to the Good Governance of Evidence*. Taylor & Francis.

Patton, Michael Quinn (2008). Utilization-Focused Evaluation. Sage publications.

Patton, Michael Quinn (2019). Transformation to global sustainability: Implications for evaluation and evaluators. *New Directions for Evaluation* 2019 (162), 103–117.

Patton, Michael Quinn (2020). Evaluation criteria for evaluating transformation: Implications for the coronavirus pandemic and the global climate emergency. *American Journal of Evaluation*, 42 (1), 53–89. https://doi.org/10.1177/1098214020933689

Paul, Regine (2020). Analyse and rule? A conceptual framework for explaining the variable appeals of ex-ante evaluation in policymaking/Ein integrierter Erklärungsansatz für die variable Hinwendung zu ex-ante Evaluierungstools in der Politikgestaltung. *dms–der moderne staat–Zeitschrift für Public Policy, Recht und Management* 13 (1), 124–142.

Ramboll (2020). Update of the European Environment Agency catalogue of available evaluations of European environment and climate policies. Available online at https://ramboll.com/-/media/files/rgr/documents/media/news/eea-evaluation-catalogue.pdf?la=en (accessed 7/28/2021).

Rist, Ray C., Stame, Nicoletta (2006). *From Studies to Streams: Managing Evaluative Systems*. Transaction Publishers.

Rog, Debra J. (2012). When background becomes foreground: Toward context-sensitive evaluation practice. *New Directions for Evaluation* 2012 (135), 25–40.

Rossi, Peter H., Lipsey, Mark W., Henry, Gary T. (2018). *Evaluation: A Systematic Approach*. Sage publications.

Sabatier, Paul A. (1988). An advocacy coalition framework of policy change and the role of policy-oriented learning therein. *Policy Sciences* 21 (2), 129–168.

Sabatier, Paul A., Weible, Christopher M. (Eds.) (2014). *Theories of the Policy Process*. Westview Press.

Sanderson, Ian (2002). Making sense of 'What Works': Evidence based policy making as instrumental rationality? *Public Policy and Administration* 17 (3), 61–75.

Sandin, Sofie, Neij, Lena, Mickwitz, Per (2019). Transition governance for energy efficiency-insights from a systematic review of Swedish policy evaluation practices. *Energy, Sustainability and Society* 9 (1), 1–18.

Schoenefeld, Jonas J. (2021). The European green deal: What prospects for governing climate change with policy monitoring? *Politics and Governance* 9 (3). https://doi.org/10.17645/pag.v9i3.4306.

Schoenefeld, Jonas J. (2023). *The Evaluation of Polycentric Climate Governance*. Cambridge University Press.

Schoenefeld, Jonas J.,Hildén, Mikael,Jordan, Andrew J. (2018). The challenges of monitoring national climate policy: Learning lessons from the EU. *Climate Policy* 18 (1), 118–128.

Schoenefeld, Jonas J., Jordan, Andrew (2017). Governing policy evaluation? Towards a new typology. *Evaluation* 23 (3), 274–293.
Schoenefeld, Jonas J., Jordan, Andrew J. (2019). Environmental policy evaluation in the EU: Between learning, accountability, and political opportunities? *Environmental Politics* 28 (2), 365–384.
Schoenefeld, Jonas J., Jordan, Andrew J. (2020). Towards harder soft governance? Monitoring climate policy in the EU. *Journal of Environmental Policy & Planning*, 22 (6), 774–786.
Schoenefeld, Jonas J., Schulze, Kai, Hildén, Mikael, Jordan, Andrew J. (2019). Policy monitoring in the EU: The impact of institutions, implementation, and quality. *Politische Vierteljahresschrift* 60 (4), 719–741.
Schoenefeld, Jonas J., Schulze, Kai, Hildén, Mikael, Jordan, Andrew J. (2021). The challenging paths to net-zero emissions: Insights from the monitoring of national policy mixes. *The International Spectator* 56 (3), 24–40.
Scriven, Michael (1991). *Evaluation Thesaurus*. Sage.
Shaw, Ian, Ronald, Greene, Jennifer C., Mark, Melvin M. (Eds.) (2006). *The SAGE Handbook of Evaluation*. Sage.
Stake, Robert E., Schwandt, Thomas A. (2006). On discerning quality in evaluation. In: Ian Shaw, Jennifer C. Greene et al. (Eds.). *The Sage Handbook of Evaluation*. Sage, 404–418.
Stame, Nicoletta (2006). Governance, democracy and evaluation. *Evaluation* 12 (1), 7–16.
Stephenson, Paul J., Schoenefeld, Jonas J., Leeuw, Frans L. (2019). The politicisation of evaluation: constructing and contesting EU policy performance. *Politische Vierteljahresschrift* 60 (4), 663–679.
Stern, Elliot (2006). Contextual challenges for evaluation practice. In: Ian Shaw, Ian Graham Ronald Shaw, Jennifer C. Greene et al. (Eds.). *The Sage Handbook of Evaluation*. Sage, 292–314.
Stern, Elliot (2009). Evaluation policy in the European Union and its institutions. *New Directions for Evaluation* 2009 (123), 67–85.
Straßheim, Holger, Schwab, Oliver (2020). Politikevaluation und Evaluationspolitik/The evaluation of politics and the politics of evaluation. *dms–der moderne staat–Zeitschrift für Public Policy, Recht und Management* 13 (1), 3–4.
Tosun, Jale (2012). Environmental monitoring and enforcement in Europe: A review of empirical research. *Environmental Policy and Governance* 22 (6), 437–448.
Toulemonde, Jacques (2000). Evaluation culture(s) in Europe: Differences and convergence between national practices. *Vierteljahrshefte zur Wirtschaftsforschung* 69 (3), 350–357.
Turnpenny, John, Radaelli, Claudio M., Jordan, Andrew, Jacob, Klaus (2009). The policy and politics of policy appraisal: emerging trends and new directions. *Journal of European Public Policy* 16 (4), 640–653.
Uitto, Juha I. (2014). Evaluating environment and development: Lessons from international cooperation. *Evaluation* 20 (1), 44–57.
United States Environmental Protection Agency (2020). Program evaluation and performance measurement at the EPA. Available online at https://www.epa.gov/evaluate/program-evaluation-and-performance-measurement-epa (accessed 7/27/2021).
Uyl, Roos M. den, Driessen, Peter P. J. (2015). Evaluating governance for sustainable development–Insights from experiences in the Dutch fen landscape. *Journal of Environmental Management* 163, 186–203.
van Voorst, Stijn, Zwaan, Pieter, Schoenefeld, Jonas J. (2023). Policy evaluation: an evolving and expanding feature of EU governance. In: Jale Tosun, Paolo Graziano (Eds.). *Elgar Encyclopedia of European Union Public Policy*. Cheltenham: Edward Elgar, 605–613.
Vedung, Evert (1997). *Public Policy and Program Evaluation*. New Brunswick, Transaction Publishers.
Weiss, Carol H. (1979). The many meanings of research utilization. *Public Administration Review* 39 (5), 426–431.
Weiss, Carol H. (1993). Where politics and evaluation research meet. *Evaluation Practice* 14 (1), 93–106.
Zwaan, Pieter, van Voorst, Stijn, Mastenbroek, Ellen (2016). Ex post legislative evaluation in the European Union: questioning the usage of evaluations as instruments for accountability. *International Review of Administrative Sciences* 82 (4), 674–693.

12

INTERNATIONAL PUBLIC ADMINISTRATIONS IN ENVIRONMENTAL GOVERNANCE

Helge Jörgens, Nina Kolleck, Alexandra Goritz, Mareike Well, Johannes Schuster, and Barbara Saerbeck

12.1 Introduction

In a study on the role and influence of the UN climate secretariat, Busch (2009a) described this international bureaucracy as a passive actor with little autonomy and only very limited influence on international climate policy outputs. The author observed that the climate secretariat had "not generated new knowledge or contributed to the scientific understanding of climate change", nor had it "played an important role in keeping climate change on the agenda" (Busch, 2009a, p. 247). Overall, Busch concluded that "the influence of the climate secretariat ha[d] been limited" and considered the secretariat to be a "technocratic bureaucracy" without "any autonomous political influence" (Busch, 2009a, p. 251). Today, this assessment seems no longer accurate. In the run-up to and during the multilateral negotiations on the Paris Agreement, the climate secretariat has turned into an autonomous actor in its own right. As the secretariat's former deputy executive secretary Richard Kinley observed:

> in the case of the Paris Conference …, the secretariat, led by a very determined and visionary Executive Secretary, considered the conference's outcome to be so important to the future of humanity that a number of efforts were made to ensure not only that governments reached agreement, but that the agreement was as meaningful and ambitious as possible with a real-world impact.
>
> *(Kinley, 2021, p. 78)*

Pointing out the "remarkable collaboration between the secretariat and the French presidency" the author concludes that the secretariat was among "the most important contributors to the successful adoption of the Paris Agreement" (Kinley, 2021, p. 94). This example shows that international public administrations (IPAs) – that is, the bureaucratic bodies of international organizations – have become an important, yet still often underestimated, actor in environmental policymaking.

DOI: 10.4324/9781003043843-14

While the role of international regimes (Young and Levy, 1999; Miles et al., 2002), multilateral treaty systems (Gehring, 2012; Chasek and Wagner, 2012), and international organizations (O'Neill, 2014; Wirth, 2016; DeSombre, 2017) in environmental governance has been extensively studied, the secretariats and bureaucratic bodies of these treaties or organizations have only recently begun to receive greater scholarly attention. In this chapter we zoom in on IPAs as a distinct class of actors in environmental policy, based on a review of the emerging literature on international environmental bureaucracies. In Section 12.2 we define IPAs as partially autonomous actors. Section 12.3 gives examples of IPA influence in environmental governance, while Section 12.4 identifies the underlying causal mechanisms of their influence. Section 12.5 lays out an agenda for the future study of international public administrations in environmental governance.

12.2 International Public Administrations as Partially Autonomous Actors in Global Governance

IPAs are distinct from international organizations. While international organizations are "associations of actors, typically states" (Martin and Simmons, 2012, p. 329), Biermann and Siebenhüner (2009b, p. 6) define international bureaucracies

> as agencies that have been set up by governments or other public actors with some degree of permanence and coherence and beyond formal direct control of single national governments (notwithstanding control by multilateral mechanisms through the collective of governments) and that act in the international arena to pursue a policy.

While the authors follow earlier characterizations of IPAs, such as Siotis's (1965, p. 178), who defined IPAs as "international bodies which have a distinct existence within a given system of multilateral diplomacy and which exercise administrative and/or executive functions, implicitly recognized or explicitly entrusted to them by the actors of the international system", their emphasis is more on the autonomy and actorness of these organizations. This focus on IPA autonomy is also taken up by Bauer et al. (2017a, p. 2) who describe international bureaucracies "as bodies with a certain degree of autonomy, staffed by professional and appointed civil servants who are responsible for specific tasks and who work together following the rules and norms" of a given international organization (see also Chapter 8 in this book). Thus, while international organizations are predominantly conceived of as institutions within which states and other actors interact (Keohane, 1984), IPAs, such as the secretariats of international organizations, are actors in their own right (Bauer et al., 2017a).

The distinction between international organizations on the one hand and their bureaucratic bodies or secretariats on the other is important for understanding the potential actorness of the latter (see also Chapter 8). However, this distinction is not always made explicit and is still far from omnipresent in the field of international relations. As Weinlich (2014, p. 39) puts it: "Most of the recent literature does not bother to make a distinction between international organisations and their bureaucracies. Often, scholars who are referring to international organisations as actors ... are actually, albeit rarely explicitly, referring to the respective bureaucracy". Similarly, Eckhard and Ege (2016, p. 967), in their systematic review on how international bureaucracies influence the policies of international organizations, conclude that only few studies "explicitly focus on the influence of IPAs as a dependent variable". Findings regarding the "bureaucratic footprint" in the policies of international organizations are "a side-product rather than the actual objective of most studies". In order to more systematically study to what extent and through which causal mechanisms international bureaucracies can shape international policy outputs, IPAs must be treated as actors in their own right, which are analytically distinct from the wider international organization or treaty system which they are a part of. Like other actors, IPAs act through international organizations. However, unlike those other actors, IPAs, at the same time, constitute a key institutional feature of these organizations. It is this dual role as a partially autonomous actor and at the same time as an integral part of the institutional architecture of international institutions that lends them considerable potential for influence. As Reinalda (2020, p. 3) puts it: "Secretariats are hierarchically organized organs of IGOs [intergovernmental organizations] that fulfil essential functions and are part of the power dynamics taking place within and around IGOs."

But understanding IPAs as potentially autonomous actors doesn't necessarily imply that they will actually influence processes of international environmental governance. Already in 1994, Sandford (1994, p. 19) argued that IPAs, especially international secretariats, invariably act in a servant-like fashion. Instead of trying to insert their own preferences into the policy outputs of their international organization or treaty system, Sandford argues that secretariats limit themselves to executing the political will of nation states in their issue area: "Underlying all secretariat activities is the notion of service. Secretariats exist to service the treaty parties." More recent research by Knill et al. (2018), however, shows that the servant-like IPA described by Sandford is just one among several possibilities. The authors develop a typology of administrative styles of IPAs and show empirically that the servant style, while still existent, is no longer the default behavior of international bureaucracies. IPAs may just as well adopt entrepreneurial or even advocacy-oriented administrative styles. While administrative styles may vary between IPAs, they may also vary across issue areas or phases of the policy cycle (Bayerlein et al., 2020; Knill et al., 2018; see also Well et al., 2020). This diversity of administrative styles indicates that – despite a lack of formal decision-making powers – IPAs often attempt to move beyond the role of passive servants in order to influence the processes and outputs of their respective international organizations or treaty systems. Against this backdrop, Trondal (2017, p. 36) sums up that "it has been shown that the task of IPAs has become increasingly that of active and independent policy-making institutions and less that of passive technical supply instruments for IGO plenary assemblies".

Most IPAs are "issue-specific" bureaucracies (Bauer, 2006, p. 28). Except for the secretariats of universal international organizations, such as the UN secretariat or the European Commission, their functions are usually limited to a specific policy domain or to a multilateral treaty. Within these issue areas, IPAs engage in activities "such as conducting studies, preparing draft decisions …, assisting states parties, and receiving reports on the implementation of commitments" (Churchill and Ulfstein, 2000, p. 627). Their tasks "typically range from generation and processing of data, information and knowledge over providing administrative, technical, legal and advisory support in intergovernmental negotiation processes to ensuring and monitoring compliance with multilateral decisions" (Busch, 2014, pp. 46–47). Consequently, their role and influence are closely linked to the specific problem structures and actor constellations of the policy domains in which they operate.

12.3 IPA Influence in Environmental Governance

Today there is little doubt that IPAs can have an autonomous influence on international environmental policy processes and outputs (Jörgens et al., forthcoming, see also Chapter 8). However, concrete examples of IPA influence are still relatively scarce. The main reason is the methodological challenges of observing the often-hidden activity of IPAs (Kolleck et al., 2023). In addition, it is often methodologically difficult to link the actions of IPAs to observed changes in the processes or outcomes of multilateral negotiations. The fact that IPAs either do not reveal their political preferences or pass them off as preferences of other actors makes it even more difficult to clearly identify IPA action (or the preferences of IPAs) as the cause of observed policy changes (Jörgens et al., 2016, pp. 981–983).

A first generation of studies, therefore, focused on qualitative in-depth case studies of different environmental secretariats. One prominent example has been the secretariat of the United Nations Convention on Biological Diversity (CBD). In a first systematic study of the biodiversity secretariat, Siebenhüner (2009, p. 272) observed that since 2000 it had been "entrusted with the drafting of decisions of the conference of the parties" and the meetings of the CBD's Subsidiary Body on Scientific, Technical and Technological Advice (SBSTTA). Siebenhüner finds that, "based on positive experiences over past years, governmental delegations consider the secretariat's position on topics under debate as neutral and accept it as a valuable input in negotiations". In more contested and polarized issue areas the secretariat drafts are usually changed or even completely redrafted by the national delegations. However, when it comes to more technical issues like the development of biodiversity indicators, the author observes that "the secretariat prepared texts that passed with minor amendments". A study by Jinnah (2011) shows that the biodiversity secretariat has played a key role in linking the topic of global biodiversity governance to the more dynamic UN climate policy regime. Jinnah observes that "the CBD Secretariat filtered, framed, and reiterated strategic representations of the biodiversity-climate change linkage in a way that aimed to shape how member states understand the biodiversity-climate interface" (Jinnah, 2011, p. 24).

A case study by Bauer (2009a) focused on the role and influence of the Secretariat of the United Nations Convention to Combat Desertification (UNCCD). According to Bauer, "the desertification secretariat was pivotal in the establishment" of a permanent subsidiary body for implementation in 2001, the Committee for the Review of the Implementation of the Convention. Its creation was pushed through against the interests of most UNCCD countries. This example shows how treaty secretariats can actively shape the architecture of those international institutions they are supposed to serve rather passively (Bauer, 2009a, p. 300). The desertification secretariat also played an active role in the creation of Regional Coordination Units. "Although these units are welcomed in affected regions, donor countries are wary of institutional duplication and question the necessity of such units" (Bauer, 2009a, p. 301). Again, this example shows how secretariats often transcend the passive role originally assigned to them and begin to actively shape "their" treaty or convention.

Focusing on the Environment Directorate of the Organization for Economic Co-operation and Development (OECD), Busch (2009b) shows that this international bureaucracy has developed or disseminated key terms and concepts in environmental governance, such as the Polluter Pays Principle (see also Bernstein, 2000) or the Pressure-State-Response Framework. In addition, the OECD has been a major driver of the cross-national diffusion of best practices in environmental policymaking (see also Jörgens, 2004; Busch and Jörgens, 2012). Through the publication of environmental performance reviews, the OECD Environment Directorate has created a constant dynamic of environmental policy innovation and diffusion across its member states (Long, 2000; OECD, 2001; on the impact of the OECD environmental performance review in New Zealand, see Bührs, 2003). Other studies pertaining to this first wave of research on international environmental bureaucracies focused on the World Bank Environment Department (Gutner, 2005; Nielson and Tierney, 2003), the secretariat of the United Nations Environment Program (UNEP) (Bauer, 2009b), the secretariat of the International Maritime

Organization (IMO) (Campe, 2009), and of course the aforementioned climate secretariat (Depledge, 2007; Busch, 2009a).

Building on this initial research, a second wave of case studies linked the study of IPAs in environmental governance to major research topics from a range of political science sub-disciplines such as international relations and international public administration. Examples are comparative studies on how different treaty secretariats deal with the institutional fragmentation of global governance (Jinnah, 2014), studies on the interplay of public and private governance at different levels of government (Chan et al., 2015; Dingwerth and Jörgens, 2015; Newell et al., 2012), and research on processes of delegation and agency in global environmental politics (Wagner and Mwangi, 2010). Focusing, for example, on institutional fragmentation, Jinnah (2012, p. 113) finds that "nearly all tools" used by the conferences of the parties of the CBD "to mandate overlap management activities can be traced back to one document produced by the Secretariat in 1995". This example shows that IPAs may bring about changes in the institutional design of international treaties, thereby establishing path dependencies which perpetuate individual instances of IPA influence over longer periods.

Another group of second-wave studies has attempted to overcome the aforementioned methodological challenge of identifying the policy preferences of international secretariats by moving beyond qualitative research designs. Applying methods of social network analysis (SNA) based on large-N surveys or Twitter data (Goritz et al., 2022b; Mederake et al., 2021; Saerbeck et al., 2020; Goritz et al., 2019; Jörgens et al., 2016), these studies have shown that treaty secretariats such as the climate or biodiversity secretariats occupy very central positions both in issue-specific cooperation networks and in issue-specific communication flows. For example, using Twitter data, Kolleck et al. (2017b) find that the United Nations Framework Convention on Climate Change (UNFCCC) secretariat occupies a central and thus potentially influential position within the communication networks focusing on climate change education during UNFCCC negotiations from 2009 to 2014. Saerbeck et al. (2020) corroborate this finding with data from an original large-N survey, showing that the climate secretariat was among the five most central organizations during the negotiations leading up to the Paris Agreement. More than other actors, it maintains strong links with a wide range of state and non-state actors, which allows it to act as a policy broker between different types of actors in global climate governance.

12.4 Causes of IPA Influence in Environmental Policymaking

What explains the policy impact of international public administrations which, according to their organizational goals and mandates, should remain politically neutral and refrain from trying to shape international environmental policy outputs? Based on the findings of a large collaborative research project, Bauer et al. (2017b, pp. 182–189) distinguish five sources of IPA influence. First, and contrary to the traditional view that conceives of IPAs as merely instrumental arrangements created to support intergovernmental cooperation, they argue that IPAs are inherently autonomous, more so than national ministerial bureaucracies (see also Bauer and Ege, 2017). Second, the authors find that IPAs do not necessarily take a passive and supportive stance vis-à-vis their political principals, but often act in a more entrepreneurial manner, for example, by using their autonomy to advocate for their own policy ideas and preferences (see also Jörgens et al., 2017; Knill et al., 2017). Third, IPAs rely more on expertise and information than on rules and formal powers. As the formal mandates and legal competencies of IPAs are rather restrictive when compared to national bureaucracies, their impact on global policy outputs relies more on the strategic use of expertise, ideas, and procedural knowledge. This cognitive and normative influence is facilitated by their often-central position in issue-specific information flows (nodality) (see also Busch and Liese, 2017). This observation is in line with previous research findings that environmental IPAs predominantly exert cognitive and normative influence, and only to a lesser degree executive influence (Biermann and Siebenhüner, 2009a). Fourth, IPAs are able to overcome budgetary restrictions, which states may use strategically to limit the autonomy of IPAs, by generating new sources of financing. Although IPAs are much more vulnerable to budgetary instability than national bureaucracies, they find ways of mobilizing "budgetary means from alternative sources in order to reduce their dependence on member state contributions" (Bauer et al., 2017b, p. 187; see also Patz and Goetz, 2017). Fifth, and finally, the authors find that IPAs are continuously shaping their organizational environment. They do so, for example, by setting up and forming structures of multilevel administration and by creating informal alliances with non-state actors at all levels of government. Within their domain-specific organizational environments and the respective issue-specific information flows IPAs then typically occupy a central position which in turn increases their potential to influence other actors in the network (see also Benz et al., 2017; Jörgens et al., 2016).

Focusing more narrowly on environmental IPAs, Wit et al. (2020) identify three general sources of IPA influence: the degree of organizational autonomy of these IPAs, their ability to deliver specific governance functions, and the way in which the complex multi-level and multi-actor structure of the international system enhances the capacity of IPAs to actively participate in processes of international environmental governance. In the following we will zoom in on two of these factors that are of particular relevance for IPA influence on environmental policy processes and outputs: autonomy and centrality in issue-specific actor networks.

12.4.1 IPA Autonomy

Verhoest et al. define autonomy as "the extent to which an agency can decide itself about matters that it considers important" (Verhoest et al., 2010, pp. 18–19; see also Chapter 8). With regard to international organizations and IPAs, Hawkins et al. (2006, p. 8) define autonomy as "the range of potential independent action available to an agent after the principal has established mechanisms of control". Thus, the autonomy of IPAs is mainly defined by the amount of discretion which the member states of an international organization or treaty system decide to grant their bureaucracy. Bauer and Ege (2016) refer to this as an IPA's *formal autonomy*.

But the initial delegation of a certain degree of autonomy to an IPA through formal mandates is not the only factor that determines the bureaucracy's range of maneuver. The way in which an IPA positions itself vis-à-vis its principals can also affect its autonomy. We refer to this autonomy resulting from an IPA's organizational strategies and development as its *organizational* autonomy. In international organizations and treaty systems principals are often less homogeneous than at the national level. Thus, IPAs are often confronted with multiple (Dehousse, 2008) or collective (Dijkstra, 2017, p. 603) principals. If an international organization or treaty system is characterized by multiple or collective principals, there is a potential for secretariats to team up with selected states with whom they share some interests against the interests of other states. Moreover, Vaubel (2006) points out that the multiple principal architectures at the international level result in longer and more complex chains of delegation. Multiple principals thus may strengthen a secretariat's organizational autonomy by leaving additional room for strategic circumvention of rules and supervision by principals. This, in turn, constitutes a potential precondition for secretariat influence beyond their formal mandate. In contrast, as Jönsson (1986, p. 44) points out, "hegemonic and polar issue structures, where issue-specific capabilities are concentrated in one or a few states, can be expected to allow less room for maneuver by international organizations than fragmented structures".

International secretariats may also attempt to increase their organizational autonomy by redefining or expanding their formal mandates (see, e.g., Barnett and Coleman, 2005). Hall (2016), for example, analyses how and to what extent the United Nations Development Program (UNDP) was successful in integrating climate adaptation into its mandate. She argues that UNDP administrators, rather than states, played a critical role in mandate expansion by deciding "whether and how to expand into a new issue-area and then lobby states to endorse this expansion" (Hall, 2016, p. 4). The study contributes to an emerging literature on how the leaderships of IPAs navigate financial, ideational, and normative opportunities to expand their bureaucracies' mandates. Michaelowa and Michaelowa (2017) argue that IPAs can also use external stimuli, such as an increase in funding, to change their mandates and expand their competencies within their respective treaty systems. The authors show that the increased revenue from the Clean Development Mechanism (CDM) within the UNFCCC both directly and indirectly strengthened the role of the climate secretariat. In particular, the study finds "that CDM staff was able to gain substantial influence over concrete policy decisions and even change the structure of relevant decision-making and consultation processes" (Michaelowa and Michaelowa, 2017, p. 247).

Another potential determinant of an IPA's organizational autonomy is salience or visibility. Already in 1974, Finkelstein (1974, p. 501) had observed that "institutional autonomy correlates with lack of salience to the powerful members". Accordingly, several studies find that international secretariats often try to maintain an image of neutrality. They do so by deliberately hiding their own policy preferences behind those of their international organization's or treaty system's member states or other actors. Thus, when IPAs attempt to influence multilateral negotiations, they often do so in an "invisible" "or behind the scenes" way (Bauer, 2006, p. 32; see also Well et al., 2020). Consequently, Mathiason (2007) refers to the political influence of international secretariats as "invisible governance". Jinnah (2014, p. 1) writes that "from the outside of an organization, office secretaries are nearly invisible". And Beach (2008, p. 220) cites an official of the General Secretariat of the Council of the European Union saying that "*le Secrétariat du Conseil n'existe pas*". But maintaining a low-key profile is not the only way in which IPAs can increase their organizational autonomy. IPA scholars increasingly observe that international secretariats step out from behind the scenes and put themselves in the spotlight of multilateral negotiations, side by side with their principals and a range of non-state and sub-state actors. A case in point is the secretariat of the UNFCCC. In 2009, Busch found that the climate secretariat was caught in a "straitjacket" of "formal and informal rules" imposed by the UNFCCC member states that "ruled out any proactive role or autonomous initiatives" and led to an "organizational culture that bars staff … from exercising any leadership vis-à-vis parties and from assuming a more independent role" (Busch, 2009a, p. 261). Today, this characterization no longer seems accurate as several scholars consider that the climate secretariat "has been able to loosen its straitjacket, demonstrating its capacity to be an autonomous actor in global climate policymaking" (Hickmann et al., 2021, p. 23). In reaction to the failure of a globally binding post-Kyoto agreement on climate change at COP 15 in 2009 in Copenhagen (Dimitrov, 2010), and confronted with a long-lasting stalemate among the formal negotiating parties, the climate secretariat no longer acts as a passive servant to the negotiating parties. Instead, it has increasingly turned its attention to other, non-party actors at different levels of government in order to gain leverage on the substance and processes of global climate governance. In particular, Hickmann et al. highlight the climate secretariat's proactive role in opening up the climate negotiations to non-state actors that are supportive of the secretariat's policy preferences through secretariat-led initiatives such as the "Lima-Paris Action Agenda" or the "Non-State Actor Zone for Climate Action" (Hickmann et al., 2021).

This changing role of international environmental treaty secretariats is reflected in new concepts of IPAs as orchestrators (Abbott et al., 2015; Abbott and Snidal, 2010; Hickmann et al., 2021) or as attention-seeking bureaucracies (Jörgens et al., 2017). For example, Bäckstrand and Kuyper (2017, p. 2) argue that "a crucial outcome of the Paris Agreement is that the UNFCCC has been consolidated as the central orchestrator of non-state actors and transnational initiatives in global climate governance". Jörgens et al. (2017) suggest that IPAs may attempt to strengthen their autonomy by deliberately attracting the attention of policymakers in order to feed their own issue-specific expertise and preferred policy recommendations into multilateral negotiations. Both concepts argue that the complex and dynamic institutional structure of multilateral agreements provides the organizations acting inside them with multiple options for strategic positioning (on the opportunity structure provided by environmental treaty systems, see Gehring, 2012). In these cases, the underlying logic of action of international bureaucracies shifts from "shirking" to "attention-seeking".

12.4.2 IPA Centrality in Issue-Specific Actor Networks

Another potential source of IPA influence in environmental governance is the central position IPAs often occupy in issue-specific policy networks. As Sandford (1994, p. 17) observed in 1994, "secretariats are the organizational glue that holds the actors and parts of a treaty system together". Jinnah (2012, p. 109) characterizes secretariats as "the operational hubs of [their] regimes". This centrality allows IPAs to interact with a wide range of actors and potentially take up a brokerage position between actors who normally don't interact directly with each other. Jönsson (1986, p. 45) refers to this as a "linking-pin position" and argues that "in order to assume an effective linking-pin position, an organization needs to have a location in the issue-specific network which allows it to reach, and to be reached by, other important organizational actors". In a similar vein, Fernandez and Gould (1994, p. 1460) argue that "organizational actors linking otherwise unconnected pairs of actors play a critical role in policy domains because they permit information to flow easily among a large and diverse set of actors, which in turn allows actors to coordinate their efforts to formulate and influence policies".

In the environmental domain, several studies have shown that treaty secretariats such as the climate or biodiversity secretariats occupy very central positions both in issue-specific communication flows and in issue-specific cooperation networks (Goritz et al., 2019; Jörgens et al., 2016; Kolleck et al., 2017b; Saerbeck et al., 2020; Well et al., 2020). This empirical focus on the centrality of IPAs in policy networks and how this centrality relates to the potential influence of IPAs on international environmental policy processes and outputs requires new methods. Against this backdrop, Jörgens et al. (2016) argue that SNA may provide a promising method for assessing the political influence of IPAs. Instead of relying on an actor's openly expressed policy preferences or on its reputation for being influential, SNA infers influence from the actor's relative position in issue-specific communication networks (Kolleck, 2016). However, descriptive techniques of SNA are only able to assess an actor's potential for becoming an influential actor in environmental governance. To study whether IPAs are actually willing and able to exploit this potential, inferential techniques of SNA as well as a combination of quantitative SNA with qualitative methods may result in more accurate accounts of secretariat influence and lead to a better/deeper understanding of the causal mechanisms through which this influence occurs. Kolleck et al. (2017b), for example, combine SNA with participant observation in their study on the role of the climate secretariat in promoting climate change education. Kolleck (2016), Kolleck et al. (2017a), and Goritz et al. (2021, 2022a) apply inferential techniques of SNA based on large datasets. Saerbeck et al. (2020) combine a survey-based SNA with insights from 33 semi-structured interviews to better understand whether and how the climate secretariat uses its brokerage position to shape issue-specific information flows.

The centrality and influence of international bureaucracies is not necessarily limited to individual issue areas. Often, they can be found to operate also at the boundary between two or more neighboring policy sub-domains or sub-systems. Based on her research on overlap management between international environmental regimes, Jinnah (2012, p. 108) argues that secretariats are able to "manage regime overlap more efficiently and effectively than other actors". "When it comes to coordination of daily, weekly, or even monthly activities between large numbers of actors across two or more international regimes, there is nobody better suited to manage the process than Secretariat staff" (Jinnah, 2012, p. 109). In a similar vein, Jönsson (1986, p. 42) suggests that at the international level "boundary-role occupants ... are typically found within the secretariat". International bureaucracies can thus be expected to occupy central positions at the intersection of different environmental issue areas. In fact, their centrality and potential for influence may turn out to be even greater if the focus is on networks of bureaucracies operating at different levels of government within a given policy domain rather than individual IPAs.

12.5 Conclusion and Future Research

In this chapter we have argued that international bureaucracies constitute a potentially influential type of actor in international environmental policy which has only recently received greater scholarly attention. From a policy perspective, research has focused mainly on the potential influence of IPAs on environmental policy processes and outputs as well as the underlying causal mechanisms. The main sources of IPA influence are:

- The formal and organizational autonomy of environmental IPAs;
- Their will and ability to move beyond a neutral stance and actively advocate for their own policy ideas and preferences;
- The strategic use of their unique expertise, normative ideas, and procedural knowledge;
- Their centrality in issue-specific policy networks and communication flows;
- Their ability to actively and deliberately shape their organizational environment and redefine their own position within these organizational fields.

After having advanced considerably in the past decade and a half, research on the role and influence of IPAs in environmental governance can now address new topics as well as conceptual and methodological challenges.

First, the observation that unelected bureaucracies beyond the nation state may actively seek to influence multilateral policies raises questions of democratic legitimacy. Focusing on two high-level orchestration efforts led by the climate secretariat, the Lima-Paris Action Agenda (LPAA) and the Non-state Actor Zone for Climate Action (NAZCA), Bäckstrand and Kuyper (2017) find that key elements of democratic governance, such as participation, deliberation, transparency, and accountability, are only weakly developed in both processes. However, the authors show that, overall, the LPAA fares better in this regard than the NAZCA. As a next step, a more comprehensive and systematic analysis could attempt to identify best practices in increasing the transparency, accountability, and participation of instances of IPA influence in environmental governance.

Second, and building on the growing literature on the interaction and networking of environmental bureaucracies at different levels of government, scholars have started to explore the possible emergence of global or transnational administrative spaces in environmental governance (Kolleck et al., 2023). Drawing on conceptual and empirical work on the European Administrative Space (Olsen, 2003; Trondal and Peters, 2015) and transnational administration (Stone and Ladi, 2015; Stone and Moloney, 2019), this emerging literature asks whether relatively stable, network-like structures, comprising and linking state and non-state actors at different levels of government, exist in different fields of environmental policy and to what extent these transnational administrative structures have an impact on the quality and effectiveness of environmental governance. In a second step, different administrative spaces could be compared with regard to parameters such as network density, the centrality of different types of actors, or the role of IPAs in establishing such structures as well as their impact on issue-specific information and communication flows within environmental administrative spaces.

Third, the study of IPAs is confronted with unique methodological challenges which require innovative methods and research designs. In particular, international bureaucracies still often seek to maintain a public image of neutrality regarding the policy outputs of their international organization or treaty system. This makes it difficult to identify the policy preferences of a given IPA, compare them to the preference of other actors, particularly nation states, and then assess to what extent IPA preferences have shaped concrete environmental policy outputs. Thus, most of the established methods to empirically infer the influence of political actors – the attributed influence method and the assessment of preference attainment (Betsill and Corell, 2008; Dür, 2008; Klüver, 2013) – are of limited use when focusing on international bureaucracies. New methods for assessing the influence of international bureaucracies that complement and go beyond the traditional combination of interviews and document analysis need to be developed. Mixed-method designs which combine quantitative and qualitative approaches and use longitudinal data are promising. For example, Michaelowa and Michaelowa (2017) combine longitudinal data on staff and budget growth with expert interviews, document analysis, and data obtained from CDM databases to infer changes in the climate secretariat's influence on the technical regulation of the CDM mechanism over time. Goritz et al. (2021) triangulate offline data from a large-N survey with online data from Twitter to examine to what extent they provide distinct theoretical and methodological insights into the role of IPAs in global governance. Goritz et al. (2022a) apply exponential random graph models (ERGMs) to data from an original large-N survey of participants of global climate negotiations to assess the climate secretariat's potential to influence global climate policy outputs. In another study on the climate secretariat, Saerbeck et al. (2020) first use SNA to examine the secretariat's relations with non-party and state stakeholders and to identify its position in the UNFCCC policy network. In a second step, the authors conducted 33 semi-structured interviews to corroborate the findings of SNA and to better understand the motivation of secretariat staff.

Finally, an increasing cross-fertilization of IPA research in different policy domains regarding concepts, research designs, methods, and findings can be observed (Knill and Steinebach, 2023). In a next step, research designs that systematically compare the role and influence of IPAs across different policy domains could be developed in order to relate IPA influence to varying problem structures and features of the policy domain.

Box 12.1 Chapter Summary

- IPAs, defined as the secretariats or bureaucratic bodies of international organizations, have become important, albeit often-neglected, actors in international environmental governance.
- To understand and study the potential actorness of IPAs, it is necessary to analytically distinguish them from international organizations or international institutions.
- There is consensus among scholars that IPAs are best conceptualized as (partially) autonomous actors with a potential to influence international environmental policy processes and outputs.
- Most IPAs are issue-specific bureaucracies whose role and influence are closely linked to the specific problem structures and actor constellations of the policy domains in which they operate.
- IPAs rely on their formal and organizational autonomy, their expertise and procedural knowledge, and their central position in issue-specific information flows to advocate for their own policy ideas and preferences. They make use of the complex multi-level and multi-actor structure of the international system to create support for their preferred policy options.
- Traditionally, IPAs have tried to stay out of the spotlight of global governance and maintain an image of neutrality and impartiality. Their influence is often hidden as they prefer to act "behind the scenes". However, more recently, secretariats have taken on a more proactive role, openly communicating their policy preferences to decision-makers and bringing non-state or sub-national actors that are supportive of the secretariat's policy preferences into multilateral environmental negotiations. This changing role of IPAs is reflected in the concepts of IPAs as orchestrators or attention-seeking bureaucracies.
- Research on IPAs in environmental governance has moved from case studies of individual bureaucracies to comprehensive mappings of their position and centrality in issue-specific policy networks, using the innovative methods of SNA.

References

Abbott, Kenneth W.; Genschel, Philipp; Snidal, Duncan; Zangl, Bernhard (Eds.) (2015): *International Organizations as Orchestrators*. Cambridge: Cambridge University Press.

Abbott, Kenneth W.; Snidal, Duncan (2010): International Regulation Without International Government: Improving IO Performance through Orchestration. *Review of International Organizations* 5 (3), pp. 315–344.

Bäckstrand, Karin; Kuyper, Jonathan W. (2017): The Democratic Legitimacy of Orchestration: The UNFCCC, Non-State Actors, and Transnational Climate Governance. *Environmental Politics* 26 (4), pp. 764–788.

Barnett, Michael; Coleman, Liv (2005): Designing Police: Interpol and the Study of Change in International Organizations. *International Studies Quarterly* 49 (4), pp. 593–620.

Bauer, Michael W.; Eckhard, Steffen; Ege, Jörn; Knill, Christoph (2017a): A Public Administration Perspective on International Organizations. In Michael W. Bauer, Christoph Knill, Steffen Eckhard (Eds.): *International Bureaucracy: Challenges and Lessons for Public Administration Research*. Basingstoke: Palgrave Macmillan, pp. 1–12.

Bauer, Michael W.; Ege, Jörn (2016): Bureaucratic Autonomy of International Organizations' Secretariats. *Journal of European Public Policy* 23 (7), pp. 1019–1037.

Bauer, Michael W.; Ege, Jörn (2017): A Matter of Will and Action: The Bureaucratic Autonomy of International Public Administrations. In Michael W. Bauer, Christoph Knill, Steffen Eckhard (Eds.): *International Bureaucracy: Challenges and Lessons for Public Administration Research*. Basingstoke: Palgrave Macmillan, pp. 13–41.

Bauer, Michael W.; Knill, Christoph; Eckhard, Steffen (2017b): International Public Administration: A New Type of Bureaucracy? Lessons and Challenges for Public Administration Research. In Michael W. Bauer, Christoph Knill, Steffen Eckhard (Eds.): *International Bureaucracy: Challenges and Lessons for Public Administration Research*. Basingstoke: Palgrave Macmillan, pp. 179–198.

Bauer, Steffen (2006): Does Bureaucracy Really Matter? The Authority of Intergovernmental Treaty Secretariats in Global Environmental Politics. *Global Environmental Politics* 6 (1), pp. 23–49.

Bauer, Steffen (2009a): The Desertification Secretariat: A Castle Made of Sand. In Frank Biermann, Bernd Siebenhüner (Eds.): *Managers of Global Change: The Influence of International Environmental Bureaucracies*. Cambridge, MA: MIT Press, pp. 293–317.

Bauer, Steffen (2009b): The Secretariat of the United Nations Environment Programme: Tangled Up in Blue. In Frank Biermann, Bernd Siebenhüner (Eds.): *Managers of Global Change: The Influence of International Environmental Bureaucracies*. Cambridge, MA: MIT Press, pp. 169–201.

Bayerlein, Louisa; Knill, Christoph; Steinebach, Yves (2020): *A Matter of Style: Organizational Agency in Global Public Policy*. Cambridge: Cambridge University Press.

Beach, Derek (2008): The Facilitator of Efficient Negotiations in the Council: The Impact of the Council Secretariat. In Daniel Naurin, Helen Wallace (Eds.): *Unveiling the Council of the European Union: Games Governments Play in Brussels*. Basingstoke: Palgrave Macmillan, pp. 219–237.

Benz, Arthur; Corcaci, Andreas; Doser, Jan Wolfgang (2017): Multilevel Administration in International and National Contexts. In Michael W. Bauer, Christoph Knill, Steffen Eckhard (Eds.): *International Bureaucracy: Challenges and Lessons for Public Administration Research*. Basingstoke: Palgrave Macmillan, pp. 151–178.

Bernstein, Steven (2000): Ideas Social Structure and the Compromise of Liberal Environmentalism. *European Journal of International Relations* 6 (4), pp. 464–512.

Betsill, Michele M.; Corell, Elisabeth (2008): Analytical Framework: Assessing the Influence of NGO Diplomats. In Michele M. Betsill, Elisabeth Corell (Eds.): *NGO Diplomacy. The Influence of Nongovernmental Organizations in International Environmental Negotiations*. Cambridge, MA: MIT Press, pp. 19–42.

Biermann, Frank; Siebenhüner, Bernd (Eds.) (2009a): *Managers of Global Change: The Influence of International Environmental Bureaucracies*. Cambridge, MA: MIT Press.

Biermann, Frank; Siebenhüner, Bernd (2009b): The Role and Relevance of International Bureaucracies: Setting the Stage. In Frank Biermann, Bernd Siebenhüner (Eds.): *Managers of Global Change: The Influence of International Environmental Bureaucracies*. Cambridge, MA: MIT Press, pp. 1–14.

Bührs, Ton (2003): From Diffusion to Defusion: The Roots and Effects of Environmental Innovation in New Zealand. *Environmental Politics* 12 (3), pp. 83–101.

Busch, Per-Olof (2009a): The Climate Secretariat: Making a Living in a Straitjacket. In Frank Biermann, Bernd Siebenhüner (Eds.): *Managers of Global Change: The Influence of International Environmental Bureaucracies*. Cambridge, MA: MIT Press, pp. 245–264.

Busch, Per-Olof (2009b): The OECD Environment Directorate: The Art of Persuasion and Its Limitations. In Frank Biermann, Bernd Siebenhüner (Eds.): *Managers of Global Change: The Influence of International Environmental Bureaucracies*. Cambridge, MA: MIT Press, pp. 75–99.

Busch, Per-Olof (2014): The Independent Influence of International Public Administrations: Contours and Future Directions of an Emerging Research Strand. In Soonhee Kim, Shena Ashley, W. Henry Lambright (Eds.): *Public Administration in the Context of Global Governance*. Cheltenham: Edward Elgar, pp. 45–62.

Busch, Per-Olof; Jörgens, Helge (2012): Governance by Diffusion: Exploring a New Mechanism of International Policy Coordination. In James Meadowcroft, Oluf Langhelle, Audun Ruud (Eds.): *Governance, Democracy and Sustainable Development: Moving Beyond the Impasse?* Cheltenham: Edward Elgar, pp. 221–248.

Busch, Per-Olof; Liese, Andrea (2017): The Authority of International Public Administrations. In Michael W. Bauer, Christoph Knill, Steffen Eckhard (Eds.): *International Bureaucracy: Challenges and Lessons for Public Administration Research*. Basingstoke: Palgrave Macmillan, pp. 97–122.

Campe, Sabine (2009): The Secretariat of the International Maritime Organization: A Tanker for Tankers. In Frank Biermann, Bernd Siebenhüner (Eds.): *Managers of Global Change: The Influence of International Environmental Bureaucracies*. Cambridge, MA: MIT Press, pp. 143–168.

Chan, Sander; Asselt, Harro van; Hale, Thomas; Abbott, Kenneth W.; Beisheim, Marianne; Hoffmann, Matthew J. et al. (2015): Reinvigorating International Climate Policy: A Comprehensive Framework for Effective Nonstate Action. *Global Policy* 6 (4), pp. 466–473.

Chasek, Pamela S.; Wagner, Lynn M. (2012): An Insider's Guide to Multilateral Environmental Negotiations since the Earth Summit. In Pamela S. Chasek, Lynn M. Wagner (Eds.): *The Roads from Rio: Lessons Learned from Twenty Years of Multilateral Environmental Negotiations*. London: Routledge, pp. 1–15.

Churchill, Robin R.; Ulfstein, Geir (2000): Autonomous Institutional Arrangements in Multilateral Environmental Agreements: A Little-Noticed Phenomenon in International Law. *American Journal of International Law* 94 (4), pp. 623–659.

de Wit, Dominique; Ostovar, Abby Lindsay; Bauer, Steffen; Jinnah, Sikina (2020): International Bureaucracies. In Frank Biermann, Rakhyun E. Kim (Eds.): *Architectures of Earth System Governance*. Cambridge: Cambridge University Press, pp. 57–74.

Dehousse, Renaud (2008): Delegation of Powers in the European Union: The Need for a Multi-principals Model. *West European Politics* 31 (4), pp. 789–805.

Depledge, Joanna (2007): A Special Relationship: Chairpersons and the Secretariat in the Climate Change Negotiations. In *Global Environmental Politics* 7 (1), pp. 45–68.

DeSombre, Elizabeth R. (2017): Global Environmental Institutions. 2nd ed. London: Routledge.

Dijkstra, Hylke (2017): Collusion in International Organizations: How States Benefit from the Authority of Secretariats. *Global Governance* 23 (4), pp. 601–618.

Dimitrov, Radoslav S. (2010): Inside UN Climate Change Negotiations: The Copenhagen Conference. *Review of Policy Research* 27 (6), pp. 795–821.

Dingwerth, Klaus; Jörgens, Helge (2015): Environmental Risks and the Changing Interface of Domestic and International Governance. In Stephan Leibfried, Frank Nullmeier, Evelyne Huber, Matthew Lange, Jonah Levy, John Stephens (Eds.): *The Oxford Handbook of Transformations of the State*. Oxford: Oxford University Press, pp. 338–354.

Dür, Andreas (2008): Interest Groups in the European Union: How Powerful Are They? *West European Politics* 31 (6), pp. 1212–1230.

Eckhard, Steffen; Ege, Jörn (2016): International Bureaucracies and their Influence on Policy-Making: A Review of Empirical Evidence. *Journal of European Public Policy* 23 (7), pp. 960–978.

Fernandez, Roberto M.; Gould, Roger V. (1994): A Dilemma of State Power: Brokerage and Influence in the National Health Policy Domain. *American Journal of Sociology* 99 (6), pp. 1455–1491.

Finkelstein, Lawrence S. (1974): International Organizations and Change: The Past as Prologue. *International Studies Quarterly* 18 (4), pp. 485–520.

Gehring, Thomas (2012): International Environmental Regimes as Decision Machines. In Peter Dauvergne (Ed.): *Handbook of Global Environmental Politics*. 2nd ed. Cheltenham: Edward Elgar, pp. 51–63.

Goritz, Alexandra; Jörgens, Helge; Kolleck, Nina (2021): Interconnected Bureaucracies? Comparing Online and Offline Networks During Global Climate Negotiations. *International Review of Administrative Sciences* 87 (4), 813–830.

Goritz, Alexandra; Jörgens, Helge; Kolleck, Nina (2022a): A Matter of Information – The Influence of International Bureaucracies in Global Climate Governance Networks. *Social Networks*. DOI: 10.1016/j.socnet.2022.02.009.

Goritz, Alexandra; Kolleck, Nina; Jörgens, Helge (2019): Education for Sustainable Development and Climate Change Education: The Potential of Social Network Analysis Based on Twitter Data. *Sustainability* 11 (19), p. 5499.

Goritz, Alexandra; Schuster, Johannes; Jörgens, Helge; Kolleck, Nina (2022b): International Public Administrations on Twitter: A Comparison of Digital Authority in Global Climate Policy. *Journal of Comparative Policy Analysis: Research and Practice* 24 (3), pp. 271–295.

Gutner, Tamar L. (2005): World Bank Environmental Reform: Revisiting Lessons from Agency Theory. *International Organization* 59 (3), pp. 773–783.

Hall, Nina (2016): *Displacement, Development, and Climate Change: International Organizations Moving Beyond their Mandates*. London: Routledge.

Hawkins, Darren; Lake, David A.; Nielson, Daniel L.; Tierney, Michael J. (2006): Delegation Under Anarchy: States, International Organizations, and Principal-Agent Theory. In Darren Hawkins, David A. Lake, Daniel L. Nielson, Michael J. Tierney (Eds.): *Delegation and Agency in International Relations*. Cambridge: Cambridge University Press, pp. 3–38.

Hickmann, Thomas; Widerberg, Oscar; Lederer, Markus; Pattberg, Philipp (2021): The United Nations Framework Convention on Climate Change Secretariat as an Orchestrator in Global Climate Policymaking. *International Review of Administrative Sciences* 87 (1), pp. 21–38.

Jinnah, Sikina (2011): Marketing Linkages: Secretariat Governance of the Climate-Biodiversity Interface. *Global Environmental Politics* 11 (3), pp. 23–43, checked on 9/1/2011.

Jinnah, Sikina (2012): Singing the Unsung: Secretariats in Global Environmental Politics. In Pamela S. Chasek, Lynn M. Wagner (Eds.): *The Roads from Rio: Lessons Learned from Twenty Years of Multilateral Environmental Negotiations*. London: Routledge, pp. 107–126.

Jinnah, Sikina (2014): *Post-Treaty Politics: Secretariat Influence in Global Environmental Governance*. Cambridge, Mass.: MIT Press.

Jönsson, Christer (1986): Interorganization Theory and International Relations. *International Studies Quarterly* 30 (1), pp. 39–57.

Jörgens, Helge (2004): Governance by Diffusion: Implementing Global Norms Through Cross-National Imitation and Learning. In William M. Lafferty (Ed.): *Governance for Sustainable Development. The Challenge of Adapting Form to Function*. Cheltenham: Edward Elgar, pp. 246–283.

Jörgens, Helge; Kolleck, Nina; Saerbeck, Barbara (2016): Exploring the Hidden Influence of International Treaty Secretariats: Using Social Network Analysis to Analyse the Twitter Debate on the 'Lima Work Programme on Gender'. *Journal of European Public Policy* 23 (7), pp. 979–998.

Jörgens, Helge; Kolleck, Nina; Saerbeck, Barbara; Well, Mareike (2017): Orchestrating (Bio-)Diversity: The Secretariat of the Convention of Biological Diversity as an Attention-Seeking Bureaucracy. In Michael W. Bauer, Christoph Knill, Steffen Eckhard (Eds.): *International Bureaucracy: Challenges and Lessons for Public Administration Research*. Basingstoke: Palgrave Macmillan, pp. 73–95.

Jörgens, Helge; Kolleck, Nina; Well, Mareike (Eds.) (forthcoming): *International Public Administrations in Environmental Governance: The Role of Autonomy, Agency and the Quest for Attention*. Cambridge: Cambridge University Press.

Keohane, Robert O. (1984): *After Hegemony. Cooperation and Discord in the World Political Economy*. Princeton: Princeton University Press.

Kinley, Richard (2021): Mission: Adoption with Ovations: The Contribution of the UNFCCC Secretariat to the Achievement of the Paris Agreement. In Henrik Jepsen, Magnus Lundgren, Kai Monheim, Hayley Walker (Eds.): *Negotiating the Paris Agreement: The Insider Stories*. Cambridge: Cambridge University Press, pp. 65–96.

Klüver, Heike (2013): *Lobbying in the European Union: Interest Groups, Lobbying Coalitions, and Policy Change*. Oxford: Oxford University Press.

Knill, Christoph; Bayerlein, Louisa; Enkler, Jan; Grohs, Stephan (2018): Bureaucratic Influence and Administrative Styles in International Organizations. *Review of International Organizations* 7 (3), p. 221.

Knill, Christoph; Enkler, Jan; Schmidt, Sylvia; Eckhard, Steffen; Grohs, Stephan (2017): Administrative Styles of International Organizations: Can We Find Them, Do They Matter? In Michael W. Bauer, Christoph Knill, Steffen Eckhard (Eds.): *International Bureaucracy: Challenges and Lessons for Public Administration Research*. Basingstoke: Palgrave Macmillan, pp. 43–71.

Knill, Christoph; Steinebach, Yves (Eds.) (2023): *International Public Administrations in Global Public Policy: Sources and Effects of Bureaucratic Influence*. London: Routledge.

Kolleck, Nina (2016): Uncovering Influence through Social Network Analysis: The Role of Schools in Education for Sustainable Development. *Journal of Education Policy* 31 (3), pp. 308–330.

Kolleck, Nina; Jörgens, Helge; Well, Mareike (2017a): Levels of Governance in Policy Innovation Cycles in Community Education: The Cases of Education for Sustainable Development and Climate Change Education. *Sustainability* 9 (11), p. 1966.

Kolleck, Nina; Jörgens, Helge; Well, Mareike; Saerbeck, Barbara; Goritz, Alexandra; Schuster, Barbara (2023): Behind the Scenes: How International Treaty Secretariats Use Social Networks to Exert Influence in the Global Climate Policy Regime. In Christoph Knill, Yves Steinebach (Eds.): *International Public Administrations in Global Public Policy: Sources and Effects of Bureaucratic Influence*. London: Routledge, pp. 199–219.

Kolleck, Nina; Well, Mareike; Sperzel, Severin; Jörgens, Helge (2017b): The Power of Social Networks: How the UNFCCC Secretariat Creates Momentum for Climate Education. *Global Environmental Politics* 17 (4), pp. 106–126.

Long, Bill L. (2000): *International Environmental Issues and the OECD 1950–2000*. Paris: OECD.

Martin, Lisa L.; Simmons, Beth A. (2012): International Organizations and Institutions. In Walter Carlsnaes, Thomas Risse, Beth A. Simmons (Eds.): *Handbook of International Relations*. 2nd ed. London: Sage, pp. 326–351.

Mathiason, John (2007): *Invisible Governance: International Secretariats in Global Politics*. Bloomfield: Kumarian Press.

Mederake, Linda; Saerbeck, Barbara; Goritz, Alexandra; Jörgens, Helge; Well, Mareike; Kolleck, Nina (2021): Cultivated Ties and Strategic Communication: Do International Environmental Secretariats Tailor Information to Increase their Bureaucratic Reputation? *International Environmental Agreements* (published online) 22, pp. 481–506.

Michaelowa, Katharina; Michaelowa, Axel (2017): The Growing Influence of the UNFCCC Secretariat on the Clean Development Mechanism. *International Environmental Agreements* 17 (2), pp. 247–269.

Miles, Edward L.; Underdal, Arild; Andresen, Steinar; Wettestad, Jørgen; Skjærseth, Jon Birger; Carlin, Elaine M. (Eds.) (2002): *Environmental Regime Effectiveness. Confronting Theory with Evidence*. Cambridge, MA: MIT Press.

Newell, Peter; Pattberg, Philipp; Schroeder, Heike (2012): Multiactor Governance and the Environment. *Annual Review of Environment and Resources* 37 (1), pp. 365–387.

Nielson, Daniel L.; Tierney, Michael J. (2003): Delegation to International Organizations: Agency Theory and World Bank Environmental Reform. *International Organization* 57 (2), pp. 241–276.

O'Neill, Kate (2014): International organizations Global and regional environmental cooperation. In Paul G. Harris (Ed.): *Routledge Handbook of Global Environmental Politics*. London: Routledge, pp. 97–109.

OECD (2001): *Environmental Performance Reviews: Achievements in OECD Countries*. Paris: OECD.

Olsen, Johan P. (2003): Towards a European Administrative Space? *Journal of European Public Policy* 10 (4), pp. 506–531.

Patz, Ronny; Goetz, Klaus H. (2017): Changing Budgeting Administration in International Organizations: Budgetary Pressures, Complex Principals and Administrative Leadership. In Michael W. Bauer, Christoph Knill, Steffen Eckhard (Eds.): *International Bureaucracy: Challenges and Lessons for Public Administration Research*. Basingstoke: Palgrave Macmillan, pp. 123–150.

Reinalda, Bob (2020): *International Secretariats: Two Centuries of International Civil Servants and Secretariats*. London: Routledge.

Saerbeck, Barbara; Well, Mareike; Jörgens, Helge; Goritz, Alexandra; Kolleck, Nina (2020): Brokering Climate Action: The UNFCCC Secretariat Between Parties and Nonparty Stakeholders. *Global Environmental Politics* 20 (2), pp. 105–127.

Sandford, Rosemary (1994): International Environmental Treaty Secretariats: Stage-Hands or Actors? In Helge O. Bergesen, Georg Parmann (Eds.): *Green Globe Yearbook of International Cooperation on Environment and Development 1994*. Oxford: Oxford University Press, pp. 17–29.

Siebenhüner, Bernd (2009): The Biodiversity Secretariat: Lean Shark in Troubled Waters. In Frank Biermann, Bernd Siebenhüner (Eds.): *Managers of Global Change: The Influence of International Environmental Bureaucracies*. Cambridge, Mass: MIT Press, pp. 265–291.

Siotis, Jean (1965): The Secretariat of the United Nations Economic Commission for Europe and European Economic Integration: The First Ten Years. *International Organization* 19 (2), pp. 177–202.

Stone, Diane; Ladi, Stella (2015): Global Public Policy and Transnational Administration. *Public Administration* 93 (4), pp. 839–855.

Stone, Diane; Moloney, Kim (Eds.) (2019): *The Oxford Handbook of Global Policy and Transnational Administration*. Oxford: Oxford University Press.

Trondal, Jarle (2017): A Research Agenda on International Public Administration. In Jarle Trondal (Ed.): *The Rise of Common Political Order: Institutions, Public Administration and Transnational Space*. Cheltenham: Edward Elgar, pp. 35–48.

Trondal, Jarle; Peters, B. Guy (2015): A Conceptual Account of the European Administrative Space. In Michael W. Bauer, Jarle Trondal (Eds.): *The Palgrave Handbook of the European Administrative System*. Basingstoke: Palgrave Macmillan, 79–92.

Vaubel, Roland (2006): Principal-Agent Problems in International Organizations. *Review of International Organizations* 1 (2), pp. 125–138.

Verhoest, Koen; Roness, Paul G.; Verschuere, Bram; Rubecksen, Kristin; MacCarthaigh, Muiris (2010): *Autonomy and Control of State Agencies: Comparing States and Agencies*. Basingstoke: Palgrave Macmillan.

Wagner, Lynn M.; Mwangi, Mawaki (2010): Be Careful What You Compromise For: Postagreement Negotiations within the UN Desertification Convention. *International Negotiation* 15 (3), pp. 439–458.

Weinlich, Silke (2014): *The UN Secretariat's Influence on the Evolution of Peacekeeping*. Basingstoke: Palgrave Macmillan.

Well, Mareike; Saerbeck, Barbara; Jörgens, Helge; Kolleck, Nina (2020): Between Mandate and Motivation: Bureaucratic Behavior in Global Climate Governance. *Global Governance: A Review of Multilateralism and International Organizations* 26 (1), pp. 99–120.

Wirth, David A. (2016): Environment. In Jacob Katz Cogan, Ian Hurd, Ian Johnstone (Eds.): *The Oxford Handbook of International Organizations*. Oxford: Oxford University Press.

Young, Oran R.; Levy, Marc A. (1999): The Effectiveness of International Environmental Regimes. In Oran R. Young (Ed.): *The Effectiveness of International Environmental Regimes. Causal Connections and Behavioral Mechanisms*. Cambridge, MA: MIT Press, pp. 1–32.

13

THE ROLE OF LITIGATION OF ENVIRONMENTAL NON-GOVERNMENTAL ORGANIZATIONS IN ENVIRONMENTAL POLITICS AND POLICY

Annette Elisabeth Töller, Paul-Philipp Schnase, and Fabio Bothner

13.1 Introduction

In 1998, the Aarhus Convention (AC) was adopted as an international treaty under the auspices of the United Nations Economic Commission for Europe (UNECE). Its three pillars – access to environmental information, public participation in environmental protection, and access to justice in environmental matters – are said to have the potential to fundamentally change the environmental governance framework (Sommermann, 2017; Kingston et al., 2021). This chapter focuses on the third pillar of the Convention, specifically on the access to justice (i.e., administrative courts) for environmental non-governmental organizations (ENGOs). While legal research is primarily concerned with the convention's implementation in the national legal systems of the Convention states, political science research asks about the use of the right by environmental associations and what effects this has.

The regulations of the AC meet highly diverse conditions in the 45 Convention states. Those 27 Convention states which are also EU members have been categorized into three different groups with regard to standing rights (i.e., the right to initiate legal proceedings) for ENGOs. However, most countries have also seen changes in their regulations over time, mainly due to Court of Justice of the European Union (CJEU) case law. Section 13.2 provides an overview of this implementation in some Convention states. Yet, accurate statistics are needed to understand the extent to which the ENGOs' standing right is actually used, the areas in which lawsuits are concentrated, how successful they are, and the patterns of lawsuit activity over time. Despite the documentation at the AC Secretariat, only a few states have collected and published reliable data on this, which we evaluate in Section 13.3.

On the other hand, a whole series of analyses addresses the question of the extent to which associations take up the right of action as an opportunity structure. Which strategies do they pursue? And what do they depend on? We provide an overview of this in Section 13.4.

 DOI: 10.4324/9781003043843-15

Finally, the effects of the right of associations to sue are analysed and discussed in research in two different aspects. For environmental policy research, the main question is whether using the standing right by environmental associations helps improve the application of the notoriously poorly implemented environmental law and thus enhances environmental quality. We address this question with regard to three different issue areas in Section 13.5. Furthermore, the effects of the access of associations to courts and its use are discussed specifically in the context of the legal order of the EU as a possible element of "Eurolegalism" (Kelemen, 2011). The issue here is, on the one hand, whether suing leads to a more adversarial style of politics, one more common in the United States. On the other hand, the right to sue by association is also discussed as an "outsourcing" of the enforcement of European environmental law. We address these debates in Section 13.6 and summarize in Section 13.7.

We deliberately do not discuss the multitude of mostly recent cases of climate litigation in this chapter. Even though some of these cases were brought by ENGOs, they are predominantly not actions before administrative courts that focus on the enforcement of existing environmental law (see Setzer & Higham, 2022).

13.2 The Transposition of the Aarhus Convention

Whereas Art. 9(1) and 9(2) of the AC secure access to (judicial) review against violations of the information and participation rights established in the first and second pillar of the Convention, Art. 9(3) goes beyond that by granting access to (judicial) review for violations of all environmental provisions of domestic law to individuals and environmental groups (third pillar). By rendering any administrative action relevant to the environment subject to judicial review, Art. 9(3) grants a general right of access to justice in environmental matters (Schmidt et al., 2014, 15; Schlacke et al., 2019, 468; Lycourgos et al., 2021, 6).

The AC needs to be implemented by all Convention parties, which includes states as well as the EU which is an independent party to the Convention. Accordingly, a "dual" implementation obligation arises for states that are both parties to the AC and Member States of the EU, since they must fulfil their individual implementation obligations and comply with the legislation adopted by the EU to implement the Convention. While Art. 9(1) and 9(2), regarding information and participation rights, were translated into European law quite swiftly by adopting the Environmental Information Directive (2003/4/EC) and the Public Participation Directive (2003/35/EC), a first draft for a "Directive on Access to Justice in Environmental Matters" failed in 2003 due to lack of approval by Member States (Altmayer, 2017, 3; Schlacke et al., 2019, 28), and no further attempt has been made to comprehensively implement Art. 9(3) to date.

Before the AC was adopted, the legal systems of states had varied considerably as to who has access to administrative courts (Prieur, 1998). These legal traditions had a strong impact on how states transposed the Convention rules into national law, especially given the lack of harmonized implementation in European law and also because the wording of Art. 9(3) itself renders standing rights subject to terms potentially set out in domestic law (Lycourgos et al., 2021, 6). Thus, the legal provisions on ENGOs' access to justice continue to differ significantly across EU Member States. On the other hand, the Convention and the decisions by the CJEU gradually changed these rules even in those states with a more reluctant stance. With regard to the legal standing of ENGOs, EU Member States can roughly be clustered into three groups (Darpö, 2013).

States with an "*actio popularis*" system allow anyone to challenge administrative acts without having to claim a violation of individual rights. This permissive form of standing rights is predominant in Portugal and Latvia (Milieu Consulting, 2019, 104) and partly realized in Spain and Slovenia (Darpö, 2013, 12). While most of these states do not define any criteria for standing (Darpö, 2013, 14), Spain, however, imposes strict requirements on the eligibility of ENGOs to sue (Sanchis-Moreno, 2007, 10).

In stark contrast to the *actio popularis* system, in the "impairment of rights system" access to administrative courts is granted only to plaintiffs that can claim an impairment of subjective rights through an act of the administration ("*Schutznormtheorie*", Darpö, 2013, 13, Schlacke et al., 2019, 443). Germany and Austria are paradigmatic of this system. Since the basic idea of giving ENGOs access to courts as stipulated in Art. 9(3) of the AC is fundamentally opposed to the subjective rights-based systems, its implementation poses the greatest challenges to legal systems of this type.

A third group is characterized as "interest-based" systems, granting standing rights to ENGOs if the organization is affected regarding a (legitimate) interest by an act of the administration (Darpö, 2013, 13). Such an interest can be understood quite broadly, as in the case of the French "*intérêt à agir*", where it is sufficient if an organization's claim relates to issues linked to their statutory objectives (Kingston et al., 2021, 148; Lycourgos et al., 2021, 12). Most European countries can be assigned to this group[1] (see Darpö, 2013, 13). Given the size of the group, there are, however, large intra-group differences. This includes countries such as the Czech Republic, which follows a rather restrictive approach to standing rights (Milieu Consulting, 2019, 103) and is therefore occasionally classified as a rights-based system (e.g. Szegedi, 2014, 128), as well as France, which has a rather generous (in application) system of standing rights (Schmidt et al., 2017, 90; Vanhala, 2016).

Whereas differentiating these groups is a good starting point to identifying how Convention states grant legal access to ENGOs, decisions by the CJEU have triggered considerable change over the years. Germany is a case in point here. The initial transposition of Art. 9(3) of the AC occurred by adopting the *Umweltrechtsbehelfsgesetz* (UmwRG) in 2006, which served to implement Directive 2003/35/EC and thus the first two paragraphs of Art. 9 (see explanatory memorandum, BT-Drs. 16/2495). According to the German legislator, this adequately fulfilled access to justice requirement for ENGOs (Schmidt et al., 2014, 27). However, this represented only a very limited transposition. In particular, ENGOs could only sue against standards that serve to protect the environment and establish individual rights (Schmidt & Zschiesche, 2018, 7); their right of action remained generally limited to projects subject to an Environmental Impact Assessment (Schlacke et al., 2019, 479 ff.). This restrictive transposition of the AC in Germany was challenged before the CJEU and also before German courts, based on CJEU rulings. An important change resulted from the CJEU's Trianel ruling (C-115/09), in which the court in 2011 held that national authorities must not limit the standing of ENGOs to appeals against norms that establish individual rights, as this would generally deprive ENGOs (which usually protect collective interests) of the rights granted to them by the AC (Lycourgos et al., 2021, 8). As a result, the German legislator abolished the restriction on the right to complain (Schlacke et al., 2019, 479 ff.). In 2013 the Federal Administrative Court ruled that ENGOs are entitled to challenge air quality plans (Töller, 2021). In its decision, the court referred to the Slovak brown bear case (C-240/09) in which the CJEU in 2011 had stated that national courts must interpret the provisions of national law on standing in a way that is to the fullest extent possible in accordance with the objective of the AC to enable ENGOs to challenge official decisions that may be in conflict with Union law (Altmayer, 2017, 5; Schmidt et al., 2017, 58). In 2017, Germany had to abolish its strict preclusion rules, following a further CJEU decision adopted in 2015 (C-139/14).

In addition to restrictions on legal standing, costs of judicial procedure are a considerable obstacle to access to environmental justice in most of the Member States (Darpö, 2013, 38). The "loser pays system", where the losing party is responsible for covering the legal fees of the other party, is a major constraint for many ENGOs (Vanhala, 2018, 385). Such systems still prevail in most of the Member States' administrative court systems,[2] even though in some countries it is common practice for the court to order each side to bear its own costs despite loser pays systems being in operation (Vanhala, 2018, 387). Only in a few member states do ENGOs have possible access to legal aid (Denmark, Hungary, Romania, Slovenia, and Spain) (Darpö, 2013, 20). In 2019 the European Commission identified 12 Member States that must improve ENGOs' access to justice without facing prohibitive costs (European Commission, 2019, 4).

Beyond standing rights and costs, the general effectiveness of the legal systems, which vary considerably between AC states, has an impact on the effectiveness of access to justice rights (Hofmann, 2019, 353). More recently, the European Commission emphasized that effective access to justice by ENGOs is a relevant requirement for successful implementation of the EU's Green Deal, and announced that it would take action to improve the access of citizens and NGOs to justice before national courts in all Member States (European Commission, 2020, 1).

13.3 Data on the Use of the Right to Stand

According to Art. 10 of the AC, parties should keep the implementation of the Convention under continuous review and present regular and publicly accessible National Implementation Reports (UNECE, 2022). In practice, however, a quick search with the UNECE's search engine reveals that the data situation on the use of the access to court right by ENGOs is desolate. Under the relevant question 30 ("Further information on the practical application of the provisions of article 9"), very few states provide useful information. Some countries (e.g. Poland) provide figures on administrative actions, but do not break them down by plaintiff. For a few countries, there is at least some overview data. For Finland, we learn that

> of the matters filed at the Supreme Administrative Court between 2011–2015, 535 were related to the environment, 305 to water resources engineering and 121 to soil. Cases falling within the sphere of implementation of the Aarhus Convention accounted for approximately 17 percent of matters resolved by the Supreme Administrative Court.
>
> *(UNECE, 2022)*

Switzerland has at least detailed annual documentation (e.g. BAFU, 2022).

However, we have a comparatively good data situation on ENGO legal actions for Germany, which is also reflected in the UNECE reports under question 30. A major evaluation was carried out in 2018 on behalf of the German Advisory Council on the Environment (SRU) (Schmidt & Zschiesche, 2018), and data were updated in 2021 (Habigt et al., 2021). Accordingly, German ENGOs utilize the right to stand extensively (Schmidt & Zschiesche, 2018) and this use has continued to increase over recent years. Compared to 140 cases (35 per year) in the 2013 to 2016 period, the number of ENGO lawsuits was 237 cases (59.2 per year) in the years 2017 to 2020. This increase is concentrated on lawsuits against the approval of wind energy plants and clean air plans: 63 cases (27 percent of all lawsuits) were related to wind energy plants and 25 (11 percent) to air quality plans (see Section 13.5). Others refer to building plans (19), street plans (18), and livestock facilities (13) (Habigt et al., 2021, 45). More than half of these lawsuits (51 percent) were fully or partly successful (Habigt et al., 2021, 19,

54). This is particularly remarkable given the fact that among *all* administrative law suits in Germany, the share of successful cases is only 12 percent (Schmidt & Zschiesche, 2018, 19, 26).

13.4 Lawsuits as a Strategy of ENGOs

For the ENGOs, the right to take legal action represents a legal opportunity structure. Thus, suing is basically a "tactic to influence political outcomes", among other strategies (Hofmann & Naurin, 2020, 1236; Vanhala, 2022). The key questions are: Which associations use the law, how, and under what conditions? Are there patterns across regulatory fields and countries? Vanhala (2012, 2013, 2016) examines the utilization of this legal opportunity structure for different countries. Taking the UK as an example, she demonstrates that associations do *not* sue just to succeed with the lawsuit. Rather, suing is seen as a way to delay proceedings, to build awareness and support, but also to establish contacts in politics and so on (Vanhala, 2012, 544–545), or as Hofmann and Naurin formulate it: "the value of a case lost … can be equal to that of a case won" (Hofmann & Naurin, 2020, 1237). In a cross-country comparative study, Vanhala finds for different types of bird conservation associations in Finland, France, Italy, and the United Kingdom that they resort to these lawsuits to varying degrees by country and type of association (Vanhala, 2018, 383). However, in terms of influencing factors, she does not identify clear patterns that apply to all countries and associations. In some cases, the self-image of the associations plays a role (Vanhala, 2018, 399). In other cases, the propensity to sue correlates with exclusion from political decision-making (Vanhala, 2018, 401–402). In turn, the density of lawyers in associations mattered in some cases but not in others (Vanhala, 2018, 403–405).

Hofmann (2022) analyses – based on data from the Comparative Interest Group Survey project for interest groups in seven medium-sized EU member states – the litigation activities of 104 ENGOs. Overall, environmental associations are more likely to sue than most other associations – only trade unions sue more frequently (Hofmann, 2022, 11). About half of all environmental associations surveyed are involved in lawsuits. Environmental associations in the Eastern European (Poland, Lithuania, and Slovenia) and Southern European (Portugal) countries tend to sue more often than in the North-Western European countries (Belgium, the Netherlands, and Sweden) (Hofmann, 2022). The factors explaining which associations go to court and which do not are identical to those in the related study on all associations (Hofmann & Naurin, 2020). Accordingly, human resources in general and legal staff in particular have an impact on the propensity to sue (Hofmann & Naurin, 2020, 1246). This is in line with theoretical expectations and is (re)confirmed by a study on environmental association lawsuits in Ireland, the Netherlands, and France (Kingston et al., 2021, 153). Furthermore, the intensity of political activity has an impact, and this is independent of the status of associations as "insiders" or "outsiders". Moreover, groups that are financed by public money sue less frequently (Hofmann & Naurin, 2020, 1247).

13.5 Do Lawsuits Improve Environmental Quality?

Yet, while it appears that ENGOs use lawsuits strategically to achieve their goals, it is quite open as to what real-world impact they have. Both politicians and scholars often assume that "increased public access to justice in environmental matters contributes to achieving the objectives of Community policy on the protection of the environment by overcoming current shortcomings in the enforcement of environmental law and, eventually, to a better environment" (European Commission, 2003, 15; SRU, 2005, 3; Kingston et al., 2021). Arguably, part of the effect is achieved by administrations better applying environmental law in anticipation, knowing that environmental associations will sue them otherwise (SRU, 2005, 5). Whether lawsuits do indeed improve the application of environmental law and environmental quality is an open question so far. Research on this question is patchy and very heterogeneous, with qualitative individual case studies and comparative case studies standing alongside a few studies with large-N and quantitative methods. We look here at three groups of lawsuits that are, on the one hand, frequent and, on the other hand, comparatively well-studied.

13.5.1 The Fight for Clean Air

Regarding air pollution, ENGOs in Europe are particularly active in taking legal action, partly due to transnational collaboration, such as the Right to Clean Air Project.[3] While ENGOs have various tools to enforce compliance with European directives, litigation before national courts appears to be particularly effective in the area of air pollution (Reiners & Versluis, 2022). Yet, while in some countries ENGOs are still fighting for the full recognition of the information and legal rights promised by the AC (e.g., in Poland and Bulgaria), there are other countries in which legal proceedings have already been used for some time to ensure compliance with the air pollution limits set by the EU Air Quality Directive (Germany, Spain, the UK, Czech Republic) (ClientEarth, 2020; Deutsche Umwelthilfe & Frank Bold, n.d.). The greatest success so far has been achieved by ENGOs in Germany, where almost 50 lawsuits were filed against local air quality plans (most of them by the Deutsche Umwelthilfe, Töller, 2021). This is due to the fact that Germany had the largest number of cities in which the air quality limit values for NO_2 were exceeded (Töller, 2021). In all of these lawsuits, the respective court found that the air quality plan did not comply with European law and thus had to be revised (Töller, 2021). In many cases, the courts suggested driving bans for diesel cars, but this highly restrictive instrument was finally adopted only in a few metropolitan areas (Töller, 2021, 493).

In the UK, ClientEarth successfully challenged the government's air quality plans several times (2015, 2017), so that the government had to revise and adapt its plans (Deutsche Umwelthilfe, 2019, 6). In the course of this process, the Ultra-Low Emission Zone was established in London, with restricted access for diesel vehicles older than Euro 6 and gasoline vehicles older than Euro 4 (Harvey, 2015). In 2018, the Spanish ENGO Ecologistas en Acción achieved their aim of the regional government of Castilla y Leon having to develop an air quality plan within one year to curb ozone exceeding EU target values in the region (Deutsche Umwelthilfe, 2019, 8). In the Czech Republic, the organization Frank Bold achieved recognition that the air pollution control plan for Ostrava was insufficient. As a result, the organization sued again in 2018 against the Czech Ministry of Environment to obtain an effective adjustment of air pollution control plans for the cities of Radvanice and Bartovice (Deutsche Umwelthilfe, 2019, 9).

Yet it is largely unknown whether these lawsuits in the field of air pollution do affect air quality. While Li et al. (2021) can show, using the example of OECD countries, that the number of ENGOs has a fundamentally positive effect on air quality, the impact of specific instruments, such as environmental lawsuits, is largely unclear. A positive effect on air quality seems particularly feasible in cases where lawsuits lead to the implementation of specific policies. For example, as mentioned above, the Ultra-Low Emission Zone was established in London in response to the lawsuits filed by ClientEarth. For this individual measure, however, Ma et al. (2021) find only a very small effect on NO_2 concentration and no significant effect on the concentration of other pollutants. While these studies only look at the effect of measures in response to lawsuits, there has been little research on whether lawsuits by ENGOs affect air quality.

One of the few studies in this area was conducted by Bothner et al. (2022). The researchers investigated whether 49 lawsuits that environmental associations filed against the air quality plans of German cities between 2011 and 2019 positively affected air quality by reducing NO_2 emissions in the respective cities. They find a small but statistically significant effect of lawsuits on air quality. The authors argue that the complaint by an ENGO already leads to anticipatory behaviour by the local decision-makers, and thus measures for air improvement are initiated even before the judgment. The study also shows that the effects of the lawsuits increase over time, which suggests that it takes time to adopt measures and that the measures adopted need time to have an effect.

13.5.2 Wolf Protection

Wolf protection is another case in point for analysing the impact of ENGOs' right to sue. The return of the wolf is a delicate issue, not only in Sweden and Finland. While protected as an endangered species, wolves are seen as a threat by parts of the rural population, thus demanding the right to hunt them (Epstein and Darpö, 2013). Sweden has already experienced ENGOs using the right to take legal action against wolf-hunting permissions in administrative courts, as early as in 2013, though only with limited success (Epstein & Darpö, 2013). The Finnish Tapiola case is generally viewed as a success story as a small group of Finnish wolf guards succeeded in challenging the Finnish Wildlife Authority's decision that wolves may be hunted (Epstein and Kantinkoski, 2020, 1).

Although wolves have been considered a protected species in Finland since 1973, the Finnish government allowed their hunting (wolf management), starting in 2007 under the pretext of stopping poaching. This led to a drastic decline in the wolf population (Epstein and Kantinkoski, 2020, 5). This development and the missing interest of existing Finnish ENGOs led to the establishment of the Tapiola Group, as Epstein and Kantinkoski (2020) demonstrate in their single-case study. This group conducted a multi-year legal fight to stop the controversial wolf management, beginning with administrative courts, followed by the Finnish Constitutional Court, finally arriving at the CJEU. As the authors describe, the AC played an important role in this process; while access to public participation and access to information was well established in Finland, access to judicial decisions in environmental matters was further improved by the AC (Epstein and Kantinkoski, 2020, 5). In the end, the activists were successful with their engagement, achieving not only some successes in regional administrative courts (Epstein and Kantinkoski, 2020, 6–7), but also in the CJEU and the Finnish Supreme Administrative Court. Both institutions criticized the practice of wolf management in their 2019 and 2020 respective rulings, allowing it only under strict conditions (Epstein and Kantinkoski, 2020, 9–10).

However, how sustainable this success is remains an open question, as the Finnish authorities announced at the end of 2021 that they would continue to control the wolf population through management. The Finnish authorities are still pursuing the goal of keeping the wolf population constant while reducing illegal hunting through legalization (Barkham, 2022). Even though hunting was suspended for 2022 after protests from ENGOs, saving the lives of 20 wolves, it is not yet clear if hunting will be permanently suspended (Lewis, 2022).

13.5.3 Fighting Wind Energy Plants: Climate versus Species Protection

Conflicts become more complicated in the next group of cases. In Germany, the largest group of lawsuits brought up by ENGOs relates to the approval of wind energy plants (Habigt et al., 2021, 19). A further expansion of onshore wind energy is badly needed to implement the German *Energiewende* (energy transformation) and meet ambitious reduction targets for greenhouse gas (GHG) emissions in the energy sector (Töller, 2022). Yet, since 2018, this expansion has experienced a dramatic decline (Deutsche Windguard, 2021). Whereas the reasons for this slump are manifold (SRU, 2022), lawsuits against the approval of plants played a role. Between 20 and 24 percent of the planned plants are subject to lawsuits challenging their approval (Töller, 2022), 61 percent of which were filed by ENGOs (Quentin, 2019, 14). As to the success of these lawsuits, the balance is more mixed, though, than in the field of air pollution. Of the 63 cases filed between 2017 and 2020, 29 percent were successful from the view of the plaintiff, 36 percent were lost, and the remainder have not yet been decided (Habigt et al., 2021, 54ff.). Given the fact that a relevant proportion of wind energy plants are the subject of lawsuits – with an increasing tendency – some observers have criticized ENGO lawsuits as slowing down the expansion of wind energy, deterring investors, and thus ultimately jeopardizing the realization of the *Energiewende* and the fight against climate change in Germany (cf. Töller, 2022). This debate demonstrates that it is not straightforward to determine whether ENGO lawsuits ultimately improve environmental quality.

At the core of most of these lawsuits is a conflict between the wind energy plant and species (primarily birds) protection. When it comes to a decision based on the Federal Emission Act to approve a wind energy plant, the protection of endangered species, based on European species protection law, is legally well established. On the other hand, the wind energy plant was evaluated like any other plant or infrastructure installation. Its specific value for climate protection by producing carbon-free energy could be assessed only under restricted conditions. Thus, when ENGOs file suits against the approval of a wind energy plant, usually putting species protection at the fore, environmental objectives, climate protection, and species protection compete, yet under quite unequal conditions (Töller, 2022). In 2022, the federal legislator fundamentally modified this legal situation by clarifying that the production of renewable energy is in the public interest and also serves public security (BMWK, 2022, 6) and thus made it easier to assert these interests against species protection concerns (§ 2 EEG).

Whereas Germany displays a huge number of ENGO legal actions bringing up conflicts between wind energy and species protection, such lawsuits are not restricted to Germany. A study by Rodela et al. (2016) shows how the Slovene bird-watching association DOPPS successfully challenged a permission given in 2006 for the installation of 33 wind energy plants before administrative courts (Rodela et al., 2016). In 2015, a wind farm in the Italian Alps could not be built due to a lawsuit filed by environmental and Alpine associations citing landscape and bird protection reasons (Italian Council of State, decision 4775/2014, UNECE, 2022).

13.6 ENGO Litigation as Part of "Eurolegalism"

Further debates on the effects of ENGO litigation rights refer to the EU, which comprises only part of the Convention states, and its specific legal and regulatory system (Kelemen, 2011; Kelemen & Pavone, 2022). These debates refer to the concept of "Eurolegalism" coined by Kelemen. With this term, Kelemen captures a longer-term development in the EU in which detailed rulemaking is accompanied by establishing rights, which are brought to bear by lawsuits from individuals, firms, and associations (Kelemen, 2011). Different questions may arise from this.

One question is whether a more confrontational interaction, especially between state and societal actors, is developing out of the increasing use of the right to sue. Töller (2020) argues that the extensive use of the right to sue by associations in Germany leads, on the one hand, to a power shift in favour of environmental associations. On the other hand, lawsuits also bring about a shift in conflicts. Due to the logic of administrative lawsuits, there is a confrontation between associations and state institutions, which certainly play a role due to the frequent lack of application of environmental law. Yet, through administrative lawsuits, the state institutions (in Germany, primarily the state administrations of the 16 *Länder*) become the scapegoat for the mistakes of economic actors (Töller, 2020, 292–294; in a similar vein Kingston et al., 2021, 160; Reiners & Versluis, 2022, 14). Those whose environmental impact is at stake in the lawsuits, as for instance the automotive industry or agriculture, are exactly *not* part of the proceedings and thus easily fall out of the public focus. In this respect, we observe a change in the *character* of the conflicts rather than environmental policy processes becoming more confrontational in the sense of a US-style adversarial legalism (similar to the results of Rehder & van Elten, 2020 and Hofmann & Naurin, 2020).

Another question discussed in the literature is what role ENGO lawsuits play in the overall context of the often precarious enforcement of EU environmental law. While in other policy sectors individuals and firms were provided with rights early on, stimulating decentralized enforcement of European law, such legal positions of individuals did not exist, for example, with regard to clean air, water (Hofmann, 2022), or species protection (Eliantonio, 2018). The standing right of ENGOs filled this gap. Hofmann (2019) argues that, on the one hand, the Commission seeks to provide environmental associations with EU-wide access to justice, but in return reduces its own activities to enforce European environmental law. He shows that in parallel with the implementation of the AC, the infringement procedures brought by the Commission before the CJEU have drastically decreased (on a general note see Kelemen & Pavone, 2022), more so in environmental policy than on average in all policy fields (p. 344). Hofmann criticizes this as a problematic "outsourcing" of the enforcement of European environmental law. He argues that the right of associations to sue cannot replace infringement procedures brought up by the Commission. This is because there are still considerable differences between countries in the actual standing rules, as well as with regard to the performance of the legal systems and the strength and differentiation of environmental associations (Hofmann, 2019, 357–358; 2022; Eliantonio, 2018; and already Slepcevic, 2009). While Hofmann recently sees "grounds for cautious optimism" (Hofmann, 2022), Kingston and co-authors are less optimistic (Kingston et al., 2021, S160).

13.7 Conclusion and Future Research

This chapter has dealt with legal actions filed by ENGOs in administrative courts, based on Art. 9(3) of the AC. There are two reasons the article displays a bias in favour of cases and data from Germany at some points. First, there seems to be a particularly high number of lawsuits filed by ENGOs in Germany as compared to other countries, even though due to the lack of reliable data such statements should be taken with due caution. This seems surprising given the history of the reluctant transposition of Art. 9(3) of the AC, but less surprising due to the relative strength of the environmental movement in Germany. Second, the cases in Germany are at least comparatively well documented, and there is some initial political science research on them, although much remains to be done.

There is a large number and range of essentially legal studies dealing in particular with the legal implementation of Art. 9(3) of the AC and the legal procedures arising from it. This is in marked contrast to the still rather limited number of studies on the effects of the standing right and its use by ENGOs. The fact that there are no regular reports for all convention states on the use of the right of action in their jurisdictions is a huge shortcoming that should be remedied quickly. The research situation is much better with regard to whether and under which conditions the environmental groups do or do not use this right. Yet, we need more studies that cover more countries, and in particular a finer differentiation of the phenomena to be explained. The question is not only which associations use the right to sue and which do not, but also what relative importance suing has for the individual association in the portfolio of possible strategies and how this portfolio has changed over time. How are lawsuits distributed across different environmental associations operating in a country? Can we identify a specialization of some associations in suing? Moreover, it would be worth asking whether the right to sue leads to a shift of resources within associations, so that more resources go into enforcing existing law but fewer into influencing policy-making. As a result, this might be detrimental to environmental protection, because even in the best case scenario, lawsuits before administrative courts can only help enforce *existing* environmental law.

We already know, on the basis of the few existing studies, that lawsuits do have the potential to improve the application of law and possibly even the quality of the environment if – as in the field of air pollution control – the law makes clear rules but these are not properly applied. Yet, when there are conflicting protected interests and the law is less clear or even biased – as in the field of wind energy plants – lawsuits can have a far weaker effect or even harm specific protected interests in favour of other protected interests.

While some studies examine the improved application of law as a result of lawsuits by ENGOs, the effect of suits on environmental quality is still a huge research gap. Here, following Bothner et al. (2022), further studies with quantitative methods would be desirable that succeed in establishing the causal relationship between lawsuits and environmental improvement with methodological certainty in other subject areas and countries (Vanhala, 2022, 110). Complementary qualitative studies are needed to help us better understand the causal mechanisms that lie between lawsuit filing and improved air quality values or improved species protection (Bothner et al., 2022, 6592; Vanhala, 2022, 109). Future research should also investigate the conditions that lead to the filing or non-filing of lawsuits. In order to do this, it is important to overcome the selection bias of current studies, which focus exclusively on those cases in which lawsuits are filed (Börzel, 2006, 129). Instead, comparable cases in which *no* lawsuits are filed must be included. Moreover, a suitable research design has to be developed to investigate the question of changes in policy styles as a result of the expansion of ENGO lawsuits.

Box 13.1 Chapter Summary

- While legal tradition shaped the implementation of the AC in Convention states, CJEU case law forced several EU Member States to extend their standing rights for ENGOs.
- For about half of ENGOs the right to take legal action is part of their strategy. Suing activities display regional and organizational patterns and depend particularly on staff, the intensity of political activity, and whether associations receive public money.
- Lawsuits filed by ENGOs can improve not only the application of environmental laws, but also environmental quality – under the condition that the law sets unambivalent rules that are clearly not being complied with.
- The European Commission's outsourced enforcement strategy appears risky, given the high variation in standing rights and overall efficiency of legal systems as well as the heterogeneity in the strength of the environmental movement.

Notes

1 Specifically: Belgium, Bulgaria, Denmark, Finland, France, Greece, Hungary, Croatia, Ireland, the Netherlands, Luxembourg, Italy, Slovakia, Slovenia, and the United Kingdom.
2 Except for Austria, Belgium, Bulgaria, Hungary, Finland, Ireland, and the Netherlands (see Darpö, 2013).
3 See www.right-to-clean-air.eu

References

Altmayer, A. (2017). Implementing the Aarhus Convention. Access to Justice in Environmental Matters. *European Parliamentary Research Service*. Available at https://www.europarl.europa.eu/thinktank/en/document/EPRS_BRI(2017)608753 (accessed 20 October 2022).

Barkham, P. (2022). Finland, Sweden and Norway to Cull Wolf Population. Available at https://www.theguardian.com/environment/2022/jan/15/finland-sweden-norway-cull-wolf-population-eu (accessed 15 October 2022).

Börzel, T. A. (2006). "Participation Through Law Enforcement: The Case of the European Union." *Comparative Political Studies*, 39(1): 128–152. https://doi.org/10.1177/0010414005283220

Bothner, F., Töller, A. E. & Schnase, P. P. (2022). "Do Lawsuits by ENGOs Improve Environmental Quality? Results from the Field of Air Pollution Policy in Germany." *Sustainability*, 14(11): 6592. https://doi.org/10.3390/su14116592

Bundesamt für Umwelt BAFU. (2022). Auswertung der abgeschlossenen Beschwerdefälle der beschwerdeberechtigten Umweltorganisationen für das Jahr 2021. Available at https://www.bafu.admin.ch/dam/bafu/de/dokumente/recht/fachinfo-daten/auswertung-der-abgeschlossenen-beschwerdefaelle-der-beschwerdeberechtigten-umweltorganisationen-fuer-das-jahr-2021.pdf.download.pdf/Auswertung_Berichterstattung_für_das_Jahr_2021_d.pdf (accessed 14 November 2022).

Bundesministerium für Wirtschaft und Klimaschutz BMWK. (2022). Überblickspapier Osterpaket. Available at https://www.bmwk.de/Redaktion/DE/Downloads/Energie/0406_ueberblickspapier_osterpaket.html (accessed 02 May 2022).

ClientEarth (2020). EU Issues Legal Warning as Bulgarian and Polish Governments Block Right to Clean Air. Available at https://www.clientearth.org/latest/press-office/press/eu-issues-legal-warning-as-bulgarian-and-polish-governments-block-right-to-clean-air/ (accessed 15 October 2022).

Darpö, J. (2013). *Effective Justice? Synthesis Report of the Study on Implementation of Articles 9.3 and 9.4 of the Aarhus Convention in the Member States of the European Union*. UNECE Task Force on Access to Justice. United Nations Economic Commission for Europe (UNECE).

Deutsche Umwelthilfe (2019). Legal Actions for Clean Air. Backgroundpaper 2019. Available at https://www.right-to-clean-air.eu/fileadmin/Redaktion/Downloads/Right-to-Clean-Air_Europe_Backgroundpaper_2019_english_final.pdf (accessed 14 November 2022).

Deutsche Umwelthilfe and Frank Bold (n.d.) (o.J.). Right to Clean Air. Poland. Lawsuits and Decisions. Available at https://www.right-to-clean-air.eu/en/lawsuits-and-decisions/poland/lawsuits-and-decisions/ (accessed 14 November 2022).

Deutsche Windguard (2021). Status des Windenergieausbaus an Land in Deutschland: Jahr 2020. Available at https://www.windguard.de/jahr-2021.html (accessed 14 November 2022).

Eliantonio, M. (2018). "The role of NGOs in Environmental Implementation Conflicts: 'Stuck in the Middle' Between Infringement Proceedings and Preliminary Rulings?" *Journal of European Integration*, 40(6): 753–767. https://doi.org/10.1080/07036337.2018.1500566

Epstein, Y. & Darpö, J. (2013). "The Wild Has No Words: Environmental NGOs Empowered to Speak for Protected Species as Swedish Courts Apply EU and International Environmental Law." *Journal for European Environmental & Planning Law*, 10(3): 250–261. https://doi.org/10.1163/18760104-01003004

Epstein, Y. & Kantinkoski, S. (2020). "Non-Governmental Enforcement of EU Environmental Law: A Stakeholder Action for Wolf Protection in Finland". *Frontiers in Ecology and Evolution*, 8. https://doi.org/10.3389/fevo.2020.00101

European Commission. (2003). *Proposal for a Directive of the European Parliament and of the Council on Access to Justice in Environmental Matters; COM/2003/0624 Final*. European Commission: Brussels, Belgium.

European Commission (2019). *Annex to the Communication from the Commission to the European Parliament, the Council, the European Economic and Social Committee and the Committee of the Regions. Environmental Implementation Review 2019: A Europe that Protects its Citizens and Enhances their Quality of Life. COM(2019) 149 final*. Brussels: European Commission.

European Commission (2020). *Improving access to justice in environmental matters in the EU and its member states. Communication from the Commission to the European Parliament the Council, the European Economic and Social Committee and the Committee of the Regions. COM(2020) 643 final*. Brussels: European Commission.

Habigt, L., Hamacher, L., Tryjanowski, A., Zschiesche, M., Schmidt, A., Heß, F & Teßmer, D. (2021). *Wissenschaftliche Unterstützung des Rechtsschutzes in Umweltangelegenheiten in der 19 in der Legislaturperiode*. Dessau-Roßlau: Umweltbundesamt (UBA).

Harvey, F. (2015). Supreme Court Orders UK to Draw Up Air Pollution Cleanup Plan. Available at https://www.theguardian.com/environment/2015/apr/29/supreme-court-orders-uk-to-draw-up-air-pollution-cleanup-plan (accessed 15 October 2022).

Hofmann, A. (2019). "Left to Interest Groups? On the Prospect for Enforcing Environmental Law in the European Union." *Environmental Politics*, 28(2): 342–364. https://doi.org/10.4324/9781003031178-8

Hofmann, A. (2022). How European environmental NGOs mobilize the law. Paper presented at the *11th Biennial Conference of the ECPR Standing Group on the European Union*, Rome, 8 – 10 June 2022.

Hofmann, A. & Naurin D. (2020). "Explaining Interest Group Litigation in Europe: Evidence from the Comparative Interest Group Survey." *Governance* 34(4): 1235–1253. https://doi.org/10.1111/gove.12556

Kelemen, R. D. (2011). *Eurolegalism: The Transformation of Law and Regulation in the European Union*. Harvard University Press.

Kelemen, R. D. & Pavone, T. (2022). Where Have the Guardians Gone? Law Enforcement and the Politics of Supranational Forbearance in the European Union. Available at SSRN: https://ssrn.com/abstract=3994918 or http://dx.doi.org/10.2139/ssrn.3994918.

Kingston, S., Alblas, E., Callaghan, M. & Foulon, J. (2021). "Magnetic Law: Designing Environmental Enforcement Laws to Encourage Us to go Further." *Regulation & Governance*, 15(1): 143–162. https://doi.org/10.1111/rego.12416

Lewis, L. (2022). Breaking! Norway & Finland Suspend This Year's Wolf Killings Sparing The Lives Of An Estimated 70 Wolves. 2022. Available at https://worldanimalnews.com/breaking-norway-finland-suspend-this-years-wolf-killing-sparing-the-lives-of-an-estimated-70-wolves/ (accessed 15 October 2022).

Li, G., He, Q., Wang, D. & Liu, B. (2021). Environmental Non-Governmental Organizations and Air-Pollution Governance: Empirical Evidence from OECD countries. *PloS one* 16(8): e0255166. https://doi.org/10.1371/journal.pone.0255166

Lycourgos, C., Vlachogiannis, A. & Yiordamli, A. (2021). Access to Justice of Environmental NGOs. A Comparative Perspective EU, France, Cyprus). Friedrich Ebert Stiftung (Climate Change, Energy and Environment).

Ma, L., Graham, D. J. & Stettler, M. E. J. (2021). "Has the Ultra Low Emission Zone in London Improved Air Quality?" *Environmental Research Letters* 16(12): 124001. https://doi.org/10.1088/1748-9326/ac30c1

Milieu Consulting (2019). *Study on EU Implementation of the Aarhus Convention in the Area of Access to Justice in Environmental Matters. Final report*. Brussels: Milieu Law & Policy Consulting.

Prieur, M. (1998). *Complaints and Appeals in the Area of Environment in the Member States of the European Union. General Report*. Study for the Commission of the European Community, Brussels.

Quentin, J. (2019). *Hemmnisse beim Ausbau der Windenergie in Deutschland: Ergebnisse einer Branchenumfrage zu Klagen gegen Windenergieanlagen sowie zu Genehmigungshemmnissen durch Drehfunkfeuer und militärische Belange der Luftraumnutzung*. Available at https://www.windenergie.de/fileadmin/redaktion/dokumente/pressemitteilungen/2019/20190719_FA_Wind_Branchenumfrage_beklagte_WEA_Hemmnisse_DVOR_und_Militaer.pdf (accessed 14 November 2022).

Rehder, B. & van Elten, K. (2020). "Klagende Verbände. Drei Logiken des justiziellen kollektiven Handelns in Deutschland." *dms – der moderne staat – Zeitschrift für Public Policy, Recht und Management*, 13(2-2020): 384–404. https://doi.org/10.3224/dms.v13i2.07

Reiners, K. & Versluis, E. (2022). "NGOs as New Guardians of the Treaties? Analysing the Effectiveness of NGOs as Decentralised Enforcers of EU law." *Journal of European Public Policy*: 1–19: https://doi.org/10.1080/13501763.2022.2084146

Rodela, R., Udovč, A. & Boström, M. (2016). "Developing Environmental NGO Power for Domestic Battles in a Multilevel Context: Lessons from a Slovenian Case." *Environmental Policy and Governance*, 27(3): 244–255. https://doi.org/10.1002/eet.1735

Sachverständigenrat für Umweltfragen SRU (2005). *Rechtsschutz für die Umwelt – die altruistische Verbandsklage ist unverzichtbar.* Available at https://www.umweltrat.de/SharedDocs/Downloads/DE/04_Stellungnahmen/2004_2008/2005_Stellung_Rechtsschutz_fuer_die_Umwelt.pdf;jsessionid=335F02EAD569964EBD5F76E4AE663BC3.intranet232?__blob=publicationFile&v=2 (accessed 10 November 2022).

Sachverständigenrat für Umweltfragen SRU (2022). Climate Protection Needs Tailwind: Towards a Reliable Expansion of Onshore Wind Energy in Germany. Available at https://www.umweltrat.de/SharedDocs/Downloads/EN/04_Statements/2020_2024/2022_05_Statement_wind_energy.pdf?__blob=publicationFile&v=6. (accessed 10 November 2022).

Sanchis-Moreno, F. (2007). "Civil Society Organizations and the Aarhus Convention in Court: Judicialization from below in Scotland?" *Representation Journal of Representative Democracy*, 49(3): 309–320.

Schlacke, S., Schrader, C. & Bunge, T. (2019). *Aarhus-Handbuch. Informationen, Beteiligung und Rechtsschutz in Umweltangelegenheiten.* Berlin: Erich Schmidt Verlag.

Schmidt, A., Schrader, C. & Zschiesche, M. (2014). *Die Verbandsklage im Umwelt- und Naturschutzrecht.* München: Beck.

Schmidt, A., Stracke, K., Wegener, B., Zschiesche, M., Zwicker, J. & Bar, M. (2017). *Die Umweltverbandsklage in der rechtspolitischen Debatte. Eine wissenschaftliche Auseinandersetzung mit Argumenten und Positionen zur Umweltverbandsklage, zugleich ein rechtvergleichender Beitrag zur weiteren Diskussion des Verbandsrechtsschutzes im Umweltbereich.* Dessau-Roßlau: Umweltbundesamt (UBA).

Schmidt, A. & Zschiesche, M. (2018). *Die Klagetätigkeit der Umweltschutzverbände im Zeitraum von 2013 bis 2016: Empirische Untersuchung zu Anzahl und Erfolgsquoten von Verbandsklagen im Umweltrecht.* Berlin: Sachverständigenrat für Umweltfragen.

Setzer, J. and Higham, C. (2022). *Global Trends in Climate Change Litigation: 2022 Snapshot.* London: Grantham Research Institute on Climate Change and the Environment and Centre for Climate Change Economics and Policy, London School of Economics and Political Science. Available at https://www.actu-environnement.com/media/pdf/news-39926-LES-rapport-contentieux-climatiques.pdf (accessed 14 November 2022).

Slepcevic, R. (2009). "The Judicial Enforcement of EU Law Through National Courts: Possibilities and Limits." *Journal of European Public Policy*. 16(3): 378–394. https://doi.org/10.1080/13501760802662847

Sommermann, K.-P. (2017). "Transformative Effects of the Aarhus Convention in Europe." *Zeitschrift Für Ausländisches Öffentliches Recht Und Völkerrecht*, 77: 321–337.

Szegedi, L. (2014). "The Eastern Way of Europeanisation in Light of Environmental Policymaking? Implementation Concerns of Aarhus Convention-related EU Law in Central and Eastern Europe." *ELTE Law Journal*, 1: 117–134.

Töller, A. E. (2020). "Das Verbandsklagerecht der Umweltverbände in Deutschland: Effekte auf Rechtsanwendung, Umweltqualität und Machtverhältnisse." *Der Moderne Staat – Zeitschrift Für Public Policy, Recht Und Management*, 13(2): 280–299. https://doi.org/10.3224/dms.v13i2.05

Töller, A. E. (2021). "Driving Bans for Diesel Cars in German Cities: The Role of ENGOs and Courts in Producing an Unlikely Outcome." *European Policy Analysis*, 7(2): 486–507. https://doi.org/10.1002/epa2.1120

Töller, A. E. (2022). Do ENGOs' Lawsuits Against Wind Energy Plants Jeopardize the German "Energiewende"? Paper presented at the *Hagen-based digital Conference "Environmental Nongovernmental Organizations' right to take legal action in EU member states: Preconditions and impacts on the application of law, policies, environmental quality and power relations"*, February 10-11, 2022.

United Nations Economic Committee for Europe UNECE (2022). *Aarhus Convention National Implementation Reports.* Available at: https://aarhusclearinghouse.unece.org/national-reports/reports (accessed 14 November 2022).

Vanhala, L. (2012). "Legal Opportunity Structures and the Paradox of Legal Mobilization by the Environmental Movement in the UK." *Law & Society Review*, 46(3): 523–556.

Vanhala, L. (2013). "Civil Society Organisations and the Aarhus Convention in Court: Judicialisation from below in Scotland?" *Representation*, 49(3): 309–320. https://doi.org/10.1080/00344893.2013.830483

Vanhala, L. (2016). "Legal Mobilization under Neo-corporatist Governance: Environmental NGOs before the Conseil d'Etat in France, 1975–2010." *Journal of Law and Courts*, 4(1): 103–130. https://doi.org/10.1086/684649

Vanhala, L. (2018). "Is Legal Mobilization for the Birds? Legal Opportunity Structures and Environmental Nongovernmental Organizations in the United Kingdom, France, Finland, and Italy." *Comparative Political Studies*, 51(3): 380–412. https://doi.org/10.1177/0010414017710257

Vanhala, L. (2022). "Environmental Legal Mobilization." *Annual Review of Law and Social Science*, 18(1): 101–117. https://doi.org/10.1146/annurev-lawsocsci-050520-104423

14

INDIGENOUS AND LOCAL KNOWLEDGE IN ENVIRONMENTAL DECISION MAKING

The Case of Climate Change

Israel Solorio, Raúl Romero, and Claudia Ros

14.1 Introduction

If there is a topic that currently dominates the environmental agenda it is, undoubtedly, climate change. Over the past decades, an overwhelming amount of scientific evidence has shown the many forms in which global warming is closely related to the way of life and production of modern societies. The ecological crisis faced by humanity has challenged our traditional understanding of concepts such as the 'environment' and, correspondingly, of environmental policies, as noted by Frank Biermann in an article published for the 30th anniversary special issue of *Environmental Politics* (Biermann, 2021). The long-established dichotomy of 'humans' versus 'environment' has become outdated, dragging with it the anthropocentric notions of sustainability as well (Seghezzo, 2009). It is in this context that critiques of the traditional paradigm of the 'environment' are growingly present in the mainstream literature. But truth be told, this idea has been floating in the air for decades among Global South academics (Leff, 2002) and, more importantly, it has been present for centuries in the cosmovision of many indigenous cultures (López Bárcenas, 2002; Walsh, 2011; Ros, 2021).

This paradigm shift in environmental politics has led to a double transformative effect. In the research field, there has been a surge of a more integrated thinking oriented towards blurring the frontiers between social and ecological systems (Biermann, 2021). Concepts such as socio-ecological systems (Young et al., 2006), social metabolism (Foster, 1999; Martinez-Alier, 2009a), or world ecology (Moore, 2011) have proliferated over the past years in the literature. In the everyday life practice of environmental policy, a greater focus has been placed on social participation as a way to look for innovative solutions to environmental challenges (Bulkeley & Mol, 2003; Berry et al., 2019), changing at the same time the modes of environmental governance (Knill & Tosun, 2012). For example, not only climate governance has been moving towards a more polycentric bottom-up approach by including non-state actors (Rayner, 2010; Jordan et al., 2015; see Chapter 5 in this volume), but also the Intergovernmental Panel on Climate Change (IPCC) has called on the employment of traditional indigenous knowledge

 DOI: 10.4324/9781003043843-16

in order to improve the range of available options to manage the planet's ecosystems (IPCC, 2018: 12). Yet, this new trend in environmental policy is not trouble-free, and the incorporation of local and indigenous knowledge in decision making and implementation has proved to be rather problematic.

Centering the attention on the case of climate change, this chapter presents the most prominent debates on the participation of indigenous peoples in environmental policy and governance. Guided by the climate justice literature, this contribution argues that the inclusion of indigenous and local knowledge in environmental decision making is hampered by a three-layered explanation: a distributive, a procedural, and an epistemic injustice. Distributive (in)justice refers to the way in which indigenous peoples carry the burden of climate change impacts and policies while rarely receiving the direct benefits. Procedural (in)justice refers to the exclusion of indigenous peoples from the main decision centers of climate policy. Finally, epistemic (in)justice captures the way in which climate policy overshadows indigenous people's capacity to communicate their own knowledge and way of relating to the environment.

The chapter is structured in the following way. The next section presents the drivers that have led to a major participation on the part of indigenous peoples in climate policy and governance. The following section presents the main findings of the literature regarding the distributive injustice placed on the shoulders of indigenous peoples. Then the procedural barriers are presented, followed by epistemic considerations. The chapter ends with a concluding section that highlights emerging areas of research, together with some brief concluding notes.

14.2 Climate Justice and Indigenous Peoples

Among the variety of academic debates that have pushed the ethical and moral issues linked to climate policy and governance, climate justice has received the spotlight (Roser and Seidel, 2016). Even when for many authors climate justice pertains more to the climate social movements than to academia, all the time more scholars are picking the threads of this debate. As traced by Schlosberg and Collins (2014), climate justice has its beginning in grassroots environmental movements in the United States back in the 1980s, with Hurricane Katrina in 2005 generally considered as the definitive development for the intersection between environmental and climate justice (Bullard and Johnson, 2009).

In brief, climate justice is "a framework that brings into view the intersection between climate change and the way social inequalities are experienced as structural violence" (Porter et al., 2020: 293). This works by recognizing the fact that social groups are differently affected by climate change and that its impacts exacerbate inequitable social conditions (Simmons, 2020). The issues that are most commonly considered through these lenses are gender (Terry, 2009), racial inequalities (Mattar et al., 2021), and the affect on indigenous peoples (Whyte, 2020). Altogether these investigations have pushed for an intersectional approach on climate change analysis (Kaijser and Kronsell, 2014), and for climate policy and governance to put more attention on "local impacts and experience, inequitable vulnerabilities, the importance of community voice, and demands for community sovereignty and functioning" (Schlosberg and Collins, 2014: 359). Some scholars have criticized the fact that climate justice has been deployed principally as an instinctive reaction to a "perceived injustice", limiting its translation into concrete changes in the governance structure (Okereke, 2010: 464).

In the international climate negotiations, climate justice has been mainly translated by means of equity considerations related to who shares the burden of climate action (Grubb et al., 1992). For example, the Paris Agreement recognizes the "principle of equity and common but differentiated responsibilities and respective capabilities, in the light of different national circumstances" (UNFCCC, 2015: 1). The IPCC recognizes that "many of the impacts of warming up to and beyond 1.5°C, and some potential impacts of mitigation actions required to limit warming to 1.5°C, fall disproportionately on the poor and vulnerable" (IPCC, 2018: 51). In other words, climate change is directly related to the inequality produced by broader socio-economic structures embedded in the political system (IPCC, 2018). This debate is most commonly found in the literature as the North–South divide (Parks & Roberts, 2008; Blicharska et al., 2017), having led some authors to hypothesize that "global inequality may be a central impediment to interstate cooperation on climate change policy" (Parks & Roberts, 2008: 621).

The fact that climate justice discussions have mainly taken place in the international arena has sidelined national and local implications. This is rather problematic because it is the states who ultimately have the rights and responsibilities of implementing climate policies and it is at the local level where climate (in)action actually impacts (Bulkeley et al., 2014). Accordingly, Harris proposed the notion of 'cosmopolitan justice', arguing that "this problem is rapidly becoming less about states and more about people" (Harris, 2010: 217). But Harris's argument in the sense that less attention has to be placed on state responsibility is more than controversial, once we consider the inequalities of global (climate) politics and governance. As pointed out by Normann (2021), states are also responsible for imposing developmental models that reproduce coloniality in intercultural relations.

Be it at the international or at the national and local level, another pending question remains related to the way of understanding climate (in)justice in regimes and the best means of addressing such inequities. Indeed, most of the economics contributions have continued adopting a narrow approach focused on distributive effects (Jafino et al., 2021), whereas political science scholars have explored broader considerations. For example, while Gupta and Bhandari (1999) considered three dimensions of climate justice (compensatory justice, distributive justice, and procedural justice), Okereke (2010) listed four dimensions, including: mitigation and burden sharing, impact and adaptation, procedural justice, and systemic justice. Other popular attempts to systematize climate justice have been made by Shue (1993) and Parks and Roberts (2006). Interestingly, only few of them come from the Global South and they don't pay sufficient attention to the domestic implications of climate justice or its impact on vulnerable groups such as indigenous peoples. As a result, in many Global South regions the debates have followed other theoretical guidance. In Latin American, for example, the idea of the 'environmentalism of the poor' proposed by Martínez Alier (2009b) enjoys an outstanding popularity among scholars and social movements alike.

Moving forward the debate, the IPCC acknowledges that the equity principle encompasses both procedural justice (i.e., participation in decision making) and distributive justice (i.e., how the costs and benefits of climate actions are distributed), also including the intergenerational, international, and national dimensions of justice (IPCC, 2018: 55). Given their socio-historical characteristics, indigenous peoples represent a paradigmatic example of climate injustices (Ford et al., 2016). Firstly, given their reliance on natural resources and ecosystems, they endure distributive effects linked to the costs of climate change impacts and policies (distributive injustice). Very frequently they are forced to abandon their lands by deforestation, sea-level rise, major infrastructure projects, and conflict arising from resource scarcity. Secondly, despite the fact that the IPCC suggests including 'traditional' indigenous knowledge as part of the range of practices to face climate change, indigenous peoples are excluded from international fora (procedural injustice). On top of that, a recent current of the literature has been pointing out that indigenous knowledge is generally not regarded as equal in relation to Western knowledge

Box 14.1 Challenges to Indigenous and Local Knowledge in Environmental Decision Making

Injustice	*Definition*	*Examples*
Distributive injustice	Refers to the way in which indigenous peoples carry the burden of climate change impacts and policies while rarely receiving the direct benefits.	• Clean Development Mechanisms • Reducing emissions from deforestation and forest degradation and the role of conservation, sustainable management of forests (REDD+) • Climate migrants • Green colonialism
Procedural injustice	Relates to the exclusion of indigenous peoples from the main decision centers of climate policy.	Internationally. • The Local Communities and Indigenous Peoples Platform (LCIPP) as a way of trying to include their voices in the international negotiations. • The World People's Conference on Climate Change and the Rights of Mother Earth as a platform to demand measures beyond financial compensation.
Epistemic injustice	Captures the way in which climate policy overshadows the indigenous people's capacity to communicate its own knowledge and way of relating to the environment.	Nationally: • The indigenous consultations and other decision-making procedures. • Decontextualization of indigenous knowledge • Cultural Western bias • Epistemic extractivism

when it comes to discussing climate solutions, making it necessary to include a third dimension of this problem: the epistemic (in)justice. In the following this chapter will go deep into these debates, which are summarized in Box 14.1.

14.3 Distributive (In)justice and Indigenous Peoples

It has been widely recognized that indigenous peoples are particularly vulnerable to climate change, being importantly affected by climate impacts and policies (OHCR, 2015; IPCC, 2018). The structural origin of this burden is quite straightforward: on the one hand, indigenous groups maintain a close link with their territories (IIPFCC, 2019); on the other, they have historically suffered from an economic marginalization that has placed them in some of the world's poorest regions (Comberti et al., 2019). This vicious circle has done nothing but reinforce their reliance on ecosystems, aggravating their vulnerability to climate change. All these elements have altered the way of life and production of indigenous peoples, making them subject to climate displacement (Escobar, 2008; Goodwin-Gill & McAdam, 2017).

Paradoxically, indigenous peoples also play a crucial role in climate mitigation and adaptation, especially considering that they inhabit the world's main ecosystems (IIPFCC, 2019). Because of this, the 1997 Kyoto Protocol conceived the Clean Development Mechanisms (CDMs) as a way to boost collaboration between developed and developing countries. The general idea was to allow developed countries to meet their greenhouse gas emissions reduction targets by promoting sustainable development in developing countries (Van Asselt & Gupta, 2009). However, the literature has demonstrated that the CDM projects have been highly problematic due to the absence of any international safeguards on human rights, this having importantly affected indigenous groups all across the Global South (Schade & Obergassel, 2014).

International climate policy has tackled distributive (in)justice mainly by means of financial transfers supposedly oriented to assisting developing countries to face climate mitigation and adaptation challenges (Ravindranath & Sathaye, 2002: 242). However, this policy has been harshly criticized in the Global South for diffusing occidental values and ways of life (Normann, 2021), having even been labeled as "green colonialism" (Van Asselt y Gupta, 2009: 337–338). In this context, Fairhead and his colleagues coined the term 'green grabbing' to refer to "the appropriation of land and resources for environmental ends" (Fairhead et al., 2012: 277). Scholars have also pointed out that these policies have mainly been centered on distributive aspects, neglecting cultural and political elements that might guarantee local and indigenous peoples' participation in the decision-making process (Satyal et al., 2021).

The development of CDMs in the Global South such as wind and solar parks have faced strong local opposition (Escobar, 2008; Mendoza, 2010), so the relation between indigenous lands and extractivism has been quite discussed in the literature (Menezes & Barbosa, 2021; Solorio et al., 2021). On the one hand, indigenous groups argue that climate policies are being sustained on the dispossession of their lands; on the other, there is a demand for national and international recognition of their identity, culture, way of life, and relation to the territory (Ulloa, 2008). This dynamic has attracted important scholarly attention, having been labeled in several ways. Whereas Martínez Alier and his colleagues have employed the term 'ecological distributive conflicts' (Scheidel et al., 2018), other popular terms that have emerged are 'socio-environmental conflicts' (Toledo et al., 2015; Paz, 2017; Pacheco-Vega, 2017) and 'socio-ecological conflicts' (Sen & Pattanaik, 2016).

Bearing in mind that millions of indigenous peoples live in the world's main forests, the instrument for Reducing Emissions from Deforestation and forest Degradation (REDD+) was acknowledged as part of the range of available options to mitigate climate change in the Cancun Agreements of the United Nations Framework Convention on Climate Change (UNFCCC) (Schroeder, 2010; Wallbott & Florian-Rivero, 2018). Its functioning and design are similar to the CDMs, consisting of developed countries giving funds to developing countries for protecting their forests sustainably (Bhullar, 2013). The implementation of REDD+ relates to the struggles of indigenous peoples around the world, but particularly in Latin America, "to strengthen their control over land and the territories they inhabit" (Aguilar-Støen, 2017: 91). Yet, the historical experiences of indigenous peoples' marginalization, their exposure to the effects of deforestation and forest degradation, and their mistrust of forest-based policy measures has led to the adoption of social safeguards (Wallbott & Florian-Rivero, 2018). But scholars have signaled the fact that "nation state responses to REDD often may not match indigenous and local community interests" (Godden & Tehan, 2016: 5), this leading to debates about how to better include indigenous peoples in policy making.

14.4 Procedural (In)justice and Indigenous Peoples

The international climate regime (as with all other environmental regimes) has been constructed with states as central authorities, whereas non-state actors have a limited role in the negotiations. In climate governance non-state actors are entitled, at the most, to an observer status – this extends also to indigenous peoples (Schroeder, 2010: 324). Certainly, this role allows for the presentation of declarations, policy recommendations, and proposals to state representatives (Ford et al., 2016), but in practice the influence in the outcomes of the negotiations is rather limited (Comberti et al., 2019).

Indigenous groups must rely on the representation of their states, where they have traditionally suffered a history of colonization, discrimination, and marginalization (Ulloa et al., 2012; Whyte, 2017). Despite the fact that indigenous groups are historical inhabitants of many of the planet's main ecosystems, they are underrepresented in the decision-making process of international climate policy and governance (Shawoo & Thornton, 2019). As contended by Ford et al. (2016: 440), the "state-centric structure frames the responsibility for climate action of sub-national populations as being in the jurisdiction of national governments".

One of the most significant efforts to integrate indigenous peoples into climate governance was the World People's Conference on Climate Change and the Rights of Mother Earth, which took place at the city of Cochabamba, Bolivia in April 2010. Promoted by the former Bolivian President Evo Morales, this event gathered more than 35,000 people from 140 countries (Kruse, 2014). Having the People's Agreement of Cochabamba as the main outcome, the participants demanded that developed countries commit to quantifiable goals of emission reduction, an acknowledgment of Mother Earth's Rights, measures beyond financial compensation, and the creation of an International Climate and Environmental Justice Tribunal (World People's Conference on Climate Change and the Rights of Mother Earth, 2010). All these proposals represented a way to promote equity in climate policy, but they were largely ignored in the Cancun Agreements during COP 16. Instead, the funding commitments for developing countries were increased both for mitigation and adaptation (Seoane, 2013). This experience represents one of the most recent and iconic examples of the exclusion of indigenous voices in climate international negotiations.

Sponsored by the United Nations, LCIPP is an attempt to facilitate the exchange of experience and the sharing of best practices between indigenous groups within climate negotiations (Shawoo & Thornton, 2019: 1). Members of the Platform have pointed out its potential for increasing the influence of indigenous peoples in negotiations, but have at the same time complained about the lack of funding and technical support for their proper functioning (LCIPP, 2018). For Riedel and Bodle (2018), who carried out one of the few studies on the LCIPP arrangement, a main challenge is the lack of a governance structure that supervises the fulfillment of its functions.

However, this experiment of transnational climate governance deserves further academic and political attention, mostly taking into consideration that it is one of the few ways in which indigenous peoples can overcome the political marginalization to which they have been condemned by national governments (Comberti et al., 2019; Shawoo & Thornton, 2019). To the best of our knowledge, the study developed by Ella Belfer and her colleagues is one of the most complete attempts at assessing the opportunities for and constraints against indigenous participation in the UNFCCC. Their findings highlight that the lack of financial resources and meaningful recognition constrain indigenous participation in climate governance, but at the same time there is a window of opportunity for resource sharing, coordination, and support among indigenous delegates (Belfer et al., 2019).

The procedural injustice faced by indigenous peoples also has national implications. Since 1989, the Indigenous and Tribal Peoples Convention of the International Labour Organization recognizes "the aspirations of these peoples to exercise control over their own institutions, ways of life and economic development" (ILO, 1989). Accordingly, it is the responsibility of national governments to protect the rights of indigenous peoples and to guarantee their integrity whenever "consideration is being given to legislative or administrative measures which may affect them directly" (ILO, 1989), bringing the right to indigenous consultation to the front of the debate. Differently to other instruments of citizen participation, the indigenous consultation is linked to the territorial and cultural rights of indigenous peoples and must comply with certain international standards, including free, prior, informed, and culturally appropriated communication and information exchange (Leifsen et al., 2017).

In practice, nevertheless, indigenous consultation has functioned as an instrument to legitimize the imposition of large-scale projects in indigenous territories (Solorio and Romero, 2021). Following several investigations, very rarely have these processes functioned to empower indigenous communities. The findings of Torres Wong are categorical: "The implementation of the right to prior consultation which was intended to include indigenous peoples in policy making has failed to deter industrialized extractivism" (Torres Wong, 2018: 137). Drawing on the debates of participatory governance and employing cases from Bolivia and Peru, Flemmer and Schilling-Vacaflor (2016) analyzed the "unfulfilled promise of the consultation approach" by focusing their attention on the power asymmetries that affect indigenous peoples. In the same vein and using the case of a contested wind park in Mexico, Dunlap (2018) referred to free, prior, and informed consent as a "bureaucratic trap", that is, a procedure that ultimately affirms state and organizational processes and agendas (Dunlap, 2018). Overall, the indigenous consultation has been seen as "the juridification of collective claims of cultural identity, self-determination, and control over territories and resources" (Rodríguez-Garavito, 2011: 275).

Although many national governments argue that the consultation might be useful to include indigenous groups in decision-making processes, many groups have claimed further recognition of their political and territorial rights. In Latin America, for example, over the past decades there has been a wide process of recognition of indigenous peoples' rights within the national constitutions (CEPAL, 2020). Yet, the difficulty of implementing these reforms has led to some experiences of self-organization and autonomy in which several policies for ecosystem protection and sustainable production have been deployed (Gudynas, 2009).

14.5 Epistemic (In)justice and Indigenous Peoples

Epistemic justice, first defined by Fricker (2007), refers to the epistemological asymmetric positions between groups that create dynamics of injustice. In environmental and climate policy making, scientific knowledge is deemed to be universal, objective, and rigorous, while indigenous and local knowledge is regarded as irrational and subjective. The former holds a position of authority over the latter, and in consequence indigenous knowledge is incorporated within the Western frames of environmental management – reaffirming a cultural Western bias (Shawoo and Thornton, 2019:3). Considering that environmental policy is based upon scientific knowledge, the representation of indigenous people within environmental discourses resonates with colonial history, reproduces relations of exclusion/appropriation, and, ultimately, helps to maintain inequalities (Ulloa, 2012). The attempt to integrate indigenous knowledge usually leads to epistemic extractivism (Guerrero Mc Manus, 2021; McGregor, 2004; Townsend & Townsend, 2021), which happens when the indigenous knowledge is decontextualized and incorporated within (Western) environmental policies without leaving behind any benefits to the communities (Klein, 2013; Grosfoguel, 2015). Whereas this debate has been neglected by the mainstream literature of environmental and climate policy, an important bulk of contributions coming from the anthropology and sociology fields have prepared the ground for discussion about epistemic (in)justice.

Even though indigenous knowledge is a way to relate to the environment (Whyte, 2017; McGregor, 2004), it has been considered that it can easily be integrated into environmental policy without the need of a broader understanding of the different epistemes involved. For example, the IPCC has suggested the use of indigenous and 'traditional' knowledge as part of the strategies of adaptation and mitigation. Due to the social collective memory accumulated within indigenous knowledge, indigenous communities hold a great adaptive capability in the management of agroecological systems. However, the IPCC also recognizes that indigenous knowledge is endangered by multiple factors such as acculturation, dispossession of land rights and land grabbing, environmental changes, and colonization (IPCC, 2018: 337). More recently, there has been a boom of academic literature discussing ways of incorporating indigenous knowledge in climate policy, yet there is no consensus or an operative definition that can be applied throughout different disciplines (Smith & Sharp, 2012).

The IPCC defines indigenous knowledge as the "understandings, skills and philosophies developed by societies with long histories of interaction with their natural surroundings" (IPCC, 2018: 552). But while there are many similar definitions that have a utilitarian approach, it has to be highlighted that the concept itself is a creation of Western knowledge. Indigenous knowledge must be seen as an imposed category given that no indigenous community would classify their worldview as "indigenous knowledge" (Smith & Sharp, 2012: 468). The field of indigenous knowledge and traditional ecological knowledge has found a niche around the environmental and resource management sectors, demonstrating further its conception as a tool to environmental policy (McGregor, 2004).

However, there is an underlying epistemic injustice that affects the incorporation of indigenous knowledge into the process of environmental decision making: taking indigenous groups (and their knowledge) as stakeholders instead of actors with rights and responsibilities affects their autonomy, while their knowledge is basically used as mere data for the decision making (Latulippe & Klenk, 2020). But indigenous knowledge cannot (and should not) fulfill the same functions as scientific knowledge (Mistry and Berardi, 2016). In the words of Tsuji and Ho (2002), indigenous and scientific knowledge are different ways of constructing knowledge and thus they should be treated as such by avoiding integration attempts and opting for a reciprocal exchange.

Much literature has focused on the relationship between scientific and indigenous knowledge (Thompson et al., 2020; Bohensky & Maru, 2011). This is related to recent calls to decolonize academia and climate research (Aikenhead & Ogawa, 2007; Wheeler & Root-Bernstein, 2020), but also connects with the attempts of indigenous groups to build bridges between ways of knowledge. For example, in 2017 the Mexican Zapatista Army for National Liberation[1] declared that the sciences have a lot to learn but also to contribute to the knowledge of indigenous groups. However, the epistemic (in)justice affecting indigenous knowledge is something to be further analyzed by environmental policy experts.

14.6 Conclusion and Future Research

By focusing on the case of climate change and following a three-layered explanation (distributive, procedural, and epistemic injustice), this chapter has assessed the main literature on the inclusion of indigenous and local knowledge in environmental decision making. The review here presented has signaled the need for more trans- and interdisciplinary research, but also for a more inclusive academia in terms of encompassing contributions from the Global South. Only in this way can climate governance be truly reformed to face the challenges posed by the environmental crises.

Several research avenues need to be traced. First and foremost, this contribution calls for climate justice to be analyzed more in national and local contexts. This will help to overcome the state-centered approach in which climate policy investigation has been stuck for decades. Second, researchers need to pay further attention to self-organization experiences of indigenous groups. Too much attention has been placed on the implementation of participative processes such as indigenous consultation, while very little consideration has been paid to the ways in which indigenous groups have tried to overcome the constraints posed by the state. Finally, environmental research needs to pick up the debates on decolonizing knowledge. Otherwise, proposals for integrating indigenous knowledge do nothing but perpetuate the injustices to which indigenous peoples have been condemned for centuries.

Note

1 The EZLN is an indigenous-based political organization established in Chiapas, Mexico, which appeared publicly in 1994. The Congreso Nacional Indígena (National Indigenous Congress – CNI) emerged in 1996 as a result of the EZLN's call for a national fight to guarantee the rights of indigenous communities.

References

Aguilar-Støen, M. (2017, January). Better safe than sorry? Indigenous peoples, carbon cowboys and the governance of REDD in the Amazon. In *Forum for Development Studies*, *44*(1), 91–108. .

Aikenhead, G. S., & Ogawa, M. (2007). Indigenous knowledge and science revisited. *Cultural Studies of Science Education*, *2*(3), 539–620.

Belfer, E., Ford, J. D., Maillet, M., Araos, M., & Flynn, M. (2019). Pursuing an indigenous platform: exploring opportunities and constraints for indigenous participation in the UNFCCC. *Global Environmental Politics*, *19*(1), 12–33.

Berry, L. H., Koski, J., Verkuijl, C., Strambo, C., & Piggot, G. (2019). *Making space: how public participation shapes environmental decision-making.* Stockholm Environment Institute.

Bhullar, L. (2013). REDD+ and the clean development mechanism: A comparative perspective. *International Journal of Rural Law and Policy*, *1*, 1–8.

Biermann, F. (2021). The future of 'environmental' policy in the Anthropocene: Time for a paradigm shift. *Environmental Politics*, *30*(1–2), 61–80.

Blicharska, M., Smithers, R. J., Kuchler, M., Agrawal, G. K., Gutiérrez, J. M., Hassanali, A., Huq, S., Koller, S., Marjit, S., Mshinda, H. M., Masjuki, H. H., Solomons, N. W., Van Staden, J., & Mikusiński, G. (2017). Steps to overcome the North–South divide in research relevant to climate change policy and practice. *Nature Climate Change*, 7, 21–27.

Bohensky, E. L., & Maru, Y. (2011). Indigenous knowledge, science, and resilience: What have we learned from a decade of international literature on "integration"?. *Ecology and Society*, *16*(4), 6.

Bulkeley, H., Edwards, G. A., & Fuller, S. (2014). Contesting climate justice in the city: Examining politics and practice in urban climate change experiments. *Global Environmental Change*, *25*, 31–40.

Bulkeley, H., & Mol, A. P. (2003). Participation and environmental governance: consensus, ambivalence and debate. *Environmental Values*, *12*(2), 143–154.

Bullard, R. D., & Johnson, G. S. (2009). *Environmental justice grassroots activism and its impact.* Environmental Sociology: From Analysis to Action, 63.

CEPAL. (2020). *Los pueblos indígenas de América Latina–Abya Yala y la Agenda 2030 para el Desarrollo Sostenible: tensiones y desafíos desde una perspectiva territorial.* Comisión Económica para América Latina y el Caribe.

Comberti, C., Thornton, T. F., Korodimou, M., Shea, M., & Riamit, K. O. (2019). Adaptation and resilience at the margins: Addressing indigenous peoples' marginalization at international climate negotiations. *Environment: Science and Policy for Sustainable Development, 61*(2), 14–30.

Dunlap, A. (2018). "A bureaucratic trap:" Free, prior and informed consent (FPIC) and wind energy development in Juchitán, Mexico. *Capitalism Nature Socialism, 29*(4), 88–108.

Escobar, E. (2008) Las mujeres indígenas: víctimas y protagonistas de la resistencia contra los megaproyectos. In A. Ulloa, E. Escobar, L. Donato, & P. Escobar (Eds.), *Mujeres indígenas y cambio climático. Perspectivas latinoamericanas*. UNAL-Fundación Natura de Colombia-UNODC, 85–103.

Fairhead, J., Leach, M., & Scoones, I. (2012). Green grabbing: a new appropriation of nature?. *Journal of Peasant Studies, 39*(2), 237–261.

Flemmer, R., & Schilling-Vacaflor, A. (2016). Unfulfilled promises of the consultation approach: the limits to effective indigenous participation in Bolivia's and Peru's extractive industries. *Third World Quarterly, 37*(1), 172–188.

Ford, J., Maillet, M., Pouliot, V., Meredith, T., & Cavanaugh, A. (2016). Adaptation and indigenous peoples in the United Nations framework convention on climate change. *Climatic Change, 139*, 429–443.

Foster, J. B. (1999). Marx's theory of metabolic rift: Classical foundations for environmental sociology. *American Journal of Sociology, 105*(2), 366–405.

Fricker, M. (2007). *Epistemic injustice: Power and the ethics of knowing*. Oxford University Press.

Godden, L., & Tehan, M. (2016). REDD+: climate justice and indigenous and local community rights in an era of climate disruption. *Journal of Energy & Natural Resources Law, 34*(1), 95–108.

Goodwin-Gill, G. S., & McAdam, J. (2017). *UNHCR and climate change, disasters and displacement*. The United Nations Refugee Agency (UNHCR).

Grosfoguel, R. (2015). Del extractivismo económico al extractivismo epistémico y ontológico. *Revista Internacional de Comunicación y Desarrollo, 4*, 33–45.

Grubb, M., Sebenius, J., Magalhaes, A., & Subak, S. (1992). Sharing the burden. In I. M. Mintzer (Ed.), *Confronting climate change: Risks, implications and responses*. Cambridge University Press, 305–322.

Gudynas, E. (2009). La ecología política del giro biocéntrico en la nueva Constitución de Ecuador. *Revista de estudios sociales, 32*, 34–46.

Guerrero Mc Manus, S. (2021). Injusticias epistémicas y crisis ambiental. *Izatapalapa Revista de Ciencias Sociales y Humanidades, 90*(42), 179–204.

Gupta, S., & Bhandari, P. M. (1999). An effective allocation criterion for CO2 emissions. *Energy Policy, 27*(12), 727–736.

Harris, P. G. (2010). Misplaced ethics of climate change: political vs. environmental geography. Ethics, *Place and Environment, 13*(2), 215–222.

IIPFCC. (2019). About the International Indigenous Peoples' Forum on Climate Change. *International Indigenous Peoples' Forum on Climate Change*. http://www.iipfcc.org/quienes-somos

ILO. (1989). C169 - Indigenous and Tribal Peoples Convention, 1989 (No. 169). *International Labour Organization*. https://www.ilo.org/dyn/normlex/en/f?p=NORMLEXPUB:12100:0::NO::P12100_ILO_CODE:C169

IPCC. (2018). *Global warming of 1.5° C: an IPCC special report on the impacts of global warming of 1.5° C above pre-industrial levels and related global greenhouse gas emission pathways, in the context of strengthening the global response to the threat of climate change, sustainable development, and efforts to eradicate poverty*. Intergovernmental Panel on Climate Change.

Jafino, B. A., Kwakkel, J. H., & Taebi, B. (2021). Enabling assessment of distributive justice through models for climate change planning: A review of recent advances and a research agenda. *Wiley Interdisciplinary Reviews: Climate Change, 12*(4), e721.

Jordan, A. J., Huitema, D., Hildén, M., Van Asselt, H., Rayner, T. J., Schoenefeld, J. J., Tosun, J., Forster, J., & Boasson, E. L. (2015). Emergence of polycentric climate governance and its future prospects. *Nature Climate Change, 5*(11), 977–982.

Kaijser, A., & Kronsell, A. (2014). Climate change through the lens of intersectionality. *Environmental politics, 23*(3), 417–433.

Klein, N. (2013, March 6). Dancing the World into Being: A Conversation with Idle No More's Leanne Simpson. https://www.yesmagazine.org/social-justice/2013/03/06/dancing-the-world-into-being-a-conversation-with-idle-no-more-leanne-simpson

Knill, C., & Tosun, J. (2012). *Public policy: A new introduction*. Red Globe Press.

Kruse, J. (2014). Reframing climate change: The Cochabamba conference and global climate politics. In Dietz, Matthias; Garrelts, Heiko (Eds.), *Routledge handbook of the climate change movement*. London: Routledge, 280–291.

Latulippe, N., & Klenk, N. (2020). Making room and moving over: knowledge co-production, Indigenous knowledge sovereignty and the politics of global environmental change decision-making. *Current Opinion in Environmental Sustainability*, *42*, 7–14.
LCIPP (2018). Report of the multi-stakeholder workshop: Implementing the functions of the Local Communities and Indigenous Peoples Platform. *Documents-LCIPP: United Nations Framework Convention on Climate Change.* https://unfccc.int/topics/local-communities-and-indigenous-peoples-platform/the-big-picture/introduction-to-lcipp/documents-lcipp
Leff, E. (2002). *Saber ambiental: sustentabilidad, racionalidad, complejidad, poder.* Siglo xxi.
Leifsen, E., Gustafsson, M. T., Guzmán-Gallegos, M. A., & Schilling-Vacaflor, A. (2017). New mechanisms of participation in extractive governance: between technologies of governance and resistance work. *Third World Quarterly*, *38*(5), 1043–1057. https://doi.org/10.1080/01436597.2017.1302329
López Bárcenas, F. (2002). Territorios, tierras y recursos naturales de los pueblos indígenas en México. In González Galván, Jorge Alberto (Ed.), *Constitución y derechos indígenas*, Universidad Nacional Autónoma de México, 121–143.
Martinez-Alier, J. (2009a). Social metabolism, ecological distribution conflicts, and languages of valuation. *Capitalism Nature Socialism*, *20*(1), 58–87.
Martinez-Alier, J. (2009b). *El Ecologismo de los pobres.* Editorial Icaria.
Mattar, S. D., Jafry, T., Schröder, P., & Ahmad, Z. (2021). Climate justice: priorities for equitable recovery from the pandemic. *Climate Policy*, 1–11.
McGregor, D. (2004). Coming full circle: Indigenous knowledge, environment, and our future. *American Indian Quarterly*, *28*(3/4), 385–410.
Mendoza, E., & Pérez, V. (2010). Energías renovables y movimientos sociales en América Latina. *Estudios Internacionales*, *165*, 109–128.
Menezes, R. G., & Barbosa Jr, R. (2021). Environmental governance under Bolsonaro: dismantling institutions, curtailing participation, delegitimising opposition. *Zeitschrift für Vergleichende Politikwissenschaft*, *15*, 229–247
Mistry, J., & Berardi, A. (2016). Bridging indigenous and scientific knowledge. *Science*, *352*(6291), 1274–1275.
Moore, J. W. (2011). Ecology, capital, and the nature of our times: Accumulation & crisis in the capitalist world-ecology. *Journal of World-Systems Research*, *17*(1), 107–146.
Normann, S. (2021). "Time is our worst enemy:" Lived experiences and intercultural relations in the making of green aluminum. *Journal of Social Issues*, *78*(1), 163–182.
OHCR. (2015). Understanding human rights and climate change. *Office of the High Commissioner for Human Rights.* https://www.ohchr.org/Documents/Issues/ClimateChange/ COP21.pdf
Okereke, C. (2010). Climate justice and the international regime. *Wiley Interdisciplinary Reviews: Climate Change*, *1*(3), 462–474.
Pacheco-Vega, R. (2017). El megaproyecto de la presa El Zapotillo como nodo centroidal de conflicto intratable. Un análisis desde la ecología política. *Espiral (Guadalajara)*, *24*(69), 193–229.
Parks, B. C., & Roberts, J. T. (2006). Globalization, vulnerability to climate change, and perceived injustice. *Society and Natural Resources*, *19*(4), 337–355.
Parks, B. C., & Roberts, J. T. (2008). Inequality and the global climate regime: breaking the north-south impasse. *Cambridge Review of International Affairs*, *21*(4), 621–648.
Paz, M. F. (2017). Luchas en defensa del territorio. Reflexiones desde los conflictos socio ambientales en México. *Acta sociológica*, *73*, 197–219.
Porter, L., Rickards, L., Verlie, B., Bosomworth, K., Moloney, S., Lay, B., Latham, B., Anguelovski, I. & Pellow, D. (2020). Climate justice in a climate changed world. *Planning Theory & Practice*, *21*(2), 293–321.
Ravindranath, N. H. & Sathaye, J. A. (2002). *Climate change and developing countries.* Kluwer Academic Publishers.
Rayner, S. (2010). How to eat an elephant: a bottom-up approach to climate policy. *Climate Policy*, *10*(6), 615–621.
Riedel, A., & Bodle, R. (2018). *Local communities and indigenous peoples platform:-potential governance arrangements under the Paris Agreement.* Nordic Council of Ministers.
Rodríguez-Garavito, C. (2011). Global governance, indigenous peoples, and the right to prior consultation in social minefields. *Indiana Journal of Global Legal Studies*, *18*(1), 263–305. https://doi.org/10.2979/indjglolegstu.18.1.263
Ros, C. (2021). Perspectivas ontológicas sobre la gobernanza ambiental en México. In I. Solorio (Coord.), *México ante la encrucijada de la gobernanza climática. Retos de participación.* Universidad Nacional Autónoma de México, 51–76.

Roser, D., & Seidel, C. (2016). *Climate justice: An introduction*. Routledge.
Satyal, P., Byskov, M. F., & Hyams, K. (2021). Addressing multi-dimensional injustice in indigenous adaptation: the case of Uganda's Batwa community. *Climate and Development, 13*(6), 529–542.
Schade, J., & Obergassel, W. (2014). Human rights and the clean development mechanism. *Cambridge Review of International Affairs, 27*(4), 717–735.
Scheidel, A., Temper, L., Demaria, F., & Martínez-Alier, J. (2018). Ecological distribution conflicts as forces for sustainability: an overview and conceptual framework. *Sustainability Science, 13*(3), 585–598.
Schlosberg, D., & Collins, L. B. (2014). From environmental to climate justice: climate change and the discourse of environmental justice. *Wiley Interdisciplinary Reviews: Climate Change, 5*(3), 359–374.
Schroeder, H. (2010). Agency in international climate negotiations: the case of indigenous peoples and avoided deforestation. *International Environmental Agreements, 10*(4), 317–332.
Seghezzo, L. (2009). The five dimensions of sustainability. *Environmental Politics, 18*(4), 539–556.
Sen, A., & Pattanaik, S. (2016). Politics of biodiversity conservation and socio ecological conflicts in a city: The case of Sanjay Gandhi national park, Mumbai. *Journal of Agricultural and Environmental Ethics, 29*(2), 305–326.
Seoane, J. (2013). Crisis climática: gestión sistémica, falsas soluciones y alternativas desde los pueblos. In J. Seoane, E. Taddei & C. Algranati (Eds.), *Extractivismo, despojo y crisis climática*, 285–315.
Shawoo, Z., & Thornton, T. F. (2019). The UN local communities and Indigenous peoples' platform: A traditional ecological knowledge-based evaluation. *Wiley Interdisciplinary Reviews: Climate Change, 10*(3), e575.
Shue, H. (1993). Subsistence emissions and luxury emissions. *Law & Policy, 15*(1), 39–60.
Simmons, D. (2020). What is 'Climate Justice'?. *Yale Climate Connections*, https://yaleclimateconnections.org/2020/07/what-is-climate-justice/
Smith, H. A. & Sharp, K. (2012). Indigenous climate knowledges. *WIREs Climate Change, 3*, 467–476.
Solorio, I., Ortega, J., Romero, R., & Guzmán, J. (2021). AMLO's populism in Mexico and the framing of the extractivist agenda: the construction of the hegemony of the people without the indigenous voices. *Zeitschrift für Vergleichende Politikwissenschaft, 15*, 249–273.
Solorio, I., & Romero, R. (2021). La gobernanza climática en México frente a las fuerzas sociales. In I. Solorio (Coord.), *México ante la encrucijada de la gobernanza climática. Retos de participación*. Universidad Nacional Autónoma de México, 19–50.
Terry, G. (2009). No climate justice without gender justice: an overview of the issues. *Gender & Development, 17*(1), 5–18.
Thompson, K. L., Lantz, T., & Ban, N. (2020). A review of Indigenous knowledge and participation in environmental monitoring. *Ecology and Society, 25*(2), 10.
Toledo, V. M., Garrido, D., & Barrera-Bassols, N. (2015). The struggle for life: socio-environmental conflicts in Mexico. *Latin American Perspectives, 42*(5), 133–147.
Torres Wong, M. (2018). *Natural resources, extraction and indigenous rights in Latin America: Exploring the boundaries of environmental and state-corporate crime in Bolivia, Peru and Mexico*. Routledge.
Townsend, D. L., & Townsend, L. (2021). Epistemic Injustice and Indigenous Peoples in the Inter-American Human Rights System. *Social Epistemology, 35*(2), 147–159.
Tsuji, L. J., & Ho, E. (2002). Traditional environmental knowledge and western science: in search of common ground. *Canadian Journal of Native Studies, 22*(2), 327–360.
Ulloa, A. (2008). Implicaciones ambientales y culturales del cambio climático para los pueblos indígenas. In A. Ulloa, E. Escobar, L. Donato & P. Escobar (Eds.), *Mujeres indígenas y cambio climático. Perspectivas latinoamericanas*. UNAL-Fundación Natura de Colombia-UNODC, 17–34.
Ulloa, A. (2012). Producción de conocimientos en torno al clima: procesos históricos de exclusión/apropiación de saberes y territorios de mujeres y pueblos indígenas. Working Paper, 21.
UNFCCC. (2015). Paris Agreement. *United Nations Framework Convention on Climate Change*. https://unfccc.int/sites/default/files/english_paris_agreement.pdf
van Asselt, H., & Gupta, J. (2009). Stretching too far? Developing countries and the role of flexibility mechanisms beyond Kyoto. *Stanford Environmental Law Journal, 28*(2), 311–379.
Wallbott, L., & Florian-Rivero, E. M. (2018). Forests, rights and development in Costa Rica: a political ecology perspective on indigenous peoples' engagement in REDD+. *Conflict, Security & Development, 18*(6), 493–519.
Walsh, C. (2011). Afro and indigenous life-visions in/and politics.(De) colonial perspectives in Bolivia and Ecuador. *Bolivian Studies Journal, 18*, 49–69.

Wheeler, H. C., & Root-Bernstein, M. (2020). Informing decision-making with Indigenous and local knowledge and science. *Journal of Applied Ecology*, *57*, 1634–1643.

Whyte, K. (2017). Indigenous climate change studies: Indigenizing futures decolonizing the anthropocene, *English Language Notes*, *55*(1), 153–162.

Whyte, K. (2020). Too late for indigenous climate justice: Ecological and relational tipping points. *Wiley Interdisciplinary Reviews: Climate Change*, *11*(1), e603.

World People's Conference on Climate Change and the Rights of Mother Earth (2010, April 22). *People's agreement of Cochabamba*. World people's conference on climate change. Cochabamba.

Young, O. R., Berkhout, F., Gallopin, G. C., Janssen, M. A., Ostrom, E., & Van der Leeuw, S. (2006). The globalization of socio-ecological systems: an agenda for scientific research. *Global Environmental Change*, *16*(3), 304–316.

15

THE SCIENCE–POLICY INTERFACE AND EVIDENCE-BASED POLICYMAKING IN ENVIRONMENTAL POLICY

Promises and Pitfalls

Adam M. Wellstead, Kei Schmidt, and Anat Gofen

15.1 Introduction

On August 9, 2021, the Working Group I Contribution to the Sixth Assessment Report of the Intergovernmental Panel on Climate Change (IPCC) released its voluminous 3,949-page report, *Climate Change 2021: The Physical Science Basis* (Intergovernmental Panel on Climate Change, 2021). News outlets and social media were quick to seize on the Report's sobering messages about the potentially devastating impacts of climate change. Notably, the IPCC's pinned tweet stated: "ClimateChange 2021: the Physical Science Basis – provides the most updated physical understanding of the climate system and #climatechange, combining the latest advances in climate science and multiple lines of evidence."

The word "evidence" played a prominent role in the full report and was mentioned 2,589 times. The same emphasis on the irrefutable evidence of climate change impacts was also found in the shorter but more digestible *Summary for Policymakers* that accompanied the main report. Since 1990, over 45 similarly themed scientific IPPC reports have been published covering a wide range of topics including climate change impacts, adaptation, vulnerability, carbon dioxide capture and storage, and the impacts on biodiversity and forestry (Intergovernmental Panel on Climate Change, 2014).

Other international and national environmental organizations tasked with investigating environmental problems and providing scientific research also stress the importance of evidence and provide similar messages. For example, in the case of plastic waste, a report commissioned by the Center of Environmental Law stated that "the evidence collected in this report is conclusive that there is an urgent need to adopt a precautionary approach to protect human health from the plastic pollution crisis" (Azoulay et al., 2019, p. 1). Evidence that the UN's 17 Sustainable

DOI: 10.4324/9781003043843-17

Development Goals (SDGs) are part of national government policies is a central concern to the Independent Evaluation Office of the United Nations Development Programme (UNDP) (van den Berg et al., 2017).

At the national level, the US Environmental Protection Agency (EPA) annually publishes an online "Report on the Environment" (ROE) which examines the air, water, land, human exposure as well as health, and ecological conditions across the USA. This report, a compilation of scientific reports, relies on over 80 indicators of peer-reviewed evidence from governmental and non-governmental sources (Environmental Protection Agency 2021).

Despite the constant output of high-quality scientific research and accompanying evidence, very few policies have emerged that significantly decrease CO_2 levels or reduce the amount of plastic in the oceans. There is a growing evidence-based policymaking literature that identifies challenges hindering the development of progressive policies.

In the policy literature, evidence-based policymaking is defined as using what is known from program evaluation or creating more knowledge to better inform future policy decisions. This approach prioritizes rigorous research findings, data, analytics, and the evaluation of innovations. Sanderson (2002) argues that there are two forms of evidence required to improve government effectiveness:

- Promoting the overall accountability of the government;
- Seeking effective policies and programs.

Accountability is understood as information based on performance management measures (e.g., indicators and targets). Evidence of effective policies and programs is qualitatively different and requires a "knowledge of how policy interventions achieve change in social systems" (Sanderson, 2002: 3). Similarly, Head (2008) highlights the need for evidence in policy development, program evaluation, and program improvement. Tied to evidence-based policy are science–policy interfaces defined "as social processes which encompass relations between scientists and other actors in the policy process" (Van den Hove, 2007: 815). Here the concept of the co-production of knowledge is critical and extends beyond simply supplying or bridging knowledge but instead facilitating it (Mass et al., 2022). Shaxon et al.'s (2012) so-called K★Knowledge Spectrum (Figure 15.1) illustrates the variety of entry points for the science–policy interface.

This chapter addresses the promises and pitfalls in the environmental science–policy interface and evidence-based policymaking literature by examining four important contributions. First, an overview of recent scholarship on the barriers and drivers of science–policy interface taxonomies is highlighted and discussed. Many of these barriers and drivers correspond to Cairney's (2016) literature review of evidence-based policymaking in the environmental field. We replicate and update Cairney's study. Second, the scholarship on barriers overlooks the fact that policymakers are unable to consider all the evidence relevant to policy problems. In particular, their bounded rationality means that the ambiguity and complexity inherent in making policy decisions are not considered critical in evidence-based policymaking. That is not to say that policymakers have completely ignored politics or policymaking. To overcome these pitfalls, particular attention needs to be given to understanding the policymaking process and the contributions of policy theory. A third promising development is the role of causal mechanisms in enhancing the barriers and the policy theory approaches. Finally, the rise of environmental policy innovation labs (PILs) as a promising development in evidence-based policymaking is discussed. They may provide a space in which novel approaches to environmental policymaking can be assessed.

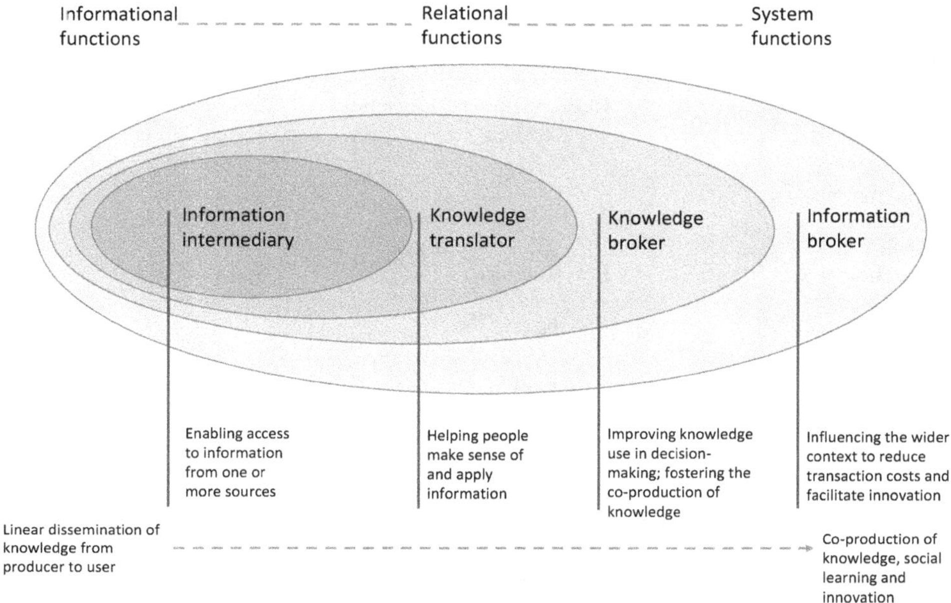

Figure 15.1 The K*Knowledge Spectrum. Based on Shaxon et al. (2012).

15.2 Environmental Evidence-Based Policymaking Scholarship: Science–Policy Barriers and Driver Taxonomies

We begin by highlighting the literature that systematically provides taxonomies of the barriers and drivers of evidence-based policymaking in the environmental sector. Barriers are impediments that can stop, delay, or divert a particular policy change, whereas drivers have the opposite effect, namely that they often trigger change (Moser and Ekstrom, 2010). Based on the studies of practitioners and policymakers, Rose et al. (2018) and Walsh et al. (2019) provide an exhaustive study of these barriers and drivers (Table 15.1). For example, Walsh et al. (2019) identified 64 possible barriers and drivers. We organized these barriers by superimposing Head's (2016) well-known three evidence lenses, namely:

- Scientific research;
- Professional practices;
- Political judgment.

In contrast to the barriers and drivers of political judgment, the barriers and drivers of scientific research and professional judgment are the most numerous and well-developed.

In his study of over 60 peer-reviewed articles, Cairney (2016) analyzed a decade's worth (2005–2015) of empirical literature, identifying the 'barriers' to the adoption of evidence by policymakers in the environmental policy field. He compared these findings with a similar search in the health sector, where such evidence-based policy studies are more prevalent. The central argument is that, in both sectors, policy insights are often influenced by the bounded rationality of policymakers along with their specific policy environments, networks, and contexts. The key point about bounded rationality is that policymakers naturally take shortcuts

Table 15.1 A taxonomy of barrier and driver categories in environmental evidence-based policymaking

Scientific research	*Professional practices*	*Political judgment*
Nature of the evidence	**Research–practitioner links**	**Decision context**
• Existence	• Research–practice links	• Decision makers
• Accessibility	• Manager–advisor relationships	• Social, political, and economic context
• Relevance and applicability	• Demand by advisors	• Implementation capacity
• Quality	• Community strategies	
• Policy relevant science		
Researchers and organizations	**Practitioners**	**Stakeholders**
• Attitudes	• Attitudes	• Stakeholder values and beliefs
• Skills and awareness	• Skills	• Policymaker–stakeholder relationship
• Researcher demands	• Personal characteristics	• Wider policy community
• Researcher culture	• Decision processes	
	• Culture of practitioners	
	• Awareness of literature	
Funding for research	**Management organizations**	**Use of research**
• Funding levels	• Capacity and resources	• Political will
	• Management structure	
	• Organizational culture	

because they cannot consider all evidence relevant to policy problems. Two shortcuts are pursued: the 'rational', by pursuing clear goals and prioritizing certain kinds and sources of information; and the 'irrational', by drawing on emotions, gut feelings, deeply held beliefs, and habits to make decisions quickly (Cairney et al., 2016).

The environmentally based studies that Cairney assessed employed a wide variety of empirically different methods (workshops, interviews, surveys), and they were from a variety of environmental fields (e.g., conservation, climate change, water). Scholars from non-policy fields often were the lead investigators. The specific barriers (and solutions) to the use of evidence in environmental policy across the studies corresponded to Rose et al. (2018) and Walsh et al. (2019) including improving: the supply of or increasing the demand for scientific evidence to or by policymakers, timing, opportunity, and policymakers' minimal understanding of science and reliance on day-to-day crisis management. Cairney found that most of the environmental evidence-based studies focus only on information shortcuts, namely problems of uncertainty and incomplete information. Very few studies referred to policy theory or made explicit ties to the evidence-based policymaking (EBPM) literature.

In March, 2022, we replicated Cairney's study's method by employing the same search parameters for peer-reviewed articles that specifically addressed challenges to evidence-based policy or decision-making. Our search netted seven peer-reviewed articles. They are summarized below.

Salomaa et al.'s (2016) study of 59 Finnish forest stakeholders focused on how evidence is utilized. They found a demand for a wider understanding of evidence to include knowledge of local experiences and their settings. Acknowledged barriers included challenges utilizing different stakeholders' knowledge, conflicting aims, and outlooks about conservation policies. They argued that locally relevant knowledge can be diffused and implemented if all stakeholders are better educated and establish trustworthy relationships.

Marshall et al.'s (2017) key informant interviews of senior policymakers and scientists examined the challenges of incorporating social science into conservation management which is relevant in incorporating the human viewpoint into the discussion on systems and effective outcomes. While some familiar solutions from Cairney's study to overcoming barriers are highlighted, they do raise the importance of acquiring policy acumen, understanding when and where science can be informative in a complex system, understanding the policy process, and how and where social scientists can help produce effective benefits.

Karam-Gemael et al.'s (2018) study of Brazil's biodiversity sector found that the biggest barriers facing policymakers were the time available to read the scientific literature, the difficulty understanding the technical nature of information, and language barriers (because many reports were written in English). From their survey of scientists, legislators, and agency managers, there was a pronounced difference in the priority ranking of the conservation biology agenda. Several proposed initiatives were suggested as a way to bridge the gap between science and policy. They were similar to Cairney's suggestion of 'translating' scientific papers into booklets with a more accessible language.

Rose et al.'s (2018) multi-country multi-phase surveys asked 131 respondents (those in policy positions, practitioners, and research scientists) to identify barriers preventing the use of conservation science in policymaking. The barriers included the lack of policy-relevant sciences, conservation not being a political issue, mismatch of timescales, complex (uncertain) problems, policymakers (scientists) not understanding science (policymaking), the lack of funding, the priority of the private sector in the policy agenda, stakeholders not being valued, and poor communication between scientists and policymakers. Suggested solutions focused generally on overcoming uncertainty such as "demonstrate benefits of conservation,", "improve policy education of scientists", "better science advocacy", "more knowledge brokers", or more "collaboration between scientists and policymakers".

The most extensive barrier and enabler typology was developed by Walsh et al. (2019). Their findings were based largely on interviews of key informant conservation practitioners and a literature review of the healthcare-based barriers and drivers. A total of 230 barriers formed their "conservation knowledge-action framework". They were included in one of eight broad categories (nature of evidence, research-practice links, decision context, researchers and research organizations, practitioners, management organizations, stakeholders, and wider community). These were further refined and are listed in Table 15.1. Importantly, in some cases, what was considered to be a barrier in one context could also be a driver in another.

Lemieux et al. (2021) surveyed 181 practitioners, mainly from government organizations, to examine the state of evidence-based decision-making in Canada's protected area organizations. They noted that over the past six years, the number of perceived barriers affecting the access and use of evidence had increased. In particular, limited financial resources, the lack of policy that prioritizes or promotes the use of empirical evidence, limited training or experience in evaluating evidence, uncertain information, and a lack of trust were identified as the most significant barriers. They did note the importance of greater collaboration with aboriginal people, recognizing them as key policy actors. They recommended a more "relevant evidence base", more collaborative forums for sharing information, a greater role for scientists in protected area organizations, and greater accountability and transparency towards knowledge management as critical policy drivers.

In the seventh and final study, Zea-Reyes et al. (2021) examined institutional barriers to climate change adaptation in Beirut, Lebanon. They found the two major influences are "political interference" and an altered view of where and how climate change should be addressed. Significantly these two barriers function in cycles, connecting and further enforcing many different barriers which operate on lower levels. They also identified 11 opportunities to confront these barriers, which include options at the "individual", "organizational", "enabling environment", and "non-government" levels.

Predictably, the offered solutions in our selected literature review were similar to those highlighted by Cairney in 2016, namely an emphasis placed on hierarchies of evidence, strategies for improving the supply of information to policymakers, and encouraging academic–practitioner workshops. Absent in the barriers and drivers literature is a consideration of policy theory, the key consideration in Head's (2008) political judgment lens for EBPM.

15.3 The Importance of Policy Theory: Addressing Ambiguity and Complexity

The sole strategy of informing decision-makers with an increasing volume of information may be unsuccessful or contribute to, at best, a slow, indirect, and cumulative effect of "knowledge creep" (Daviter, 2015). It is part of their nature that scientists avoid the topic of ambiguity. Cairney et al. (2016) points to two reasons. First, many try to draw an artificial and misleading line between facts and values or science and politics. They prefer to see themselves as playing the role of an "honest broker" rather than advocating for any particular issue or policy (see Pielke, 2007; Douglas, 2009; Smith and Stewart, 2017). Second, they do not understand the policy process. Their exposure to the policymaking process is centered on the heuristic policy cycle (Oxman et al., 2009; Cairney, 2016). In professional circles, policymaking is viewed as a mysterious "black box" containing irrational policymakers who lack the necessary "political will" or courage and that is responsible for unpredictable and suboptimal outcomes (Biesbroek et al., 2015). More problematic "is categorizing any factor or process as a barrier reduces complex and highly dynamic decision-making processes into simplified, static and metaphorical statements about why current outcomes are 'incorrect'" (Biesbroek et al., 2015, p. 494). Important concepts such as political values and power are often overlooked, as are debate, coalition formation, the role of persuasion, and the ability to frame policy problems to reduce ambiguity.

It is often these political factors that the demand for scientific evidence comes from rather than the constant stream of studies and data. Cairney and Oliver (2017) suggest that those advocating scientific evidence engage in such strategies as forming coalitions or telling stories (narratives) that weave in the evidence to overcome the emotional or ideological bias of policymakers. These strategies provide advice about how to intervene in practice to secure better outcomes. Over the past quarter-century, policy theories and frameworks have been developed from which concepts can be employed to explain other sources of ambiguity and complexity. Table 15.2 outlines some possible avenues for future research.

Table 15.2 Examples of policy theory concepts influencing evidence-based policymaking

Multiple Streams Approach (MSA)	*Advocacy Coalition Framework (ACF)*	*Punctuated Equilibrium Theory (PET)*	*Institutional Analysis & Development Framework (IAD)*	*Narrative policy Framework (NPF)*
Ambiguity of policymaking (policy streams)	Policy-oriented beliefs	Policy feedback	Co-production	Narratives
Policy windows	Coalition formation	Venue shifts	Rule-making	Storytelling

15.4 Moving beyond Black and Gray Boxes: A Call for Causal Mechanisms

Employing policy theory concepts to understand ambiguity in the science–policy interface adds to the understanding of environmental EBPM in terms of barriers and drivers. Policy process theories also contribute to a causal understanding of why evidence often does not contribute to policy change. However, these approaches do not go far enough in that they present a 'gray boxed' approach to causality. That is, the causal claims are not fully explicitly made. A causal mechanism approach, building on Cairney and Oliver's work above, to EBPM is another promising avenue of research. In fact, their proposed strategies of coalition formation and 'storytelling' which emerge from policy theories are well developed causal mechanisms. We will elaborate on this concept.

A mechanism is defined as "a process in a concrete system, such that it is capable of bringing about or preventing some change in the system as a whole or in some of its subsystems" (Bunge, 1997, p. 414). Elster also argues that "roughly speaking, mechanisms are frequently occurring and easily recognizable causal patterns that are triggered under generally unknown conditions or with indeterminate consequences. They allow us to explain, but not to predict" (Elster, 2007, 36). Beisbroek et al. (2014) provide further elaboration and state that mechanisms consist of

> entities (with their properties) and the activities that these entities engage in either by themselves or in concert with other entities. These activities bring about change, and the type of change brought about depends on the properties of the entities and the way in which they are linked to one another.
>
> *(p. 109)*

Since mechanisms are portable concepts, they can be applied by policy researchers to investigate the long-term nature of EBPM (Falleti and Lynch, 2009).

A mechanism-based perspective creates the possibility for generalizing about the cause–effect relationships that observing barriers and drivers cannot do. Examples of well know social science causal mechanisms include self-fulfilling prophecies, spill-over effects, and dialogues of the deaf. Providing an evidence-based policy causal explanation requires careful consideration of the interaction between the mechanism(s) and the contextual conditions within which the mechanism operates. Figure 15.2 illustrates how an initial set of conditions play a key role in determining if, when, and how certain mechanisms are triggered and how they might only play out under certain contextual conditions. Causal mechanisms help explain how and

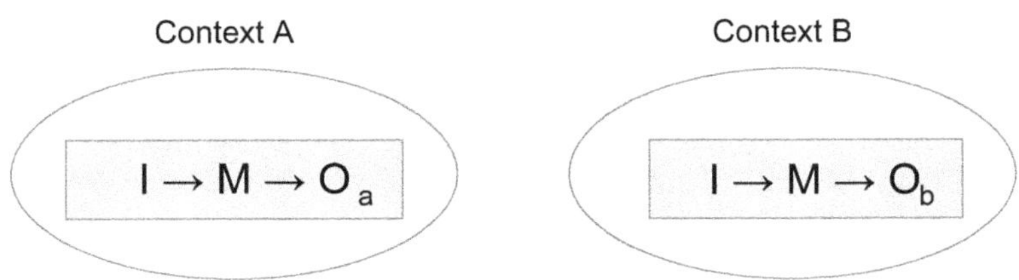

Figure 15.2 I → M → O model in different contexts.

Source: Falleti and Lynch (2009).

why a hypothesized cause, in a given context, contributes to a particular outcome. Context is important as it allows for formulating more refined hypotheses by specifying under which conditions certain mechanisms are most likely to occur or produce a certain effect, which is critical for evidence-based approaches.

Here, observed patterns of (un)intended outcomes can be explained by identifying the plausible causal set of mechanisms within the situational context of the process. By adopting a causal mechanisms approach to EBPM, careful attention to the inputs (e.g., actors, nature of the evidence) and context (e.g., policy style, administrative traditions) is required.

For several decades, the causal mechanism literature has been employed in the program evaluation literature (see Pawson, 2000). While the calls for understanding causality in policy theory are not new (Sabatier, 1991), recently there has been a growing interest in seriously considering the role of policy-related causal mechanisms in the area of policy design (Capano et al., 2019). Based on a large bibliometric study, van der Heijden et al. (2021) examined identified causal mechanisms in papers applying five popular policy process frameworks.[1] They concluded that a majority of studies, including those with an environmental focus, failed to explicitly engage in a mechanism-based perspective. In Table 15.3, we list some of the barriers listed in Table 15.1 and policy process concepts from Table 15.2. For each case, possible causal mechanisms are listed with examples from the environmental policy literature. The portability of causal mechanisms and their prevalence in the social science literature offers promising directions in understanding EBPM.

Stemke et al. (2012) consider how mixed-scanning can address uncertainties in long-term environmental and climate change planning and design. Overland and Sovacool (2020) examine how research funding transparency and coordination mechanisms would improve

Table 15.3 Possible evidence-based policy causal mechanisms

Evidence lenses	*Barriers/policy theories*	*Possible causal mechanisms*	*Examples from the environmental policy literature*
Scientific	Accessibility of evidence	Information/mixed scanning	Stremke et al. (2012): Environmental planning
	Research funding	Agency funding procedures	Overland & Sovacool (2020): Climate change research funding
Professional	Decision processes	Protecting turf	Kinley (2021): US government climate change planning
	Organizational culture	Blame avoidance	Howlett (2014): Climate change policy innovation
	Attitudes	Groupthink	Kennedy (1988): US natural resource agencies
Political	Policy windows (MSA)	Strategic behavior of policy entrepreneurs	Arnold (2021): US Fracking
	Coalition formation (ACF)	Coalition coordination	Hysing & Olsson (2008): Swedish forest policy
	Policy feedback (PET)	Venue shifting	Carter & Jacobs (2014): UK environmental and energy policy
	Rulemaking (IAD)	Negotiation	Wilkes-Allemann et al. (2022): Recreational policy (Austria, Germany, Switzerland)
	Policy narratives (NPF)	Angel/devil shift	Pattison et al. (2022): Fracking

communication between climate change researchers and policymakers. Kinley et al. (2021) argue that individual US federal government departments often protected their own turf, a common professional mechanism, which often led to a dispersed and weak government-wide response to climate change. Howlett (2014) developed a blame avoidance framework based on intentionality and intensity that allows for hypothesis testing of the propensity for climate change policy innovation. Groupthink is a well-known mechanism that Kennedy (1988) examined within the context of the US natural resource agencies. Specifically, he found the illusion of morality, shared stereotypes, 'mind guarding', and the illusion of invulnerability impacted how these agencies operated.

From the policy theory literature, Arnold (2021) found that the strategic behavior of policy entrepreneurs was critical in advancing fracking policy in the state of New York. The coordination of the Swedish environmental coalition was central to developing Sweden's forestry Environmental Quality Objectives (Hysing & Olsson, 2008). Applying the punctuated equilibrium theory (PET), Carter and Jacobs (2014) found four significant venue shifts in the UK's environmental and energy policy in the early 2000s during the Labour government. Negotiation processes (or action situations) between mountain bike users and other recreational users in Austria, Germany, and Switzerland was examined by Wilkes-Allemann et al. (2022). Pattison et al. (2022) examined the devil and angel shifts, concepts from behavioral economics in Twitter debates regarding fracking. The devil shift is misperceiving an opponent's intentions as more malicious or evil than would be expected.

In addition to identifying the scientific, professional, and political policy evidence-based causal mechanisms, the policy research and analysis communities should be interested in testing policy mechanisms. There are several under-utilized qualitative methods where policy analysts can collect and diagnose evidence and test hypotheses from which they can provide narratives explaining how a particular outcome or set of events came about. Of these process tracing is the most promising method that can assist in investigating complex interactions and empirical fingerprints that the mechanism makes and unpacking it into its constituent parts (Meyfroidt, 2016; Trampusch and Palier, 2016). As a qualitative method, a variety of information sources such as archival documents, interviews, reports, and memos can be utilized.

15.5 Policy Labs: A Venue to Address Policy–Science Challenges?

In scarcely a decade, a global 'labification' trend has taken hold, whereby the search for innovative policy solutions is embedded within scientific experiment-like structures. Often referred to as policy innovation labs (PILs), they can be found within government agencies, universities, and not-for-profit organizations.[2] Each seeks to address a pressing social or economic issue. Their rapid growth has led to claims that PILs "are on the path to becoming a pervasive part of the social infrastructure of modern public organizations" (Carstensen and Bason, 2012, 5). PILs share similarities and resemble well-known organizations, including think tanks, research institutes, or policy shops. They build on new knowledge and skills that are often recruited from other parts of an organization (such as is the case with government-based policy labs) or by autonomous or semi-autonomous organizations. Often there are established teams (or organizations, or institutes) set up specifically for innovative activities in physical spaces to conduct workshops or other stakeholder activities. Flexibility, adaptation, and creativity to deal with environments characterized by uncertainty, ambiguity, and information overload are some of their key features that enable them to produce innovative products and services, adapt quickly to new opportunities, and build emergent strategies (Lindquist and Buttazzoni, 2021).

Table 15.4 Policy innovation lab focus

Environment only	30
Multi-sector labs that include environmental issues	64
Total	94

Table 15.5 Location of environmental policy innovation labs

Inter-organizational partnerships	17
UN	4
National governments	9
City/municipal governments	12
Universities	27
Private organizations	19
Not-for-profit organizations	6

PILs tend to share three distinctive features. (1) The use of design-thinking methodology (e.g. McGann et al., 2018), which originated in industrial and product and service design. (2) A focus on innovation through the application of experimental approaches and the emulation of scientific methodologies to test and measure the efficacy of various public policies and programs, thus drawing on experiments, often as pilots or prototypes. By seeking to emulate scientific methodologies, PILs attempt to test and measure the efficacy of various public policies and programs as well as to provide evidence for evidence-based design (Bason, 2017). (3) A user-centric approach whereby target populations actively engage in the design process (Lee and Ma, 2020). Guided by user-centric approaches and drawing on experiments as pilots, policy labs aim to address implementation gaps (e.g. Gassner and Gofen, 2018) and noncompliance (Gofen, 2014; 2015) by enhancing the notion of evidence-based design. The collaborative engagement of stakeholders and experimentation have been suggested as a strategy for the diffusion of sustainability issues.

Based on a global search, we identified 94 active environmentally oriented PILs. Some PILs are solely environmentally focused while others address a suite of issues including environmental ones (Table 15.4). An example of the latter category would be the Philadelphia Behavioral Science Initiative, whose goal is to "help the City harness the best ideas and methods from economics, psychology, political science, and sociology, and other fields, to address the needs of Philadelphia's residents and businesses in creative and effective ways" and has three projects which are specifically focused on the environment (Philadelphia Behavioral Science Initiative, 2021). Common environmental areas examined include climate change, the circular economy, green urban development, the energy transition, clean transportation, waste disposal, agriculture and food, environmental conservation, and clean water. A majority (68) of the environmental PILs focus on more than one of these areas, and on average the labs have nearly eight individual projects occurring within a variety of areas. Environmental PILs are found within governments and in arm's length groups (Table 15.5). However, for these groups, government funding is common.

15.6 Conclusion and Future Research

In this chapter, we have discussed four promising areas in the environmental EBPM field. Important empirical research has been conducted that maps out the barriers and drivers in the environmental science–policy interface. Here, the focus has been on challenges to scientific objectivity and challenges to technical or professional expertise. Cairney et al. (2016) argue that little attention has been paid to how scientists and experts engage in policy and political advocacy. This engagement is important because relaying mostly on information, policymakers will struggle through inaccessible and equivocal scientific evidence. This is a common story throughout the environmental sector as illustrated in Cairney's (2016) and our review of case studies. Cairney et al. (2016) and Cairney and Oliver (2017) argue for the need to address ambiguity and complexity. In doing so, they turn to policy theory which we have elaborated upon in the environmental policy context. Thirdly, we have argued that the barrier–driver typology literature and policy theory insights to EBPM would be further enhanced by understanding the causal mechanisms underpinning these concepts. Additionally, qualitative methods such as process tracing can be utilized which would allow researchers to understand cause and effect in EBPM. Finally, we have suggested that the emergence of PILs in the environmental sector might offer promise to enhancing EBPM particularly with their emphasis on experimental approaches and stakeholder engagement.

Algorithmic policymaking is a research area that has received relatively little scholarly attention in the environmental policy field but will have a significant impact on EBPM. Algorithmic decision-making is embedded with other disruptive technologies including artificial intelligence, machine learning, autonomous vehicles, big data, block chains, quantum computing, and the internet-of-things. Genomics and synthetic biology are two relatively new environmental policy areas. More importantly, the growing importance of algorithms will have an impact of EBPM. Algorithms are computational sets of rules that computer programmers input into technologies to 'design in' particular patterns of decision-making, which generate outputs that technologies are designed to solve. Algorithms underpin many of the attempts at 'automating' decisions in contemporary public policy. Combined with big data and machine learning, scientists will have a greater predictive modeling capacity, such as predicting greenhouse gas concentrations, sea ice loss, risks to wildlife, and forest degradation. As discussed earlier and despite the call for data-driven policymaking, more and improved information will not lead to policy change. A more significant future area of research is how algorithms consistently and subtly shape human behavior and our influence on the world's landscapes, oceans, air, and ecosystems. The critical ethical question is rather: On the basis of what decision, or what assumption, is complex reality to be reduced to an algorithmic function that is not neutral but embodied? The critical consideration in 'digital era governance' is the context of our choices and how they are framed. The imperative of what appears to be objective quantifiable data yielding the possibility of prediction requires responsible engagement with the frame that is reinforced by technological practices that are not neutral but rather frame our habitual perceptions, generating and/or reinforcing implicit biases. This new level of complexity associated with EBPM only amplifies the importance of the earlier themes highlighted in this chapter and points to promising areas of research.

Box 15.1 Chapter Summary

- There is growth in the scholarship developing taxonomies of barriers and drivers of EBPM.
- Most of the barriers to evidence in environmental policy centers on the lack of information or uncertainty by policymakers.
- Ambiguity related barriers are often overlooked. Policy process concepts can help explain the challenges associated with the science–policy interface.
- Policy-process-related causal mechanisms improve the understanding of EBPM.
- Environmentally based PILs are a promising avenue for evidence-based policy.

Notes

1 The Streams Approach, Advocacy Coalition Framework, Punctuated Equilibrium Theory, Narrative Policy Framework, and Institutional Analysis and Development Framework.

2 PILs are also referred to as "public innovation labs", "public sector innovation labs", "government innovation labs", "organizational innovation labs", "policy innovation labs", "innovation labs", "public policy labs", "social innovation labs", "systems change labs", "design labs", and "policy labs" (Whicher, 2021; Hinrichs-Krapels et al., 2020).

References

Arnold, G., 2021. Does entrepreneurship work? Understanding what policy entrepreneurs do and whether it matters. *Policy Studies Journal*, 49(4), pp. 968–991.

Azoulay, D., Villa, P., Arellano, Y., Gordon, M.F., Moon, D., Miller, K.A. and Thompson, K., 2019. Plastic & health: the hidden costs of a plastic planet. *CIEL*. https://www.ciel.org/plasticandhealth/

Bason, Christian. 2017. *Leading Public Design*. Bristol: Policy Press.

Biesbroek, G.R., Termeer, C.J., Klostermann, J.E. and Kabat, P., 2014. Rethinking barriers to adaptation: Mechanism-based explanation of impasses in the governance of an innovative adaptation measure. *Global Environmental Change*, 26, pp. 108–118.

Biesbroek, R., Dupuis, J., Jordan, A., Wellstead, A., Howlett, M., Cairney, P., Rayner, J. and Davidson, D., 2015. Opening up the black box of adaptation decision-making. *Nature Climate Change*, 5(6), pp. 493–494.

Bunge, M., 1997. Mechanism and explanation. *Philosophy of the Social Sciences*, 27(4), pp. 410–465.

Cairney, P., 2016. *The Politics of Evidence-based Policy Making*. Springer.

Cairney, P. and Oliver, K., 2017. Evidence-based policymaking is not like evidence-based medicine, so how far should you go to bridge the divide between evidence and policy?. *Health Research Policy and Systems*, 15(1), pp. 1–11.

Cairney, P., Oliver, K. and Wellstead, A., 2016. To bridge the divide between evidence and policy: reduce ambiguity as much as uncertainty. *Public Administration Review*, 76(3), pp. 399–402.

Capano, G., Howlett, M., Ramesh, M. and Virani, A., 2019. *Making Policies Work*. Edward Elgar Publishing.

Carstensen, Helle Vibeke and Bason, Christian. 2012. Powering collaborative policy innovation: Can innovation labs help? *The Innovation Journal: The Public Sector Innovation Journal*, 17(1), article 4.

Carter, N. and Jacobs, M., 2014. Explaining radical policy change: the case of climate change and energy policy under the British Labour government 2006–10. *Public Administration*, 92(1), pp. 125–141.

Daviter, F., 2015. The political use of knowledge in the policy process. *Policy Sciences*, 48(4), pp. 491–505.

Douglas, Heather E. 2009. *Science, Policy, and the Value-Free Ideal*. Pittsburgh, PA: University of Pittsburgh Press.

Elster, Jon. 2007. *Explaining Social Behavior: More Nuts and Bolts for the Social Sciences*. Cambridge: Cambridge University Press.

Falleti, T.G. and Lynch, J.F., 2009. Context and causal mechanisms in political analysis. *Comparative Political Studies*, 42(9), pp. 1143–1166.

Gassner, Drorit, and Gofen, Anat. 2018. Street-level management: A clientele-agent perspective on implementation. *Journal of Public Administration Research and Theory* 28(4), pp. 551–568.

Gofen, A. 2015. Citizens' entrepreneurial role in public service provision. *Public Management Review* 17(3), pp. 404–424.

Gofen, Anat. 2014. Mind the gap: Dimensions and influence of street-level divergence. *Journal of Public Administration Research and Theory* 24(2), pp. 473–493.

Head, B. W. 2008. Three lenses of evidence-based policy. *Australian Journal of Public Administration* 67(1), pp. 1–11.

Head, B.W., 2016. Toward more "evidence-informed" policy making?. *Public Administration Review*, 76(3), pp. 472–484.

Hinrichs-Krapels, Saba, Jocelyn Bailey, Harriet Boulding, Bobby Duffy, Rachel Hesketh, Emma Kinloch, Alexandra Pollitt, et al. 2020. Using policy labs as a process to bring evidence closer to public policy-making: A guide to one approach. *Palgrave Communications* 6(1): 1–9.

Howlett, M., 2014. Why are policy innovations rare and so often negative? Blame avoidance and problem denial in climate change policy-making. *Global Environmental Change*, 29, pp. 395–403.

Hysing, E. and Olsson, J., 2008. Contextualising the advocacy coalition framework: theorising change in Swedish forest policy. *Environmental Politics*, 17(5), pp. 730–748.

Intergovernmental Panel on Climate Change. 2014. List of Major IPCC Reports. In *Climate Change 2014 – Impacts, Adaptation and Vulnerability: Part B: Regional Aspects: Working Group II Contribution to the IPCC Fifth Assessment Report* (pp. 1783–1786). Cambridge: Cambridge University Press.

Intergovernmental Panel on Climate Change. 2021. *Climate Change 2021: The Physical Science Basis.* Contribution of Working Group I to the Sixth Assessment Report of the Intergovernmental Panel on Climate Change. Cambridge: Cambridge University Press.

Karam-Gemael, M., Loyola, R., Penha, J. and Izzo, T., 2018. Poor alignment of priorities between scientists and policymakers highlights the need for evidence-informed conservation in Brazil. *Perspectives in Ecology and Conservation*, 16(3), pp. 125–132.

Kennedy, J.J., 1988. Legislative confrontation of groupthink in US natural resource agencies. *Environmental Conservation*, 15(2), pp. 123–128.

Kinley, R., Cutajar, M.Z., de Boer, Y. and Figueres, C., 2021. Beyond good intentions, to urgent action: Former UNFCCC leaders take stock of thirty years of international climate change negotiations. *Climate Policy*, 21(5), pp. 593–603.

Lemieux, C.J., Halpenny, E.A., Swerdfager, T., He, M., Gould, A.J., Carruthers Den Hoed, D., Bueddefeld, J., Hvenegaard, G.T., Joubert, B. and Rollins, R., 2021. Free Fallin'? The decline in evidence-based decision-making by Canada's protected areas managers. *FACETS*, *6*(1), pp. 640–664.

Lindquist, Evert A. and Buttazzoni, Michael. 2021. The ecology of open innovation units: Adhocracy and competing values in public service systems. *Policy Design and Practice* 4(2), pp. 212–227.

Marshall, N., Adger, N., Attwood, S., Brown, K., Crissman, C., Cvitanovic, C., De Young, C., Gooch, M., James, C., Jessen, S. and Johnson, D., 2017. Empirically derived guidance for social scientists to influence environmental policy. *PLoS One*, *12*(3), p. e0171950.

McGann, Michael, Blomkamp, Emma and Lewis, Jenny M. 2018. The rise of public sector innovation labs: Experiments in design thinking for policy. *Policy Sciences* 51(3), pp. 249–267.

Meyfroidt, Patrick. 2016. Approaches and terminology for causal analysis in land systems science. *Journal of Land Use Science* 11(5), pp. 501–522.

Moser, Susanne C. and Ekstrom, Julia A. 2010. A framework to diagnose barriers to climate change adaptation. *Proceedings of the National Academy of Sciences of the United States of America* 107(51), pp. 22026–22031.

Overland, I. and Sovacool, B.K., 2020. The misallocation of climate research funding. *Energy Research & Social Science*, 62, p. 101349.

Pattison, Andrew, Cipolli, William and Marichal, Jose. 2022. The devil we know and the angel that did not fly: An examination of devil/angel shift in twitter fracking "debates" in NY 2008–2018. *Review of Policy Research* 39(1), pp. 51–72.

Pawson, Ray. 2000. Middle-range realism. *European Journal of Sociology* 41(2), pp. 283–325.

Philadelphia Behavioral Science Initiative. 2021. *What We Do*. https://phillybsi.org/what-we-do

Pielke, Roger A. Jr. 2007. *The Honest Broker: Making Sense of Science in Policy and Politics*. Cambridge: Cambridge University Press.

Rose, D.C., Sutherland, W.J., Amano, T., González-Varo, J.P., Robertson, R.J., Simmons, B.I., Wauchope, H.S., Kovacs, E., Durán, A.P., Vadrot, A.B. and Wu, W., 2018. The major barriers to evidence-informed conservation policy and possible solutions. *Conservation Letters*, 11(5), p. e12564.

Sabatier, Paul A. 1991. Toward better theories of the policy process. *Political Science and Politics* 24(2), pp. 147–156.

Salomaa, A., Paloniemi, R., Hujala, T., Rantala, S., Arponen, A. and Niemelä, J., 2016. The use of knowledge in evidence-informed voluntary conservation of Finnish forests. *Forest Policy and Economics*, 73, pp. 90–98.

Sanderson, Ian. 2002. Evaluation, policy learning and evidence-based policy making. *Public Administration* 80 (1), pp. 1–22.

Shaxon, L., Bielak, A., Ahmed, I., Brien, D., Conant, B., Middleton, A., Fisher, C., Gwyn, E., Klerkx, L., Morton, S. and Pant, L., 2012. Expanding our understanding of K*(KT, KE, KTT, KMb, KB, KM, etc.), K* Conference.

Stremke, S., Van Kann, F. and Koh, J., 2012. Integrated visions (part I): Methodological framework for long-term regional design. *European Planning Studies*, 20(2), pp. 305–319.

Trampusch, Christine and Palier, Bruno. 2016. Between X and Y: How process tracing contributes to opening the black box of causality. *New Political Economy* 21(5), pp. 437–454.

van den Berg, R., Naidoo, I., and S. Tamondong, eds. 2017. *Evaluation for Agenda 2030: Providing Evidence on Progress and Sustainability*. Exeter, UK: IDEAS.

Van den Hove, S., 2007. A rationale for science–policy interfaces. *Futures*, 39(7), pp. 807–826.

van der Heijden, J., Kuhlmann, J., Lindquist, E. and Wellstead, A., 2021. Have policy process scholars embraced causal mechanisms? A review of five popular frameworks. *Public Policy and Administration*, 36(2), pp. 163–186.

Walsh, J.C., Dicks, L.V., Raymond, C.M. and Sutherland, W.J., 2019. A typology of barriers and enablers of scientific evidence use in conservation practice. *Journal of Environmental Management*, 250, p. 109481.

Whicher, A. 2021. Evolution of policy labs and use of design for policy in UK government. *Policy Design and Practice* 4(2), pp. 252–270.

Wilkes-Allemann, J., Ludvig, A., Gobs, S., Lieberherr, E., Hogl, K. and Selter, A., 2022. Getting a grip on negotiation processes: Addressing trade-offs in mountain biking in Austria, Germany and Switzerland. *Forest Policy and Economics*, 136, p. 102683.

Zea-Reyes, L., Olivotto, V. and Bergh, S.I., 2021. Understanding institutional barriers in the climate change adaptation planning process of the city of Beirut: vicious cycles and opportunities. *Mitigation and Adaptation Strategies for Global Change*, 26(6), pp. 1–24.

PART III

Environmental Policy Change

16

POLICY CHANGE AND POLICY ACCUMULATION IN THE ENVIRONMENTAL DOMAIN

Causes and Consequences

Christoph Knill

16.1 Introduction

Policy experts, NGOs, and citizens regularly criticize the insufficient nature and status quo orientation of political action (Roberts, 2019). This frustration has been particularly prominent in the context of climate change politics, where incremental policy responses have been strongly criticized (Luce, 2020) and described as a "failure of democracy" (Shearman and Smith, 2007). Yet, widespread concerns about absent, insufficient, or delayed governmental responses can be found not only for the challenge of climate change, but for almost any environmental problem. This is not to say that governments have not been effective in addressing certain pressing environmental problems, but rather that a lot of problems persist and/or have even worsened over time. To name just a few examples: the growing use of fertilizers in agriculture poses a growing threat to water quality; pesticides and deforestation pose increasing challenges to flora and fauna; and growing traffic and urbanization come with growing air pollution. Taken together, these considerations suggest that governmental policy responses are generally not sufficiently effective to deal with many of the currently pressing environmental problems (Adam et al., 2022).

From the viewpoint of theories of policy change, this statement is hardly surprising. As emphasized by Punctuated Equilibrium Theory (PET), policy-makers suffer from bounded rationality and cannot pay attention to all demands for policy adjustment at the same time. Typically, the existence of policy subsystems allows for the parallel processing of several demands and compensates for these cognitive limitations. At the same time, however, the processing of new information by subsystems typically reinforces the policy status quo rather than promoting policy change. This is because subsystems tend to provide a fixed venue in which stable actor constellations interact and, over time, create a dominant policy image or paradigm. This policy image will influence the processing of new incoming information and thereby attenuate the perceived demands for policy adjustment. But when information is able to attract attention beyond the relevant subsystem, this negative feedback mechanism that attenuates the perceived demand for change can turn into a positive feedback mechanism amplifying the perceived demand for adjustment. These mechanisms of negative and positive feedback mean that political

DOI: 10.4324/9781003043843-19

systems tend to process new information disproportionately. They "shift from underreacting to overreacting to information" (Jones and Baumgartner, 2012, p. 7). In turn, one central claim of PET is that policy output reflects a pattern of punctuated equilibrium where policy stability is disrupted every now and then by substantial punctuations. In other words, the steady flow of political demands is typically translated into patterns of enduring policy stability and incrementalism interrupted by rare instances of major change (Baumgartner and Jones, 1993; Baumgartner et al., 2014).

Yet, this pattern of rare policy breakthroughs is in sharp contrast to the findings of studies that have analyzed aggregate developments of environmental policy portfolios over time. Despite certain variation across countries, the dominant picture is one of highly dynamic policy accumulation in modern democracies, with a continuous and steep increase in policy outputs over the last five decades, which strongly exceeds accumulation rates in other policy areas (Adam et al., 2019). In other words, the strong growth in the size of environmental policy portfolios over time seems to be barely compatible with general accounts of policy stability and limited policy effectiveness. Rather aggregate assessments reveal the emergence of an impressive stock of highly diversified environmental policies (Fernández-i-Marín et al., 2021), hinting at considerable progress and development in national governance repertoires in order to effectively tackle environmental policy problems.

How do these contrasting assessments go together? The main argument advanced in this chapter is that both assessments do not contradict, but partially reinforce each other. On the one hand, the phenomenon of policy accumulation is only weakly related to concerns for improving policy effectiveness. The main driver of accumulation is political responsiveness, which does not necessarily mean that these responses are actually effective in tackling environmental problems. On the other hand, policy accumulation creates additional problems that undermine overall policy effectiveness. These problems refer in particular to the overburdening of implementation capacities and growing policy complexity as a result of multiple interactions between an ever-increasing number of policy elements. In view of these considerations, the main challenge is to better align political responsiveness and policy effectiveness; that is to develop mechanisms that ensure that the growth in policy outputs actually improves overall policy effectiveness.

In order to present these arguments in closer detail the remainder of this chapter is structured as follows. The next section provides an empirical overview of policy accumulation in the environmental domain based on data on the long-term development of environmental policy stocks in OECD countries. Based on this assessment, we then turn to a discussion of the potential causes that are responsible for policy development and also its variation across countries (Section 16.3). In Section 16.4, we turn to the question of how policy accumulation undermines the potential for developing effective policy responses. The resulting question of how to better reconcile policy accumulation and policy effectiveness is addressed in Section 16.5. Section 16.6 concludes.

16.2 Policy Accumulation in the Environmental Domain: Empirical Overview

Policy accumulation relates to aggregate developments of sectoral policy stocks. In the comparative public policy and public administration literature, the phenomenon of policy accumulation has appeared on the analytical radar only recently. The dominant focus of research has been on describing and explaining instances of individual policy change and implementation (Knill and Tosun, 2020). Despite this general assessment, there is broad scholarly agreement that policy stocks in modern democracies are continuously growing, regardless of the policy area or country under study. There are a number of publications pointing to the presence of accumulation patterns, emphasizing widespread and densely populated "policyscapes" (Mettler, 2016), as well as layered and increasingly complex mixes of policy instruments and targets (Hacker, 2004; Thelen, 2004). In short, policy accumulation is an undisputed phenomenon. It occurs whenever the rate of policy production exceeds the rate of policy termination (Adam et al., 2019).

Policy accumulation hence reflects the result of a continuous addition of new policy elements to existing policy portfolios without the compensatory reduction of already existing policy elements. In this context, a policy element constitutes the combination of (1) a policy target and (2) a policy instrument. While policy targets define *what or who is being addressed* by a new policy, policy instruments define *how the target is being addressed* (Adam et al., 2019). For measuring the size of sectoral policy portfolios, we are not interested in the restrictiveness or generosity of a certain policy, that is the setting of the policy instruments, since changes in instrument settings do not contribute to the size of the policy portfolio. Instead, we are interested in the introduction of new policy instruments and/or the widening of the scope of existing policy instruments to new policy targets.

For instance, consider a law that amends a country's policy portfolio by introducing a maximum value for lead emissions into continental surface water. In our approach, this legal provision represents one policy element as it brings together one policy instrument, namely the setting of an obligatory standard of a maximal limit for emissions, and one policy item, that is the emissions of lead into surface waters. Similarly, we count the first-time regulation of a maximal value for the emissions of phosphates into continental surface water as one additional policy element within the national policy portfolio. While such a regulation includes the same policy instrument – an obligatory standard – the instrument does address a new policy target, that is the emission of phosphates into continental surface waters. A reform that only implies a lower maximal limit on phosphate emissions into surface waters does affect the restrictiveness of a certain policy but leaves the size of a country's policy portfolio completely unchanged.

The differentiation between policy targets and instruments leaves us with a two-dimensional portfolio space (Adam et al., 2017; Fernández-i-Marín et al., 2021). Based on this portfolio space, we can calculate a standardized measure of the sectoral portfolio size that can range from 0 (no policy instrument for any of the targets) to 1 (all policy instruments for all the targets). We measure policy accumulation with reference to a predefined benchmark of a maximum number of policy targets and policy instruments for each policy field under study. If one policy item or instrument is adopted in any of the countries in our sample at any point in time, it is included in the list of items that can potentially be addressed.

Based on this approach, we can identify 48 policy targets across the three subfields of clean air, water, and nature conservation policy. Policy targets are mostly pollutants like ozone, carbon dioxide, or sulfur dioxide in the air, but also comprise other substances like lead content in gasoline, sulfur content in diesel, nitrates, and phosphates in continental surface water, as well as environmental objects like native forests, endangered plants, or endangered species. We distinguish 12 types of policy instruments (plus one residual category) that range from hierarchical forms of governing, such as obligatory policy standards and technological prescriptions, to economic incentives through taxes, subsidies, and other forms of market intervention. Combining all available policy instruments and items results in a maximal size of the environmental rule stock consisting of 624 spaces (each of the 48 items can be addressed by up to 13 instruments).

Figure 16.1 displays the growth of environmental policy portfolios for 21 OECD countries between 1976 and 2018. The data reveal a general pattern of impressive growth of

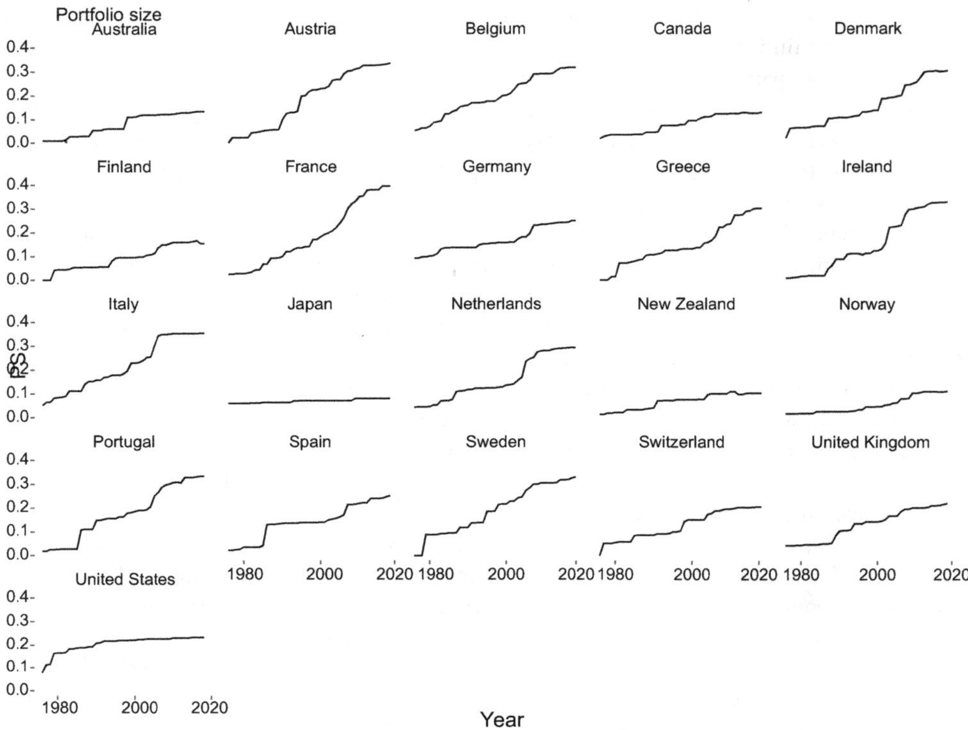

Figure 16.1 Growth of environmental policy portfolios in 21 OECD countries.

environmental policy outputs. As shown by Adam et al. (2017, 2019), these rates of environmental policy accumulation by far exceed the growth rates observed for other policy areas, such as social policy or morality policy. Yet, despite the overall trend, there is considerable variation in policy accumulation rates across countries. While some countries, like Austria, France, Ireland, Italy, Portugal, or Sweden, display very high growth rates and have filled around 40 percent of the maximum portfolio space over the investigation period, rates of policy accumulation remain at somewhat lower levels for other countries. Yet, also in these countries, we nevertheless observe pronounced increases in policy stocks, with only Japan standing out as a pattern of relative stagnation.

As we will discuss in the following sections, the degree of environmental policy accumulation as such does not tell us very much about actual policy performance. Larger policy portfolios do not necessarily mean that countries are more effective in handling problems of environmental pollution or climate change. Rather even the opposite might be the case in certain constellations.

16.3 Causes of Environmental Policy Accumulation: Responsiveness to National and International Demands

Although there is considerable country variation in environmental policy accumulation, the overall trend is clear. Over the last 45 years, OECD countries have strongly expanded their environmental policy portfolios. Yet, as will be argued in this section, this does not by any means ensure that these increases are actually matched by corresponding improvements in policy performance, that is the overall effectiveness of the measures taken in order to address environmental problems. The main reason for this loose coupling between policy accumulation and policy effectiveness lies in the fact that effectiveness per se is not the central driver of accumulation. Policy accumulation is driven by a range of factors, which are primarily related to political responsiveness and international influences rather than concerns about policy performance.

16.3.1 The Political Economy of Policy Accumulation

First, the most central driver of policy accumulation is the responsiveness of political actors to societal demands (Adam et al., 2019). Policies pile up in modern democracies for primarily political reasons: vote-seeking politicians demonstrate responsiveness to their citizens' demands by constantly proposing new policies in the form of laws, regulations, or programs. Policies, in turn, create expectations and dependence in beneficiaries and thus are difficult to terminate or dismantle once established. Political incentive structures hence entail that over time, and regardless of the exact policy sector in question, governments typically adopt more policies than they eliminate.

Power-seeking politicians thus have strong incentives to demonstrate their responsiveness to societal demands by constantly proposing new policies. Yet, this interest in policy production does not extend to the subsequent challenge of policy implementation. As political responsibilities for implementation success are often unclear, there are weak electoral incentives for politicians to invest in implementation capacities. Although such capacity increases improve policy effectiveness, the attribution of such improvements to the actions of particular political actors is difficult for voters. This is reinforced by the fact that implementation effectiveness is not only a matter of administrative capacities, but also affected by a range of other factors (Harding and Stasavage, 2013). This weakens the incentives for political actors to engage in costly improvements of implementation capacities, while the fruits of gaining political credit by announcing new ambitious policies hang much lower (Fernández-i-Marín et al., 2021a). It is only in a constellation in which voters are directly affected by certain policies and are hence highly aware of the effectiveness of these measures that merely demonstrating responsiveness might not be sufficient for vote-seeking politicians. They rather need to ensure that these policies actually work (Dasgupta and Kapur, 2020).

But what happens if administrative resources do not keep pace with accumulating policy stocks? In a recent comparative study on the link between policy accumulation and policy effectiveness in the environmental field, Limberg et al. (2021) show that the adoption of new policies enhances policy effectiveness only as long as increases in the sectoral policy stock are backed by an expansion of implementation capacities. In the absence of such an expansion, further policies either remain largely ineffective or make things even worse, that is when the over-burdened bureaucracy opts to prioritize and selectively implement only those policies and cases that are easy to control but may not lead to substantive improvement (Tummers et al., 2015).

From a mere political logic, we should expect a 'toxic combination' of strong policy accumulation and stagnating or even declining implementation capacities. Although the urgency of this problem might vary across countries and sectors in view of differences in accumulation rates and variation in levels of implementation capacity, the nature of the challenge remains the same: if we assume that policy-making is exclusively driven by politics, we would expect an ever-growing 'burden-capacity gap,' with ever more policies undermining rather than strengthening overall policy effectiveness.

16.3.2 International Drivers of Policy Accumulation

Many national policies originate from international influences, either via legal obligation to comply with international rules or policy diffusion driven by trade relations or intensified transnational communication between countries (Holzinger and Knill, 2005; Holzinger et al., 2008). As a consequence, environmental policy accumulation to a considerable extent can be interpreted as domestic response to international influences.

It has long been acknowledged that many environmental problems, most notably issues related to climate change, are transnational in nature and hence require the development of coordinated policy approaches at the international level. In this regard, the European Union (EU) has developed an impressive policy stock since the early 1970s (Knill and Liefferink, 2007), although this dynamic has been considerable reduced since the 2008 financial crisis (Steinebach and Knill, 2018; Burns et al., 2019). EU member states are legally obliged to transpose these measures by adopting necessary policies. As a consequence, compliance with EU environmental law and to a similar extent also with agreements at the level of international organizations should result in domestic policy accumulation.

In addition to domestic policy adoptions triggered by compliance obligations, the literature on environmental policy diffusion (Busch et al., 2003; Verdolini and Bosetti, 2017) hints at the role of trade dependence between countries that might facilitate policy adoptions in the environmental domain. Close trade connections seem to stimulate the adoption of additional policies. The most likely reasons for this connection are the harmonization of domestic regulatory standards in the case of environmental policy (Holzinger et al., 2008; Vogel, 1995). Moreover, the mere fact that countries exchange information about the policy approaches adopted in other countries might constitute a driver of domestic policy adoptions. Such effects of transnational communication might not only be driven by legitimacy concerns (i.e. the striving of governments to increase their legitimacy by embracing forms and practices that are valued within the broader environment), but also because countries transfer policy solutions from other countries they consider particularly successful in addressing certain problems (Rose 1991; Dolowitz and Marsh, 2000; Evans, 2017).

At first glance, there seems to be a closer connection between the adoption of internationally required or inspired policies and policy effectiveness. Many environmental problems can only be effectively addressed by internationally agreed solutions, and in particular practices of transferring successful policies developed elsewhere should basically help to improve overall sectoral policy performance. Yet, this argument overlooks the fact that policies of international origin simply contribute to the extension of existing national policy stocks. Hence, while the size of environmental policy portfolios continues to grow, this does not mean that this growth is actually matched by corresponding increases in implementation capacities. The large literature on the implementation problems of EU environmental policy (Knill and Lenschow, 2000; Thomann and Sager, 2017; Börzel, 2021) shows that there is no reason to assume that governments take more effort to implement policies of European rather than national origin. Moreover, as mentioned above, legitimacy concerns might be a central driver of policy diffusion. While adopting policies that are already in place in many other countries might strengthen governmental legitimacy, this does not in anyway imply that governments are actually interested in effectively implementing these policies. In short, similar to national policies, international policies also lead to domestic policy accumulation and hence potentially create similar challenges with regard to implementation capacities and hence policy effectiveness.

16.3.3 The Ambiguous Nature of Policy Accumulation

We have seen so far that the reasons for ever-growing environmental policy stocks are much less related to attempts of increasing sectoral policy performance as one might intuitively expect. Concerns over policy effectiveness are of subordinate relevance in the strategic calculations of political actors. Policy accumulation is hence in the first place nothing more than the result of political decisions; there is no guarantee that these decisions are followed by corresponding action in the sense of effective implementation (Brunsson, 2002).

Yet, it is important to emphasize that decisions and hence the adoption of new policies are a necessary condition to effectively tackle environmental problems. And we should also not deny that on the basis of such decisions a broad range of problems has been successfully addressed in the past. The steep rise in environmental policy stocks in this respect of course meant significant progress in national and international attempts to deal with environmental policy challenges. In many ways, the accumulation of environmental policy measures is the hard-fought result of democratically led battles on how to mitigate pressing environmental problems. While many people forget about the important benefits of regulation and government intervention, we can generally assume that most people are happy not to live in a country that still trusts in the environmental policy portfolio of the 1950s. Accumulating public policies have achieved substantial improvements in public health, the water quality of rivers and lakes, and many areas of nature protection (Adam et al., 2019).

And yet the continuous expansion of environmental policy portfolios is a highly ambivalent process. On the one hand, it is the essential prerequisite for tackling environmental challenges. On the other hand, effective problem-solving is not the major impetus behind policy accumulation. We should hence be careful to conclude that growing policy stocks comes with better policy performance.

Box 16.1 Causes and Consequences of Policy Accumulation

The phenomenon of environmental policy accumulation is defined by the constant growth of environmental policy portfolios, i.e. the number of policy targets and policy instruments is growing over time.

Causes of policy accumulation

- Strong political incentives for policy production (democratic responsiveness);
- Low political incentives to improve implementation effectiveness of existing policies;
- International factors (EU harmonization, international organizations, policy diffusion).

Consequences of policy accumulation

- Policies are important tools to address environmental problems; however
- Policy accumulation creates policy complexity;
- Policy accumulation bears the risk of bureaucratic overload, hence undermining implementation effectiveness.

16.4 Consequences of Environmental Policy Accumulation: Policy Complexity and Bureaucratic Overload

These doubts about a direct coupling between accumulation and policy performance are reinforced by the fact that policy accumulation comes with two additional challenges that might undermine sectoral policy effectiveness. These problems associated with policy accumulation refer to policy complexity and bureaucratic overload. While policy complexity creates problems for effective policy design, bureaucratic overload undermines the implementation effectiveness of sectoral policies.

16.4.1 Policy Complexity

A major reason why policies may not achieve their intended objectives is flawed policy design. Yet, avoiding such design flaws is far from trivial. Governments need to solve multiple issues at the same time, even when it comes to individual policy areas such as environmental or climate policy. Here, governments not only need to deal with air pollutants and greenhouse gas emissions from both stationary and nonstationary sources, they also need to curb water pollution, provide protection to endangered plants and animals, and attend to other relevant environmental concerns. To address these diverse targets, governments must develop new policies that specify concrete instruments for the different issues at stake. What sounds obvious in theory, however, is difficult in practice, as the choice of suitable instruments is far from self-evident. For instance, whether command-and-control measures, economic incentives, mere information provision, or a combination of these instruments will work more effectively in a certain context is often subject to intense political and academic debate. The fact that governments face many such decisions only reinforces the underlying challenges of designing effective public policies (Fernández-i-Marín et al., 2021).

The study of policy design (Howlett and Mukherjee, 2018); seeks to identify what makes public policies more or less effective and then, on the basis of these findings, to inform and improve policy-making efforts and outcomes. Crucial here are studies that focus on different types of policy instruments, their advantages and disadvantages, as well as on the processes involved in their selection and implementation (Peters, 2018). Through these efforts, scholars have identified several abstract principles characterizing well-designed policies – in particular, the consistency, coherence, and congruence of policy targets and instruments (Howlett and Rayner, 2013). Although the respective terms are often defined quite ambiguously, they essentially imply that a government's multiple policy targets and instruments should be logically connected and mutually reinforce, rather than work against, one another (Gunningham and Sinclair, 1999).

Yet, what sounds plausible in theory is an increasingly difficult undertaking in practice because the continuous accumulation of policy portfolios makes it more and more difficult to learn about policy effects (Adam et al., 2018). Policy portfolios are not merely complicated systems that comprise a large (and increasing) number of independent policy targets and instruments, but also – and to an increasing extent – reflect complex systems characterized by the interactions between different policy targets and instruments (Howlett and Rayner, 2013). Complexity is a direct consequence of continuous policy accumulation: as national policy mixes combine a growing number of interacting policy targets and instruments, the degree of policy complexity increases. Accordingly, the number of interactions that have to be considered when trying to relate outcomes to existing policy designs also grows. This complicates any attempts to evaluate the effectiveness of public policies in a way that will be able to feed back into policy-making. In short, policy accumulation makes policy portfolios increasingly complex. Policy complexity, in turn, hampers the design of effective policies.

16.4.2 Bureaucratic Overload

As discussed in the previous section, the political drivers of policy accumulation might come with the problem of an increasing gap between accumulating implementation tasks and restricted administrative capacities for implementation. If administrative resources (in particular staff, expertise, funding) are not expanded in line with growing tasks, then growing bureaucratic overload will lead to ineffective implementation and hence negatively affect policy performance.

To deal with policy accumulation under substantial resource constraints, implementers develop coping strategies (Lipsky, 2010) that come along with different types of implementation deficits (Oberfield, 2010). First, implementers can opt for an incremental piecemeal approach leading to implementation backlogs. They carry out their tasks in a routine way while taking some measures to reduce their case and workload. This can include limiting clients' demands by making themselves unavailable to contacts or by forcing clients to wait (Tummers et al., 2015). Second, implementers can opt for prioritization and hence selective implementation. Regardless of the rationale guiding implementers to prioritize some cases or tasks ahead of others, this coping strategy ultimately leads to patterns of selective implementation, with some policies or parts of policies being systematically preferred to others while some are not or not sufficiently put into practice (O'Brien and Li, 1999; Winter and Nielsen, 2008).

Implementation research so far has not systematically recognized these problems of accumulation-induced bureaucratic overload. Implementation research typically uses detailed case studies to analyze the implementation effectiveness of individual policies. A whole research branch has developed in recent decades, analyzing the variation of implementation effectiveness and underlying causes (Winter, 2012). Although these studies advance our understanding of individual implementation processes, we lack a holistic perspective on the prevalence of implementation deficits in a whole policy sector that systematically takes account of interdependencies across different policies. In view of constrained implementation capacities, effective implementation of a newly adopted policy 'A' might come with the deficient implementation of already existing policies 'B' or 'C,' as implementers shift their priorities. Regardless of whether implementation is analyzed top-down or bottom-up (Hill and Hupe, 2009), the focus on potential deficits remains case- rather than prevalence-oriented. Related to the above problem is the fact that, so far, no systematic attempts have been made to study the extent to which the accumulation of implementation burdens contributes to the prevalence of deficient implementation.

Yet, while data on policy accumulation are still relatively easy to collect, aggregate measures of environmental implementation capacities pose considerable challenges in terms of data collection. Any precise assessment would entail extremely time-intensive research on changes in the budgets and manpower of administrative bodies in charge of implementing environmental policies. Moreover, in many cases, such data might not be available for longer periods of time. To deal with these problems, Fernández-i-Marín et al. (2021a) developed a measure of implementation capacities by using a combination of different data sources, such as information on the quality of national public administration provided by the V-Dem 10 dataset (Coppedge et al., 2019) and the World Bank data on tax revenues and administrative renumeration (2020), the 'Weberianness index' by Rauch and Evans (2000), the index of 'information capacity' as generated by Brambor et al. (2019), as well as the environmental institutional capacity index (Jahn, 2016). Figure 16.2 displays the development of implementation capacities based on this assessment for the same sample of countries we have analyzed above. Without going too much into detail, it becomes clearly obvious that implementation capacities – across all countries – have developed much less dynamically than has been the case for their environmental policy portfolios. Implementation capacities overall display a pattern of high stability over time. This development points to a potentially increasing gap between accumulating policies and restricted implementation capacities.

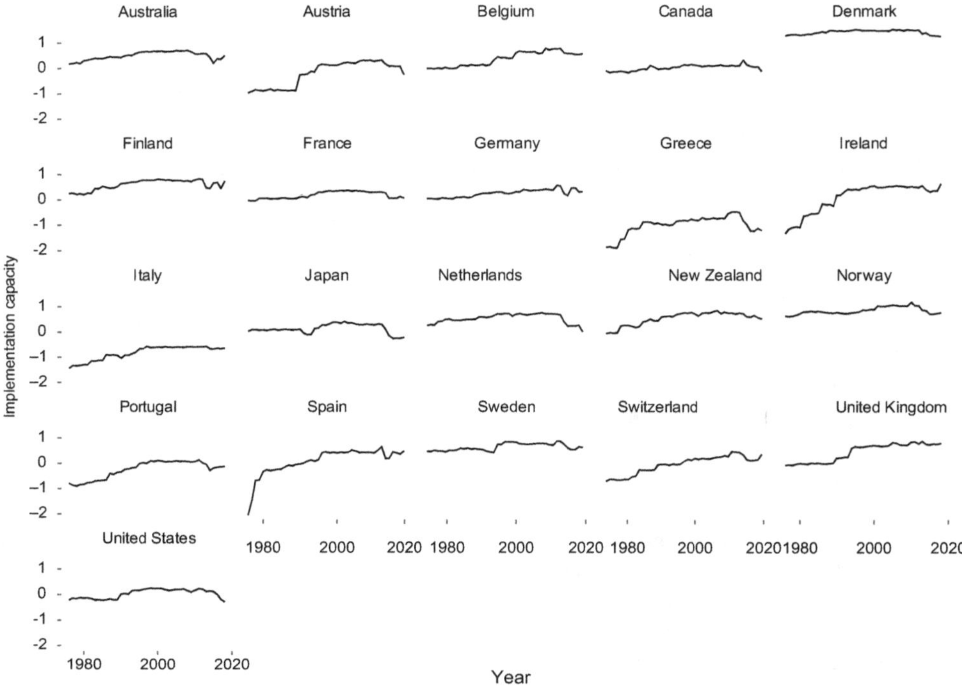

Figure 16.2 Development of implementation capacities in 21 OECD countries.

16.5 The Challenge of Sustainable Policy Accumulation

We have seen that there are potential trade-offs between policy responsiveness and policy effectiveness. The more governments respond to societal demands and environmental problems via policy accumulation, the more they might overburden implementation bodies with ever-more and increasingly complex policies (Limberg et al., 2021). This, in turn, has the potential to undermine the long-term support for governmental intervention.

Theoretically, governments might avoid this "responsiveness trap" (Adam et al., 2019) by expanding administrative capacities along with growing implementation burdens emerging from policy accumulation. Yet, most governments face fundamental fiscal and ideological constraints for public sector expansion in times of globalized financial markets, austerity, and the still reverberating ideas of New Public Management (Lobao et al., 2018). A more realistic option is that governments manage to keep policy accumulation at a sustainable level. Sustainable accumulation means that governments respond responsibly to the constant demand for new policies. In concrete terms, this implies that policy growth rates are kept at a moderate level while at the same time existing implementation capacities are allocated as efficiently and purposefully as possible (Knill et al., 2020).

Achieving this objective and avoiding excessive accumulation is everything but easy. The central challenge is that, in many instances, there is a division of labor between the sectoral bureaucracies in charge of policy formulation and those in charge of policy implementation. On the one hand, there are bureaucracies – typically at the ministerial level – that are responsible for drafting new policies in response to political or societal demands. It is in the very interest of these bureaucracies to expand their competences and strengthen their institutional status by constantly producing new policy proposals. On the other hand, and in contrast to the ministerial level where policies are off the table once they are adopted, the burden to implement these policies accumulates at the 'street-level,' that is at the desks of people working in implementing bodies and executive agencies.

Structural capacities for integrating the processes of policy production and implementation play a decisive role in bridging the gap between responsive and effective policy-making. These patterns of vertical policy-process integration (VPI) (Knill et al., 2020) might not only vary across countries, but also across policy sectors. There are two dimensions of institutionalized feedback and exchange through which VPI might contribute to avoiding excessive policy accumulation. The first dimension – *bottom-up integration* – captures the volume of the 'voice' that implementation bodies have in policy formulation, in particular with regard to their experience with the design flaws of existing policies. This way, VPI reduces the risk that ever-new policies are needed to compensate for the deficits of the policies already in place. The second dimension – *top-down integration* – refers to the extent to which the implementation costs are internalized in the process of policy formulation. To what extent do bureaucracies that produce policies also have to bear the cost for implementing these measures? In short, well-integrated bureaucracies should not only produce fewer policies, but also need fewer new policies than administrations that lack effective VPI.

In a recent study on the impact of VPI on the development of the gap between accumulating implementation tasks and implementation capacities in 21 OECD countries, Fernandez-í-Marín et al. (2021a) provide strong support for the above arguments. The more countries have institutionally integrated processes of environmental policy formulation and policy implementation, the lower is their 'burden-capacity gap.' The lower the gap, the higher in turn is the potential that new policies are actually effective and improve sectoral policy performance (Limberg et al., 2021).

16.6 Conclusion and Future Research

While established theories of policy change, in particular PET, expect high policy stability over time that is rarely interrupted by punctuations of major change, an aggregate view on the development of environmental policy stocks reveals rather dynamic patterns of continuous growth and policy accumulation. Yet, policy accumulation should not be confused with improved policy performance: more policies need not necessarily enhance policy effectiveness, but might even undermine it. In this regard, it makes no difference whether the growth of the policy stock is driven by a constant flow of incremental additions or by more abrupt large-scale reforms.

On the one hand, the adoption of new policies is an essential condition for tackling the broad array of environmental policy problems governments are confronted with. Policy accumulation in many ways reflects progress in the problem-solving capacities of modern democracies. On the other hand, policy accumulation comes with several problems that might undermine rather than improve sectoral policy performance. Policy accumulation makes environmental policy portfolios not only more complex, but also comes with the risk of bureaucratic overload, leading to growing implementation deficits. As a consequence, new policies might even make things worse if they are not backed by corresponding expansions of implementation capacities.

Tackling environmental problems is therefore a balancing act. Governments need to reconcile political responsiveness and policy effectiveness. Although this sounds pretty obvious, political incentive structures indicate potential contradictions between drivers of policy accumulation and policy performance. Recent research findings suggest that those countries that dispose of well-developed institutional arrangements for integrating the formulation and implementation of environmental policies are in a much better position to achieve 'sustainable policy accumulation,' implying that policy responses are backed by necessary implementation capacities. Future research should further investigate trade-offs between the adoption of new policies and potential disadvantages emerging from policy growth with regard to policy complexity and bureaucratic overburdening. Research indicates that institutional arrangements facilitating the vertical integration of policy production and policy implementation might help to avoid these problems associated with policy accumulation. These arguments need to be explored more systematically in future research.

References

Adam, C., S. Hurka, C. Knill, & Y. Steinebach (2019). *Policy Accumulation and the Democratic Responsiveness Trap*. Cambridge University Press.

Adam, C., S. Hurka, C. Knill, & Y. Steinebach (2022). On democratic intelligence and failure: The Vice and virtue of incrementalism under political fragmentation and policy accumulation. *Governance*, online first, 35, 525–543.

Adam, Christian, Christoph Knill, & Xavier Fernandéz-i-Marín. (2017). Rule growth and government effectiveness: Why it takes the capacity to learn and coordinate to constrain rule growth. *Policy Sciences* 50, 241–268.

Adam, C., Y. Steinebach, & C. Knill (2018). Neglected challenges to evidence-based policy-making: the problem of policy accumulation. *Policy Sciences* 51(3), 269–290.

Baumgartner, F. R., & B. D. Jones (1993). *Agendas and Instability in American Politics*. Chicago University Press.

Baumgartner, F. R., B. D. Jones, & P. B. Mortensen (2014). Punctuated equilibrium theory: Explaining stability and change in public policymaking. In P. A. Sabatier & C. M. Weible (Eds.), *Theories of the Policy Process* (pp. 25–58). Westview Press.

Börzel, Tanja A. (2021). *Why Non-Compliance? The Politics of Law in the European Union*: Cornell: Cornell University Press.

Brambor, Thomas, Agustín Goenaga, Johannes Lindvall, & Jan Teorell (2019). The lay of the land: Information capacity and the modern state. *Comparative Political Studies* 53, 175–213.

Brunsson, N. (2002). *The Organization of Hypocrisy: Talk, Decisions and Actions in Organizations*. Copenhagen: Copenhagen Business School Press.

Busch, Per-Olof, Helge Jörgens, & Kerstin Tews (2003). The diffusion of new environmental policy instruments. *European Journal of Political Research* 41(2), 569–600.

Burns, C., V. Gravey, A. Jordan, & A. Zito (2019). De-europeanising or disengaging? EU environmental policy and brexit. *Environmental Politics* 28(2), doi:10.1080/09644016.2019.1549774.

Coppedge, Michael, John Gerring, Carl Henrik Knutsen, Staffan I. Lindberg, Jan Teorell, David Altman, Michael Bernhard, M. Steven Fish, Adam Glynn, Allen Hicken, Anna Lührmann, Kyle L. Marquardt, Kelly McMann, Pamela Paxton, Daniel Pemstein, Brigitte Seim, Rachel Sigman, Svend-Erik Skaaning, Jeffrey Staton, Steven Wilson, Agnes Cornell, Lisa Gastaldi, Haakon Gjerløw,&Nina Ilchenko. (2019). *V-Dem [Country-Year/Country-Date] Dataset v9. Varieties of Democracy (V-Dem) Project.*

Dasgupta, Aditya, & Devesh Kapur (2020). The political economy of bureaucratic overload: Evidence from rural development officials in India. *American Political Science Review* 114(4), 1316–1334.

Dolowitz, D. P., & D. Marsh (2000). Learning from abroad: The role of policy transfer in contemporary policy making. *Governance* 13(1), 5–24.

Evans, M. (2017). *Policy Transfer in Global Perspective*. London: Routledge.

Fernandéz-i-Marín, Xavier, Christoph Knill, & Yves Steinebach (2021). Studying policy design quality in comparative perspective. *American Political Science Review*, 115(3), 931–947.

Fernandéz-i-Marín, Xavier, Christoph Knill, Yves Steinebach, & Christina Steinbacher (2021a). *Bureuacratic Quality and the Gap Between Policy Accumulation and Implementation Capacities*. Unpublished Manuscript.

Gunningham, Neil, & Darren Sinclair (1999). Regulatory pluralism: Designing policy mixes for environmental protection. *Law & Policy* 21(1), 49–76.

Hacker, Jacob S. (2004). Privatizing risk without privatizing the welfare state: The hidden politics of social policy retrenchment in the United States. *American Political Science Review* 98, 243–260.

Harding, Robin, & David Stasavage (2013). What democracy does (and doesn't do) for basic services: School fees, school inputs, and African elections. *The Journal of Politics* 76(1), 229–245.

Hill, Michael, & Peter Hupe (2009). *Implementing Public Policy: An Introduction to the Study of Operational Governance*. London, UK: SAGE Publication.

Holzinger, Katharina, & Christoph Knill (2005). Causes and conditions of cross-national policy convergence. *Journal of European Public Policy* 12(5), 775–796.

Holzinger, Katharina, Christoph Knill, & Thomas Sommerer (2008). Environmental policy convergence: The impact of international harmonization, transnational communication and regulatory competition. *International Organization* 62(3), 553–587.

Howlett, Michael, & Ishani Mukherjee (2018). *Routledge Handbook of Policy Design*. London: Routledge.

Howlett, Michael, and Jeremy Rayner (2013). Patching vs packaging in policy formulation: Assessing policy portfolio design. *Politics and Governance* 1, 170–182.

Jahn, Detlef (2016). *The Politics of Environmental Performance*. Cambridge, UK: Cambridge University Press.

Jones, Bryan D., & Frank R. Baumgartner (2012). From there to here: Punctuated equilibrium to the general punctuation thesis to a theory of government information processing. *Policy Studies Journal* 40(1), 1–20.

Kaufmann, Daniel, Aart Kraay, & Massimo Mastruzzi (2011). The worldwide governance indicators: Methodology and analytical issues. *Hague Journal of the Rule of Law* 3, 220–246.

Knill, Christoph, & Andrea Lenschow (2000). *Implementing EU Environmental Policy. New Directions and Old Problems*. Manchester: Manchester University Press.

Knill, Christoph, & Duncan Liefferink (2007). *Environmental Politics in the European Union: Policy-making, Implementation and Patterns of Multilevel Governance*. Manchester: Manchester University Press.

Knill, Christoph, Christina Steinbacher, & Yves Steinebach (2020). Balancing trade-offs between policy responsiveness and effectiveness: The impact of vertical policy-process integration on policy accumulation. *Public Administration Review* 81, 157–160.

Knill, C., & J. Tosun (2020). *Public Policy: A New Introduction*. Basingstoke: Palgrave Macmillan.

Limberg, Julian, Yves Steinebach, Louisa Bayerlein, & Christoph Knill (2021). The more the better? Rule growth and policy impact from a macro perspective. *European Journal of Political Research* 60, 438–454.

Lipsky, M. (2010). *Street-Level Bureaucracy: Dilemmas of the Individual in Public Services*. New York, NY: Russel Foundation.

Lobao, Linda, Mia Gray, Kevin Cox, & Michael Kitson (2018). The shrinking state? Understanding the assault on the public sector. *Cambridge Journal of Regions, Economy and Society* 11, 389–408.

Luce, E. (2020). Democracies are ill-suited to deal with climate change. *Financial Times*.

Mettler, Suzanne (2016). The policyscape and the challenges of contemporary politics to policy maintenance. *Perspectives on Politics* 14, 369–390.

O'Brien, K. J., & L. Li (1999). Selective policy implementation in rural China. *Comparative Politics* 31(2), 167–186.

Oberfield, Z. (2010). Rule following and discretion at government's frontlines: Continuity and change during organization socialization. *Journal of Public Administration Research and Theory* 20(4), 735–755.

Peters, B. Guy. 2018. *Policy Problems and Policy Design*. Northampton, MA: Edward Elgar Publishing.

Rauch, James, & Peter B. Evans (2000). Bureaucratic structure and bureaucratic performance in less developed countries. *Journal of Public Economics* 75, 49–71.

Roberts, D. (2019). The Green New Deal and the case against incremental climate policy. *vox.com*. https://www.vox.com/energy-and-environment/2019/3/28/18283514/green-new-deal-climate-policy

Rose, R. (1991). Lesson drawing across nations. *Journal of Public Policy* 11(1), 3–30.

Shearman, D., & J. W. Smith (2007). *The Climate Change Challenge and the Failure of Democracy*. Praeger Publishers.

Steinebach, Yves, & Knill, Christoph (2018). Social policy during economic crises: An analysis of cross-national variation in crisis-coping strategies in Europe from 1980 to 2013. *Journal of European Public Policy* 25(11), 1566–1588.

Thelen, Kathleen (2004). *How Institutions Evolve: The Political Economy of Skills in Germany, Britain, the United States and Japan*. New York: Cambridge University Press.

Thomann, E., & F. Sager (2017). Toward a better understanding of implementation performance in the EU multilevel system. *Journal of European Public Policy* 24(9), 1385–1407.

Tummers, Lars L. G., Victor Bekkers, Evelien Vink, & Michael C. Musheno (2015). Coping during public service delivery: A conceptualization and systematic review of the literature. *Journal of Public Administration Research and Theory* 25, 1099–1126.

Verdolini, E., & V. Bosetti (2017). Environmental policy and the international diffusion of green and brown technologies. *Environmental and Resource Economics* 66, 497–536.

Vogel, David (1995). *Trading Up: Consumer and Environmental Regulation in a Global Economy*. Cambridge: Harvard University Press.

Winter, S. C., & V. L. Nielsen (2008). *Implementering af Politik*. Aarhus: Gyldendal.

Winter, Søren C. (2012). Implementation perspectives: Status and reconsideration. In B. Guy Peters & Jon Pierre (Eds). *Handbook of Public Administration* (2nd ed, pp. 265–278). London, UK: SAGE Publication.

17

LEADERS, PIONEERS, AND FOLLOWERS IN ENVIRONMENTAL GOVERNANCE

Duncan Liefferink, Mikael Skou Andersen, Jana Gheuens, Paul Tobin, Diarmuid Torney, and Rüdiger K. W. Wurzel

17.1 Introduction

Leadership is important for enacting change. The vast general literature on leadership ranges from the analysis of hegemonic power in world politics to individuals like Julius Caesar and United States (US) Presidents (e.g. Burns, 1978; Rhodes and Hart, 2014). An entire branch of the Management Studies literature seeks to support inspiring, sustainable and effective leadership – preferably all of those together. This chapter focuses on the specific body of literature that deals with the dynamics of *leadership in environmental politics and policy*. Leaders in environmental governance initiate new policies, set examples and push for wider action. For example: the European Union (EU) has consistently sought to act as a global climate leader, such as through its 2019 European Green Deal; the 2015 Paris Conference would not have been concluded in its present form without the diplomatic efforts of the host country France (Bocquillon and Evrard, 2017); cities annually compete for being awarded the title 'Green Capital of Europe' (Kern, 2019); and companies, such as mobile phone manufacturer Fairphone (Biedenkopf et al., 2019), seek to pioneer more sustainable business modes. What makes certain actors environmental leaders? How does leadership manifest itself?

Although some studies of environmental leadership focus on the individual or organisational level (e.g. Evans et al., 2015), the literature mostly has had a strong focus on leadership by states. We thus start at the global level, by discussing environmental leadership in the International Relations (IR) literature. Next, we show how the leadership theme was taken up by the environmental Comparative Politics (CP) literature. Inspired particularly by the politics of climate change (an area pre-eminently involving all levels of policy-making, from the global to the local), increasingly links have been established between the IR and CP literatures. More recent theorising has attempted to incorporate the roles of sub-national and non-state actors as leaders in environmental policy.

A useful analytical distinction can be made between pioneers, who actively promote stringent policies for mainly domestic purposes, and leaders that also seek to attract followers in other jurisdictions (Liefferink and Wurzel, 2017). As we will see, this basic distinction can be

DOI: 10.4324/9781003043843-20

worked out in various, more fine-grained, typologies. Nevertheless, we will employ 'leadership' as the overarching term. This term subsumes 'pioneership', which does not primarily contain the aim of attracting followers. Furthermore, 'environmental leadership' almost inevitably has normative dimensions. An actor can lead the way to a more sustainable future, but it can also lead the resistance against environmental action. The latter is usually not regarded as 'environmental leadership' (see Tobin et al., 2022). This understanding is in line with both the general leadership literature (e.g. Burns, 1978; Rhodes and Hart, 2014) and the environmental leadership literature (e.g. Young, 1991; Underdal, 1994), which associate leadership with actions that lead to the improvement of the 'common good'.

17.2 The International Relations Perspective

The conceptualisation of *environmental* leadership started in the IR literature. In 1991, Oran Young published a now seminal article in which he analysed the role of leadership in international regime formation (Young, 1991). His focus was on individuals acting as leaders mainly on behalf of states. According to Young (1991: 285), leaders strive 'to solve or circumvent the collective action problems that plague the efforts of parties to reap joint gains in processes of institutional bargaining'. Leaders pursue a collective goal and not – or at least not primarily – self-interest (see also Malnes, 1995: 94–95). Young distinguishes the following three types of leadership, which still form the basis of many leadership conceptualisations:

- *Structural leadership* is based on material resources. In realist IR theory, this primarily involves military power (Nye, 2008), but generally in environmental politics economic power – having financial resources or being able to grant access to a large market – is more important. Also, a state's relative contribution to the problem at stake can be seen as structural power (Liefferink and Wurzel, 2017). Without the biggest greenhouse gas emitters – currently the US and China – involved, long-term climate governance is unlikely to be effective.
- *Intellectual* or *cognitive leadership* entails knowledge production and dissemination about problems, causalities, and solutions. This leadership type is particularly important in a 'technical' policy area, such as the environment. Cognitive leadership helps to frame problems and identify the range of practical solutions.
- *Entrepreneurial leadership* is primarily about diplomacy. It relies on negotiating skills, with a view to integrative bargaining and brokering compromises. The importance of this type can be seen by comparing Denmark's relatively ineffective role in forging the 2009 Copenhagen Accord (Andersen and Nielsen, 2016) with France's highly effective entrepreneurial leadership during the preparation of the 2015 Paris Agreement (Bocquillon and Evrard, 2017).

The three types of leadership distinguished by Young (1991) do not mutually exclude each other and can be combined in different ways in practice.

17.3 The Comparative Politics Perspective

The theme of environmental leadership was soon taken up by the CP literature. While one line of research concentrated on comparing domestic factors facilitating environmental leadership by states, another line focused on strategies employed by environmental leader states in the international arena. In recent years, more systematic attention to the role of followers has emerged.

17.3.1 Domestic Factors

Martin Jänicke and Helmut Weidner investigated the capacities necessary for developing successful national environmental policies and becoming an international trendsetter or pioneer (e.g. Weidner and Jänicke, 2002). With slight variations over the years, they distinguish: (1) *country-specific factors*, such as the degree of economic development and various political and institutional factors; (2) *issue-specific factors*, such as the visibility and salience of specific problems and solutions; (3) *situative factors*, or 'policy windows', that may change due to, for instance, economic fluctuations, elections or shock events; and (4) *strategic factors*, or the 'will and skill' to act as a pioneer (Jänicke, 2005).

Jänicke observes that 'the most important characteristic of "green" pioneer countries is their high degree of economic development' (Jänicke, 2005: 136). In line with this definition, Tanja Börzel (2002) identifies economic development – which underpins both policy preference and action capacity – as the key domestic factor explaining leadership. An innovative aspect of Börzel's contribution was that it focused not only on 'pace-setters', but also on 'fence-sitters' and 'foot-draggers' (i.e. followers and laggards), and the political dynamic between them.

17.3.2 Strategies

The second strand of research within the CP perspective examined the strategies of environmental leader states. With its primary focus on the international process rather than domestic factors, this research line reconnects with the IR perspective.

For example, when Sweden, Finland and Austria became members of the EU in 1995, these countries were expected to join the EU's traditional environmental leaders Germany, Denmark and the Netherlands, known at the time as the 'green trio'. While investigating the strategies of these supposedly 'green' Member States, Liefferink and Andersen (1998) observed that leaders may channel their efforts either directly into environmental policy-making, for instance by pushing at the EU level new issues or more stringent standards, or indirectly via the introduction of national environmental policies that have an impact on the EU's internal market policies. In addition, they argued that 'green' states may act either with the explicit aim of affecting EU policies or in a more incremental way, starting from policies initially developed mainly for domestic purposes. Liefferink and Andersen (1998) thus already differentiated environmental leaders, who want to attract followers, from those actors who adopt ambitious domestic environmental policies without the explicit aim of recruiting followers. Liefferink and Wurzel (2017) subsequently used the analytical term *leaders* only for environmentally progressive actors who attempt to attract followers and *pioneers* for actors who do not do so. Note that the pioneer role deviates from Young's (1991) and Malnes's (1995) assumption that leaders always pursue a collective goal. Liefferink and Andersen (1998) found that cooperation between the 'green' Member States is by no means obvious, as strategies regarding specific issues often differ.

17.3.3 Followers

Building on Börzel's (2002) identification of 'fence-sitters' and 'foot-draggers', the topic of followership has gained increased attention within the environmental leadership literature from the late 2010s. Although Börzel's typology distinguished between leaders, followers and laggards, it did not explicitly consider the role of followers and followership.

Torney (2019) defines climate followership as the adoption of a policy, idea, institution, approach, or technique for responding to climate change by one actor by subsequent reference to its previous adoption by another actor. Note that there must be intentionality on the part of the follower but not the leader/pioneer.

Torney (2019) goes on to consider the following three central questions: (1) who follows? (2) through what pathways does followership emerge? and (3) what conditions facilitate or hinder followership? First – and in line with the increasing attention to leaders other than states (see below) – a wide variety of actor types, including businesses, NGOs, individuals, and epistemic communities, can be followers. However, leader–follower relations are more likely to emerge between actors of the same type. As regards the second question, followership can materialise through four principal pathways, which can be differentiated according to the logic of social interaction at play (March and Olsen, 1998). In the case of a 'logic of consequences', followership can result from coercion or provision of incentives. In the case of a 'logic of appropriateness', followership can emerge from learning, persuasion or the wish to raise the legitimacy of domestic action (DiMaggio and Powell, 1983). Third, with a view to conditions, followership in the case of a 'logic of consequences' can be enabled or constrained by the extent and type of the leader's material power, ability and willingness to mobilise resources to attract followers. By contrast, the leader's perceived legitimacy and credibility matter more in the case of a 'logic of appropriateness'. In a further development, Busby and Urpelainen (2020) have developed a typology of potential follower types based on motivation and capacity, distinguishing between 'enthusiasts', 'pliables', 'reluctants' and 'hard nuts'.

Empirical investigation of followership is challenging because intentionality can be difficult to identify: a challenge common also to the literature on policy learning and transfer (e.g. Rose, 1993; Dolowitz and Marsh, 1996). Different actors may implement similar policy solutions independently of each other rather than as a result of leader–follower dynamics. Notwithstanding these difficulties, climate followership has been studied empirically in a number of contexts, including in international climate negotiations (Karlsson et al., 2011; Parker et al., 2015), and within the USA (Wang, 2012), Europe (Kammerer et al., 2021; Tobin & Schmidt, 2021) and Asia (Urban et al., 2021), though a significant research terrain remains to be explored.

17.4 Combining the International Relations and Comparative Politics Perspectives

From the late 1990s, the issue of climate change increasingly came to dominate the global environmental agenda. One part of the rapidly expanding literature on global, European and comparative climate politics focused on climate leadership, particularly the EU's role as an alleged global climate leader. Moreover, addressing both the EU's external role and the internal dynamics between the EU institutions and the Member States established connections between the IR and the CP perspectives discussed above.

17.4.1 Climate Leadership

In the 1970s, the USA was considered *the* global environmental leader, setting the norm for many product standards for consumer and environmental protection, for example on auto emissions and chemical policy, partly due to the sheer size of its economy (Vogel, 1995; Le Cacheux and Laurent, 2015). However, with European integration and the creation of the single European market, the EU has frequently replaced the USA's leadership status. At the United Nations' (UN) annual climate Conferences of the Parties (COPs), the EU has explicitly been seeking to press deliberations forward, although it has not always been successful in taking the lead.

An extensive literature rooted in IR has shown how the EU's ambitious pursuit of global climate leadership has evolved from its pre-2009 Copenhagen emphasis on structural leadership (Eckersley, 2020). During the 2010s, from mainly offering material resources, the EU became more entrepreneurial, building alliances and skilfully exploiting diplomatic avenues, continuously underpinned by cognitive leadership, and culminating with the 2015 Paris Agreement (Bäckstrand and Elgström, 2013; Parker et al., 2017; Wurzel et al., 2017a; Oberthür and Dupont, 2021). Yet, among Global South countries, the EU's leadership role remains disputed, with China emerging as a more powerful player next to the USA, though both great powers have occasionally leaned more towards an obstructive role (Parker et al., 2015).

As a basis for understanding the international climate negotiations, the CP literature has explored the domestic factors that can explain how and why one or more of the 'big three' (the EU, the USA and China) have been able to take the lead – or to undermine agreement. Analyses of EU positions are especially rich, due to the confederative nature of the Union, with a dynamic subset of Member States pushing hard for leadership while others are relatively unengaged if not outright sceptical (Carter et al., 2019). The mechanisms within the EU reflect traits conducive to both IR and CP perspectives on leadership. Once the black box is opened up, it becomes clear that Member States' domestic politics can help us to understand both leader and laggard positions. The significance of domestic 'spoilers' to leadership was evident with the Trump administration of the USA and to a lesser degree with regard to attempts by some Member States, especially Poland, to undermine the EU's climate leadership ambitions (Bang and Schreurs, 2017; Jankowska, 2017; Selin and VanDeveer, 2021; and for India: Jayaram, 2018). A better understanding of the interplay between negotiation games that are played simultaneously at the global and domestic level, and how 'win-sets' are identified and become accepted (Putnam, 1988), calls for more and better integration of IR and CP perspectives in future research.

17.4.2 Towards a More Comprehensive Conceptual Approach

As explained above, the burgeoning literature particularly on climate leadership led to a variety of different approaches to environmental/climate leadership, often combining IR and CP elements. In an attempt to bring together some of the main lines of enquiry, Liefferink and Wurzel (2017) proposed a model encompassing different *positions*, *types* and *styles* of environmental and climate leadership. With some limitations, the model can be applied also to sub-national governmental and non-state actors (Wurzel et al., 2019a, 2019b).

The starting point of the model by Liefferink and Wurzel (2017) is four *positions* that actors might adopt based on their internal and external ambitions (see Table 17.1). The level

Table 17.1 Ambitions and positions of environmental leaders and pioneers

	Internal ambitions	
External ambitions	*Low*	*High*
Low	(a) Laggard	(b) Pioneer
High	(c) Symbolic leader	(d) Pusher • constructive • conditional

Adapted from Liefferink and Wurzel, 2017: 954.

of ambition (high or low) is measured in relation to other actors rather than compared to, for example, a scientific standard that might change over time. High *external* ambitions refer to efforts to influence other actors, that is to ('genuine') leadership. High *internal* ambitions can refer to an actor either being the first to adopt a certain environmental measure or formulating the highest environmental standard – or both at the same time, that is 'first in class' *and* 'best in class' (Liefferink et al., 2009). A *pusher* (field d) combines high external ambitions with high internal ambitions. Within the pusher category, a further distinction can be made between pushers whose internal ambitions are conditional on the policies of others (conditional pusher), and those who carry on regardless (constructive pusher). When high external ambitions are not underpinned by high internal ambitions, we can speak of *symbolic leadership* (field c). An actor pursuing high internal ambitions in combination with no or limited ambition to attract followers can be seen as a *pioneer* (field b). A combination of low internal and low external ambitions, finally, is characteristic of a *laggard*. In short, rather than assessing leadership on effectiveness or goal achievement, the model examines the intention of actors to attract followers and the domestic policy output that lends credibility to the leadership position.

Subsequently, leaders and pioneers can use different types and styles of leadership. Building on existing conceptualisations of leadership *types* (in particular: structural, entrepreneurial and cognitive; see Young, 1991), Liefferink and Wurzel (2017) added exemplary leadership as a fourth type to assess how leaders and pioneers exert leadership. Due to their high internal ambitions without the intention to attract followers, pioneers can be seen as mainly using unintentional exemplary leadership. Constructive pushers, in contrast, are likely to set intentionally an example while relying on structural, entrepreneurial or cognitive leadership or a combination of these.

Finally, including the *styles* of leadership as developed by Burns (1978, 2003) allows for an assessment of the degree of change of actors' positions over time (Liefferink and Wurzel, 2017; Wurzel et al., 2019b). Whereas transactional leadership is focused on achieving short-term, incremental goals, transformational leadership sets out to establish more profound, long-term change. While in principle the two styles can be combined with all types of leadership, and, thus, with the four positions, the time horizon might vary based on the position. For example, if symbolic leaders try to employ transformational leadership on the international level without the capacity or ambition of backing it up with domestic action, it will most likely remain short-term focused leadership because such a strategy lacks credibility. For instance, the EU's ambiguous role in international climate politics in the early 1990s amounted to little more than symbolic leadership (Grubb and Gupta, 2000; Wurzel et al., 2017c). In contrast, a pioneer consistently using a transactional approach may achieve durable long-term change, for example Danish pioneership on nitrate pollution (Andersen, 1997) and Costa Rica's on payments for ecosystem services and forestry (Urban et al., 2021), both in the 1990s. The credibility of an actor is therefore essential for assessing the long-term impact of leadership.

17.5 Bringing in Other Actors

The early literature on leadership in environmental politics and policy was rooted in IR and CP and focused on states. However, the literature has increasingly acknowledged that actors other than states, such as sub-national governments, businesses and civil society organisations, can act as environmental leaders too.

17.5.1 Sub-National Actors

The parallel trends of greater devolution and the increasing complexity of many environmental problems (e.g. climate change) have amplified attention paid to sub-national levels. Indeed, Betsill and Bulkeley (2006: 141) argue that 'it is only by taking a multilevel perspective that we can fully capture the social, political, and economic processes that shape global environmental governance'. Alongside being important actors worthy of study in their own right, local actors can elevate other localities' and even national-level environmental performance. The 'California effect' (Vogel, 1995) describes how California's high environmental standards for automobiles led national standards to follow the state's lead, as manufacturers pursued the large and wealthy California market's preferences anyway. This can be seen as a case of structural leadership (through market power) by a sub-national authority. Sub-national leadership is especially important when national policy-makers do not offer environmental leadership or even try to veto action, such as US President Trump (Selin and VanDeveer, 2021).

However, the ratcheting upwards of sub-national standards is not a foregone conclusion. Vogel (1995) additionally identified a 'Delaware effect', also known as the 'race to the bottom' (Porter, 1999; Holzinger and Sommerer, 2011), involving the competitive lowering of sub-national environmental standards to attract businesses, including high polluters. Casado-Asensio and Steurer (2016) find that in Switzerland federalism produced a 'lowest common denominator' approach regarding climate mitigation in the building sector. Regarding German federalism and also the EU, the 'joint decision trap' can lead to suboptimal (environmental) policy solutions to avoid a policy being vetoed (Scharpf, 1988). According to Scharpf (1996), joint decision traps can be overcome more easily for product standards (e.g. emission limits for automobiles) than for process standards which stipulate how products can be manufactured, aligning with Vogel's (1995) California effect.

Much empirical research on local level environmental leaders/pioneers has focused upon large and/or wealthy cities which are often seen as laboratories for environmental innovation that can be scaled up. Bulkeley and Castán Broto (2013) examine 627 urban climate change experiments in 100 'global cities', finding that experiments create new political spaces that blur public and private authority. Smaller localities have fewer resources for such experimentation, and institutional capacity is necessary if a locality is to engage with and inspire others, especially internationally, through cognitive leadership in such networks as C40 (Lee and Koski, 2014). Disappointingly, the level of change proposed by C40 cities has been found to support often little more than the status quo (Heikkinen et al., 2019). However, there are instances of environmental leadership and pioneership in poorer locations including 'structurally disadvantaged cities', characterised by high unemployment, social deprivation and other problems (Wurzel et al., 2019b).

While actors in the Global South have also displayed environmental leadership, for instance with the creation of the now defunct Clean Development Mechanism of the Kyoto protocol, they have remained under-researched. Keiner and Kim (2007: 1389) place African and South American cities highly in their rankings of transnational sustainability networks. While we often associate local environmental governance as representing 'bottom-up' leadership, some localities outside of the Global North have been found to tackle city-level action differently. Li (2021) explores multiple low-carbon pioneers at the city level in China, where the initial imperative for action is top-down (i.e. central government-initiated), in contrast to the bottom-up California effect. Much city-level climate change research in the Global South has focused on adaptation rather than mitigation (Chu et al., 2016).

17.5.2 Non-State Actors

Non-state actors, such as businesses, charities, campaigning organisations and religious communities can act as important 'nodes' with governance networks. Indeed, independent, overlapping, non-state actors are at the heart of polycentric governance (Ostrom, 2010; Jordan et al., 2018; see Chapter 6 of this volume). However, it is not clear to what degree actors in polycentric governance structures are able to influence 'higher' governance levels and/or engender 'upscaling' of ambition. Proponents of polycentricism tend to understate the significance of power relations: structural power matters, especially if strong power asymmetries exist between leaders and followers (Wurzel et al., 2019b).

Environmental Non-Governmental Organisations (ENGOs) have been at the forefront of environmental action at different governance levels. In Europe, many ENGOs possess offices in Brussels for lobbying EU legislators in the pursuit of higher standards and transformative change (Fitch-Roy et al., 2020). Well established ENGOs such as Greenpeace and Friends of the Earth (FOE) have long been identified as offering environmental leadership (e.g. Wurzel et al., 2017b). In recent years, new ENGOs have sprung up that have tried to offer leadership especially on climate related issues. For example, young people have set up new ENGOs or movements (e.g. Friday for Futures) and/or joined local Youth Climate Councils (Wurzel et al., 2019b). YOUNGO is an official observer at UN climate conferences and seeks to initiate greater ambition at the global level by lobbying diplomats (Thew et al., 2021). The literature remains ambivalent about businesses' ability to act as environmental leaders or pioneers (e.g. Grant, 2017). One case of exemplary business leadership is the social enterprise Fairphone, which produces more sustainable mobile phones (Biedenkopf et al., 2019).

Environmental leadership/pioneership from non-state actors may also be taken up by groups for which the environment is not the main *raison d'être*. Bomberg and Hague (2018) find that Christian congregations in Scotland mobilise on climate change around rituals and symbols that are distinct from other climate groups. Moreover, religious communities have founded several of the 'big players' in the aid, trade and development sector, which in turn seek to drive climate ambition at the global level (Saunders, 2008). Yet, some evangelical groups especially in the USA have been at the forefront of opposing climate action (Veldman, 2019). In recent years, ENGOs and development groups have worked more closely together, such as in the run up to the 2015 Paris COP21 (Wurzel et al., 2017b). Many trade unions are increasingly rejecting the 'jobs versus environment' dichotomy in favour of sustainable development which puts equal emphasis on economic, social and environmental concerns (Räthzel and Uzzell, 2011). In short, this section has provided a snapshot of how almost all sub-national and non-state actors are *potential* environmental leaders or pioneers.

17.6 Conclusion and Future Research

The study of leadership in environmental governance started in the IR and CP literatures. Increasingly, the two perspectives have been brought together. This interconnection has enabled the development of several typologies describing the different positions environmental leaders may take and the ways in which leadership can be exerted. At the most basic level, a distinction can be made between leaders, who actively seek to attract followers, and pioneers, who promote stringent policies mainly for domestic purposes but whose actions may nevertheless energise other actors. Leaders and pioneers may exert structural, entrepreneurial, cognitive or exemplary leadership, or combinations of these. Furthermore, leadership implies followership, a theme that has also been taken up in recent years. Parallel to this conceptual development, the scope of attention to environmental leadership has been widened to sub-national and non-state actors, such as cities, businesses and various types of NGOs, and to environmental leadership and pioneership in the Global South.

Although our understanding of environmental leadership has advanced over the past decades, several issues remain understudied. First, while the work on followership has only just started, the dynamics between leaders and followers deserve more attention: when, why and under which circumstances do actors follow or refrain from doing so even in the case of active leadership efforts? These questions warrant a focus on mechanisms. To add greater clarity, the logics of consequences and appropriateness that may be at work merit attention, as does the policy diffusion and transfer literature (e.g. Busch and Jörgens, 2005; Elkins and Simmons, 2005). Such mechanisms include lesson-drawing (where actors actively use experiences elsewhere to solve domestic problems), transnational problem-solving (where policies are developed in, for instance, transnational elite networks or epistemic communities) and emulation (where policies are copied for normative or legitimacy-driven reasons) (Busch and Jörgens, 2005; Holzinger and Knill, 2008; Graham et al., 2013).

Second, leadership by non-state actors and, to a lesser degree, sub-national actors (such as cities) is still under-researched; their roles need to be more systematically investigated. The notion of polycentricity may provide a fruitful basis for understanding local action, experimentation, learning and diffusion (Jordan et al., 2018). Under conditions of polycentricity, leadership and pioneership exercised by, for instance, corporations or civil society actors may take on even more diverse forms than under more traditional, hierarchical conditions. Hierarchical governance relations may nevertheless remain helpful for widening the potential range of followers beyond the confines of relatively autonomous polycentric units (Liefferink and Wurzel, 2018).

Finally, studies of environmental leadership – like most other sub-fields of environmental policy studies – tend to focus on rich, industrialised countries, although there are notable exceptions (e.g. Clapp and Swanston, 2009; Jayaram, 2018; Theys and Rietig, 2020; various contributions to Wurzel et al., 2021). Widening the scope to the Global South should be an urgent priority. Here the notion of polycentricity may be even more relevant, as it allows for a more fine-grained analysis of the dynamics between global and regional pioneers and leaders.

Box 17.1 Chapter Summary

- While leaders actively seek to attract followers, pioneers focus mainly on stringent internal policies.
- Leaders and pioneers may exert structural, entrepreneurial, cognitive or exemplary leadership, or combinations of these.
- Leadership implies followership. The initial focus on states as leaders has been gradually widened to include the role of followers.
- Increasing attention is being paid to leadership by sub-national and non-state actors.
- Leadership in and by the Global South has so far remained largely unexplored.

References

Andersen, M. S., 1997. Denmark: The shadow of the green majority. In: M.S. Andersen and D. Liefferink, eds., *European Environmental Policy: The Pioneers*. Manchester: Manchester University Press, 251–286.

Andersen, M.S. and Nielsen, H.Ø., 2016. Small state with a big voice and bigger dilemmas. In R.K.W. Wurzel, J. Connelly and D. Liefferink, eds., *The European Union in International Climate Change Politics*. London: Routledge, 83–97.

Bäckstrand, K. and Elgström, O., 2013. The EU's role in climate change negotiations: From leader to 'Leadiator'. *Journal of European Public Policy*, 20(10), 1369–1386.

Bang, G. and Schreurs, M.A., 2017. The United States: The challenge of global climate leadership in a politically divided state. In R.K.W. Wurzel, J. Connelly and D. Liefferink, eds., *The European Union in International Climate Change Politics*. London: Routledge, 239–253.

Betsill, M. and Bulkeley, H., 2006. Cities and the multilevel governance of global climate change. *Global Governance* 12, 141–160.

Biedenkopf, K., Van Eynde, S. and Bachus, K., 2019. Environmental, climate and social leadership of small enterprises: Fairphone's step-by-step approach. *Environmental Politics*, 28(1), 43–63.

Bocquillon, P. and Evrard, A., 2017. French climate policy: Diplomacy in the service of symbolic leadership. In R.K.W. Wurzel, J. Connelly and D. Liefferink, eds., *The European Union in International Climate Change Politics*. London: Routledge, 98–113.

Bomberg, E. and Hague, A., 2018. Faith-based climate action in Christian congregations: Mobilisation and spiritual resources. *Local Environment*, 23(5), 582–596.

Börzel, T.A., 2002. Pace-setting, foot-dragging and fence-sitting. *Journal of Common Market Studies*, 40(2), 193–214.

Bulkeley, H. and Castán Broto, V., 2013. Government by experiment? Global cities and the governing of climate change. *Transactions of the institute of British geographers*, 38(3), 361–375.

Burns, J.M., 1978. *Leadership*. New York: Harper & Row.

Burns, J.M., 2003. *Transforming Leadership*. New York: Grove Press.

Busby, J.W. and Urpelainen, J., 2020. Following the leaders? How to restore progress in global climate governance. *Global Environmental Politics*, 20(4), 99–121.

Busch, P.O. and Jörgens, H. (2005). The international sources of policy convergence: Explaining the spread of environmental policy innovations. *Journal of European Public Policy*, *12*(5), 860–884.

Carter, N., Little, C. and Torney, D., 2019. Climate politics in small European states. *Environmental Politics*, 28(6), 981–996.

Casado-Asensio, J. and Steurer, R., 2016. Mitigating climate change in a federal country committed to the Kyoto Protocol: How Swiss federalism further complicated an already complex challenge. *Policy Sciences*, 49(3), 257–279.

Chu, E., Anguelovski, I. and Carmin, J., 2016. Inclusive approaches to urban climate adaptation planning and implementation in the Global South. *Climate Policy*, 16(3), 372–392.

Clapp, J. and Swanston, L., 2009. Doing away with plastic shopping bags: International patterns of norm emergence and policy implementation. *Environmental Politics*, 18(3), 315–332.

DiMaggio, P. J. and Powell, W. W., 1983. The iron cage revisited: Institutional isomorphism and collective rationality in organizational fields. *American Sociological Review*, 48(2), 147–160.

Dolowitz, D. and Marsh, D., 1996. Who learns what from whom: A review of the policy transfer literature. *Political Studies*, 44(2), 343–357.

Eckersley, R., 2020. Rethinking leadership: understanding the roles of the US and China in the negotiation of the Paris Agreement. *European Journal of International Relations*, *26*(4), 1178–1202.

Elkins, Z. and Simmons, B. 2005. On waves, clusters, and diffusion: A conceptual framework. *The Annals of the American Academy of Political and Social Science*, *598*(1), 33–51.

Evans, L.S., Hicks, C.C.,. Cohen, P.J., Case, P., Prideaux, M. and Mills, D.J., 2015. Understanding leadership in the environmental sciences. *Ecology and Society*, 20(1), 50.

Fitch-Roy, O., Fairbrass, J. and Benson, D. 2020. Ideas, coalitions and compromise: Reinterpreting EU-ETS lobbying through discursive institutionalism. *Journal of European Public Policy*, 27(1), 82–101.

Graham, E.R., Shipan, C.R. and Volden, C. 2013. The diffusion of policy diffusion research in political science. *British Journal of Political Science*, *43*(3), 673–701.

Grant, W., 2017. Business: Greening at the edges. In R.K.W. Wurzel, J. Connelly and D. Liefferink, eds., *The European Union in International Climate Change Politics*. London: Routledge, 207–220.

Grubb, M. and Gupta, J., 2000. Leadership. In J. Gupta and M. Grubb, eds., *Climate Change and European Leadership*. Springer: Dordrecht, 15–24.

Heikkinen, M., Ylä-Anttila, T. and Juhola, S., 2019. Incremental, reformistic or transformational: What kind of change do C40 cities advocate to deal with climate change?. *Journal of Environmental Policy & Planning*, 21(1), 90–103.

Holzinger, K. and Knill, C., 2008. Theoretical framework: Causal factors and convergence expectations. In Holzinger, K., Knill, C. and Arts, B., eds., *Environmental Policy Convergence in Europe. The Impact of International Institutions and Trade*. Cambridge: Cambridge University Press, 30–63.

Holzinger, K. and Sommerer, T., 2011. Race to the bottom'or 'race to Brussels'? Environmental competition in Europe. *Journal of Common Market Studies*, 49(2), 315–339.

Jänicke, M., 2005. Trend-setters in environmental policy: the character and role of pioneer countries. *European Environment*, 15(2), 129–142.

Jankowska, K., 2017. Poland's clash over energy and climate policy: green economy or grey status quo? In R.K.W. Wurzel, J. Connelly and D. Liefferink, eds., *The European Union in International Climate Change Politics*. London: Routledge, 145–158.

Jayaram D, 2018. From "Spoiler" to "Bridging Nation": The reshaping of India's climate diplomacy. *Revue internationale et stratégique*, 109(1), 181–190.

Jordan, A., Huitema, D., Van Asselt, H. and Forster, J., eds., 2018. *Governing Climate Change: Polycentricity in Action?* Cambridge: Cambridge University Press,.

Kammerer, M., Ingold, K. and Dupuis, J., 2021. Switzerland: International commitments and domestic drawbacks. In R.K.W. Wurzel, M.S. Andersen and P. Tobin, eds., *Climate Governance Across the Globe: Pioneers, Leaders and Followers*. London: Routledge, 235–256.

Karlsson, C., Parker, C.F., Hjerpe, M. and Linnér, B-O., 2011. Looking for leaders: Perceptions of climate change leadership among climate change negotiation participants, *Global Environmental Politics*, 14(1), 89–107.

Keiner, M., & Kim, A. 2007. Transnational city networks for sustainability. *European Planning Studies*, 15(10), 1369–1395.

Kern, K., 2019. Cities as leaders in EU multilevel climate governance: Embedded upscaling of local experiments in Europe. *Environmental Politics*, 28(1), 125–145.

Le Cacheux, J. and Laurent, E., 2015. The EU as a global ecological leader. In *Report on the State of the European Union*. London: Palgrave Macmillan, 125–138.

Lee, T. and Koski, C., 2014. Mitigating global warming in global cities: Comparing participation and climate change policies of C40 cities. *Journal of Comparative Policy Analysis: Research and Practice*, 16(5), 475–492.

Li, X., 2021. China: Emerging low-carbon pioneers at city level. In R.K.W. Wurzel, M.S. Andersen and P. Tobin, eds., *Climate Governance Across the Globe*. London: Routledge, 23–44.

Liefferink, D. and Andersen, M.S., 1998. Strategies of the "green" member states in EU environmental policy-making. *Journal of European Public Policy*, 5(2), 254–270.

Liefferink, D., Arts, B., Kamstra, J. and Ooijevaar, J., 2009. Leaders and laggards in environmental policy. *Journal of European Public Policy*, 16(5), 677–700.

Liefferink, D. and Wurzel, R.K.W., 2017. Environmental leaders and pioneers: Agents of change? *Journal of European Public Policy*, 24(7), 651–668.

Liefferink, D. and Wurzel, R.K.W., 2018. Leadership and pioneership: exploring their role in polycentric governance. In Jordan, A., Huitema, D., Van Asselt, H. and Forster, J., eds., *Governing Climate Change. Polycentricity in Action?* Cambridge: Cambridge University Press, 135–151.

Malnes, R., 1995. 'Leader' and entrepreneur' in international negotiations: A conceptual analysis. *European Journal of International Relations*, 1(1), 87–112.

March, J.G. and Olsen, J. P., 1998. The institutional dynamics of international political orders. *International Organization*, 52(4), 943–969.

Nye Jr, J., 2008. *The powers to lead*. Oxford: Oxford University Press.

Oberthür, S. and Dupont, C., 2021. The European Union's international climate leadership: Towards a grand climate strategy? *Journal of European Public Policy*, online. https://doi.org/10.1080/13501763.2021.1918218

Ostrom, E., 2010. Polycentric systems for coping with collective action and global environmental change. *Global Environmental Change*, 20, 550–557.

Parker, C.F., Karlsson, C. and Hjerpe, M., 2015. Climate change leaders and followers: Leadership recognition and selection in the UNFCCC negotiations. *International Relations*, 29(4), 434–454.

Parker, C.F., Karlsson C. and Hjerpe, M., 2017. Assessing the European Union's global climate change leadership: From Copenhagen to the Paris Agreement, *Journal of European Integration*, 39(2), 239–252. https://doi.org/10.1080/07036337.2016.1275608

Porter, G., 1999. Trade competition and pollution standards: "race to the bottom" or "stuck at the bottom". *Journal of Environment & Development*, 8(2), 133–151.

Putnam, R., 1988. Diplomacy and domestic politics: The logic of two-level games. *International Organization*, 42(3), 427–460.

Räthzel, N. and Uzzell, D., 2011. Trade unions and climate change: The jobs versus environment dilemma. *Global Environmental Change*, 21(4), 1215–1223.

Rhodes, R.A.W. and Hart, P.T., eds., 2014. *The Oxford Handbook of Political Leadership*. Oxford: Oxford University Press.

Rose, R., 1993. What is lesson-drawing. *Journal of Public Policy*, 11(1), 3–30.

Saunders, C., 2008. The stop climate chaos coalition: Climate change as a development issue. *Third World Quarterly*, 29(8), 1509–1526.

Scharpf, F., 1988. The joint decision trap. Lessons from German federalism and European integration. *Public Administration*, 66, 239–278.

Scharpf, F., 1996, Negative and positive integration in the political economy of European welfare states. In G. Marks et al., eds, *Governance in the European Union*, London: Sage, 15–39.

Selin, H. and VanDeveer, S.D., 2021. Climate change politics and policy in the United States: Forward, reverse and through the looking glass. In R.K.W. Wurzel, M.S. Andersen and P. Tobin, eds., *Climate Governance across the Globe*. London: Routledge, 123–141.

Thew, H., Middlemiss, L. and Paavola, J., 2021. Does youth participation increase the democratic legitimacy of UNFCCC-orchestrated global climate change governance? *Environmental Politics*, 1–22. https://doi.org/10.1080/09644016.2020.1868838

Theys, S. and Rietig, K., 2020. The influence of small states: How Bhutan succeeds in influencing global sustainability governance. *International Affairs*, 96(6), 1603–1622.

Tobin, P. and Schmidt, N.M., 2021. European Union leadership before, during and after the Paris Conference of the Parties. In R.K.W. Wurzel, M.S. Andersen and P. Tobin, eds., *Climate Governance Across the Globe: Pioneers, Leaders and Followers*. London: Routledge, 142–159.

Tobin, P., Torney, D. and Biedenkopf, K. 2022. Leadership: Global and domestic dimensions. In T. Rayner, K. Szulecki, A. Jordan and S. Oberthür, eds., *Handbook on European Union Climate Change Policy and Politics*. Cheltenham: Edward Elgar (forthcoming).

Torney, D., 2019. Follow the leader? Conceptualising the relartionship between leaders and followers in polycentric climate governance. *Environmental Politics*, 28(1), 167–186.

Underdal, A., 1994. Leadership theory: Rediscovering the arts of management. In W.I. Zartman, ed., *International Multilateral Negotiation*. San Francisco: Jossey-Bass, 178–197.

Urban, F., Siciliano, G., Villalobos, A., Anh, D.N. and Lederer, M., 2021. Costa Rica and Vietnam: Pioneers in green transformations. In R.K.W. Wurzel, M.S. Andersen and P. Tobin, eds., *Climate Governance across the Globe*. London: Routledge, 61–81.

Veldman, R.G., 2019. *The Gospel of Climate Skepticism: Why Evangelical Christians Oppose Action on Climate Change*. Oakland: University of California Press.

Vogel, D., 1995. *Trading up. Consumer and Environmental Regulation in a Global Economy*. Cambridge, MA.: Harvard University Press.

Wang, R., 2012. Leaders, followers, and laggards: Adoption of the us conference of mayors climate protection agreement in California. *Environment and Planning C: Government and Policy*, 30(6), 1116–1128.

Weidner, H. and Jänicke, M., eds., 2002. *Capacity Building in National Environmental Policy: A Comparative Study of 17 Countries*. Berlin: Springer.

Wurzel, R.K.W., Andersen, M.S. and Tobin, P., eds., 2021. *Climate Governance Across the Globe. Pioneers, Leaders and Followers*. London: Routledge.

Wurzel, R.K.W., Connelly, J. and Liefferink, D., eds., 2017a. *The European Union in International Climate Change Politics*. London: Routledge.

Wurzel, R. K.W., Connelly, J. and Monaghan, E., 2017b. Environmental NGOs: Pushing for leadership. In R.K.W. Wurzel, J. Connelly and D. Liefferink, eds., *The European Union in International Climate Change Politics*. London: Routledge, 221–236.

Wurzel, R.K.W., Liefferink, D. and Connelly, J., 2017c. Conclusion: Re-assessing European Union climate leadership. In R.K.W. Wurzel, J. Connelly and D. Liefferink, eds., *The European Union in International Climate Change Politics*. London: Routledge, 287–302.

Wurzel, R.K.W., Liefferink, D. and Torney, D., eds., 2019a. Pioneers, leaders and followers in multilevel and polycentric climate governance, *Environmental Politics (special issue)*, 28(1), 1–186.

Wurzel, R.K.W., Moulton, J.F., Osthorst, W., Mederake, L., Deutz, P. and Jonas, A.E., 2019b. Climate pioneership and leadership in structurally disadvantaged maritime port cities. *Environmental Politics*, 28(1), 146–166.

Young, O.R., 1991. Political leadership and regime formation: on the development of institutions in international society. *International Organization*, 45(3), 281–308.

18

CONVERGENCE AND DIFFUSION OF ENVIRONMENTAL POLICIES

Christoph Knill, Yves Steinebach, and Xavier Fernandéz-i-Marín

18.1 Introduction

One of the key issues of globalization research in the social sciences is the question of whether globalization leads to the convergence of political institutions, policies, legal orders, and societal structures. Is the world becoming ever more similar as a result of globalization and Europeanization as the "world society approach" (Meyer et al., 1997) implies? Does the strong growth of economic and institutional interlinkages between nation states as well as the presence of transnational problems exceeding the scope of national borders lead to increasingly similar policy measures and institutions across countries (see, e.g., Boushey, 2010; Butler et al., 2017; Holzinger et al., 2008a, 2011; Popp et al., 2011)? In short, to what extent are national policy responses driven by the global diffusion of policy solutions and hence an overall trend of policy convergence?

In this chapter, we will take a closer look at both the extent to which such dynamics are present in the environmental domain and the consequences of these developments for governments' problem-solving capacities. In so doing, we proceed in three steps. First, we introduce the concepts of policy diffusion and policy convergence, including a discussion of how these concepts are related and how they can be measured empirically. In a second step, we turn to the factors that have been identified as central drivers of environmental policy diffusion and policy convergence. Third, we provide an empirical overview of the extent to which diffusion and convergence dynamics can be observed in the environmental domain. To what extent can we observe a global spread of environmental policies? Are there some policies that are more widely adopted than others? Moreover, while there is some research suggesting that cross-national diffusion of environmental policy innovation regularly occurs (e.g. Busch et al., 2005), we are still left with the question as to whether this really represents an overall trend leading to greater similarity or homogeneity of national environmental policies. To what extent do national environmental policies converge over time? Fourth, we take a look at the broader consequences of diffusion and convergence dynamics in terms of policy ambitions. What is the direction of the potential convergence trends? Can we observe an overall tendency of a strengthening or weakening of regulatory requirements or is it rather the case that only "bad policies spread," while the "good ones don't" (Shipan and Volden, 2021)?

DOI: 10.4324/9781003043843-21

18.2 Policy Diffusion and Policy Convergence: Different, Yet Related Concepts

In the literature, the concepts of policy diffusion and policy convergence are often mentioned jointly in order to emphasize cross-national policy dynamics. Yet, this does not mean that diffusion and convergence actually mean the same thing in practice. Although the concepts are closely related they describe distinct phenomena. While diffusion captures the *process* of globally spreading policies, convergence refers to the (end) *result or effects* of this process, namely changes in cross-national policy similarity over time (Knill, 2005).

18.2.1 Policy Diffusion

Policy diffusion typically refers to processes that might result in increasing policy similarities across countries, hence leading to policy convergence (Elkins and Simmons, 2005, 36). Diffusion is generally defined as the socially mediated spread of policies across and within political systems, including communication and influence processes which operate both on and within populations of adopters (Gilardi and Wasserfallen, 2019). Most of the diffusion literature is characterized by this approach. Diffusion studies typically start out from the description of adoption patterns for certain policy innovations over time. In a subsequent step, they analyze the factors that account for the empirically observed spreading process. From this perspective, policy diffusion is not restricted to the operation of specific 'mediation mechanisms,' but rather includes all conceivable channels of influence between countries, reaching from the voluntary adoption of policy models that have been communicated in the international system, diffusion processes triggered by legally binding harmonization requirements defined in international agreements or by supranational regulations, to the imposition of policies on other countries through external actors.

In contrast to this definition, however, some authors suggest a narrower focus of the concept, explicitly restricting diffusion to processes of voluntary policy transfer (Busch and Jörgens, 2005; Kern, 2013). In consequence, diffusion is conceived as a distinctive causal factor that drives international policy convergence rather than a general process that is caused by the operation of varying (both voluntary and coercive) influence channels. Following this approach, Busch and Jörgens distinguish three mechanisms of policy convergence: international harmonization (legal obligation from international or supranational agreements deliberately agreed by the involved countries in multilateral negotiations), imposition of policies, and policy diffusion (where national policy-makers voluntarily adopt policy models that are communicated internationally).

We are thus confronted with two different conceptions of policy diffusion. On the one hand, the concept describes the process of spreading policies across countries with the possible result of cross-national policy convergence, regardless of the causal factors that are driving this development (e.g. regulatory competition, international harmonization, imposition). On the other hand, diffusion is conceived as a distinctive causal factor leading to policy convergence by voluntary (in contrast to obliged or imposed) transfer of policy models. Both conceptions of diffusion are analytically well-grounded and applied in the literature. Yet, most scholarly contributions rely on the former conception of diffusion as a general process rather than as a specific causal mechanism of policy convergence (e.g. Boushey, 2010; Shipan and Volden, 2021).

Policy diffusion thus rests on the assumption that governments do not learn about policy practices randomly, but tend to do so through common affiliations, negotiations, and institutional membership (Simmons and Elkins, 2004). Diffusion processes hence require that actors are informed about the policy choices of others (Strang and Meyer, 1993: 488). Departing from this perspective, the central dependent variable in diffusion research refers to the general patterns that characterize the spread of innovations within or across political systems. The diffusion literature focuses more on the spatial, structural, and socioeconomic reasons for particular adoption patterns rather than on the reasons for individual policy adoptions as such (Jordana and Levi-Faur, 2005). Diffusion studies often reveal a rather robust adoption pattern, with the cumulative adoption of a policy innovation over time following an S-shaped curve (Gray, 1973). Relatively few countries adopt innovation during the early stages. Over time, the rate of adoption increases until the process gets closer to saturation, and the rate ultimately slows down again.

18.2.2 Policy Convergence

Policy convergence differs from policy diffusion in several ways. First, differences exist with respect to the underlying analytical focus. While diffusion studies are concerned with process patterns, convergence studies place particular emphasis on effects. Diffusion thus reflects processes which under certain circumstances might result in policy convergence. This does not imply, however, that the empirical observation of converging policies must necessarily be the result of diffusion (Drezner, 2001). It is well conceivable that policy convergence is the result of similar but relatively isolated domestic events. Second, the concepts differ in their dependent variables. Convergence studies typically seek to explain changes in policy similarity over time, while the focus of diffusion research is on the explanation of adoption patterns over time (Elkins and Simmons, 2005).

Following the above considerations, policy convergence can be defined as any increase in the similarity between one or more characteristics of a certain policy (e.g. policy objectives, policy instruments, policy settings) across a given set of political jurisdictions (supranational institutions, states, regions, local authorities) over a given period of time. Policy convergence thus describes the end result of a process of policy change over time towards some common point, and this regardless of the causal processes (Knill, 2005).

In the literature on policy convergence, we find different options for how convergence can be measured and evaluated empirically (Heichel et al., 2005; Holzinger et al., 2011). First, the most basic way of assessing policy convergence is to analyze the extent to which the policies of countries have become more similar to each other over time. This convergence dimension captures the *homogeneity* of the policy repertoire across countries. Here, scholars are typically interested in the extent to which the domestic policy repertoires (the portfolio of environmental policy measures of a country) and levels of regulatory standards have become more similar over time. The second convergence dimension refers to the *direction* of domestic policy changes. Have environmental policies become stricter over time and have the countries under investigation developed in a similar way? Third, the dimension of *mobility* refers to the extent to which laggard countries have caught up with leaders or even have overtaken former front-runners over time.

Depending on the exact convergence dimension investigated, empirical results might be interpreted very differently. Evidence of growing policy homogeneity, for instance, does not necessarily mean that there is also convergence in terms of mobility, with laggards catching up and overtaking the leaders. In a similar vein, evidence of mobility does not imply that there must also be an increase in homogeneity. The fact that laggard countries change more fundamentally than leader countries is not a sufficient condition for a decrease in variance across all countries. Moreover, neither mobility nor homogeneity growth provides us with information about the overall direction of policy ambitions. Especially when comparing empirical results from different studies, it is therefore crucial to be clear about the specific type of convergence that has been investigated.

18.3 Theoretical Discussion: Drivers of Policy Diffusion and Policy Convergence

Which factors affect the extent to which policies spread across countries and why are some policies or countries more infectious than others? And which factors make the occurrence of cross-national policy convergence more or less likely? The literature on convergence and diffusion offers a broad range of causal factors that might be of relevance in this regard. At a very general level, these factors can be grouped into two categories: (1) causal mechanisms triggering the convergent policy changes across countries and (2) moderating factors that affect the strength of the respective mechanisms (Holzinger and Knill, 2005; Knill, 2005).

18.3.1 Causal Mechanisms

With regard to causal mechanisms, both diffusion and convergence can be generally expected to increase with the extent of *transnational communication* between (groups) of countries. Under this heading, several mechanisms are summarized which all have in common that they purely rest on communication and information exchange among countries (Dobbins and Knill, 2017; Holzinger and Knill, 2005; Howlett and Joshi-Koop, 2011; Strunz et al., 2018). They include lesson-drawing (where countries deliberately seek to learn from successful problem-solving activities in other countries) (Rose, 1991), joint problem-solving activities within transnational elite networks or epistemic communities (Haas, 1992), the promotion of policy models by international organizations with the objective of accelerating and facilitating cross-national policy transfer as well as the emulation of policy models.[1]

It is well conceivable that transnational problem-solving in elite networks can prepare the ground for the subsequent activities of international harmonization. This holds especially true for problems characterized by strong interdependencies. At the same time, however, it is emphasized that international organizations play an important role in forging and promulgating transnational epistemic communities (Simmons and Elkins, 2004). In other words, regular negotiations and discussions on problems subject to harmonization provide the ground for joint problem-solving in related areas that do not necessarily require a joint solution through international law. This argument is supported by the findings of Kern (2013: 144) who shows that international organizations play an important role in accelerating and facilitating cross-national policy transfer. They constitute important channels for multilateral communication and policy diffusion. Kern shows that – compared to policy exchange resting on bilateral and horizontal communication between countries – policy models spread much broader and faster if these countries are members of the same international organization (see also Shipan and Volden, 2021).

Second, *competition* emerging from the increasing economic integration of European and global markets has been identified as an important factor that drives the mutual adjustment of policies across countries and the global spread of public policies. Competition presupposes economic integration among countries. Especially with the increasing integration of European and global markets and the abolition of national trade barriers, the international mobility of goods, workers, and capital puts competitive pressure on the nation states to redesign domestic market regulations in order to avoid regulatory burdens restricting the competitiveness of domestic industries (Gilardi and Wasserfallen, 2019). The pressure arises from (potential) threats of economic actors to shift their activities elsewhere, inducing governments to lower their regulatory standards. This way, regulatory competition among governments may lead to a race to the bottom in policies, implying policy convergence (Drezner, 2001: 57–9; Simmons and Elkins, 2004). Theoretical work, however, suggests that there are a number of conditions that may drive policy in both directions (Vogel, 1997), including, amongst others, the type of policy concerned (product or process standards) or the presence of interests other than business in national politics (Ostry, 2018).

Third, emphasis is placed on the *harmonization* of national policies through international or supranational law. Countries are obliged to comply with international rules on which they have deliberately agreed during multilateral negotiations (Arbolino et al., 2018; Strunz et al., 2018; Zhou et al., 2019). Harmonization refers to a specific outcome of international cooperation, namely to constellations in which national governments are legally required to adopt similar policies and programs as part of their obligations as members of international organizations. International harmonization and more generally international cooperation presuppose the existence of interdependencies or externalities which push governments to resolve common problems through cooperation within international organizations, hence sacrificing some independence for the good of the community (Drezner, 2001: 60). Once established, institutional arrangements will constrain and shape domestic policy choices, even as they are constantly challenged and reformed by their member states. This way, international institutions are not only the object of state choice, but at the same time consequential for subsequent governmental activities (Martin and Simmons, 1998: 743). However, as member states voluntarily engage in international cooperation and actively influence corresponding decisions and arrangements, the impact of international harmonization on national policies constitutes no hierarchical process, but rather can be interpreted as negotiated diffusion or convergence (Holzinger and Knill, 2005).

Fourth, several studies emphasize the convergence and diffusion process effects stemming from *coercion*, and hence the imposition of policies. Coercion refers to constellations where countries or international organizations force other countries to adopt certain policies by exploiting asymmetries in political or economic power. This presupposes asymmetry of power. Often, there is an exchange of economic resources for the adoption of policies (Gilardi and Wasserfallen, 2019; Zhou et al., 2019). There are two typical cases: the unilateral imposition of a policy on a country by another country, and conditionalities imposed by international institutions. The first case might, for example, occur after a war. This is rare and does not lead to far-reaching diffusion and convergence, as it does not usually involve many countries. The second case is more prevalent and usually involves a greater number of countries. Moreover, the policies which form the content of the conditionality – typically economic policies or human rights – are already applied in wider parts of the international community (Holzinger and Knill, 2005).

18.3.2 Moderating Factors

As mentioned above, the strength of the effect size of the different causal mechanisms might be affected by a range of moderating variables that are discussed in the relevant literature. More specifically, emphasis is placed on two groups of factors (Knill, 2005). The first group in that respect refers to characteristics, or more precisely, the similarity of the countries under investigation. It is argued that converging policy developments and policy diffusion are more likely for and across countries that are characterized by high institutional similarity. Policies are transferred and properly implemented only insofar as they fit with existing institutional arrangements (see, for instance, Knill, 2005). Moreover, cultural similarity plays an important role in facilitating cross-national policy diffusion. In their search for relevant policy models, decision-makers are expected to look to the experiences of those countries with which they share an especially close set of cultural ties (Baldwin, 2018; Sommerer and Tallberg, 2019). Finally, similarity in socio-economic structures and development has been identified as a factor that facilitates the diffusion of policies across countries (Arbolino et al., 2018; Holzinger et al., 2011).

The second group of moderating variables that can be analytically distinguished is composed of the characteristics of the underlying policies. In this context, the type of policy is identified as a factor that influences the likelihood of convergence and diffusion. The expectation is that policies involving high distributional conflicts between domestic actor coalitions will diffuse and hence converge to a lesser extent than regulatory policies with comparatively small redistributional consequences (Tews and Busch, 2002). A second argument about the impact of policy characteristics on convergence and diffusion concentrates on different policy dimensions. Holzinger et al. (2008a), for instance, distinguish between policy targets and policy instruments and settings, arguing that policy convergence and policy diffusion are more likely for abstract policy elements (such as policy targets), while it is less likely (and more demanding from a convergence or diffusion perspective) that countries not only adopt similar policy targets, but also rely on the same instruments to achieve these targets and even calibrate these instruments in exactly the same way (see also Steinebach and Knill, 2018).

18.4 Empirical Findings: Diffusion and Convergence of Environmental Policy Portfolios

In this section, we take a closer look at the empirical patterns of environmental policy diffusion and convergence. In so doing, we adopt a more holistic approach capturing these patterns for sectoral policy portfolios rather than individual policies. This latter focus has been prevalent in particular with regard to studies on environmental policy diffusion that analyzed, for instance, the global spread of single policy measures such as green taxation policies or environmental audit schemes (e.g. Busch and Jörgens, 2005).

On environmental policy convergence, by contrast, more systematic data covering 24 countries (21 European states as well as the USA, Mexico, and Japan) are available. The empirical findings of the ENVIPOLCON project strongly point to the occurrence of environmental policy convergence in Europe in the 1970 to 2000 period. This conclusion is valid in two senses. Environmental policies have generally grown more alike over time (growing homogeneity), but at the same time they have moved in an 'upward' direction, thus becoming also stricter over time. Hence, a 'race to the bottom' due to regulatory competition – i.e. a weakening of environmental standards by countries as a consequence of engaging in competitive markets, as often predicted in the literature – has not taken place (Holzinger et al., 2008a, 2008b; Jörgens et al., 2014). Moreover, empirical findings also hint at considerable mobility in policy developments across countries, with laggard countries partially catching up with the leaders in terms of environmental policy adoptions and the strictness of regulatory levels (Holzinger et al., 2010). Overall, these findings indicate also that – contrary to Shipan and Volden's (2021) conclusion – those policies spread that come with higher regulatory ambitions.

In theoretical terms, the findings of the ENVIPOLCON project point to the strong impact of international harmonization (both at the level of the EU and international organizations) and transnational communication. The latter is particularly pronounced for those policies that are not subject to some binding form of supranational or international regulation. By contrast, competition plays a minor role in accounting for the pronounced convergence developments. Apart from failing to produce the often predicted 'race to the bottom,' there is no effect of regulatory competition that goes beyond the effects of harmonization or communication. Moreover, the impact of trade is not more pronounced for trade-related policies than for those not directly related to trade. Finally, domestic factors also contribute to the explanation of environmental policy convergence. Among the factors controlled for in the ENVIPOLCON project, the effects of income are most pronounced, whereas political demand exerted by green parties and environmental problem pressure shows weaker effects.

In the following, we complement these findings in two ways. First, we rely on a more recent dataset, capturing sectoral policy developments over more than 40 years (1976 to 2020) (Knill et al., 2020). Second, we take a more holistic approach by studying the diffusion and convergence of environmental policy portfolios, capturing the subfields of clean air policy, water policy, and nature protection (Adam et al., 2017, 2019; Fernández-i-Marín et al., 2021). Policy portfolios are typically composed of two dimensions: policy targets and policy instruments. *Policy targets* are all issues addressed by the government. *Policy instruments* are the means that governments have at their disposal to address policy targets.

Overall, the data identify 48 policy targets, which are mostly pollutants like ozone, carbon dioxide, or sulfur dioxide in the air, but also comprise other substances like lead content in gasoline, sulfur content in diesel, nitrates, and phosphates in continental surface water, as well as environmental objects like native forests, endangered plants, or endangered species. In addition to their substantive area, the targets differ in the extent to which they entail economic costs for the target groups, that is the actors that are addressed by the policy in question. For our purposes, we distinguish between two categories. First, policies can entail concentrated costs for specific target groups. This applies to all policies that contain specific requirements that need to be fulfilled by a clearly identifiable group of actors, for example emission or product policies targeted at certain industries. Second, the costs of policy targets can spread in more diffuse patterns across a broad range of actors. This applies, for instance, to policies that address general air quality or water quality conditions. In these cases, the target group is not clearly identifiable from the outset.

Policy instruments can be grouped into three major categories, including (1) regulation (ranging from hierarchical forms of governing, such as obligatory policy standards and technological prescriptions); (2) market-based instruments, such as taxes, subsidies, and other forms of market intervention; and (3) soft instruments, capturing voluntary, information-based, and planning instruments (Holzinger et al., 2006).

18.4.1 Patterns of Policy Diffusion

Based on these distinctions, we can now take a closer look at the extent to which patterns of policy diffusion vary across the different categories of policy targets and policy instruments. Are some policies and instruments more 'infectious' than others and to what extent does policy diffusion vary between EU and non-EU countries? To study policy infectiousness, we can further differentiate between two aspects. First, adoption rates capture the number of countries that have been 'infected' by a given policy. Second, policies might vary in the speed at which they spread across countries.

Figure 18.1 provides an overview of the average adoption rates we observed for different policy targets and policy instruments at the end of our observation period. The data reveal that classical instruments based on regulatory approaches have been adopted much more widely

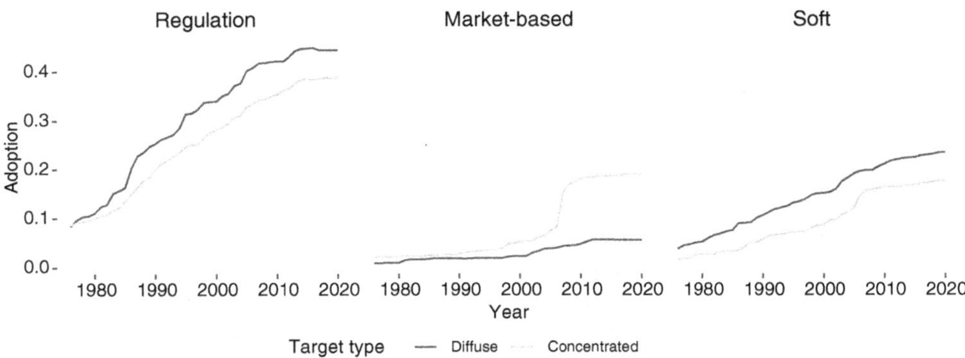

Figure 18.1 Policy dimensions and adoption rates.

across our country population than soft- and market-based instruments. This finding is well in line with studies analyzing the adoption of different environmental policy instruments at the EU level (Holzinger et al., 2006). Contrary to political rhetoric emphasizing the need to replace classical regulatory intervention by market-based instruments and softer forms of intervention, regulatory instruments have on average much higher adoption rates (around twice as high) as the two other instrument categories.

At the same time, adoption rates are generally higher for policy targets that entail diffuse rather than concentrated costs for the actor groups addressed. It seems easier to adopt policies whose costs are distributed broadly across society rather than those targeted on specific groups. As pointed out by Wilson (2019), concentrated costs facilitate the mobilization and organization of powerful private interests, such as industry associations, that try to block the adoption of such policies (see also Madden, 2014). It is only for market-based instruments that policies entailing concentrated costs diffuse more widely than policies entailing diffuse costs. This can be explained by the fact that it is in the very nature of market-based approaches that they come with clear allocations of costs or benefits for targeted actors; that is market-based instruments with diffuse costs are overall very rare and thus of only subordinate relevance.

To what extent are these patterns also observable, when we shift our focus from adoption rates to the speed of policy diffusion? Are these targets and instrument types that are most widely adopted also spreading more quickly across our country's population? To answer this question, we calculate the average adoption speeds for the different combinations of target and instrument types identified above. Average adoption speed is calculated as the average of the annual increases in the adoption rates.

Table 18.1 ranks our target and instruments dimensions with regard to the average adoption speed. Obviously, there seems to be a strong correlation between adoption rates and adoption speeds. Regulatory instruments overall diffuse more quickly than soft and market-based instruments. Moreover, targets entailing diffuse costs spread faster than targets entailing concentrated costs for clearly specified target groups. Yet, it is important to emphasize that the consideration of average speed does not fully capture different speed dynamics over time. The latter can be observed in particular for market-based instruments applied to achieved policy targets that entail concentrated costs (see again Figure 18.1). In this case, a long period of stagnation has been interrupted by a short stage of very fast diffusion, which again is followed by a period of slowed down diffusion. Obviously, market instruments at least so far display a lower potential to spread across a larger number of countries, notwithstanding a steep, short-term diffusion dynamic.

Table 18.1 Target and instrument types ranked by average adoption speed

Target type	*Instrument type*	*Average adoption speed*
Diffuse costs	Regulation	0.8202
Concentrated costs	Regulation	0.6851
Diffuse costs	Soft	0.4454
Concentrated costs	Market-based	0.3865
Concentrated costs	Soft	0.3628
Diffuse costs	Market-based	0.3628

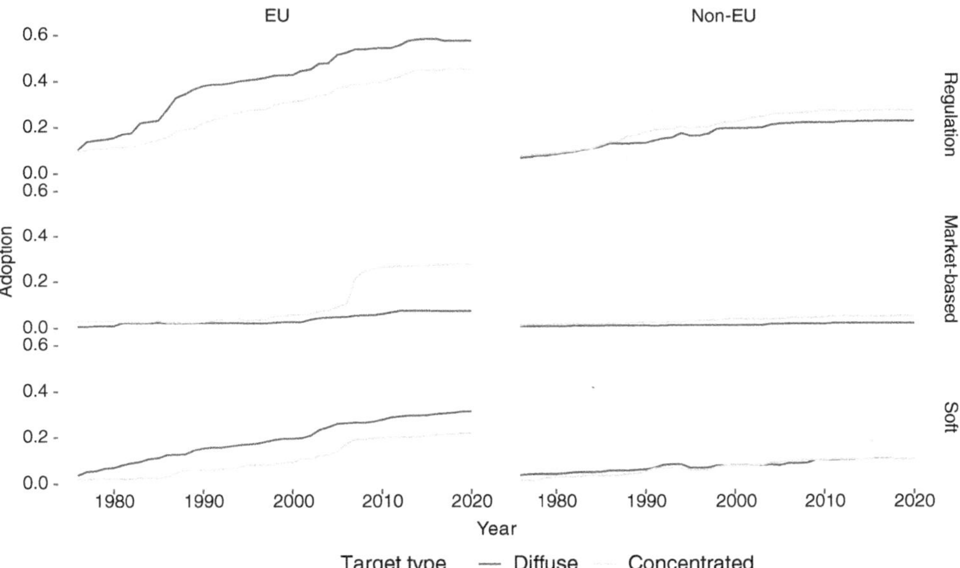

Figure 18.2 Diffusion dynamics in EU member states and non-EU countries.

Overall, our empirical findings indicate that diffusion dynamics (both in terms of adoption rates and speeds) vary across policy targets and policy instruments. In the next step, we take a closer look at the extent to which these dynamics vary in relation to the extent of international harmonization. To capture these differences, we compare diffusion dynamics between EU member states and non-EU countries (Figure 18.2). We see that diffusion dynamics are positively affected by EU membership. This is most pronounced for regulatory instruments. We also see that the steep increase in market-based instruments is mainly the result of diffusion dynamics across EU member states. It also becomes apparent that both soft and market instruments only started to spread across EU countries over the last two decades, while regulatory instruments reveal more linear spreading patterns throughout the whole observation period. This observation is well in line with studies analyzing the emergence of so-called new policy instruments in EU environmental policy (Bailey, 2017; Holzinger et al., 2006; Jordan et al., 2005). Overall, these findings provide further support to the fact that it is not only soft instruments and targets with diffuse costs implications that are spreading, but also policies that are based on regulatory intervention and concentrated costs for specific addresses. In short, there is a clear indication that ambitious policies are spreading at least as broadly and quickly as less ambitious approaches.

18.4.2 Patterns of Policy Convergence

Previous studies analyzed the changes in policy homogeneity for different policy elements (targets, instruments, instrument calibration). We complement these measurements by focusing – in a more encompassing way – on changes in the similarity of environmental policy portfolios over time. To what extent have countries filled similar spaces in their environmental policy portfolios over the observation period? To assess this development, we rely on the Jaccard coefficient, which has been developed as a key figure to analyze the similarity of sets. It is calculated as the number of common elements (the intersection of filled portfolio spaces) divided by the size of the union set (the maximum portfolio space). The closer the Jaccard coefficient is to 1, the greater the similarity of the sets. The minimum value of the Jaccard coefficient is 0.

Figure 18.3 displays the convergence of environmental policy portfolios between the countries in our sample from 1976 to 2020. The figure shows considerable increase in average portfolio similarity, ranging from almost complete dissimilarity at the beginning of the observation period to a score of over 30 percent in 2020. This increase, however, mainly occurred before the early 2000s, with only minor additional similarity changes occurring over the last two decades. This finding is well in line with the previous insight that policy ambitions at the EU level generally stagnated in the aftermath of the Great Recession (Burns and Tobin, 2020; Knill et al., 2020a; Steinebach and Knill, 2017).

In order to classify these findings more broadly, we have included data on changes in the similarity of social policy portfolios for our country sample. A comparison of both policy areas reveals that the dynamics of international policy convergence are much more pronounced for environmental policy. This difference might be explained by several factors. First, convergence for policies entailing bigger redistributional conflicts between societal groups is less likely than for regulatory policies. Second, convergence might be negatively affected by policy maturity and implied saturation effects. In contrast to social policy, environmental policy is a younger

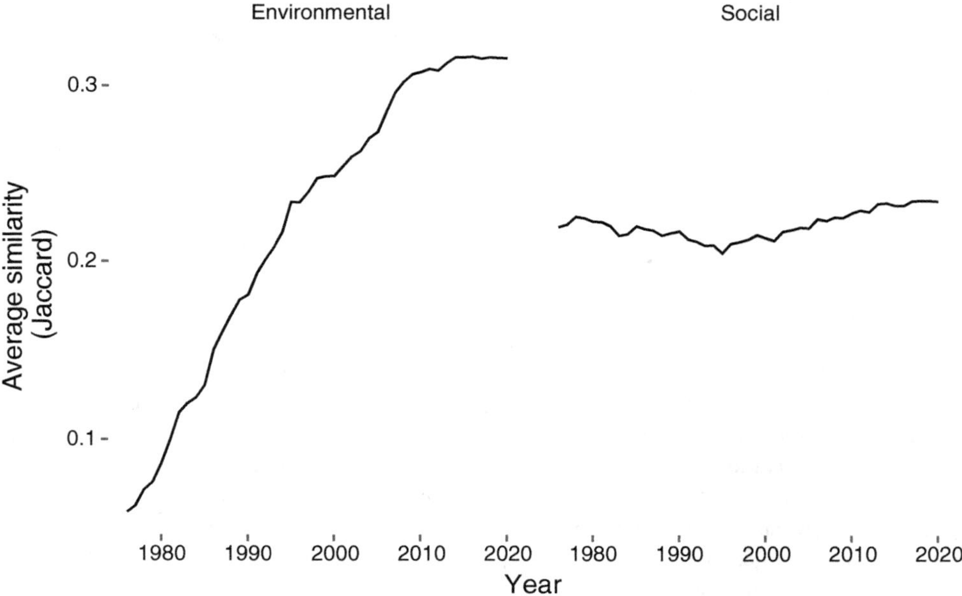

Figure 18.3 Convergence of environmental policy portfolios.

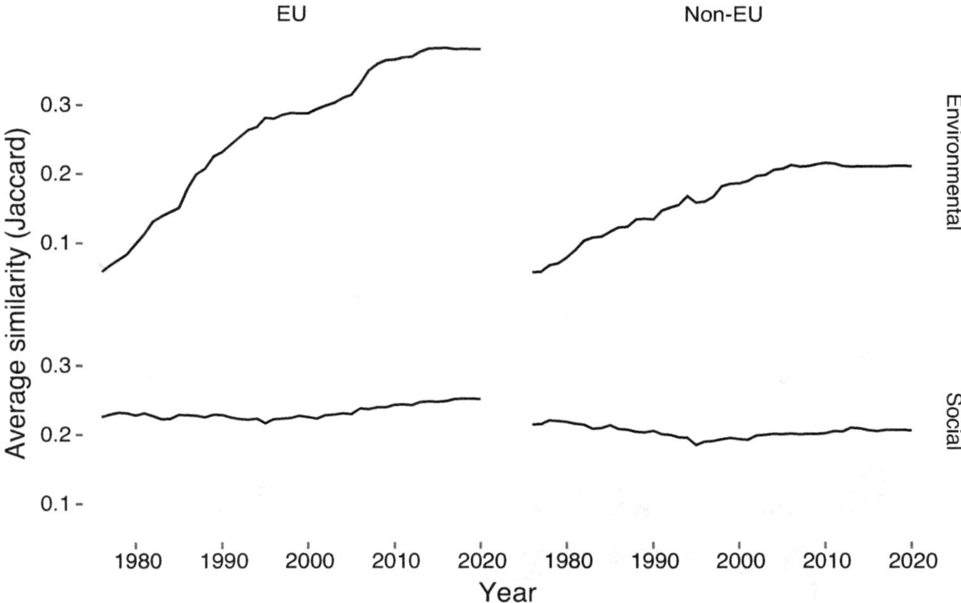

Figure 18.4 Convergence of environmental policy portfolios in EU and non-EU countries.

and more dynamic policy area that took off from the 1970s onwards. Consequently, there is a higher probability that the expansion of national policy portfolios comes with increasing policy similarity. Third, convergence might be strongly driven by international harmonization, and in particular the effects of EU legislation. EU competences are very strongly developed for environmental, but not for social, policy. This latter aspect is further illustrated when comparing policy convergence between EU and non-EU countries (Figure 18.4).

Figure 18.4 reveals that the EU is a pronounced driver of environmental policy convergence. Portfolio similarity which increases over time is much higher for EU countries, while for non-EU-members convergence is less dynamic. At the same time, however, the overall pattern of a stagnation in convergence developments applies to both country groups. Figure 18.4 also shows that the EU makes no strong difference to the convergence of social policy portfolios. This finding is rather plausible in view of the EU's comparatively weak competencies in this policy field.

A further illustration of the strong impact of the EU on environmental portfolio convergence is provided by taking a closer look at the country pairs that represent the most similar (respectively most different) portfolios in our sample (Table 18.2). The five country pairs with the highest similarity values are EU member states. Figures 18.5 and 18.6 provide a graphical illustration of two country pairs with a very high (Belgium and the Netherlands) and very low portfolio similarity (Belgium and Japan).

18.5 Conclusion and Future Research

In this chapter, we have discussed to what extent national environmental policy dynamics are affected by processes of international policy diffusion and policy convergence. While diffusion captures the scope and speed of environmental policy adoption across countries, convergence

Table 18.2 Top five country pairs with the most and the least similar environmental portfolios

Most similar pairs	*Jaccard coefficient*	*Least similar pairs*	*Jaccard coefficient*
Ireland/Italy	0.732	Australia/Japan	0.068
France/Italy	0.698	Japan/Sweden	0.094
Ireland/Netherlands	0.695	Australia/Norway	0.102
Austria/Italy	0.690	Denmark/Mexico	0.105
Austria/Ireland	0.690	Sweden/Mexico	0.112

Figure 18.5 Example of a country pair with very similar environmental portfolios in 2020.

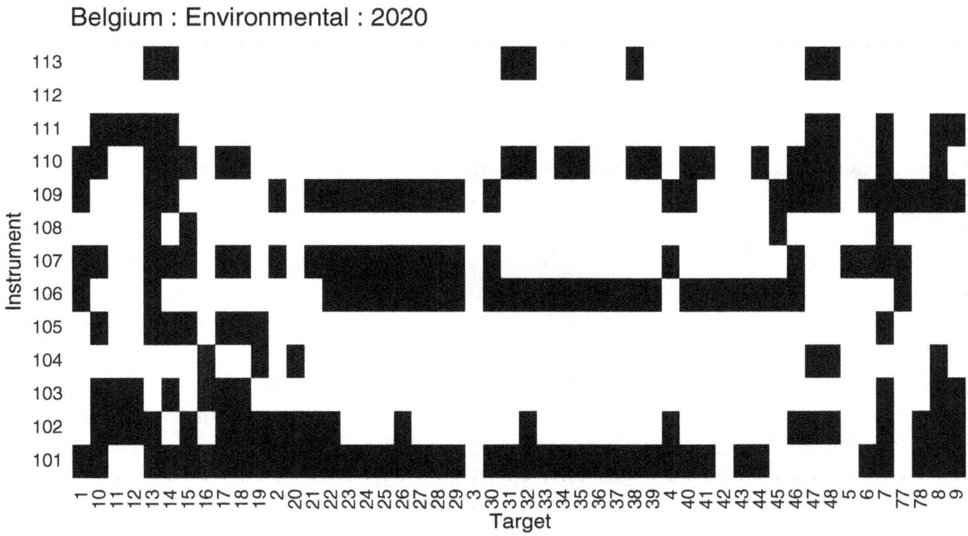

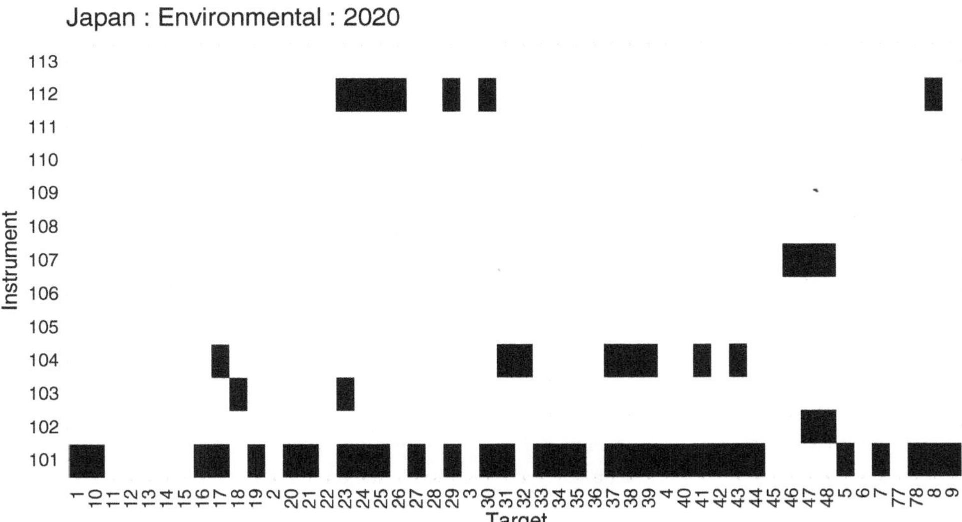

Figure 18.6 Example of a country pair with very different environmental portfolios in 2020.

refers to the changes in the similarity of environmental policy portfolios over time. Diffusion and convergence are driven by a number of factors, in particular international interdependencies between countries that emerge from cooperation and communication at the level of international and supranational organizations as well economic interlinkages in globalized markets.

From the outset it is not clear if processes of diffusion and convergence – when they actually occur – are a good or a bad thing with regard to the design of effective policy solutions to environmental problems. In particular, we do not know how convergence and diffusion affect overall policy ambitions in terms of the size and regulatory stringency of national policy portfolios. To take a closer empirical look at these effects, we complemented existing empirical evidence by studying the spread of different types of environmental policy targets and policy instruments and changes in the similarity of national environmental policy portfolios. Overall, our assessment reveals three issues.

First, it seems that ambitious policies (in terms of regulatory rather than soft instruments and concentrated rather than diffuse cost implications) spread at least as widely and fast across countries as do less ambitious policies. Second, this development is accompanied by a considerable increase in national policy portfolios over time. This becomes apparent in particular when comparing convergence patterns to other policy areas such as social policy. Third, both convergence and diffusion trends are strongly amplified by EU membership. In conclusion, these findings indicate that diffusion and convergence should general help to strengthen the capacity of national governments to address environmental problems. Moreover, the EU (and international cooperation more generally) constitutes a crucial factor reinforcing this relationship.

While there is the overall finding that diffusion and convergence contribute to the overall strengthening of environmental policies, more research is needed on the conditions under which such scenarios are more or less likely. In this regard, particular emphasis should be placed on the temporal sequence between different drivers of policy diffusion and policy convergence, in particular with regard to economic drivers and drivers linked to international harmonization and cooperation.

Box 18.1 Chapter Summary

- Policy diffusion and policy convergence capture related yet different phenomena. While diffusion refers to the scope and speed of environmental policy adoption across countries, convergence refers to the changes in the similarity of environmental policy portfolios over time.
- Diffusion and convergence are driven by a number of factors, in particular international interdependencies between countries that emerge from cooperation and communication at the level of international and supranational organizations as well economic interlinkages in globalized markets.
- To analyze the effects of convergence and diffusion on environmental problem-solving, we focused on (1) the spread of different types of environmental policy targets and instruments and (2) changes in the similarity of national environmental policy portfolios over time.
- Our findings indicate that diffusion and convergence should generally help to strengthen the capacity of national governments to address environmental problems.
- Moreover, the EU (and international cooperation more generally) constitutes a crucial factor reinforcing this relationship.

Note

1 As argued above, in the terminology used by Busch and Jörgens (2005), the mechanisms summarized under transnational communication would be referred to as policy diffusion.

References

Adam, C., Hurka, S., Knill, C., & Steinebach, Y. (2019). *Policy accumulation and the democratic responsiveness trap*. Cambridge University Press.

Adam, C., Knill, C., & Fernández-i-Marín, X. (2017). Rule growth and government effectiveness: Why it takes the capacity to learn and coordinate to constrain rule growth. *Policy Sciences*, *50*(2), 241–268.

Arbolino, R., Carlucci, F., De Simone, L., Ioppolo, G., & Yigitcanlar, T. (2018). The policy diffusion of environmental performance in the European countries. *Ecological Indicators*, *89*, 130–138.
Bailey, I. (2017). *New environmental policy instruments in the European Union: Politics, economics, and the implementation of the packaging waste directive*. Taylor & Francis.
Baldwin, R. (2018). *The great convergence*. Harvard University Press.
Boushey, G. (2010). *Policy diffusion dynamics in America*. Cambridge University Press.
Burns, C., & Tobin, P. (2020). Crisis, climate change and comitology: Policy dismantling via the backdoor? *JCMS: Journal of Common Market Studies*, *58*(3), 527–544.
Busch, P. O., & Jörgens, H. (2005). The international sources of policy convergence: Explaining the spread of environmental policy innovations. *Journal of European Public Policy*, *12*(5), 860–884.
Butler, D. M., Volden, C., Dynes, A. M., & Shor, B. (2017). Ideology, learning, and policy diffusion: Experimental evidence. *American Journal of Political Science*, *61*(1), 37–49.
Dobbins, M., & Knill, C. (2017). Higher education governance in France, Germany, and Italy: Change and variation in the impact of transnational soft governance. *Policy and Society*, *36*(1), 67–88.
Drezner, D. W. (2001). Globalization and policy convergence. *International Studies Review*, *3*(1), 53–78.
Elkins, Z., & Simmons, B. (2005). On waves, clusters, and diffusion: A conceptual framework. *The Annals of the American Academy of Political and Social Science*, *598*(1), 33–51.
Fernández-i-Marín, X., Knill, C., & Steinebach, Y. (2021). Studying policy design quality in comparative perspective. *American Political Science Review*, 1–17.
Gilardi, F., & Wasserfallen, F. (2019). The politics of policy diffusion. *European Journal of Political Research*, *58*(4), 1245–1256.
Gray, V. (1973). Innovation in the states: A diffusion study. *American political science review*, *67*(4), 1174–1185.
Haas, P. M. (1992). Introduction: Epistemic communities and international policy coordination. *International Organization*, *46*(1), 1–35.
Heichel, S., Pape, J., & Sommerer, T. (2005). Is there convergence in convergence research? An overview of empirical studies on policy convergence. *Journal of European public policy*, *12*(5), 817–840.
Holzinger, K., & Knill, C. (2005). Causes and conditions of cross-national policy convergence. *Journal of European public policy*, *12*(5), 775–796.
Holzinger, K., Knill, C., & Arts, B. (2008b). *Environmental policy convergence in Europe: The impact of international institutions and trade*. Cambridge University Press.
Holzinger, K., Knill, C., & Schäfer, A. (2006). Rhetoric or reality?'New governance'in EU environmental policy. *European Law Journal*, *12*(3), 403–420.
Holzinger, K., Knill, C., & Sommerer, T. (2008a). Environmental policy convergence: The impact of international harmonization, transnational communication, and regulatory competition. *International Organization*, *62*(4), 553–587.
Holzinger, K., Knill, C., & Sommerer, T. (2011). Is there convergence of national environmental policies? An analysis of policy outputs in 24 OECD countries. *Environmental Politics*, *20*(1), 20–41.
Holzinger, K., Knill, C., Heichel, S., & Sommerer, T. (2010). *Theorie und Empirie internationaler Politikkonvergenz: Eine vergleichende Analyse der Umweltpolitik zwischen 1970 und 2000*. Verlag Barbara Budrich.
Howlett, M., & Joshi-Koop, S. (2011). Transnational learning, policy analytical capacity, and environmental policy convergence: Survey results from Canada. *Global Environmental Change*, *21*(1), 85–92.
Jordan, A., Wurzel, R. K., & Zito, A. (2005). The rise of 'new'policy instruments in comparative perspective: Has governance eclipsed government?. *Political Studies*, *53*(3), 477–496.
Jordana, J., & Levi-Faur, D. (2005). The diffusion of regulatory capitalism in Latin America: Sectoral and national channels in the making of a new order. *The Annals of the American Academy of Political and Social Science*, *598*(1), 102–124.
Jörgens, H., Lenschow, A., & Liefferink, D. (Eds.). (2014). *Understanding environmental policy convergence: The power of words, rules and money*. Cambridge University Press.
Kern, K. (2013). *Die diffusion von politikinnovationen: Umweltpolitische innovationen im mehrebenensystem der USA*(Vol. 17). Springer-Verlag.
Knill, C. (2005). Introduction: Cross-national policy convergence: Concepts, approaches and explanatory factors. *Journal of European Public Policy*, *12*(5), 764–774.
Knill, C., Steinbacher, C., & Steinebach, Y. (2020). Sustaining statehood: A comparative analysis of vertical policy-process integration in Denmark and Italy. *Public Administration*.
Knill, C., Steinebach, Y., & Fernández-i-Marín, X. (2020a). Hypocrisy as a crisis response? Assessing changes in talk, decisions, and actions of the European Commission in EU environmental policy. *Public Administration*, *98*(2), 363–377.

Madden, N. J. (2014). Green means stop: Veto players and their impact on climate change policy outputs, *Environmental Politics*, *23*(4)), 570–589.
Martin, L. L., & Simmons, B. A. (1998). Theories and empirical studies of international institutions. *International Organization*, *52*(4), 729–757.
Meyer, J. W., Frank, D. J., Hironaka, A., Schofer, E., & Tuma, N. B. (1997). The structuring of a world environmental regime, 1870–1990. *International Organization*, *51*(4), 623–651.
Ostry, S. (2018). *National diversity and global capitalism*. Cornell University Press.
Popp, D., Hafner, T., & Johnstone, N. (2011). Environmental policy vs. public pressure: Innovation and diffusion of alternative bleaching technologies in the pulp industry. *Research Policy*, *40*(9), 1253–1268.
Rose, R. (1991). What is lesson-drawing?. *Journal of Public Policy*, *11*(1), 3–30.
Shipan, C. R., & Volden, C. (2021). Why bad policies spread (and Good Ones Don't). *Elements in American politics*. Cambridge University Press.
Simmons, B. A., & Elkins, Z. (2004). The globalization of liberalization: Policy diffusion in the international political economy. *American Political Science Review*, *98*(1), 171–189.
Sommerer, T., & Tallberg, J. (2019). Diffusion across international organizations: Connectivity and convergence. *International Organization*, *73*(2), 399–433.
Steinebach, Y., & Knill, C. (2017). Still an entrepreneur? The changing role of the European Commission in EU environmental policy-making. *Journal of European Public Policy*, *24*(3), 429–446.
Steinebach, Y., & Knill, C. (2018). Social policies during economic crises: An analysis of cross-national variation in coping strategies from 1980 to 2013. *Journal of European Public Policy*, *25*(11), 1566–1588.
Strang, D., & Meyer, J. W. (1993). Institutional conditions for diffusion. *Theory and Society*, *22*(4), 487–511.
Strunz, S., Gawel, E., Lehmann, P., & Söderholm, P. (2018). Policy convergence as a multifaceted concept: The case of renewable energy policies in the European Union. *Journal of Public Policy*, *38*(3), 361–387.
Tews, K., & Busch, P. O. (2002). Governance by diffusion? Potentials and restrictions of environmental policy diffusion. In *Proceedings of the 2001 Berlin Conference on the Human Dimension of Global Environmental Change" Global Environmental Change and the Nation State* (pp. 168–182).
Vogel, D. (1997). Trading up and governing across: Transnational governance and environmental protection. *Journal of European Public Policy*, *4*(4), 556–571.
Wilson, J. Q. (2019). *Bureaucracy: What government agencies do and why they do it*. Basic Books.
Zhou, Y., Dong, F., Kong, D., & Liu, Y. (2019). Unfolding the convergence process of scientific knowledge for the early identification of emerging technologies. *Technological Forecasting and Social Change*, *144*, 205–220.

19

POLICY DESIGN FOR SUSTAINABLE ENERGY AND THE INTERPLAY OF PROCEDURAL AND SUBSTANTIVE POLICY INSTRUMENTS

Ishani Mukherjee

19.1 Introduction

The formulation of policy instruments to address growing public problems linked with environmental sustainability and climate change, especially in pertinent sectors such as energy production, has recently come under sharp focus. According to the most recent contribution to the Sixth Assessment Report of the Intergovernmental Panel on Climate Change (IPCC, 2021), climate change has accelerated to the point where it is already too late to avoid an increase in temperatures, and the best response now necessitates aggressive policy efforts to globally reach net-zero emissions by 2050, and negative carbon dioxide emissions beyond that point. Electricity and heat production, coupled with linked land use change remains the largest contributor of these emissions, calling now for an unavoidably critical focus on formulating effective policies in these sectors.

At the same time, contemporary research in the policy sciences places *effectiveness* as the central goal of policy design. This emphasis permeates both micro-level considerations for specific policy calibrations (such as increasing existing tariffs for siting coal plants) as well as more meso-level policy tools and tool mixes (such as combining regulation and financial incentives to galvanize a switch from fossil fuels to more renewable sources of energy, coupled with vitally ramping up the diffusion of carbon capture and storage technologies). Such research inquiries within the modern policy design literature involve the study of both traditional 'substantive' instruments of policymaking, such as regulation and fiscal policy, as well as lesser recognized 'procedural' tools that are deployed for policy administration (Hood, 2007; Salamon, 1981; Salamon, 2002). The latter include tools related to governments' design and implementation of network management mechanisms and public participation, activities linked to the delivery of organizational services such as the formation of advisory

DOI: 10.4324/9781003043843-22

committees to regulatory agencies, and policy processes more generally, such as the creation of public information commissions, government data portals and repositories, and judicial review processes (Howlett et al., 2000).

When conceptualizing policy instruments, the general tendency in policy sciences has been to view them as specific depictions of policy formulation. They represent the most concrete manifestations of government intent and are the technical alternatives that policy formulators design and deploy to meet stated goals. As such, there is a general agreement in policy sciences regarding what a policy tool conceptually indicates, while the processes of designing it may in practice be diverse. Howlett (2000), for example, defines policy tools as the means of government "to deliberately affect the natures, types, quantities and distribution of goods and services provided in a society" (p. 415). Vedung et al. (1998) have similarly alluded to government *action* that is embodied in a policy instrument that represents "a set of techniques by which government authorities wield their power in attempting to ensure support and affect or prevent social change" (p. 21).

At their most fundamental level, then, it is understood that tools of policy design can be broadly typified based on their primary purpose. The first relates to the actual *content* of how public action is to be influenced. The second targets government's own *processes* of administration, and seeks to affect political or policymaking behaviour during the setting of policy aims and means (Howlett and Mukherjee, 2018). Policy instruments in policy design, then, become those specific techniques that give effect to policy goals and which also differ according to their main purpose: falling into 'substantive' and/or 'procedural' categories as above (Howlett et al., 2000; Hood and Margetts, 2007).

While substantive policies, their analysis, and categorization have remained at the forefront of studies of policy instruments for sustainability, procedural policy means have in comparison enjoyed less academic exploration. Nevertheless, there is broad consensus that procedural policy components are fundamental to the formulation and effective functioning of substantive tools, and markedly so in sectors linked with environmental sustainability. And yet, the relationship that exists *between* these two categories remains scarcely explored in the literature on policy instruments and design (Howlett et al., 2000; Howlett and Mukherjee, 2018). Unlike substantive tools that directly affect the production, consumption, and distribution activities of public goods and services, procedural instruments are envisioned as instead affecting governments' own internal actions related to the policymaking process. For example, network management and governing the behaviours and interactions of policymaking agents is one significant function of procedural tools (Klijn et al., 1995). And these tools, beyond their administrative function, can also shape the substantive policy decisions that follow from governments' process-oriented actions. Examples include governments forming an advisory committee of select citizens or experts to help timely deliberations on sensitive issues such as nuclear energy regulation, or the creation of freedom-of-information or access-to-information laws that facilitate citizen access to government legislation and records of pollution and effluent permits. Additionally, reorganizing an administration's own internal structure can impact the effective formulation of policies, as occurs when new regulating agencies are created by merging personnel from energy and environmental ministries, forcing the two to adopt a new collaborative operating arrangement to address intra-sectoral issue areas such as the water-energy nexus.

These observations indicate that procedural and substantive elements of policy tools interact frequently during policy design for sustainability; and the impact of this interaction warrants further research. With a focus on sustainable energy policy, this chapter offers an initial exploration as to how exactly procedural elements support the substantive within sustainable energy policy design. By using illustrative examples from global experience with renewable energy policy formulation, the chapter highlights the role and interactions of procedural elements in the design of policy tools and their eventual on-the-ground calibrations.

19.2 Procedural Components of Policymaking

Procedural tools are fundamental to the overall resilience of policy design, especially in sectors such as energy, which are increasingly defined by volatile resource endowments and erratic global trends – so much so that understanding the interaction between the substantive and the procedural aspects of policy design is now deemed critical towards determining how best to proceed with effective energy policy formulation (Braathen, 2007; Del Río, 2010). And it is only in recent policy design literature where there is now growing evidence of considering both substantive and procedural policy aspects of policy mixes together in the interest of upholding design effectiveness and long-term resilience (Capano and Woo, 2018; Howlett, 2019; Nair and Howlett, 2017).

In policy domains such as renewable energy in particular, procedural elements have been observed to strengthen the substantive purpose of policy tools working within dedicated climate policy mixes. For example, regular reviews and sunset clauses in the design of national clean energy plans are specifically deployed to mitigate against problems arising from the haphazard layering of policy elements, which can derail the original intent of design. Such tools can make use of processes of 'smart layering' or 'patching' to keep policy elements flexible and aligned with changing contexts (Capano and Woo, 2018; Gunningham and Sinclair, 1998; Howlett and Rayner, 2013). Furthermore, procedural tools help towards estimating fluctuations in the policy context, thereby informing the success of the substantive contribution that a policy tool makes Leung et al., 2015; Lang, 2019). Such processual activities necessitate the earmarked allocation of human capital, financial, and administrative resources at the outset that can make available and deploy managerial and procedural support for a policy instrument during changing contexts that may often require a significant recalibration or rethinking of substantive elements (Luthar and Cicchetti, 2000; Doz and Kosonen, 2014).

Further to supporting the policy formulation process, procedural tools play a significant organizational role by affecting the actions, locations, and behaviour of policy formulators, which in turn directly impact the design of substantive tools. While this research agenda is still emergent within policy sciences, some evidence from studies on regulatory policy design reveals that the deliberate adjustment of institutional frameworks faced by regulators can lead to the formulation of better welfare-enhancing (substantive) policies (Dudley and Xie, 2020). This developing line of inquiry highlights that the variation in how substantive tools and regulations are designed can be explained not just by varying endowments of professional training, experience, and information, but also in terms of the cognitive biases of regulators, subject-matter experts, and analysts in government agencies, which can all be influenced by procedural policy decisions taken early on in the design process (Linder and Peters, 1989). On the other hand, such factors that uphold durability and perpetuate dominant styles of policy design may not always lead to a favourable outcome as procedural mechanisms can be used to essentially 'lock in' the use of substantive instruments that become the "lowest common denominator" of gaining the support of multiple policy actors, but might be objectively redundant or counter-productive to addressing the problem at hand (Fernández-i-Marín et al., 2021, p. 10).

The contemporary understanding of effective policy design heavily emphasizes the temporal tenacity of policy mixes (Peters et al., 2018). Perpetuating an environment that favours effective policy mixes over time deeply relies on intentional planning and the sequencing of underlying procedural elements that uphold the stability of substantive tools in the long run. However, this temporal dimension of procedural tools has rarely been comparatively explored in the policy design literature (Howlett, 2019). The lessons that are emerging about the effect that procedural tools have on policy stability over time in sectors such as environmental policy and climate regulation studies build on the ongoing theoretical development of concepts such as path dependence and policy feedback (e.g. Mahoney, 2000; Béland, 2010), in order to guide the formulation of policy strategies that endure by securing long-term political support (Fouquet, 2016; Bernstein and Hoffman, 2018). Rosenbloom et al. (2019), for example, outline four main lessons from their study of path dependency for the formulation of climate policies, lessons which indicate procedural processes such as accounting for sunk costs and the accumulation of technological and political experience that 'lock in' the instrumental elements of policy design. This has commonly been the case for fossil-fuel subsidies that, as a category of policy instruments, have generally been resistant to reform. In an era of net-zero commitments and low-carbon transition plans under the Paris Agreement, progress towards meeting climate goals has been markedly slow, often owing to established public financial flows that perpetuate the reliance on fossil fuels (Gençsü et al., 2020). Such trends now call for better-designed procedural instruments that improve transparency about the support being provided for phasing out fossil subsidies, annual and regular reporting of progress against National Energy and Climate Plans (NECPs), and more comprehensive reporting of timelines towards progress.

Such findings suggest that one direct effect that procedural tools can have on the effective formulation of environmental policy instruments is by determining the 'stickiness' of implemented policies or policy mixes by checking reversibility, engraining political support, and expanding buy-in over time. In their exposition on what makes climate policy instruments durable by design, Jordan and Moore (2020) similarly emphasize that "such policy fosters and sustains its own political support base over time, triggering legacy effects that endure even after the waning of the political forces that generated the policy's original enactment" (p. 4). In other words, the investment that policy makers provide to the initial substantive design of often ambitious climate policies is not self-enduring without the backing of underlying procedural support, since "some recalibration of a policy's implementing instruments is likely if the policy as a whole is to remain on course to achieve its goals" (Jordan and Moore, 2020, p. 7).

The above examples reveal insights into the fundamental questions about when and how far procedural and substantive elements of policy tools interact during the process of policy design and signal the need for a perspective on procedural tools that is sensitive to their unique role, especially within environmental policy design.

Subsequent to the new turn in policy design studies of the early 2000s, a disconnect has arisen between the scholarly discourse on policy formulation and how well it vouches for sound implementation. This disconnect is indicative of a growing need to foster a dialogue about creating effective implementation plans through procedural decisions made early in the policy process, alongside those of policy design, as "it is often a highly rarefied world and policy design occurs far from places where policy implementation happens" (Mintrom and Luetjens, 2017, p. 1). And yet, there is also scholarly cautioning about the mismatched expectations and misinformation regarding contexts, which can exist between the design of the content of policy instruments and strategies for its on-the-ground implementation (Howlett and Rayner, 2007).

A focus on procedural policy tools can help to amplify this connection between the choice of policy instrument content and how the tool is applied on the ground. Cashore and Howlett's original work on policy design (2007) on the topic distilled six "elements or components" (p. 535) of ends and means of policy design occupying three levels of analysis. Subsequent developments towards this line of thinking (Howlett, 2009; Howlett Mukherjee and Rayner,

2014; Peters et al., 2018) have solidified an understanding that choices about different types of policy tools occur at the 'meso' level of policy operationalization, while concerns about how chosen tools are enacted or deployed surround their more 'micro' level, on-the-ground calibrations. While distinct from each other at two different levels of analysis, design decisions for tools and calibrations are often taken together as illustrated by the examples herein from the renewable energy policy domain. Such a perspective that accommodates design decisions traversing *meso–micro* analytical divides necessitates a multi-level lens that looks beyond just the substantive content of tools and includes more procedural deliberations regarding its design and implementation. Especially so as to uphold consistency and coherence within a policy toolkit, effective design merits that equal consideration also be given to the coordination of substantive and procedural elements of tools working within any instrument mix (Howlett et al., 2014; Peters et al., 2018).

In other words, bringing procedural considerations into policy design frameworks and uniting them with the substantive allows for a more consolidated understanding of how meso-level policy tools and their micro-level calibrations and applications are constructed. Procedural means support and uphold the aims of more substantive elements of policy instrument formulation; and they make mutually supportive yet distinct contributions to the process of policy design. The discourse also points to how the decisions about the choice and content of tools and tool mixes are linked to those dealing with their on-the-ground calibrations. Uniting these multi-level considerations of policy design, allows for a framework to be drawn about how tools and their enactment are constructed as a whole, and embody the union or 'compound' of four distinct design components (Figure 19.1).

Firstly, at the level of tools, there is the identifying substantive component (cell 1, Figure 19.1), which is what the policy instrument is 'known for' and which the tools literature traditionally organizes in terms of the resources that the government puts behind its creation (Bemelmans-Videc et al., 1998; Howlett et al., 2000; Hood, 2007). Secondly, to support the

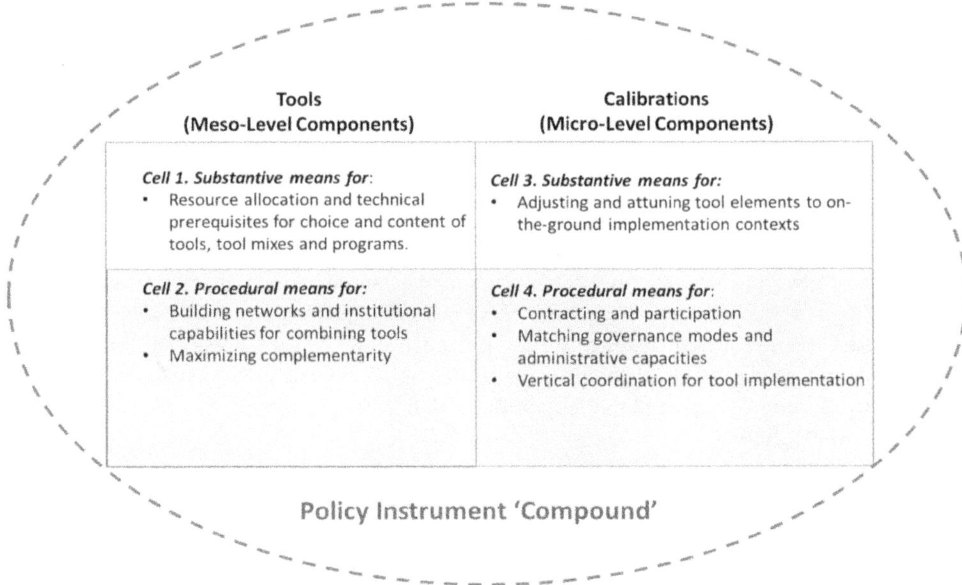

Figure 19.1 Substantive and procedural components in policy instrument "compounds."

formulation of the substantive means of policy tools through decisions related to the allocation of governance resources, the articulation of its institutional framework, as well as maximizing the complementarity of policy elements, there are procedural means that also need to be deployed concurrently (cell 2, Figure 19.1) (Howlett et al., 2017). At the level of calibration then, work on effective policy design shows that any formulation activity focused only on the characteristic features of policy instruments is incomplete without parallel consideration given to the means for implementing them (Mintrom and Luetjens, 2017; Adam et al., 2019). And so, the third component of an instrument 'compound' (cell 3, Figure 19.1) entails those substantive design considerations that are explicitly aimed at adjusting and attuning policy tool elements to the contexts within which they operate. Lastly, to address these calibrations of policy tools that govern their on-the-ground implementation, procedural means support the effectiveness of implementation planning decisions that directly implicate matters related to participation, matching of governance contexts to existing policy capacities, as well as the vertical coordination of governments' design work (cell 4, Figure 19.1). In highlighting the role of the procedural elements in policy instrument design, cells 2 and 4 of Figure 19.1 are elaborated below through examples drawn from three major categories of renewable energy policy, namely renewable portfolio standards, feed-in tariffs, and net metering.

19.3 Designing Renewable Energy Policy Mixes: An Illustrative Example

As of 2019, over 135 countries worldwide have formulated and implemented dedicated renewable energy policies in the power sector with Feed-in-Tariffs (FITs), net metering (also known as 'smart metering'), and renewable energy mandates or targets as the leading instruments within this category (REN21, 2019). Together, these instruments are classified by the sector as broadly 'regulatory' even if this generalization covers distinct tools such as quotas, mandates, and pricing instruments (REN21, 2019). Furthermore, it is understood that to ensure effectiveness from the outset of policy design, "a robust framework to monitor and penalise non-compliance is needed" (REN21, 2019, p. 56). FITs are most often designed or revised in conjunction with net-metering technology to align with national targets for renewable energy generation. To allude to Hood's (1986) categorization, this policy mix unites the functions of three distinct substantive instruments: a financial tool (FIT), an organizational tool encapsulating the direct provision of a good (smart metering) and a service (smart grid), and their ultimate coordination with an authority-based instrument (national renewable energy standard or targets). The first two are also often packaged together as a form of 'pull' policy to incentivize certain actions, complementing renewable energy mandates that direct or 'push' those actions. Most governments pursuing renewable energy targets make explicit this connection between the differing substantive means of these tools. For example, the UK Department of Energy and Climate Change states that "the objective of FITs is to contribute to the UK's 2020 renewable energy target through greater take-up of electricity generation at the small scale and to achieve a level of public engagement that will engender widespread behavioural change" (DECC, 2009).

Taken together as major policies that are usually deployed within countries' renewable energy policy 'toolkits', the general design of these instruments displays discernible procedural components that are separate from the substantive. Reflecting the elements listed under cells 2 and 4 of the framework depicted in Figure 19.1, the main procedural components pertaining to the design of tools and calibrations of renewable energy targets, FITs, and smart metering are shown in Box 19.1.

Box 19.1 Procedural Means of Renewable Energy (RE) Policy Design

Tools (Meso-Level Components)	*Calibrations (Micro-Level Components)*
Tool: RE Targets and Quotas	
• Devising clear objectives and sequencing of targets. • Outlining and customizing signals specific to consumers and industry based on type of quota; time frame; technology; obligated entities and compliance rules. • Designing federal and state level targets for compliance. • Coordinating renewable portfolio standards amendments to allow/disallow energy uses (e.g. allowing renewable heat to qualify towards quota requirement). • Creating multi-actor, expert working groups to set initial baseline and target. • Creating forums for public consultation and input during RE target deliberations.	• Facilitating market creation and moderation of tradeable energy certificate. • Accreditation and registration of renewable energy certificates or credits. • Creation of specialized division of public utilities commission or the creation of new renewable energy regulator. • Decisions regarding staged, step-wise increases or adjustments to target. • Establishing framework to monitor compliance and penalize non-compliance by central and state regulators. • Regular reporting by designated nodal agencies on compliance by obligated entities. • Devising system of collecting penalties (e.g. alternative compliance payments).
Tool: Feed-in-Tariffs (FITs)	
• Aligning RE capacity allocation to match program objectives. • Allocation of funds based on FIT programme's lifetime and/or technological advancements. • Allocation of FIT host organization rights and responsibilities (examples). • At the municipal level, the local utility administers entire FIT program including applications, interconnection, contract administration, and production payments. • At state and provincial levels, the programme is sponsored by independent purchasing agent, and national energy producers act as intermediaries between RE producers and local utilities.	• Assigning duties to national power producers' association to act as contractual intermediary between individual producers and local utilities. • Transferring capacities to local utilities to maintain original retail billing relationships and accept new FIT production payment relationships with participants. • Transferring capacities to local utilities to manage grid interconnection. • Empower province level power agency to authorize/conduct long-term planning and procurement. • Assigning responsibility to FIT agency to control and constrain total uptake (e.g. enforce a cap on the number of participants). • Administrative setting of price policies and their continuous adaptation to changing market adjustments and costs of technologies.

(Continued)

Tools (Meso-Level Components)	*Calibrations (Micro-Level Components)*
Tool: Net Metering	
• Installation of sensing technology and equipment (controls, automation, and computing systems) to existing electrical grid. • Ascertaining eligibility requirements, the value of excess electricity, additional taxes and fees, consumption entitlement period. • Raising buy-in from power consumers through consultations, training workshops, outreach forums to install two-way monitoring technology. • Implementing dynamic pricing mechanism for energy use/supply.	• At the level of federal states working with a common target for efficiency improvement or power reduction, a new association of regional governors is formed to jointly formulate transmission siting initiatives. • Alignment of existing regulatory bodies to coordinate the overseeing of new interconnection, dispatch, and tariff policies to incentivize decentralized producers to feed power into smart grid. • Regional coordination to link member states' centralized and decentralized power storage solutions. • Building in capacities to implement dynamic pricing and pilot projects to showcase and 'test' new regulatory policies for decentralized power generation.

Sources: REN21, 2016, 2017, 2018, 2019; Del Río 2010, 2012; DECC, 2009; Sun and Nie, 2015; Kent and Mercer, 2006; Lipp, 2007; Shammin and Bullard, 2009; Bertoldi and Huld, 2006; Menanteau et al., 2003; Butler and Neuhoff 2008; Zhang and Wang, 2017; UK, 2019; DOE, 2019.

19.3.1 Renewable Energy Targets or Quotas

Renewable energy quotas or targets outline mandated national (and often sub-national) goals for the development and deployment of renewables in the power sector. Often aligned to a country's national greenhouse gas (GHG) emission reduction targets, these obligations require that a set percentage of electricity must be generated from renewable energy sources and often these targets correspond with higher level national targets for sustainability or climate change mitigation. Such quotas are also known as renewable portfolio standards (RPS) or renewable purchase obligations (RPOs) and are devised to provide high-level policy signals that can then cascade down into more specific (technology or industry based) objectives (Lipp, 2007; Sun and Nie, 2015). There can be various substantive modifications that are possible in the design of this instrument, including those that specify a preferred technology (such as natural gas) or more general goals for energy resource diversification (that are non-binding for any particular form of renewable energy).

The procedural components of the design of such quotas are oriented heavily towards building administrative frameworks for monitoring compliance and penalizing non-compliance during implementation. These activities are generally carried out by either central or state-level regulators who manage rosters of the issuance of clean energy certificates (or allowances). In several instances (such as in Australia) there is also the establishment of new agencies dedicated to overseeing the implementation of national GHG emission strategies, and the role of RPS therein, that are distinct from those agencies created for informing the design of the substantive content of the RPS. As noted in the Australian example by Kent and Mercer (2006, p. 104) the latter took the form of the Renewables Target Working Group that "comprised representatives of the Commonwealth, State and Territory Governments, the electricity supply sector, electricity user groups and the renewable energy industry" and was responsible for the coordinating consultation forums and public submissions to all preliminary white papers.

As shown in Box 19.1 (and reflected through cell 2 in Figure 19.1), these instances indicate meso-level procedural means for building networks of support and analytical capacities that are put in place first, for helping to combine RPS with subsequent tools (such as FITs, among others) that are designed to meet RPS-set targets. As such, maximizing complementarity with new and emerging technological developments remains a significant procedural priority for the effective design of RPS and quotas, as their administrative components must be deliberately designed to prevent premature lock-in of any one or dominant technology. Furthermore, these considerations reflect a significant procedural effort to build market-based institutions through the design of RPS to encourage renewable energy diversity. As reflected through the case of the United States RPS, empirical evidence points to notable interaction effects between the general targets and what provisions are made therein for technology-specific designs (Kim and Tang, 2020).

At the more micro-level (Table 19.1 and cell 4 in Figure 19.1) of how the quotas are calibrated for implementation, procedural tasks geared towards enhancing participation and vertical coordination for tool implementation include establishing a system of regular reporting by designated nodal agencies on compliance or non-compliance by obligated entities. In cases where the RPS includes the concurrent creation of a tradeable energy certificate, procedural aspects of tool design also include the means of accreditation and registration of renewable energy projects, as well as the creation and moderation of trading activities among the policy targets (Lipp, 2007; Shammin and Bullard, 2009) and setting up of mechanisms that match administrative capacities for facilitating trade within predominant modes of governance (Bertoldi and Huld, 2006). Often, these procedural tasks are taken up directly by public utilities commissions (such has been done in the state of California in the United States) instead of first having to go through a national level regulatory agency (REN21, 2019).

19.3.2 Feed-in-Tariffs

The substantive means of the design of Feed-in-Tariffs (FITs) as a particular category of tools reflect elements that combine regulation-based instruments governing the rules and rates of the tariff, as well as organizational resources dedicated to the formulation of government-sanctioned contracts and a regulated market between power purchasers and producers. In the first instance, the design of substantive means concerning the financial structure of FITs can include policy choices regarding the level of the tariff (whether fixed or flexible in the form of degressions), support linked to the price of electricity over the duration of the FIT, and the setting of maximum and minimum limits on the electricity price (Menanteau et al., 2003; Butler and Neuhoff, 2008).

For almost every example of the design of FITs, the decision to create a scheme and the initial timeline of its launch are articulated in a government's overall national renewable energy target or quota. At the outset, there is a temporal dimension to the design of FITs wherein there is initially greater emphasis put on procedural means being put in place before designing the substantive elements. For example, in the UK, FITs were first introduced in the 2008 Energy Act that gave the regulatory bodies enabling power to set up the tariffs through the passing of an initial statutory instrument. This procedural aspect of design qualified relevant regulatory agencies of the government to responsibilities such as setting up the accreditation system, calculating FIT payment and levelization rates to keep up with market rates, creating a central FIT register, and mandating internal reviews of the scheme prior to the implementation of the instrument. Additionally, this statutory mechanism articulated administrative functions, such as the publication and dissemination of national FIT guides, annual reports, list of licensees, and all legal notices pertaining to any RE supplier entering into a FIT arrangement (UK, 2019). Similarly, as shown in Table 19.1, procedural means in the initial design of their respective FIT schemes have helped to lay the groundwork for the support of the substantive aspects of FITs as and when they are rolled out and perpetuated.

At the meso-level (cell 2, Figure 19.1) of the design of FITs, enabling capacities for analysis and institution-building underpin their success or failure as a viable instrument for promoting renewable energy development. For example, a lack of technical expertise for designing suitable auction mechanisms alongside the FIT policy itself has been shown to severely hinder the commercial roll-out of projects to meet national targets in Kenya (at the scale of above 10 MW). At the same level, lacking procedural legitimacy during FIT design undermines greater institutionalization of community-based renewable energy projects. For instance, a survey of such projects in the UK has recently concluded that RE technology and structural legitimacy gaps emerge without sufficient support built into FIT programs to uphold planning permissions, without eliciting stakeholder involvement, relationship-building with project partners or accessing expertise, or when developing specific renewable energy technologies such as onshore wind-turbine installation, which can be seen as being undesirable without due support from local authorities during the planning phase (Genus and Iskandarova, 2020).

Procedural aspects to support the calibration of FITs at the micro-level on the ground (cell 4, Figure 19.1) include internal mechanisms that allow national regulatory authorities to set up the initial design of a FIT and thereafter adjust the FIT rate to keep it aligned with existing renewable energy legislation (such as laws or new ministerial decrees). For example in Japan, soon after the Fukushima nuclear disaster, the government issued new legislation and a concurrent FIT scheme to ramp up support for alternative renewable energy, such as solar photovoltaic (PV) and wind, which required the creation of a special parliamentary committee to oversee progress towards the 2021 targets (REN21, 2019). As FITs must typically remain stable during the guaranteed payment period, this necessitated building in analytical capacities into relevant committees as well as review mechanisms that allowed rates to be adjusted at predetermined intervals or as deemed necessary by market realities.

19.3.3 Net Metering and Smart Grids

Net metering mainly involves the supply, installation, monitoring, and functioning of a metering apparatus that measures the relative power consumption of an individual or household or organization that also produces its own power from its own renewable energy facilities. In other words, consumers can 'feed' some part of the energy they produce into the grid and receive payment for it. Most of the cases of net metering are concentrated on decentralized solar PV technologies that can be incorporated within the consumer's own facility. The interaction between the procedural and substantive elements of instrument design is visible quite profoundly in net-metering mechanisms as they most closely reflect a nodality- or information-based policy tool that must work in conjunction with a more financial tool such as a FIT (Zhang and Wang, 2017). The meter is bidirectional or a combination of two unidirectional meters that measure the net energy produced by these facilities. A smart grid indicates a network of such meters and facilities and is defined as "the digital technology that allows for two-way communication between the utility and its customers, and the sensing along with the transmission lines" (USDOE, 2020).

At the meso-level of tool design (cell 2, Figure 19.1) fostering partnerships with the administration is a cornerstone of effective application of smart grids that, at the outset, relies on "new interconnection, dispatch and tariff policies to address market barriers and incentivize decentralized generators" (REN21, 2019, p. 184). At the level of cities or federal states (such as in the United States), procedural functions like devising permitting and interconnection procedures, reducing installation and other fees for end-users, overall 'rule making' for implementing legislative goals and even conflict management and arbitration are carried out by public utilities commissions that are appointed at the jurisdictional level and supported by the legislature (Hess and Lee, 2020). The deliberate designing of these initial procedural functions can both enable as well as limit how the substantive tool is able to subsequently address its stated objectives (e.g. to increase the share of distributed PV in the national energy mix).

At the level of calibrations (cell 4, Figure 19.1) the procedural means to facilitate the implementation of net metering and the uptake of smart grids rely heavily on the coordinating capacity of administrators and regulators to jointly enable transmission and siting initiatives (Table 19.1). Maintaining partnerships with key stakeholders (such as households producing energy and state utilities purchasing this at a favourable rate) falls under the purview of the administration that must continuously calibrate payment structures as the supply of decentralized renewable energy grows, as has been shown in the experience with net metering in the United States as well as member states of the European Union (Hess and Lee, 2020; Iliopoulos et al., 2020). Procedural roles of regulators are additionally detrimental to the vertical coordination between suppliers and utilities in the substantive implementation of net-metering policy. Such activities include, for example, setting up accounting systems for monetary/energy credits and outlining accumulation rules for energy credits (e.g. the number of billing periods and rules for 'rolling over' credits as well as defining quantity and quality restrictions on the supplied energy) (Soto et al., 2019).

19.4 Conclusion and Future Research

Until recently, procedural policies had not garnered as much systematic research attention as substantive policy tools have in modern policy studies. Ongoing and future research work in this area of policy design is shifting the focus towards understanding not only how these different design components work individually, but how they can be deliberately planned to interact and work together in policy mixes to address complex policy problems. Theorization about the particular relationship that procedural means share within the design of policy mixes – whether with each other or with more substantive elements of tools – remains a promising area of future research, especially in light of what design considerations can offer towards planning for sound and sustainable implementation in a climatically volatile future.

Furthermore, procedural policies also implicate a discussion of what cognitive and behavioural aspects of policymakers cause them to favour one design over another, and this discourse can have profound implications for what is being revealed about governments' preferred policy styles in the environmental policy arena (Howlett and Tosun, 2018). Thinking of the procedural along with the substantive during policy design, and especially environmental policy design, can offer richer insights on both choices made at a single moment in time and over a period of years or decades (Hughes and Urpelainen, 2015). In a time when we can no longer assume that sustainability will automatically always result from effective environmental policy blueprints, it is now all the more important to directly examine how good design can bring about the sound application of policies in sectors such as renewable energy and energy efficiency. This is an especially pertinent future research question for policymaking in sustainable energy, where repeated coordination between multiple substantive instruments and the implementation and capacity constraints on the ground, can itself lead to the proliferation of a wide range of procedural provisions not just at the micro-meso design level of policy tools, but also more prominently at the macro-level of inter-policy coordination. Other important avenues for future inquiry have to do with the temporality and sequencing of procedural means during policy design, which this broad examination of renewable energy has alluded to and which can help to set the direction towards this necessary and pertinent aim of modern policy studies.

The conceptualization of tool 'compounds' as introduced here proposes two main advances to the scholarship of procedural tools within policy design. Firstly, it provides an analytical space within existing frameworks of policy design to focus on questions about how substantive and procedural policy elements support and interact with each other during the design of policy tools and their calibrations for effective deployment on the ground. By doing so, the concept addresses considerations of temporality in the process of policy formulation, a topic in the policy sciences that has so far disjointedly dealt with substantive and procedural policy tools. Secondly, coupling meso- and micro-level considerations of policy design closely reflects how major classes of policy tools include planning for implementation into their initial design. This allows for the discussion on policy tool design to go beyond explorations of governance resource-based classifications of tool origins and to explore hypotheses of tool choice that are linked with variables such as policy capacity, organizational coordination, and the behavioural motivations of policy designers. The various behaviours and avenues of participation of different policy actors in policy implementation can be notably affected by procedural tools that are used to manipulate policy processes and have a bearing on how multiple tools can be expected to work towards a broader common goal, such as the development of renewable sources of energy.

With respect to stakeholder and public participation, this can occur 'naturally' or with more regulatory encouragement in the renewable energy domain, but it is also carefully devised by governments through the sequential deployment of procedural tools that enhance or curtail specific kinds of activities. These instruments, including the creation and legitimization of independent power producers and local regulators, or the provision of funding and differential tax treatments for users and producers of alternative energy, invisibly yet profoundly impact policymaking behaviour in articulating, developing, choosing, or supporting particular policy solutions over time and through changing environmental realities (Thacher and Rein, 2004).

References

Adam, Christian, Hurka, Steffen, Christoph Knill, B. Peters, Guy, & Steinebach, Yves (2019). Introducing vertical policy coordination to comparative policy analysis: The missing link between policy production and implementation. *Journal of Comparative Policy Analysis: Research and Practice*, 1–19.

Béland, D. (2010). Reconsidering policy feedback: How policies affect politics. *Administration & Society* 42 (5), 568–590.

Bemelmans-Videc, M.-L., Rist, R.C., & Vedung, E. (Eds.) (1998). *Carrots, Sticks, and Sermons: Policy Instruments and Their Evaluation*. New Brunswick: Transaction.

Bertoldi, P., & Huld, T. (2006). Tradable certificates for renewable electricity and energy savings. *Energy Policy*, 34(2), 212–222.

Braathen, N. A. (2007). Instrument mixes for environmental policy: how many stones should be used to kill a bird? *International Review of Environmental and Resource Economics*, 1(2), 185–236.

Butler, L., & Neuhoff, K., 2008. Comparison of feed-in tariff, quota and auction mechanisms to support wind power development. *Renewable Energy*, 33, 1854–1867.

Capano, Giliberto, & Woo, Jun Jie (2018). Designing policy robustness: Outputs and processes. Policy and Society, 37(4), 422–440.

Cashore, B., & Howlett, M. (2007). Punctuating which equilibrium? Understanding thermostatic policy dynamics in Pacific Northwest forestry. *American Journal of Political Science*, 51(3), 532–551.

Del Río, P., Silvosa, A. C., & Gómez, G. I. (2011). Policies and design elements for the repowering of wind farms: A qualitative analysis of different options. *Energy Policy*, 39(4), 1897–1908.

Del Río, Pablo. (2010). Analysing the Interactions between renewable energy promotion and energy efficiency support schemes: The impact of different instruments and design elements. *Energy Policy*, 38(9), 4978–4989.

Department of Energy & Climate Change (2009). *The UK Renewable Energy Strategy*. London: TSO.

Doz, Yves, & Kosonen, Mikko. (2014). Governments for the future: Building the strategic and agile state. Sitra Studies, 80, 18.

Dudley, S.E., & Xie, Z. (2020): Designing a choice architecture for regulators. *Public Administration Review* 80 (1), 151–156.

Fernández-i-Marín, X., Knill, C., & Steinebach, Y. (2021). Studying policy design quality in comparative perspective. *American Political Science Review*, 1–17.

Fouquet, R. (2016). Path dependence in energy systems and economic development. *Nature Energy* 1 (8), 1–5.

Gençsü, I., Whitley, S., Trilling, M., van der Burg, L., McLynn, M., & Worrall, L. (2020). Phasing out public financial flows to fossil fuel production in Europe. *Climate Policy* 20(8), 1010–1023.

Genus, A., & Iskandarova, M. (2020). Transforming the energy system? Technology and organisational legitimacy and the institutionalisation of community renewable energy. *Renewable and Sustainable Energy Reviews* 125.

Gunningham, Neil, & Sinclair, Darren. (1998). Designing smart regulation. In *Economic Aspects of Environmental Compliance Assurance*. OECD Global Forum on Sustainable Development.

Hess, D. J., & Lee, D. (2020). Energy decentralization in California and New York: Conflicts in the politics of shared solar and community choice. *Renewable and Sustainable Energy Reviews*, 121, 109716.

Hood, C. (1986). *The Tools of Government* (1st American ed.). Public policy and politics. Chatham, NJ: Chatham House.

Hood, C. (2007). Intellectual obsolescence and intellectual makeovers: Reflections on the tools of government after two decades. *Governance*, 20(1), 127–144.

Hood, C. C., & Margetts, Helen Z. (2007). *The Tools of Government in the Digital Age*. Macmillan International Higher Education.

Howlett, M. (2009). Governance modes, policy regimes and operational plans: A multi-level nested model of policy instrument choice and policy design. *Policy Sciences* 42(1), 73–89.

Howlett, M. (2019). *The Policy Design Primer: Choosing the Right Tools for the Job*. London: Routledge.

Howlett, M., & Mukherjee, I. (Eds.) (2018). *Routledge Handbook of Policy Design*. New York: Routledge.

Howlett, M., & Mukherjee, I. (2020). Designing public participation in the policy process: a critical review of procedural instrument theory. *Elgar Encyclopedia of Environmental Law*, 88–100.

Howlett, M., Mukherjee, I., & Rayner, J. (2014). The elements of effective program design: A two-level analysis. *Politics and Governance* 2(2), 1–12.

Howlett, M., Mukherjee, I., & Rayner, J. (2017). The elements of effective program design: a two-level analysis. In: *Handbook of Policy Formulation*. Edward Elgar Publishing, pp. 129–144.

Howlett, M., & Tosun, J. (2019). *Policy Styles and Policy-Making*. London: Routledge.

Howlett, M. (2000). Managing the 'Hollow State': Procedural policy instruments and modern governance. *Canadian Public Administration*, 43(4), 412–431.

Howlett, M., & Rayner, Jeremy (2013). Patching vs packaging in policy formulation: Assessing policy portfolio design. *Politics and Governance*, 1(2), 170–182.

Howlett, M., & Tosun, J. (Eds.). (2018). *Policy Styles and Policy-Making: Exploring the Linkages*. Routledge.

Howlett, M., & Mukherjee, Ishani (2018). Introduction: The importance of policy design: Effective processes, tools and outcomes. In *Routledge Handbook of Policy Design*, 3–19. Routledge.

Howlett, Michael, Rayner, Jeremy (2007). Design Principles for Policy Mixes: Cohesion and Coherence in 'New Governance Arrangements'. *Policy and Society* 26 (4), 1–18.

Hughes, L., & Urpelainen, J. (2015). Interests, institutions, and climate policy: Explaining the choice of policy instruments for the energy sector. *Environmental Science & Policy*, 54, 52–63.

Iliopoulos, T. G., Fermeglia, M., & Vanheusden, B. (2020). The EU's 2030 climate and energy policy framework: How net metering slips through its net. *Review of European, Comparative & International Environmental Law*, 29(2), 245–256.

IPCC (Intergovernmental Panel on Climate Change). (2021). *Climate Change 2021: The Physical Science Basis. Contribution of Working Group I to the Sixth Assessment Report of the Intergovernmental Panel on Climate Change*. Cambridge: Cambridge University Press.

Jordan, A.J., & Moore, B. (2020). *Durable by Design? Policy Feedback in a Changing Climate*. Cambridge: Cambridge University Press.

Kent, A., & Mercer, D. (2006). Australia's mandatory renewable energy target (MRET): An assessment. *Energy Policy*, 34(9), 1046–1062.

Kim, J. E., & Tang, T. (2020). Preventing early lock-in with technology-specific policy designs: The Renewable Portfolio Standards and diversity in renewable energy technologies. *Renewable and Sustainable Energy Reviews*, 123, 109738.

Klijn, Erik-Hans, Koppenjan, Joop, & Termeer, Katrien (1995). Managing networks in the public sector: A theoretical study of management strategies in policy networks. *Public Administration*, 73(3), 437–454.

Lang, Achim (2019). Collaborative governance in health and technology policy: The use and effects of procedural policy instruments. *Administration & Society*, 51(2), 272–298.

Leung, Wanda, Noble, Bram, Gunn, Jill, & Jaeger, Jochen A. G. (2015). A review of uncertainty research in impact assessment. *Environmental Impact Assessment Review*, 50, 116–123.

Linder, S.H., & Peters, B. G. (1989). Instruments of government: Perceptions and contexts. *Journal of Public Policy* 9 (1), 35–58.

Lipp, J. (2007). Lessons for effective renewable electricity policy from Denmark, Germany and the United Kingdom. *Energy Policy*, 35(11), 5481–5495.

Luthar, S. S., & Cicchetti, D. (2000). The construct of resilience: implications for interventions and social policies. *Development and Psychopathology* 12(4), 857–885.

Mahoney, J. (2000). Path dependence in historical sociology. *Theory and Society* 29(4), 507–548.

Menanteau, P., Finon, D., & Lamy, M. L. (2003). Prices versus quantities: Choosing policies for promoting the development of renewable energy. *Energy policy*, 31(8), 799–812.

Mintrom, M, & Joannah Luetjens (2017). Creating public value: Tightening connections between policy design and public management. *Policy Studies Journal* 45(1), 170–190.

Nair, Sreeja, & Howlett, M. (2017). Policy myopia as a source of policy failure: Adaptation and policy learning under deep uncertainty. *Policy & Politics* 45(1), 103–118.

Peters, B.G., Capano, G., Howlett, M., Mukherjee, I., Meng-Hsuan, C, Ravinet, P. (2018). *Designing for Policy Effectiveness: Defining and Understanding a Concept*. Cambridge University Press.

REN21 (2016). *Renewables 2016: Global Status Report*. Paris: REN21.

REN21 (2017). *Renewables 2017: Global Status Report*. Paris: REN21.

REN21 (2018). *Renewables 2018: Global Status Report*. Paris: REN21.

REN21, Renewables. (2019). "Global Status Report, Paris, 2019".

Rosenbloom, D., Meadowcroft, J., & Cashore, B. (2019). Stability and climate policy? Harnessing insights on path dependence, policy feedback, and transition pathways. *Energy Research & Social Science*, 50, 168–178.

Salamon, L.M. (1981): The goals of reorganization: A framework for analysis. *Administration & Society* 12(4), 471–500.

Salamon, Lester M. *The Tools of Government. A Guide to the New Governance*. New York: Oxford University Press, 2002.

Shammin, M. R., & Bullard, C. W. (2009). Impact of cap-and-trade policies for reducing greenhouse gas emissions on US households. *Ecological Economics*, 68(8–9), 2432–2438.

Soto, D. D. L., Mejdalani, et al. (2019). *Advancing the Policy Design and Regulatory Framework for Renewable Energies in Latin America and the Caribbean for Grid-Scale and Distributed Generation* (Vol. 785). Inter-American Development Bank.

Sun, P., & Nie, P. (2015). A comparative study of feed-in tariff and renewable portfolio standard policy in renewable energy industry. *Renewable Energy* 74, 255–262.

Thacher, D., & Rein, M. (2004). Managing value conflict in public policy. *Governance*, 17(4), 457–486.

UK, United Kingdom. GOV.UK. Feed-in Tariffs: Get Money for Generating Your Own Electricity. Accessed January 17, 2020. https://www.gov.uk/feed-in-tariffs.

USDOE. United States Department of Energy. Tax Credits, Rebates & Savings. *Energy.gov*. Accessed January 17, 2020. https://www.energy.gov/savings/dsire-page.

Zhang, X., & Wang, Y. (2017). How to reduce household carbon emissions: A review of experience and policy design considerations. *Energy Policy*, 102, 116–124.

20
SECURITIZATION, CLIMATE CHANGE, AND ENERGY

Maria Julia Trombetta

20.1 Introduction

This chapter explores the transformation of climate change and energy into security issues. The framework for the analysis is provided by securitization theory. Both climate change and access to energy have been presented as security issues as part of a process of securitization. I argue that the process is not just about the emergence and recognition of relevant threats that need to be added to the security agenda. The transformation of climate change and energy into security reflects specific discourses about which threats are considered and whose security is prioritized. This is part of a rearticulation of what counts as security and questions existing security practices. The process is relevant for the difficulties of integrating environmental and energy security discourses. By considering how distinct discourses linking climate change and energy with security have emerged and became entangled, I address the paradoxes caused by their separation and consider how their slow convergence is part of a process of rearticulation of security in the Anthropocene.

Discussing the securitization of climate change and energy in one chapter seems an obvious choice given that the two are deeply related. Existing energy systems, largely based on fossil fuels, are the main contributors to carbon dioxide emissions that cause climate change (see Trombetta, 2018). Paradoxically, however, climate change and energy policies have been considered as two separate domains in practice. Different actors and priorities have characterized them. Only recently has the importance of an integrated approach emerged (Kuzemko, 2013), and while environmental policy, both at international and national levels, takes into account energy, discussions of energy security mainly have remained at the national level and have tended to ignore climate issues (Nyman, 2018a).

This separation contributes to a set of paradoxes. Attempts to secure existing energy systems are causing other forms of insecurity. As mentioned above, fossil fuels contribute to climate change. In total, the global energy sector is responsible for more than two-thirds of all greenhouse gas emissions (IEA, 2015). The tension runs deeper as existing energy systems are unsustainable, both in terms of existing resources and environmental impacts. As Mayer and Schoulten put it: "the term 'security' appears inappropriate in 'energy security' as pursuing it threatens its very aim" (2012, 13). The argument has been echoed by Nyman (2018b), who highlights the extent of energy security paradoxes. Access to energy sources, often located

DOI: 10.4324/9781003043843-23

in distant places, can cause geopolitical tensions. Energy systems are highly unequal. Despite recent improvements, more than 770 million people have no access to electricity[1] and yet maintaining the existing level of consumption and providing access to energy services for those excluded, without rapid decarbonization, would impact deeply on the climate and on the environment. Even if the shale revolution has made oil and gas more abundant, large parts of the existing reserve of fossil fuels will need to be kept in the ground if climate change targets are to be met. Climate change is not the only environmental impact of fossil fuels. Burning coal contributes to air pollution with levels of PM2.5 often exceeding WHO recommended levels for cities. Fracking, the technology that unlocked shale gas and oil, has high environmental impacts, not to mention accidents like oil spills. Energy security has been conceived narrowly, and for long, like security of supply, downplayed other dimensions (Dyer and Trombetta, 2013).

Security discourses are relevant in shaping policies, yet what counts as energy or climate security is often taken for granted. In this context, Ciută (2010, 124) polemically notes that in traditional energy security studies "there simply is no need to debate what energy security is, because we know both *that* energy is a security issue and *what* security is". Such a perspective does not consider the way the link between climate and security or energy and security is conceptualized and established, and that the process reflects political decisions about whose security is prioritized, what needs to be protected, and against what threats and by what means. As Kester points out, it is necessary to consider "how energy is understood as a security issue, and what these understandings do" (Kester, 2017: 230). The challenge is not only about incorporating externalities, like concerns for emissions, into energy security discourse, but also about understanding how different problems, like climate change, are made thinkable in security terms (and with them all the other concepts like carbon budgets, Hothouse Earth, and abrupt climate change that allow us to visualize, make sense of, and speak of threats and vulnerabilities) and the policies these discourses legitimize. Critical security studies, considering how threats are constructed, and the implications of transforming an issue into a security issue provide relevant insights; and the debate about environmental security has anticipated many of the issues involved (Dalby, 2013) in climate security discussions. In the case of energy, the debate is more recent (Ciută, 2010, 124; Dyer and Trombetta, 2013; Mayer and Schouten, 2012; Nyman, 2018a; Kester, 2017) and yet more pervasive, given the importance of energy in all aspects of human life. Addressing climate change requires reconsidering energy security. If climate change is emerging as the greatest threat facing humankind, the conceptualization of energy security and security needs to reflect that.

The remainder of the chapter proceeds as follows. After a brief introduction of securitization theory, its relevance and limitations, and an appeal to apply it flexibly, the chapter provides a review of the securitization of climate change and energy. It points at how specific threats and discourses have been prioritized and how that made it difficult to develop an integrated approach, resulting in the creation of paradoxes. The final part explores the challenges of moving beyond securitization and rethinking how we understand security. The analysis is based on a critical review of existing literature and debates.

20.2 Concepts of Securitization

Securitization theory, originally formulated by the so-called 'Copenhagen School' (McSweeney, 1996), adopts a constructivist perspective that suggests that no objective threats are waiting to be discovered and counteracted (Buzan et al., 1998). Threats are constructed as part of a discursive process that selects and prioritizes them. Presenting an issue as a security issue is a powerful process that legitimizes both actors and measures. As the Copenhagen School emphasizes, there is a specific political tradition associated with security and with what security allows. Security, evoking existential threats, legitimizes exceptional measures and the use of all means necessary to block them. Security is not a value or a condition but refers to a specific form of social practice that transforms the way of handling an issue, lifting it above politics. In other words, while for the Copenhagen School threats are constructed, the logic of security is rather fixed and involves exceptional measures, urgency, governing by decree rather than by democratic decisions (McDonald, 2008; Trombetta, 2008).

This allows us to consider the political implications of transforming an issue into a security issue. For the Copenhagen School, transforming an issue into a security issue is a problematic development, and one that is best to avoid[2] (Wæver, 1995, 94) as it may bring about the practice outlined above. The logic of security is the logic of war. It legitimizes the breaking of accepted rules and creates enemies, making cooperation and political debate difficult. Others have pointed out that transforming a problem into a security issue may be a way to mobilize and prioritize actions and resources and that the consequences cannot be judged a priori (Floyd, 2011). Yet others have shown that in the process of threats construction or the practices and logic of security can be transformed as well as translated and adapted to new sectors and problems like environmental ones (Berling et al., 2021; Stritzel, 2011; Trombetta, 2008). The possibilities offered by the latter perspective, and the resilience of the logic of security described by the Copenhagen School, is what this chapter focuses on.

In this chapter, securitization will be used to provide an account of the process that transforms a problem into a security issue and of its political nature. It will help us to identify which threats have been selected, whose security is a stake, and to explore the consequences of attaching the label 'security' to an issue. At the same time, rather than taking for granted the logic described by the Copenhagen School that an issue is transformed into a security issue, the chapter will consider how the process of threats construction can contribute to transform security logics and practices. At the empirical level the emphasis on different discourses will allow us to consider how discourses about climate change and energy security emerged and followed independent trajectories, despite their overlapping character. At the conceptual level it will outline the existence of different security discourses and practices, suggesting that the issue is not just whether an issue is securitized or not but how security is understood in specific contexts and whether security logics and practices are changing. As will be argued there is a rearticulation of security logics toward resilience; in more and more sectors precautionary measures aimed at promoting resilience are emerging when security is evoked, though the logic depicted by the Copenhagen School is still relevant.

20.3 The Climate Security Storyline

Global environmental problems started to emerge in the 1970s and 1980s. Problems like the depletion of the ozone layer or global warming were pointing at the global consequences of actions carried out in different places on the planet. "Our Earth is one, our world is not" warned the Brundtland Commission (WCED, 1987), pointing at both the necessity of cooperation but also at the difficulties of achieving it.

Traditional approaches, dominant during the cold war period, tended to downplay environmental issues as low politics, not very relevant from a security perspective, which was concerned with high politics and problems of war and peace. In International Relations (IR), security has traditionally been understood as the protection of the state from external, often military, threats. The threats environmental problems posed were rather different from the traditional security ones. As Deudney (1990) pointed out the environment was not a security issue, as it did not fit in the dominant security discourse, and so it was not opportune to frame it as such as it could have mobilized "nationalist and militarist mindsets closely associated with national security", which could have jeopardized collaboration (p. 475). Yet, as Deudney continued, environmentalism was a threat to a specific and problematically narrow approach to security.

Commentators called for a re-understanding of security to include issues that really matter for people (Mathews, 1990) or using the security argument to advocate environmental action, often mentioning indirect threats to national security such as the spectre of millions of environmental refugees (Brown, 2018; Myers, 1997). "The overall narrative was that environmental degradation, resource depletion and pollution were threatening the global commons, from the oceans to the atmosphere and thus the security of states" (Trombetta, 2022, 226) 'Trombetta, Maria Julia. "10 Environmental Security in the Anthropocene". Global Security in an Age of Crisis, Edinburgh: Edinburgh University Press, 2022, pp. 220–243. The argument was also used to urge a more cooperative approach to security, starting with cooperation on low politics issues to build up the basis for cooperation on high politics ones, like disarmament.

Despite the enduring relevance of national security considerations, the narrative was promoting a more cooperative approach that characterized the years immediately after the end of the cold war and the preparatory work for the 1992 United Nations Conference on Environment and Development (UNCED) in Rio de Janeiro, where three relevant conventions to preserve the global commons (biodiversity, the climate and forests) were meant to be signed. As Mikhail Gorbachev stated at the United Nations General Assembly: "the threat from the sky is no longer a missile but global warming" (quoted in Myers, 1996). In the case of climate change, the construction of the threats went hand in hand with the representation of the climate as a stable yet complex system and the assumption that maintaining anthropogenic transformation within limits would have provided the possibility of adaptation to this climate change and thus security (Trombetta 2022).

As the difficulties in cooperation emerged and the optimism of the end of the cold war faded, the discourse was replaced by a focus on threats to global order (Trombetta 2012). Climate change and environmental degradation were seen as causing global instability and chaos in the peripheries, causing conflict and destabilizing fragile states (Kaplan, 1994). Environmental conflicts were analysed by extensive and well-funded research projects both in Europe and in North America (Homer-Dixon, 2010). Kaplan popularized his argument in an article in *The Atlantic Monthly*, which was quite influential on the Clinton administration (Kaplan, 1994; Matthew, 2002).

The emphasis on an external threat was calling for defensive measures – as in traditional security discourses – legitimizing new roles for the military, even if the need for precautionary approaches to prevent problematic developments was also emerging. The way environmental conflicts were conceptualized as objects of knowledge and threats reflected that climate and specific assumptions about national security, scarcity, and conflict. This holds although empirical research was rather cautious in establishing causal links between environmental degradation and conflicts. Critics questioned the methodology and the assumptions underlying environmental conflict research (Trombetta, 2012).

In this context, the concept of human security gained relevance and shifted the focus from the state and the military to individual wellbeing and human rights. Scholars and international organizations embraced it to point to the transnational character of environmental threats and to the links between development, the environment, and national security. While the environmental conflict narrative focused on a state-centric understanding of security, a more people-centred one that considered "freedom from fear and freedom from want" gained relevance at the international level, as part of the UN development agenda (Hughes, 2018, 72–73).

The narrative on environmental conflicts contributed to marginalizing climate change as it was not considered a main cause of environmental conflict. Moreover, other global threats like terrorism distracted attention from the climate, even though environmental considerations had been included in major national security strategies since the early 1990s. The securitization of climate change gained new momentum after 2004. Different actors contributed to a variety of securitizing moves. The Fourth Intergovernmental Panel on Climate Change (IPCC) report unequivocally acknowledged the anthropogenic increase in global temperature and warned about the possibility of abrupt climate change. The scientific community played a central role in detecting and representing the complex, non-linear dynamics of a vulnerable climate system (Mayer, 2012). This was echoed by a set of reports by influential think tanks. Among the most alarming ones, there was the report originally prepared for the Pentagon by Schwartz and Randall forecasting an abrupt climate change with Siberian temperatures in most of North America and Europe (Schwartz and Randall, 2003). Nick Mabey, writing for the Royal United Services Institute (RUSI) warned about a failure to acknowledge climate change security threats (Mabey, 2008). The Stern Review (Tyndall Centre for Climate Change Research and Adger, 2006) presented the economic costs of procrastinating action on climate change. Concerns were popularized by Hollywood with the movie *The Day After Tomorrow*.

In this context, the year 2007 was particularly relevant as the United Nations Security Council (UNSC) had its first meeting on climate change on a British initiative. It was an attempt to discuss not only climate change and security but also energy. In the same year, the Nobel Peace Prize was awarded to Al Gore, who had promoted climate action (and the securitization process) with the alarming documentary *An Inconvenient Truth*, and to the IPCC whose reports had provided a common ground for public and scientific debate.

Since then, climate change has been in and out of the security debate. The open meetings at the UNSC provide a sense of the ebb and flow of securitization and suggest how the security card has been evoked to mobilize action when other organizations like the United Nations Framework Convention on Climate Change (UNFCCC) and its Conferences of the Parties (COPs) failed to move negotiations forward. A second UNSC open debate was hosted in 2011 after the failure at renegotiating the Kyoto protocol, and again in 2018 when the USA withdrew from the Paris Agreement. Since then, meetings have been organized annually, reflecting increased concerns, the new context provided by the Paris Agreement, and the positional shift by key players such as China (Meartens and Trombetta, forthcoming).

The questioning of the mechanistic understanding of the relations between climate change and conflict and the growing representation of the climate system as complex and non-linear have contributed to the emergence of a discourse that rearticulates climate security discourses in terms of complexity and resilience. Commentators have noticed a shift from a logic of defence to one of risk management and resilience (Corry, 2014; Oels, 2012; Trombetta, 2008). Responding to complexity and uncertainty emphasizes the importance of preparedness, decentralization, and empowerment. Corry (2014) outlined the differences between a logic of defence, which characterizes traditional IR approaches (and securitization), and one of risk and resilience. In contrast to the logic of defence that often requires short-term actions, risk and resilience promote the long-term thinking that is required by climate change. They allow a reflexive moment that questions the rigidity of defence thinking in terms of security practices. Finally, risk and resilience do not depend on a friend–enemy logic and the problematic antagonism that it creates. For Corry, in addition to processes of securitization, processes of riskification need to be considered (2014) in which problems are transformed into risk and managed accordingly (Corry, 2014).

Resilience has been embraced as a way to cope with complexity and non-linearity. However, as it focuses on the ability of the system to respond to complex socio-ecological drivers of insecurity it has also contributed to questioning top-down, state-driven policies as they may limit the self-adaptive potential of the system. Resilience stresses practices such as the adaptation to risk and a shared responsibility to achieving human security; as such it has been embraced as an empowering strategy (Boas and Rothe, 2016). At the same time the responsibility for adaptation is left to individuals, with little consideration for the most vulnerable who cannot adapt. In consequence, as commentators have pointed out, political antagonism is shifted to another level, as the vulnerable community fails to become resilient and, in turn, can become a threat to both states and the global order (Bettini, quoted in Boas and Rothe, 2016).

More recently the term "climate emergency" has been used[3] and in 2019 the *Oxford English Dictionary* declared "climate emergency" the so-called "Word of the Year" (McHugh et al., 2021). The term "emergency" suggests a securitization, aimed at mobilizing urgent action before it is too late, and legitimizes exceptional measures. Yet, such measures are not materializing while attempts to increase resilience dominate the debate.

Commentators have pointed out that the securitization of climate change has failed as it did not mobilize exceptional measures. Others have seen this as a process of transformation in which climate considerations and practices (often based on precaution and promoting resilience) are included in more and more aspects of the security discourse and provisions, challenging and transforming the existing security logic (Trombetta, 2008). Scholars have described this process as a "climatization" of security (Maertens, 2023; Oels, 2012). In the process, energy is also considered, even if attempts to integrate it have been more challenging. A "climatization" of energy is still in the making, despite the relevance of energy policies for addressing the issue of climate change.

20.4 The Energy Security Storyline

To understand the challenges of the securitization of energy it is necessary to focus on the characteristics of modern energy systems. Since the Industrial Revolution energy systems have become increasingly reliant on fossil fuels – coal first, oil and natural gas later. Shifting from coal to oil and gas implied importing these fuels from distant places as oil and gas are unevenly distributed across the globe. From this perspective, the security of supply has become a concern for many states. The second characteristic is the centralized nature of the energy system created in the aftermath of World War II that created large, public companies, dealing with all aspects of the provision of energy services, from acquiring resources to distributing electricity or gas. Protecting these companies was easily a matter of national interest. Energy security emerged as the national security of supply, concerned with ensuring access to energy sources imported from distant places and, especially in Europe, the protection of centralized energy systems, often considered as a technical rather than a political problem (despite some debate on nuclear power or on the safety of large hydroelectric power plants).

The oil crisis in the 1970s brought about a process of securitization (McGowan, 2011). Security of supply was threatened, and exceptional measures were required to deal with the short-term effects, like including rationing and restrictions on the use of private cars. Yet, institutional arrangements and successful security mechanisms, like strategic oil reserves, organized by the newly created International Energy Agency (IEA) in the aftermath of the crisis, ensured the security of supply against short-term shocks and the main threats that emerged. As a result, energy apparently disappeared from the security agenda, and it was considered an economic issue. Yet, this security arrangement ignored and contributed to exacerbating other threats related to access to energy or environmental impacts. The traditional understanding of security in International Relations, as protection from external threats and the emphasis on the state, has shaped the way energy security has been conceptualized. Realists focused on a strategic, neo-mercantilist approach that perceives of energy security as a zero-sum game, with winners and losers, where energy security is provided by ensuring supply, via state-owned enterprises, energy diplomacy, and investments in exporting countries. Liberal approaches, in turn, focus on the market and on international organizations like the IEA to ensure energy security (Correljé and van der Linde, 2006). Ultimately, however, what is at stake is the security of states, including OECD countries.

In the early 2000s, there was a new process of securitization and the term "energy security" reappeared. The structural crisis due to unsustainable energy systems both in terms of existing resources and environmental impacts became evident. The spectre of growing demands from BRICS was echoed by growing concerns about climate change. China engaged in oil diplomacy to acquire equity oil across the globe and ensure the security of supply. In the USA the strategic relevance of energy independence contributed to the development of shale gas. The EU used the security and environmental card to develop a common energy policy and combine environmental and energy considerations.

Over the last ten years the narrative of peak oil has faded to be replaced by one of problematic abundance. The so-called "shale revolution" has substantially increased the extent of existing reserves, as well as oil and gas availability. Concerns for climate change are gaining relevance and with them the argument that existing reserves need to be kept on the ground to reduce emissions and meet the Paris Agreement targets.

While there is a growing interest in resources, including energy ones, as part of a return of power politics, the academic debate and political discourses are attentive to the emerging multi-dimensional aspects of energy politics. On the one hand, the narrative of oil wars (Kelanic, 2020; Meierding, 2020) is questioned to provide a more nuanced picture, attentive to different dimensions and typologies of conflicts. On the other hand, the potential for new conflict has emerged, both due to the distribution of the losses caused by the decrease of resource rents due to decarbonization and the increased competition for new scarce resources which are essential for low carbon systems. While the potential for solar, hydro, or wind energy is dispersed over the globe, critical materials such as rare earth elements and metals required in low-carbon technologies, such as lithium and cobalt, are concentrated in only a few countries. In consequence, geopolitical competition is turning into a scramble for resources. Back in 2008, China, which holds about one-third of existing reserves and is the main supplier to both the EU (98%) and the USA (80%), has issued restrictions on the sale of rare earth elements to foreign buyers and supported FDI to mitigate the risk to its own mineral supply chain, with both the USA and the EU trying to mitigate their dependency (Blondeel et al., 2021). Once again what is securitized is access to scarce resources. Interestingly, however, the language used is that of increasing resilience. This is reflected, for instance, in the 2020 Action Plan of the European Commission, titled "Critical Raw Material Resilience".

Moreover, the carbon budget is emerging as a concept and is turning into a new aspect to be secured as part of the attempt to not overburden the planet's carrying capacities. The prospect of unburnable carbon and of reserves that need to be kept on the ground to meet the climate targets is creating new insecurities with regard to the distribution of the losses and the capability of producing countries to cope with the decreasing revenues (Blondeel et al., 2021).[4]

On the academic side, there are attempts to reconsider energy security, calling for a more inclusive approach, questioning whose security needs to be prioritized and what needs must be protected (Kester, 2017; Nyman, 2018b). This points at the different meanings of energy security and at the need to integrate different aspects. Yet, it also implies the need to move beyond the divide between technical, material, and ideational aspects that characterize many energy security discourses and disciplinary divides.

In this regard, Kuzemko has stressed the importance of considering the climate–energy security nexus (Kuzemko, 2013). More recent attempts have tried to also include water energy and food security into the nexus to broaden the debate beyond the internalization of different kinds of externalities and to promote equity and justice considerations. Aleh Cherp and Jessica Jewell, for instance, define energy security as the "low vulnerability of vital energy systems" where both vital energy systems and their vulnerabilities are not only objective phenomena, reflecting "objective properties of energy stocks, flows, infrastructure, markets and prices", but also "political constructs" defined and prioritized by various social actors, who reflect "institutional interests, memories and distinct perspectives on the future" (2014, 420). This approach resembles efforts to secure ecosystems as part of attempts to rearticulate security as ecological security in order to move away from the narrow perspective of the security of states or even that of human security (McDonald, 2018). Yet, the emphasis on the technical dimension may remain a way to depoliticize energy governance, taking energy decisions away from the political debate while questions about the national dimension of vital energy systems remain open.

The energy sector reveals the limitations of a narrow focus and of a specific security logic. On the one hand, a narrow focus on security of supply has contributed to marginalizing other issues, like those related to security of demand, which are relevant for producer countries, and access to energy services, which point at the equity dimension and environmental impact. On the other hand, different security discourses and logics characterize the energy sector. Energy security means different things to engineers and to politicians: not all the discourse about energy security evokes the defensive, antagonist, exceptionalist logic that securitization is associated with. The exceptionalist logic of security is relevant to mobilizing action, but this action often involves technological expertise and the design of specific institutional arrangements that favour depoliticization.

As part of a climatization of the energy security discourse, climate expertise and actors are gaining relevance and new referent objects are emerging: normative contributions point to the need for overcoming narrow conceptualizations, moving beyond the incorporation of externalities like carbon emissions or pollution to a reconsideration of technological and political perspectives that incorporate a broader set of values to be protected, though energy governance and relevant security discourses have still remained rather state-centric, with few results in the promoting of much needed forms of global energy governance.

20.5 Beyond Securitization?

Applying the securitization concept to the sectors of energy and climate reveals the political implications of the discourses in both areas and warns about a process of depoliticization, in which options may be precluded, in the name of urgency, or technological considerations and challenges. It also reveals the selectivity in identifying threats and specifying whose security is considered.

While securitization, at least in the way it was originally formulated, identifies one specific security formation and its logic, and warns about the implications of applying it, a more flexible approach opens up a process of transformation. This transformation is about a re-articulation of the way of thinking and providing security.[5] Transforming an issue into a security issue is not only about the construction of threats, but also about the construction and transformation of referent objects, moving away from states or even individuals to considering different entities, be they ecosystems or vital energy systems, which points to the importance of moving beyond an anthropocentric perspective. It is also about the reworking of the very logic of security. In the process, the logic is re-articulated, moving from the defensive, reactive one described by the Copenhagen School, to one that considers risks and precautionary principles.

The processes of securitization and the re-articulation of security occur in a specific context in which views of the future and different perspectives on the possibility of handling crises play a relevant role. Much of the debate is characterized by a divide between catastrophists and technological optimists. This echoes the long-term divide between neo-Malthusian and Cornucopian views, though it has gained new momentum as part of the debate on the Anthropocene (Rothe, 2020), a new geological era in which the destiny of humankind is in human hands and the distinction between human and nature does not hold anymore. This is seen as both terrifying and empowering. While the ontological and epistemological perspectives behind those approaches are complex and beyond the scope of this overview chapter, the different perspectives have relevance for the applicability and persistence of the exceptionalist logic described by the Copenhagen School and its implications.

As I have elaborated elsewhere (Trombetta 2022), catastrophists refer to limited resources and warn about existential threats. Often, they call for radical transformation, like Extinction Rebellion. At the same time, however, this approach can be a way to legitimize the endurance of the exceptionalist logic of security that allows exceptional measures in the name of the survival of existing structures rather than their transformation. As in the case of environmental conflicts or migration, securitization can imply that the consequences rather than the root causes of the problems are addressed in a short-term perspective that focuses on emergencies and crises. Optimists, by contrast, point at how human ingenuity will be able to solve the problem and at how crises that have been announced have been regularly postponed. Fracking has unleashed abundant resources providing energy independence to North America. Geoengineering is considered as the way to stop and reverse climate change. In terms of securitization these discourses may suggest a depoliticization through appeals to security. Rather than exceptional measures, the measures promoted are of a technical nature, lifting an issue *not* above politics but away from politics in a process of technification (Hansen and Nissenbaum, 2009). As with the process of securitization, the process of technification can be problematic, especially given the high level of uncertainty and the huge consequences of these "technical" decisions. In this sense, technical decisions are everything but apolitical.

In between there is a process that focuses on adaptation and on coming to terms with the extent to which climate change is already happening. It is within this perspective that much of the re-articulation of security practices and logics is occurring. The rise of resilience, both in the climate and energy security discourse, reflects a broader material and ideational context as well the action of practitioners in the climate and energy security field.

20.6 Conclusion and Future Research

The chapter has considered considered how climate change and energy security have been incorporated into security discourses. The process has been highly selective, prioritizing some problems, while excluding others, adopting a fragmented approach that reflects traditional discourse often based on national security, even if paradoxes emerge. As a result, a focus on securing the *supply* of energy has downplayed aspects like security of demand or environmental aspects, while environmental conflict discourses have contributed to reframing environmental threats within the familiar language of national security. An approach like securitization that focuses on the construction of threat and the problematic implications of a specific security discourse allows us to navigate the way in which climate change and energy have been linked to security, considering whose security is at stake and which threats are considered. However, a focus on a specific security formation and its logic downplays a broader process of re-articulation concerning the way security is conceptualized and provided.

Future research on climate security discourses needs to explore how different discourses are assembled together, which elements are prioritized, and which are excluded, specifically what are the implications for security provision and for security logics and practices. It needs to move beyond a process that incorporates externalities like emissions or pollution in the case of energy security discourses or to point at new nexuses like the water energy security one to explore how different elements are selected and weaved together in an apparently coherent way. Exclusions in many cases are more relevant than inclusions.

Second, as new security practices and ways of governing through security, like risk or resilience, are gaining relevance, it becomes more urgent to explore their potential and limits. As critical approaches have pointed at the negative implications of the process of securitization, resilience, initially welcomed as an alternative, has been criticized for maintaining the status quo rather than promoting transformation or for implementing neoliberal perspectives that ask individuals to become resilient, leaving the most vulnerable behind to then be reconsidered as, for instance, the resilience of communities (Chandler, 2020). As in the case of security, a contextualized approach that is open to transformation and more empirical analysis is necessary.

Box 20.1 Chapter Summary

- Addressing climate security and energy security are not just about identifying a set of objective threats and developing policies to deal with them.
- The chapter has emphasized the discursive construction of climate and energy security, which identifies threats to be considered, the entities to be protected, and the means to be employed.
- Climate change and energy security are deeply related; however, they have been constructed as two distinct discourses. This creates tension and increases insecurity.
- The chapter has analysed the evolution of climate and energy security discourses and shown how they started to be integrated.
- Considering climate and energy security as part of the same discourse has implications for security practices and for climate and energy policies.

Notes

1 IEA (2021): www.iea.org/reports/sdg7-data-and-projections/access-to-electricity, accessed October 1, 2021.

2 The Copenhagen School determined a sort of 'Copernican revolution' in security studies, transforming security from a positive value and a desirable condition into a problematic practice that it is better to avoid.

3 This describes "a situation in which *urgent* action is required to reduce or halt climate change and avoid potentially *irreversible* environmental damage resulting from it" (*Oxford English Dictionary*, emphasis added).

4 While green activism is targeting fossil fuel infrastructures to maintain an on the ground campaign, a "green paradox" (Fattouh, 2021) has already been identified with producing countries trying to sell as much fossil fuel as possible, anticipating reduction in future demand and long-term price decline, making the transition to low carbon system slower and more difficult.

5 Securitization has opened up the discussion and, by focusing on a specific, resilient logic of security, warns about the risk of applying it; the literature has contributed to crystalizing that. At the same time, taking seriously the original ambition of the Copenhagen School of distilling the meaning of security from its usage, and recognizing that the usage can be different in a different context, provide relevant insights, especially in the energy sector.

References

Berling, Trine Villumsen, Ulrik Pram Gad, Karen Lund Petersen, and Ole Wæver. 2021. *Translations of security: A framework for the study of unwanted futures*. Abingdon: Routledge.

Blondeel, Mathieu, Michael J. Bradshaw, Gavin Bridge, and Caroline Kuzemko. 2021. "The Geopolitics of Energy System Transformation: A Review." *Geography Compass* 15(7), e12580.

Boas, Ingrid and Delf Rothe (2016) From conflict to resilience? Explaining recent changes in climate security discourse and practice. *Environmental Politics*, 25(4), 613–632.

Brown, Lester R. 1977. "Redefining National Security". *Worldwatch Paer 14*. https://eric.ed.gov/?id=ED147229.

Brown, Lester R. 2018. "Redefining National Security." In *Green Planet Blues: Critical Perspectives on Global Environmental Politics*, eds. Geoffrey Dabelko and Ken Conca. Abington: Routledge.

Buzan, Barry, Waever, Ole, and deWilde, Jaap. 1998. *Security: A New Framework for Analysis*. Boulder, CO: Lynne Rienner.

Chandler, David 2020. "Security Through Societal Resilience: Contemporary Challenges in the Anthropocene." *Contemporary Security Policy* 41(2): 195–214.

Cherp, Aleh, and Jessica Jewell. 2014. "The Concept of Energy Security: Beyond the Four As." *Energy Policy* (75):415–421.

Ciută, Felix. 2010. "Conceptual Notes on Energy Security: Total or Banal Security?" *Security Dialogue* 41(2): 123–144.

Correljé, Aad, and Coby van der Linde. 2006. "Energy Supply Security and Geopolitics: A European Perspective." *Energy Policy* 34(5): 532–543.

Corry, Olaf. 2014. "From Defense to Resilience: Environmental Security beyond Neo-Liberalism." *International Political Sociology* 8(3): 256–274.

Dalby, S. (2013). Biopolitics and Climate Security in the Anthropocene. *Geoforum* 49: 184–192.

Deudney, Daniel. 1990. "The Case against Linking Environmental Degradation and National Security." *Millennium – Journal of International Studies* 19(3): 461–476.

Dyer, Hugh, and Maria Julia Trombetta. 2013. "The Concept of Energy Security: Broadening, Deepening, Transforming." In *International Handbook of Energy Security*, eds. Hugh Dyer and Maria Julia Trombetta. Cheltenham: Edward Elgar Publishing, 3–16. http://www.elgaronline.com/view/9781781007891.00009.xml.

Fattouh, B. (2021). *Saudi oil policy: Continuity and change in the era of the energy transition* (No. 81). OIES Paper: WPM.

Floyd, Rita. 2011. "Can Securitization Theory Be Used in Normative Analysis? Towards a Just Securitization Theory." *Security Dialogue* 42(4–5): 427–439.

Hansen, Lene, and Helen Nissenbaum. 2009. "Digital Disaster, Cyber Security, and the Copenhagen School." *International Studies Quarterly* 53(4): 1155–1175.

Homer-Dixon, Thomas F. 2010. *Environment Scarcity, and Violence*. Princeton: Princeton University Press.

Hughes, B. B. (2018). *International Futures: Choices in the Face of Uncertainty*. Abingdon: Routledge.

IEA (2015). *World Energy Outlook*. International Energy Agency.

Kaplan, Richard. 1994. "The Coming Anarchy: How Scarcity, Crime, Overpopulation and Disease Are Rapidly Destroying the Social Fabric of Our Planet." *The Atlantic Monthly*.

Kelanic, Rosemary A. 2020. *Black gold and blackmail: Oil and great power politics*. Ithaca: Cornell University Press.

Kester, Johannes. 2017. "Energy Security and Human Security in a Dutch Gasquake Context: A Case of Localized Performative Politics." *Energy Research and Social Science* 24: 12–20.

Kuzemko, Caroline. 2013.*The energy security-climate nexus: Institutional change in the UK and beyond*. Ithaca: Cornell University Press.

Mabey, Nick. 2008. "Delivering Climate Security: International Security Responses to a Climate Changed World. Whitehall Paper 69, The Royal United Services Institute for Defence and Security Studiesitle."

Maertens, Lucile. 2023. Climatizing the UN Security Council. In Stefan Aykut, Lucile Maertens (Eds.): *The Climatization of Global Politics*. Cham: Springer, pp. 143–163.

Mathews, Jessica Tuchman. 1990. "Environment, Development, and International Security." *Bulletin of the American Academy of Arts and Sciences* 43(7), 10–26.

Matthew, Richard. 2002. "In Defense of Environment and Security Research." *Environmental Change and Security Project Report* 8(Summer).

Mayer, Maximilian. 2012. "Chaotic Climate Change and Security." *International Political Sociology* 6(2): 165–185.

Mayer, Maximilian, and Peer Schouten. 2012. "Energy Security and Climate Security under Conditions of the Anthropocene." In *Energy Security in the Era of Climate Change*, eds Jonathan Symons and Luca Anceschi. London: Palgrave Macmillan UK, 13–35. http://link.springer.com/10.1057/9780230355361_2.

McDonald, Matt. 2008. "The Copenhagen School and the Construction of Security." *European Journal of International Relations* 14(4): 1–7364.

McDonald, Matt. 2018. "Climate Change and Security: Towards Ecological Security?" *International Theory* 10(2): 153–180.

McGowan, Francis. 2011. "Putting Energy Insecurity into Historical Context: European Responses to the Energy Crises of the 1970s and 2000s." *Geopolitics* 16(3): 486–511.

McHugh, Lucy Holmes, Maria Carmen Lemos, and Tiffany Hope Morrison. 2021. "Risk? Crisis? Emergency? Implications of the New Climate Emergency Framing for Governance and Policy." *Wiley Interdisciplinary Reviews: Climate Change* 12(+): e736.

McSweeney, Bill 1996. "Identity and security: Buzan and the Copenhagen school." *Review of International Studies*, *22*(1), 81–93.

Meierding, Emily. 2020. *The Oil Wars Myth*. Ithaca: Cornell University Press.

Myers, N. 1997. "Environmental Refugees." *Population and Environment* 19(2), 167–182.

Myers, Norman. 1996. *Ultimate Security: The Environmental Basis of Political Stability*. Washington: Island Press.

Nyman, Jonna. 2018a. "Rethinking Energy, Climate and Security: A Critical Analysis of Energy Security in the US." *Journal of International Relations and Development*, 21(1).

Nyman, Jonna. 2018b. *The Energy Security Paradox: Rethinking Energy (in)Security in the United States and China*. Oxford: Oxford University Press.

Oels, Angela. 2012. ""From 'Securitization" of Climate Change' to "Climitization" of the Security Field: Comparing Three Theortical Perspectives." In *Climate Change, Human Security and Violent Conflict: challenges for societal stability*, eds. Jürgen Scheffran, Michael Brzoska, Hans Günter Brauch, Peter Michael Link, and Janpeter Schilling. Heidelberg: Springer, 185–205.

Rothe, Delf. 2020. "Governing the End Times? Planet Politics and the Secular Eschatology of the Anthropocene." *Millennium: Journal of International Studies* 48(2), 143–164.

Schwartz, Peter, and Doug Randall. 2003. "An Abrupt Climate Change Scenario and Its Implications for United States National Security." *Jet Propulsion Laboratory* (October).

Stritzel, Holger. 2011. "Security, the Translation." *Security Dialogue* 42(4–5): 343–355.

Trombetta, Maria Julia. 2008. "Environmental Security and Climate Change: Analysing the Discourse." *Cambridge Review of International Affairs* 21(4): 585–602. http://www.tandfonline.com/doi/abs/10.1080/09557570802452920.

Trombetta, Maria Julia. 2012. "Climate Change and the Environmental Conflict Discourse." In *Climate Change, Human Security and Violent Conflict*, eds. Jürgen Scheffran, Michael Brzoska, Hans Günter Brauch, Peter Michael Link, and Janpeter Schilling. Heidelberg: Springer, 151–164.

Trombetta, Maria Julia. 2018. "The Politics of Energy and the Environment." In *Global Environmental Politics* Second edition, eds. Gabriela Kutting and Kyle Herman. New York: Routledge, 178–197 | "First edition published by Routledge 2011"–T.p. verso.: Routledge. https://www.taylorfrancis.com/books/9781351716642/chapters/10.4324/9781315179537–11.

Trombetta, Maria Julia. 2022. "Environmental Security in the Anthropocene." In *Global Security in an Age of Crisis*, ed. Aiden Warren. Edinburgh: Edinburgh University Press, pp. 220–243.

Tyndall Centre for Climate Change Research, & Adger, W. N. (2006). *The Stern review on the economics of climate change*. Tyndall Centre for Climate Change Research.

Wæver, Ole. 1995. "Securitisation and Desecuritisation." In *On Security*, ed. Ronnie D. Lipschutz. New York: Columbia University Press, 46–86.

World Commission on Environment and Development WCED. 1987. *Our Common Future*. Oxford: Oxford Universiy Press.

21

ENVIRONMENTAL POLICY DYNAMICS IN SOUTHEAST ASIA

Two Steps Forward, One Step Back

Sreeja Nair, Ishani Mukherjee, Michael Howlett, and Benjamin Cashore

21.1 Introduction

Southeast Asia presents several unique environmental features as it straddles major regions of ecological richness such as those traversed by the Wallace Line, containing a wide variety of endangered species and landscapes, and as home to four of the world's 34 biodiversity hotspots. Southeast Asia additionally contains 15% of the world's tropical forests and the world's largest area of tropical peatland ecosystems, both of which make an important contribution to regional and global climate systems as carbon sinks (Hughes, 2017; Estoque et al., 2019).

All of this biodiversity, however, is under threat from habitat degradation, land-use change, deforestation, and the spread of alien invasive species (ASEAN, 2017b). Tropical forests in the region face among the highest rates of deforestation in the world, in addition to being adversely impacted by excessive mining and dam construction (Hughes, 2017). Over the last two decades, logging, unsustainable land-use practices, the conversion of wild spaces and wetlands to industrial plantations, drainage, the degradation of peatlands, and recurrent fires have contributed to large-scale ecological problems. The burning and clearing of forests and peatlands continues to contribute to transboundary haze, especially from Indonesia, severely deteriorating the air quality of neighboring nations (Dohong et al., 2017). Southeast Asia also covers nearly 31% of the world's most biodiverse mangrove forests, and these too are at threat of degradation and loss due to increased aquaculture and rice farming activity and palm-oil plantation expansion (Friess et al., 2016).

While ecologically vital at a global scale, Southeast Asia is also one of the most highly and densely populated regions in the world. Socio-economic indicators across the Association of Southeast Asian Nations (ASEAN) display uneven patterns of growth in the region from "first world" Singapore and Taiwan to low capacity Cambodia and Laos, and these differences, in turn, have impacts on the natural resource management and conservation practices followed in different jurisdictions (Morand et al., 2018). With much of the rise in the region's population projected to be in urban areas, pressures on urban infrastructure and surrounding natural habitats are likely to increase as a consequence (Padawangi, 2018; Elliott, 2011).

DOI: 10.4324/9781003043843-24

Despite increasing urbanization, most countries in Southeast Asia rely upon ample natural resource endowments from timber to minerals for economic growth and social and economic development, as well as upon oil and gas and various kinds of plantation-led intensive agriculture (Hirsch, 2020). Environmental policies in the region are thus closely intertwined and embedded within the socio-economic profile of its constituent countries and often dominated by different national developmental priorities. As a result, many of these countries face environmental policy inconsistencies as these clash with economic and social policies designed to address pressures from rising populations and rapid resource-intensive patterns of economic growth (Padawangi, 2018).

In addition to these current problems, future challenges from climate change also loom large in the region, as highlighted by the recent Sixth Assessment Report (AR6) of the Intergovernmental Panel on Climate Change (IPCC, 2022). Despite generally low per capita contributions of the region to greenhouse gas emissions, future projections for emissions growth make it imperative for policymakers to deliberatively and concertedly address both current environmental mitigation as well as future-oriented climate change adaptation (Elliott, 2011).

In discussing the trajectory and implications of these trends, this chapter offers insights into the nature of environmental policymaking within the unique socio-economic and environmental context of Southeast Asia. It provides a comprehensive commentary on how governments of the region are responding to current environmental stressors, supported with examples of environmental governance from Brunei, Cambodia, Indonesia, Laos, Malaysia, Myanmar, the Philippines, Singapore, Thailand, and Vietnam.

21.2 The Environmental Record by Country and Sector

Southeast Asia is geographically both diverse and large, with predominant styles of governance that are also different across constituent nations, making the generalization of policymaking patterns difficult at the regional level. This chapter focuses on different national approaches to key environmental challenges: air pollution, forestry and biodiversity (including fisheries), food security, and water resource management in the region.

21.2.1 Air Pollution, Including Transboundary Haze

As a policy problem, air pollution impacts within and outside national borders, far beyond the source of pollution itself. Conventionally policy responses for air pollution have seen diverse market-based instruments and command-and-control measures including pollution limits and fuel standards being deployed (Pacheco-Vega, 2020). Although most Southeast Asian countries have such measures in place, monitoring and enforcement of these measures remains an ongoing challenge (Boer, 2016).

Additionally, when national air quality standards and regulations for ambient air quality do exist, not all meet the guidelines set by the World Health Organization (WHO). In Malaysia and Singapore, comprehensive national air quality-related standards and regulations that meet WHO standards have been set and implemented. Though similar standards exist in Brunei, the Philippines, Vietnam, Indonesia, and Thailand these are masked by implementation and enforcement shortfalls. Laos and Myanmar lack any dedicated air quality policy or ambient air quality standards. In Myanmar air pollution is considered in a broad-brush manner under a generic Environmental Conservation Law, with poor enforcement. In Cambodia, similarly, enforcement remains a challenge, though a sub-decree on air pollution control was issued in

2000, including air quality standards for ambient air quality and emission limits (UNEP, 2017). Linkages of these policies to larger developmental issues, such as urban planning and development including transport and industrial growth among others, makes addressing air pollution a less than straightforward issue to tackle.

The transboundary nature of air pollution is one of the key challenges that nations in this region grapple with (Quah, 2002). Apart from the monitoring of transboundary air pollution, this phenomenon also blurs accountability measures and makes cooperative agreements between polluting and impacted nations difficult (Lee et al., 2016; Nurhidayah et al., 2015). Transboundary haze occurs regularly in Indonesia, Malaysia, Thailand, and Myanmar during drier months due to illegal burning and logging to clear new land and old plantations for agri-industrial products such as palm oil, acacia, and sugar. Occurrences of haze episodes have been observed to be more pronounced during El Niño years (Nguitragool, 2010).

While ASEAN has worked on the creation of binding transboundary haze agreements for the region, these efforts have to date only met with moderate success in curbing illegal land clearing practices. An example of more nationally led measures is Singapore's Transboundary Haze Pollution Act (THPA) that adopts statutes and mechanisms for penalizing companies owned or based in the country for their activities outside Singapore's boundaries that lead to haze (Lee et al., 2016). The THPA, however, represents a unique case of sustainability policy that has necessitated overcoming individual state jurisdictional boundaries to address regional and indeed global policy priorities requiring both mandatory and voluntary participation of the nations involved (Mukherjee, 2018).

21.2.2 Forestry and Biodiversity Management, Including Fisheries

Many communities in Southeast Asia are dependent on forests for livelihoods. Indonesia and the Philippines also have the highest rate of mining in the tropics, while Laos has a high number of large dams that are under construction. Some countries have already lost more than half of their original forest cover with projections of further losses in the coming decade (Hughes, 2017). On average the proportion of land area covered by forests has remained almost unchanged at 45% between 2005 and 2015 throughout ASEAN member states. Some of this stability could be attributed to reforestation efforts in Indonesia, Malaysia, Thailand, and Vietnam. There are also legal and administrative measures in Brunei and Vietnam to maintain a high forest cover similar to any other environmental standard (ASEAN, 2017a). Some (new) forest cover also results from the transition of peatlands to palm plantations.

These plantations have sustainability implications of their own, however, regarding those linked to palm oil biodiesel, especially in Indonesia, Malaysia, and Thailand (Mukherjee and Sovacool, 2014). Key issues are around the transparency of sustainable palm oil governance and consideration of the land rights of local communities (Howlett et al., 2017). Stronger government partnerships with NGOs, industry, and communities are needed to deal with some of these problems though these are often difficult to create and organize (Ivancic and Koh, 2016).

Cambodia has made notable advancements in meeting the United Nations Millennium Development Goal targets on reversing environmental losses (including loss of forests and fisheries). Thailand has developed a national plan for sustainable development to minimize loss of biodiversity and natural resources and environmental degradation. Implementation challenges, however, still undermine efforts for the monitoring and evaluation of changes in forest cover in many Southeast Asian countries, and capacity limitations on government conservation efforts and efforts to protect the rights of forest-dependent communities are many (ASEAN, 2017a).

Overfishing and illegal and destructive fishing practices continue to threaten the diversity of the marine ecosystems in Southeast Asia. A lack of strong fishing regulations at the national level and weak cooperation among countries in the region has led to rampant illegal, unreported, and unregulated fishing (Pomeroy et al., 2016). Efforts to sustainably boost aquaculture has also negatively affected the diversity of fish stocks, often replacing indigenous species with non-native and invasive ones (Elliott, 2011).

21.2.3 Food Security

Food security remains a major challenge in the region and has been exacerbated by the changing climate. Despite policy efforts to enhance food security, about 60 million people in the region remain undernourished (OECD, 2017). Policy decisions around agriculture are critical in many communities and have a ripple effect on agrarian economies that depend on the sector for livelihoods and sustenance. Climate variability has affected crop yields and soil productivity in most Southeast Asian countries and are expected to do so increasingly in future. Extremes of temperature, floods, and droughts are further negatively affecting many crop cycles and causing economic, demographic, and social disruption and dislocation (Raghavan et al., 2019). The IPCC AR6 (2022) highlights that the unanticipated food production losses from both agriculture and fisheries have intensified in the region, with food insecurity and nutritional deficits witnessed particularly by small-scale food producers. Food security finds a place in the highest risk severity category, and the impacts faced by poor communities in parts of Asia are mooted to be disproportionately high as per the current IPCC projections.

Publicly provided supportive agri-food policies in the region have typically included a range of measures towards food self-sufficiency and price stabilization for crops, together with public stockholding schemes to boost domestic crop production. In some cases, however, these have proven counterintuitive and damaging to food security in the region, especially when countries have had international trade restrictions imposed on them for reasons that can range from ethnic tensions to *coup d'etat*. ASEAN has focused a lot of its efforts in this area, from its Integrated Food Security Framework, to ASEAN Cooperation in Food, Agriculture and Forestry, ASEAN Plus Three Emergency Rice Reserve, and others (OECD, 2017). The role of investments in agricultural technologies, digital agriculture infrastructure, and climate services will be key aspects to reduce long-term risks to the agriculture sector from the impacts of climate change (Cenacchi et al., 2021).

21.2.4 Water Resources Management

Closely tied to these issues are those related to water and water use. Heavy pollution exists in many of the rivers in the region, damaging fisheries and impeding aqua- and agriculture (Elliott, 2011). And, of course, competing demands for water in the region exist, including those from domestic, industrial, and agricultural sources in many different countries. Southeast Asia boasts large river systems crossing multiple country boundaries, such as the Mekong, Red River, Irrawaddy, and Salween Rivers, and ending in major delta systems that support large populations. Rising populations, urbanization, and dam construction for hydropower development all impact the sustainability of these river systems and riverine and delta communities. Valuation of the ecosystem services provided by these river systems is an ongoing challenge which has generally failed to be included in environmental policies (Meynell et al., 2021).

Populations in Southeast Asia are also vulnerable to water-related natural hazards and climatic disasters. Populations in megacities especially those such as Jakarta and Manila are at very high risk but so are those living in smaller centers which face many of the same problems. These existing problems are all impacted by sea level rise, which leads to groundwater depletion and spoilage of freshwater resources. Old infrastructure and capacity issues in urban areas – which lead to disruption of water and sanitation services, coupled with urban areas which have very large informal settlements unable to access government services such as water pipelines or clean water – remains a major issue in much of Southeast Asia. Water security outcomes across this region are likely to be skewed and a combined result of a country's climate, economic situation, and urbanization. Water quality is likely to be a major challenge for both upper-middle-income and lower-middle-income countries, not only due to an increased likelihood of disasters in the region but also due to water demands placed by rapid urbanization and industrial growth (Lorenzo and Kinzig, 2020). The IPCC AR6 also projects that 90% of the additional people living in urban areas by 2050 will be in Asia and Africa and that 60% of Asia's population by that time will be urban. In countries such as Indonesia and the Philippines, rainfall extremes are anticipated to impact freshwater availability (and subsequently food production), health, as well as industrial output. Coastal cities in Southeast Asia are projected to be among the most vulnerable to economic losses by 2050 due to flooding.

Changes in water availability owing to transboundary projects such as dam construction also has had direct impacts on riverine communities. Hydropower projects are also critical for geopolitical power balances in the region as well as moves, beyond water resource management, to cross paths with other environmental concerns, such as energy security and biodiversity impacts (Tran and Suhardiman, 2020; Grundy-Warr and Lin, 2020). This has led to some innovation. (Yong, 2020) discusses a case of Thailand's Chiang Khong district to detail how community-driven adaptive strategies have helped to cope with the changes to water availability due to excessive dam construction along the Mekong.

21.3 Challenges to Finding and Implementing Solutions

As the above demonstrates, the region features many interconnected environmental problems, which are generally growing worse. Not every country, of course, suffers from the same problems or to the same degree, and some have been successful in dealing with a number of these issues. However, notwithstanding these differences, there are three major problems related to environmental policymaking which all these countries face. These three aspects – the level of uncertainty about future challenges, difficulties knowing the kinds of policy tools governments in the region should adopt, and the problems each sector and country has with the constellation of actors and interests governments face in creating policies to prevent or mitigate the worst kinds of environmental damage – are crucial to understanding how the countries in Southeast Asia differ in their approaches to designing effective policies and the results they have had from these efforts (Peters et al., 2018).

Policy uncertainty is an ongoing problem in policymaking which affects all policy areas, including the environment (Howlett and Leong, 2021). Environmental policymaking, in particular, is challenged by mismatches between the timescales over which policy planning is typically considered (ranging from years to decades) and over which the impact of these policy decisions might pan out (ranging from decades to a century) (Bai et al., 2010).

Designing policies that are proportionate to anticipated changes in the future policy context is thus an ongoing challenge (Nair, 2020). In formulating environmental policies policymakers typically face several uncertainties owing to an incomplete understanding of the biophysical and social systems affecting and being affected by the environmental processes and changes in these systems over time. Such uncertainties could wrongly lead to an overestimation or underestimation of policy problems (Jänicke and Jörgens, 2000; Nair and Howlett, 2017). Some policy problems are "wicked" in nature owing to the high complexity, uncertainty, and divergence of actor perspectives regarding the characteristics that define the problem and also solutions in response (Head, 2018).

Climate change is an archetypal example of this kind of uncertainty affecting the region (Levin et al., 2012). Designing effective policies for issues such as climate change is also affected by institutional barriers for gaining consensus, combining the judgment of multiple experts, and integrating multiple perspectives around the problem of climate change and proposed/desired solutions (Webster, 2003).

A second concern is centered on the *choice of policy tools* to design policies to address environmental problems in Southeast Asia. Many of the environmental policy issues such as water management, air pollution, deforestation, and land degradation are wicked problems that are poorly defined with solutions that are either not available or not well known. The choice of policy tool to address environmental problems is thus correspondingly uncertain. Gathering more information in these cases often does not reduce this kind of uncertainty as this is linked less to knowledge deficits than to future probabilities and in fact often exacerbates ambiguity and consequently comes in the way of problem solving (Daviter, 2017). Nevertheless, different countries do prefer different types of instruments – such as public enterprises or large or small subsidies to producers and consumers – and often utilize them despite little evidence of their effectiveness in specific policy areas (Howlett and Tosun, 2019, 2021). And once in place these choices are often difficult to change, such as eliminating fuel or food subsidies to consumers or producers.

Adaptive policies are needed to avoid future failures (Nair and Howlett, 2016), but how this adaptiveness is to be brought into policies in the region remains unclear. In the forestry sector, for example, new tools such as third-party forest certification have been placed within a broader set of forest governance institutions and innovations with some success (van der Ven and Cashore, 2018). Such legality verification as a means to address global forest degradation could potentially be a useful and innovative institutional solution to global forest governance given the challenges of other efforts to build a legally binding forest convention (Cashore and Stone, 2012). Similarly, the development of a "good governance norm complex" – such as in the case of Cambodia's forest sector to identify how policy designers might better anticipate the effects of the EU's Forest Law Enforcement, Governance and Trade program on ongoing problems – has had some success as well (Cashore and Nathan, 2020).

However, there remain challenges of policy coherence between new goals and old instruments. Examples from Thailand, Indonesia, Vietnam and the Philippines on efforts to better integrate climate policies using fuel economy policy instruments and changes in the transport sector have shown they can be effective. And the same is true of Vietnam's efforts to provide better fuel efficiency labeling for new passenger cars, and Thailand and Indonesia's incentives for smaller cars, while protecting local car manufacturing. But some policies may work at cross-purposes, such as vehicle size and fuel restriction measures, which encounter perverse incentives in the form of low fuel taxes (Bakker et al., 2017).

The third aspect concerns the difficulties of interacting with, and managing, the complex subsystems of *policy actors and networks for environmental policymaking* in the region. The politics of decisions around food, water, and energy in each country and in the region involve a set of actors with very unequal power relations. A nexus approach has been pushed by many agencies – such as the Asian Development Bank (ADB) and intergovernmental entities such as the Mekong River Commission (Middleton et al., 2015) – in which efforts are made to enhance meetings and cooperation between producer and government groups and countries. However, there are many agencies involved in environmental policymaking nationally, which often take on a narrow, clientelist, sectoral perspective that is protective of specific industries or interests. And some environmental issues such as mangrove conservation and management of urban lakes can have multiple stakeholders which affects the creation and distribution of responsibility and accountability for resource management and environmental protection in the area (Ng et al., 2018). The focus while addressing such problems is thus often on enhancing communication and negotiation between diverse stakeholders to work towards acceptability of solutions rather than to try to establish technically superior policies based on facts and research (Head, 2018).

In general, a trend towards decentralized approaches to governance can be observed in many jurisdictions in the region which have moved from earlier purely legalistic bureaucratic administrative regimes towards ones more amenable to market and civil-society-oriented policy solutions. This has been true since the 1990s, for example, in Vietnam, Thailand, Indonesia and the Philippines. Despite the range of experiences with the devolved governance of environmental resources in Asia, this movement has resulted in many programs emerging with a common implementation logic surrounding the creation of policies that address environmental as well as poverty alleviation goals through compensation mechanisms (Mukherjee and Howlett, 2016).

Considering the nature of the transboundary problems throughout the region there have also been calls for more and different forms of environmental policymaking and governance beyond the nation-state. This has led to a call for environmental governance to not be limited to jurisdictions but to consider users in an equitable, inclusive, and sustainable manner which moves beyond geographic boundaries (Miller et al., 2020).

Some of this can be seen in international actions such as the efforts to Reducing Emissions from Deforestation and Degradation (REDD+) (Boer, 2016) and improve forest governance led by international organizations such as the Food and Agriculture Organization (FAO). These efforts in Southeast Asia have focused on REDD+ "readiness" these nations. However, conversations around the community impact of REDD+ implementation, protocols for Free Prior and Informed Consent, and guidelines around REDD+ implementation and Access and Benefit Sharing are still in their infancy (Joseph et al., 2013), and many tensions within local groups, and the commercialization and privatization of natural resources, exist, for example in Tonle Sap, Cambodia (Sithirith, 2017).

In addition, there have been other efforts to deal more at the international than regional level. But problems continue to plague international efforts to promote instruments such as Payment of Ecosystem Services (PES), the creation of Marine Protected Areas, and enhancing the implementation of the Convention on Biological Diversity (Friess et al., 2016). Barriers for mainstreaming the concept of ecosystem services into policy measures in Thailand, Cambodia, and Vietnam, for example, are troubled by unclear and often overlapping institutional mandates and insufficient human and financial capacity. Singapore in comparison has been able to incorporate ecosystem service concepts better owing to a more uniform policymaking structure and public–private partnerships (Loc et al., 2020). An incomplete understanding of the needs and priorities of various stakeholders has also limited the effective functioning of both political actors and administrators, while also limiting stakeholder participation. This can be seen in PES programs in the Philippines and Indonesia (Mukherjee and Howlett, 2016).

As a result, much policymaking in the region takes place in or through regional bodies, the most prominent of which is ASEAN. However, ASEAN faces criticism in negotiating mostly top-down, non-binding agreements and being dominated by certain countries and governmental agencies. The ability to create only non-legally binding laws has hindered ASEAN from ensuring enforcement of the decisions, especially for regional environmental management and transboundary disputes (Boer, 2016). As a result, there continue to be calls for more "networked regionalism" with greater involvement of civil society including NGOs in transgovernmental networks. Whether this necessarily brings more transparency and inclusivity to balance the otherwise skewed power relationships for environmental governance in the region is, however, debatable (Elliott, 2011).

21.4 Conclusion and Future Research

Several conclusions can be drawn from this brief survey of the environment and environmental policymaking in Southeast Asia.

First, *the development of the data and indicators needed to overcome environmental uncertainty and manage the area is missing*. There is a need for "sustainability disclosures" which, although accepted in principle by all Southeast Asian countries, have only been incorporated into environmental and social risk management systems in Singapore and the Philippines.[1]

Second, in terms of policy tools and instruments, in many cases *consistent creation and implementation of targeted, measurable policies is missing*. That is, a consistent trend in terms of enhanced government attention to environmental protection is not seen. This is true in terms of government expenditure for environmental protection, for example. Although Southeast Asia has seen a rise in climate-change-related regulations launched each year (Huang and Xu, 2019) it is unclear how, or if, this translates into enforceable and measurable action. Green growth strategies are important in the region (OECD, 2017) but it is often cheaper to continue business-as-usual practices.

Third, and similar, *existing networks of actors and interests involved in the creation of old policies continue to block or impede many new ones*. This can be seen in the creation of new private sustainability standards for palm oil, such as the Roundtable for Sustainable Palm Oil (RSPO) certification versus ASEAN's efforts in addressing the political-economic drivers of fires/haze in Indonesia. However, these often run foul of "regulatory entrepreneurship by private players" including INGOs, philanthropic organizations and investors, and palm oil traders who would like to set stricter standards versus ASEAN's regional governance efforts which remain mired in clientelist domestic politics which resist change (Nesadurai, 2016).

In general, the trend for environmental policymaking seen in this region is thus one of "two steps forward, one step back". Despite the increased number of bills and acts, however, the implementation of environmental laws remains far from optimal in the region. This can mostly be attributed to the lack of monitoring for compliance at the local level, corruption, lack of political will when at crossroads with economic growth priorities, and the limited power of civil society around environmental protection laws. Challenges remain around policy capacity and competence. Provision for enforcement of environmental legislation by civil society is limited, with only Indonesia, the Philippines and Timor Leste including provisions to bring environment-related actions into the courts (Boer, 2016).

But, not all of this record has been one of failure. There have been some successes with new models of governance being demonstrated and tested. Hybrid governance models around ecological rather than geographic boundaries, for example, show some promise. Astuti (2020) explores the ways in which hybrid governance regimes comprising diverse stakeholders in Riau, Indonesia, enacted commons policies to minimize the damage generated by the drainage and conversion of peat swamp forests into agricultural plantations (Astuti, 2020) (see also ASEAN's Strategic Plan on Environment (ASPEN) and its Community Vision 2025 on environmental cooperation; Elliott, 2011).

Opportunities for increased action of this sort are likely to continue to emerge but there are still many problems in low-capacity states such as Cambodia and Laos and in backsliders such as Myanmar, the Philippines and Thailand, with new military or authoritarian governments less interested in environmental protection than in the past. There are also concerns over the potential environmental impacts of China's Belt and Road Initiative, one of the largest infrastructure development projects in recent times in Southeast Asia, including many dam, mineral, and other resource-based mega-projects and initiatives. Balancing sustainability concerns with such development is critical but remains highly problematic in countries interested both in the environment and in investment and public works.

Box 21.1 Chapter Summary

- Environmental policy in Southeast Asia reflects a mixed pattern of successes and failures.
- Consistent creation and implementation of targeted, measurable environmental policies is missing for most countries in the region.
- Challenges remain around policy capacity and competence for implementation of environmental laws. Provision for enforcement of environmental legislation by civil society is limited
- While promising models of environmental governance are emerging there are particular problems in low-capacity states (Cambodia, Laos) and backsliders (Myanmar, the Philippines, Thailand) that are yet to be satisfactorily addressed.

Note

1 Risk and Regulatory Outlook 2021. Key developments in Southeast Asia: Environment, social and governance (ESG). Retrieved from https://www.pwc.com/sg/en/insights/assets/docs/risk-regulatory-outlook-2021-esg.pdf

References

ASEAN. 2017a. ASEAN Statistical Report on Millennium Development Goals 2017. In *Jakarta (ID) ASEAN Secretariat.*

ASEAN. 2017b. *Fifth ASEAN State of the Environment Report* Jakarta: ASEAN Secretariat.

Astuti, Rini. 2020. "Fixing flammable Forest: The scalar politics of peatland governance and restoration in Indonesia." *Asia Pacific Viewpoint* 61 (2):283–300. doi: https://doi.org/10.1111/apv.12267.

Bai, Xuemei, Ryan RJ McAllister, R Matthew Beaty, and Bruce Taylor. 2010. "Urban policy and governance in a global environment: Complex systems, scale mismatches and public participation." *Current Opinion in Environmental Sustainability* 2 (3):129–135.

Bakker, Stefan, Kathleen Dematera Contreras, Monica Kappiantari, Nguyen Anh Tuan, Marie Danielle Guillen, Gessarin Gunthawong, Mark Zuidgeest, Duncan Liefferink, and Martin Van Maarseveen. 2017. "Low-carbon transport policy in four ASEAN countries: Developments in Indonesia, the Philippines, Thailand and Vietnam." *Sustainability* 9 (7):1217.

Boer, Ben. 2016. "Environmental law in Southeast Asia." In *Routledge Handbook of the Environment in Southeast Asia*, 133–150. Routledge.

Cashore, Benjamin, and Iben Nathan. 2020. "Can finance and market driven (FMD) interventions make "weak states" stronger? Lessons from the good governance norm complex in Cambodia." *Ecological Economics* 177. doi: 10.1016/j.ecolecon.2020.106689.

Cashore, Benjamin, and Michael W. Stone. 2012. "Can legality verification rescue global forest governance?" *Forest Policy and Economics* 18:13–22. doi: 10.1016/j.forpol.2011.12.005.

Cenacchi, Nicola, Shahnila Dunston, Timothy B Sulser, Keith Wiebe, and Dirk Willenbockel. 2021. "The future of diets and hunger in South East Asia under climate change and alternative investment scenarios." CG Space: A Repository of Agricultural Research Outputs.

Daviter, Falk. 2017. "Policy analysis in the face of complexity: What kind of knowledge to tackle wicked problems?" *Public Policy and Administration* 34 (1):62–83. doi: 10.1177/0952076717733325.

Dohong, Alue, Ammar Abdul Aziz, and Paul Dargusch. 2017. "A review of the drivers of tropical peatland degradation in South-East Asia." *Land Use Policy* 69:349–360. doi: 10.1016/j.landusepol.2017.09.035.

Elliott, Lorraine. 2011. "ASEAN and environmental governance: Rethinking networked regionalism in Southeast Asia." *Procedia - Social and Behavioral Sciences* 14:61–64. doi: 10.1016/j.sbspro.2011.03.023.

Estoque, Ronald C., Makoto Ooba, Valerio Avitabile, Yasuaki Hijioka, Rajarshi DasGupta, Takuya Togawa, and Yuji Murayama. 2019. "The future of Southeast Asia's forests." *Nature Communications* 10 (1):1829. doi: 10.1038/s41467-019-09646-4.

Friess, Daniel A, Benjamin S Thompson, Ben Brown, A Aldrie Amir, Clint Cameron, Heather J Koldewey, Sigit D Sasmito, and Frida Sidik. 2016. "Policy challenges and approaches for the conservation of mangrove forests in Southeast Asia." *Conservation Biology* 30 (5):933–949.

Grundy-Warr, Carl, and Shaun Lin. 2020. "The unseen transboundary commons that matter for Cambodia's inland fisheries: Changing sediment flows in the Mekong hydrological flood pulse." *Asia Pacific Viewpoint* 61 (2):249–265. doi: 10.1111/apv.12266.

Head, Brian W. 2018. "Forty years of wicked problems literature: Forging closer links to policy studies." *Policy and Society* 38 (2):180–197. doi: 10.1080/14494035.2018.1488797.

Hirsch, P. (Ed.). (2017). *Routledge Handbook of the Environment in Southeast Asia*. Routledge.

Hirsch, Philip. 2020. "Scaling the environmental commons: Broadening our frame of reference for transboundary governance in Southeast Asia." *Asia Pacific Viewpoint* 61 (2):190–202.

Howlett, Michael, and Ching Leong. 2021. "The "Inherent Vices" of policy design: Uncertainty, maliciousness, and noncompliance." *Risk Analysis*.

Howlett, Michael, Ishani Mukherjee, and Joop Koppenjan. 2017. "Policy learning and policy networks in theory and practice: The role of policy brokers in the Indonesian biodiesel policy network." *Policy and Society* 36 (2):233–250. doi: 10.1080/14494035.2017.1321230.

Howlett, Michael, and Jale Tosun. 2019. *Policy Styles and Policy-Making: Exploring the Linkages, Routledge Textbooks in Policy Studies*. London: Routledge.

Howlett, Michael, and Jale Tosun. 2021. *The Routledge Handbook of Policy Styles*. London: Routledge.

Huang, Bihong, and Yining Xu. 2019. Environmental performance in Asia: Overview, drivers, and policy implications. In *ADBI Working Paper 990. Tokyo: Asian Development Bank Institute*. Tokyo.

Hughes, Alice C. 2017. "Understanding the drivers of Southeast Asian biodiversity loss." *Ecosphere* 8 (1):e01624.

IPCC, 2022. "Working group II. Sixth assessment report. Impacts, vulnerability and adaptation." https://www.ipcc.ch/report/ar6/wg2/downloads/report/IPCC_AR6_WGII_FinalDraft_FullReport.pdf

Ivancic, Helena, and Lian Pin Koh. 2016. "Evolution of sustainable palm oil policy in Southeast Asia." *Cogent Environmental Science* 2 (1):1195032.

Jänicke, Martin, and Helge Jörgens. 2000. "Strategic environmental planning and uncertainty: A cross-national comparison of green plans in industrialized countries." *Policy Studies Journal* 28 (3):612–632.

Joseph, Shijo, Martin Herold, William D Sunderlin, and Louis V Verchot. 2013. "REDD+ readiness: Early insights on monitoring, reporting and verification systems of project developers." *Environmental Research Letters* 8 (3):034038.

Lee, Janice Ser Huay, Zeehan Jaafar, Alan Khee Jin Tan, Luis R Carrasco, J Jackson Ewing, David P Bickford, Edward L Webb, and Lian Pin Koh. 2016. "Toward clearer skies: Challenges in regulating transboundary haze in Southeast Asia." *Environmental Science & Policy* 55:87–95.

Levin, Kelly, Benjamin Cashore, Steven Bernstein, and Graeme Auld. 2012. "Overcoming the tragedy of super wicked problems: Constraining our future selves to ameliorate global climate change." *Policy Sciences* 45 (2):123–152.

Loc, Ho Huu, Kim N Irvine, Asan Suwanarit, Pakorn Vallikul, Fa Likitswat, A Sahavacharin, C Sovann, and L Ha. 2020. "Mainstreaming ecosystem services as public policy in South East Asia, from theory to practice." *Sustainability and Law*: 631–665.

Lorenzo, Theresa E, and Ann P Kinzig. 2020. "Double exposures: Future water security across urban Southeast Asia." *Water* 12 (1):116.

Meynell, Peter-John, Marc Metzger, and Neil Stuart. 2021. "Identifying Ecosystem Services for a Framework of Ecological Importance for Rivers in South East Asia." *Water* 13 (11):1602.

Middleton, Carl, Jeremy Allouche, Dipak Gyawali, and Sarah Allen. 2015. "The rise and implications of the water-energy-food nexus in Southeast Asia through an environmental justice lens." *Water Alternatives* 8 (1).

Miller, Michelle A., Jonathan Rigg, and David Taylor. 2020. "Governing transboundary commons in Southeast Asia." *Asia Pacific Viewpoint* 61 (2):185–189. doi: 10.1111/apv.12285.
Morand, Serge, Claire Lajaunie, and Rojchai Satrawaha. 2018. "Biodiversity Conservation in Southeast Asia." Routledge.
Mukherjee, Ishani. 2018. "Policy design for sustainability at multiple scales: The case of transboundary haze pollution in Southeast Asia." In *The Palgrave Handbook of Sustainability*, 37–51. Springer.
Mukherjee, Ishani, and Michael Howlett. 2016. "An Asian perspective on policy instruments: Policy styles, governance modes and critical capacity challenges." *Asia Pacific Journal of Public Administration* 38 (1):24–42. doi: 10.1080/23276665.2016.1152724.
Mukherjee, Ishani, and Benjamin K. Sovacool. 2014. "Palm oil-based biofuels and sustainability in Southeast Asia: A review of Indonesia, Malaysia, and Thailand." *Renewable and Sustainable Energy Reviews* 37:1–12. doi: 10.1016/j.rser.2014.05.001.
Nair, S., and M. Howlett. 2017. "Policy myopia as a source of policy failure: Adaptation and policy learning under deep uncertainty." *Policy & Politics* 45 (1):103–118. doi: 10.1332/030557316x14788776017743.
Nair, Sreeja. 2020. "Designing policy pilots under climate uncertainty: A conceptual framework for comparative analysis." *Journal of Comparative Policy Analysis: Research and Practice* 22 (4):344–359.
Nair, Sreeja, and Michael Howlett. 2016. "From robustness to resilience: Avoiding policy traps in the long term." *Sustainability Science* 11 (6):909–917.
Nesadurai, Helen E. S. 2016. "ASEAN environmental cooperation, transnational private governance, and the haze: Overcoming the 'Territorial Trap' of state-based governance?" *TRaNS: Trans -Regional and -National Studies of Southeast Asia* 5 (1):121–145. doi: 10.1017/trn.2016.25.
Ng, Wun Jern, Sreeja Nair, KB Shameen N Jinadasa, and Evelyn Valencia. 2018. *Saving Lakes-The Urban Socio-cultural And Technological Perspectives*. World Scientific.
Nguitragool, Paruedee. 2010. *Environmental Cooperation in Southeast Asia: ASEAN's Regime for Transboundary Haze Pollution*: Routledge.
Nurhidayah, Laely, Shawkat Alam, and Zada Lipman. 2015. "The influence of international law upon ASEAN approaches in addressing transboundary haze pollution in Southeast Asia." *Contemporary Southeast Asia* 183–210.
OECD. 2017. *Building Food Security and Managing Risk in Southeast Asia*: OECD.
Pacheco-Vega, Raul. 2020. "Environmental regulation, governance, and policy instruments, 20 years after the stick, carrot, and sermon typology." *Journal of Environmental Policy & Planning* 22 (5):620–635.
Padawangi, Rita. 2018. *Routledge Handbook of Urbanization in Southeast Asia*. Routledge.
Peters, B. G., Capano, G., Howlett, M., Mukherjee, I., Chou, M. H., & Ravinet, P. (2018). *Designing for policy effectiveness: Defining and understanding a concept*. Cambridge University Press.
Pomeroy, Robert, John Parks, Kitty Courtney, and Nives Mattich. 2016. "Improving marine fisheries management in Southeast Asia: Results of a regional fisheries stakeholder analysis." *Marine Policy* 65:20–29.
Quah, Euston. 2002. "Transboundary pollution in Southeast Asia: The Indonesian fires." *World Development* 30 (3):429–441.
Raghavan, Srivatsan V, Jiang Ze, Jina Hur, Liu Jiandong, and Nguyen Ngoc. 2019. "ASEAN Food Security under the 2 C-4 C Global Warming Climate Change Scenarios." *Towards a Resilient ASEAN* 1:37–52.
Sithirith, Mak. 2017. "Water governance in Cambodia: From centralized water governance to farmer water user community." *Resources* 6 (3):44.
Tran, Thong Anh, and Diana Suhardiman. 2020. "Laos' hydropower development and cross-border power trade in the Lower Mekong Basin: A discourse analysis." *Asia Pacific Viewpoint* 61 (2):219–235. doi: 10.1111/apv.12269.
UNEP. 2017. The South East Asia Regional Air Quality report.
van der Ven, Hamish, and Benjamin Cashore. 2018. "Forest certification: The challenge of measuring impacts." *Current Opinion in Environmental Sustainability* 32:104–111. doi: 10.1016/j.cosust.2018.06.001.
Webster, Mort. 2003. "Communicating climate change uncertainty to policy-makers and the public." *Climatic Change* 61 (1–2):1.
Yong, Ming Li. 2020. "Reclaiming community spaces in the Mekong River transboundary commons: Shifting territorialities in Chiang Khong, Thailand." *Asia Pacific Viewpoint* 61 (2):203–218. doi: 10.1111/apv.12257.

PART IV

Transformation of Environmental Policies

Paradigmatic Challenges

22
THE CHALLENGE OF LONG-TERM ENVIRONMENTAL POLICY

Detlef F. Sprinz

22.1 Introduction

Why is the world, at large, not successful at reigning in climate change impacts, curbing biodiversity loss, and still losing forested lands despite repeated attempts to do so? Why does it appear to take so long to achieve so little? Many academic, political, and other observers might ask themselves these questions. If these were not long-term environmental policy challenges, they would, most likely, already be "solved" or very substantial headway would have been made in desirable directions. By and large, the inability to easily make rapid progress is linked to the very nature of long-term environmental policy challenges which I will address in this chapter. While I will demonstrate the appreciable challenges, I remain optimistic that, with more advanced research, political innovators and entrepreneurship, as well as appropriate resource allocation, there remains substantial scope for avoiding extremely unfortunate environmental outcomes.

After defining long-term environmental policy challenges (Section 22.2), I will review select methods used to study this class of challenges (Section 22.3) before turning to the policy options to cope with them (Section 22.4). The outlook (Section 22.5) is geared toward inspiring future research.

22.2 Definition

Long-term policy (LoPo) challenges set themselves apart from shorter-term challenges. LoPo can be defined as

> public policy issues that last at least one human generation, exhibit deep uncertainty exacerbated by the depth of time, and engender public goods aspects both at the stage of problem generation as well as at the response stage.
>
> *(Sprinz, 2009, 2)*

The definition has three constituent components. First, a "human generation" relates to problems that remain unabated for a quarter century or longer, or, alternatively, to policy interventions that need a quarter century or longer to reach their long-term policy goal. Iteratively "muddling through" (Lindblom, 1959) is very unlikely to solve LoPo challenges. For example,

DOI: 10.4324/9781003043843-26

halting deforestation by 2030, such as agreed at the UN Framework Convention Climate Change's 26th Conference of the Parties (UNFCCC COP-26) at Glasgow in November 2021,[1] echoes the 2014 promise to (a) reduce deforestation by 50% until 2020 and (b) achieve the 2030 goal of halting deforestation.[2] In view of the data on forest loss or recent greenhouse gas emissions, achieving the purported environmental goals appears beyond challenging.[3]

Second, deep uncertainty refers to

> a situation where the system model and the input parameters to the system model are not known or widely agreed on by the stakeholders to the decision.
>
> *(Lempert, 2002, 7309)*

Deep uncertainty represents considerable scientific and/or perceptual uncertainty about where policy interventions (or the lack thereof) may lead to over time. This is particularly aggravated by political and administrative turnover. In particular, a "new" political or administrative leadership may deviate in its expectation about non-intervention or the effects of specific policy interventions from its predecessors even if the scientific knowledge does *not* change. The assessment of climate change by the US Trump administration provides a stark contrast to the preceding Obama administrations and the succeeding Biden administration, although the science of climate change did not fundamentally change during these three governments.

Third, lopo challenges continue to generate intertemporal negative externalities (such as greenhouse gas emissions that remain effective in the atmosphere for decades with ensuing impacts) if unchecked, thus shifting the costs of present-day inaction to future generations. Conversely, the challenge to assemble sufficiently large and effective groups of actors to curb such challenges over time shows that the logic of collective action characterizes LoPo challenges (Olson, 1971). The Great Green Wall[4] is an attempt to stop the further southward advancement of the Sahel desert and to create livelihoods, protect the environment, as well as to further a very broad array of UN sustainability goals across that latitude in Africa. It appears that only a small fraction of the 2030 ambition has yet come to fruition.[5]

Overall, LoPo challenges should be seen as a conundrum similar to "wicked" (Rittel & Webber, 1973) and "super-wicked" (Levin et al., 2012) policy challenges. Levin et al. suggest that

> [s]uper wicked problems comprise four key features: time is running out; those who cause the problem also seek to provide a solution; the central authority needed to address them is weak or non-existent; and irrational discounting occurs that pushes responses into the future.
>
> *(Levin et al., 2012, 124)*

In conjunction with the definition provided earlier, it becomes apparent that not all environmental policy challenges are long term, yet what we call LoPo environmental challenges are far from trivial and may often be associated with policy failure, defined as low effectiveness in coping with them (Helm & Sprinz, 2000; Sprinz & Helm, 1999).

The most vexing challenge associated with LoPo is the challenge of time-consistent decision-making. Kydland and Prescott (1977) raised this issue in their Nobel Prize winning work by showing that optimal policies adopted at one point in time may not be optimal at a later point in time, given political or other perturbations that intervene in the meantime. In the same vein, Elster (2000, 24), quoting Cukierman, defines time or dynamic inconsistency as "when the best policy currently planned for some future period is no longer the best when that period arrives." This also reflects intertemporal discrepancies in preference between governments and voters on LoPo challenges. The mere possibility of governments (and individuals) reneging on long-term promises, either by way of inconsistency over time in the behavior of the same person (hyperbolic discounting) or due to strategic interaction among various actors (Elster, 2000), led Kydland and Prescott (1977) to call for "rules rather than discretion", that is, rule-based decision-making by a third party (delegation) not subject to constant political pressures.[6] We will turn to this and other policy options following a brief overview of methodological approaches.

22.3 Methods

Methods inform us as to how we can study phenomena. A range of methods lend themselves to the study of LoPo challenges. Given the brevity of this chapter, a complete review of methodologies is not feasible, yet prominent approaches are briefly highlighted,[7] including

- Storytelling, Delphi methods, and foresight exercises;
- Comparative case studies;
- Statistical approaches;
- Game-theoretic models and negotiation simulation;
- Robust decision-making; and
- Agent-based models.

Perhaps the oldest approach in research may be storytelling, yet Delphi methods and foresight exercises have given way to structured exchanges about analyses, expectations, and policy interventions to reduce or even solve long-term environmental policy challenges (Georghiou et al., 2008; Gordon & Helmer, 1964; Lempert et al., 2009, 107). Conceptually, comparative case study designs also lend themselves to LoPo analyses (Bennett, 2004; George & Bennett, 2005), yet they have not been employed rigorously in the context of LoPo, potentially because comparative case studies are mostly employed retrospectively rather than prospectively.

The latter characteristic is also shared by statistical approaches to the study of LoPo (Lempert et al., 2009; Sprinz, 2004), except if they are used for predictions. While econometric forecasting has proven useful for short periods of time, it lacks the ability to cope with structural changes over long periods of time.

By contrast, game-theoretic models help us to understand strategic interaction among stakeholders, especially in a dynamic context under various assumptions about information (e.g., Hovi & Areklett, 2004; Kilgour & Wolinsky-Nahmias, 2004). Besides improving our understanding of strategic choice, game theory's perhaps most pivotal use has been its incorporation into multi-party negotiation software, such as the Predictioneer's Game and the DECIDE models (Bueno de Mesquita, 2009a, 2009b; Dijkstra et al., 2008; Sprinz et al., 2016; Stokman & Van Oosten, 1994). Only rarely have such models, however, been used to make predictions far into the future (for exceptions, see Bueno de Mesquita, 2009a; Bueno de Mesquita et al., 1985; Bueno de Mesquita et al., 1996).

Robust decision-making (RDM) models build on existing simulation models, such as climate, water, or integrated assessment models.

> RDM treats uncertainty with multiple representations of the future, as opposed to a single (probabilistic) forecast, and uses robustness, as opposed to an optimality condition, Thus, … to evaluate alternative strategies that might be pursued by policymakers … RDM also adopts key concepts from scenario planning.
>
> *(Lempert et al., 2009, 116)*

Grossly simplified, RDM relates near-term interventions to groups of desirable and undesirable long-term outcomes and helps policy-makers to identify strategies whose good performance is relatively insensitive to key uncertainties and to characterize the key tradeoffs among such strategies (Groves & Lempert, 2007; Lempert et al., 2009, 116; Lempert et al., 2003).

Similar to the simulation models used by RDM modelers, agent-based models allow for a wide plethora of structured "what if" questions to be computed in efficient ways, for example, the coalition formation process for greenhouse-gas-emission-reducing "climate clubs" or the assessment of the usefulness of the architecture of the 2015 Paris Agreement on Climate Change to hold global mean temperature change since the onset of industrialization below 2 °C (Dimitrov et al., 2019; Hovi et al., 2019, 2020; Sælen, 2020; Sælen et al., 2020; Sprinz et al., 2018).

Overall, scholars of long-term policy have a broad range of methods to choose from. Models calibrated with high-quality observed data in combination with a structured approach to explicitly treat uncertainty are likely to be most useful for practical applications.

22.4 Policy Options

While non-environmental fields, such as the quest for democratic accountability, the rule of law, and humanitarian law, provide a rich history for LoPo challenges, the environmental LoPo challenges are largely, but not exclusively, a post-World War II phenomena. Much has been written about relevant policies available in the abstract sense (Ascher, 2009; Hovi et al., 2005, 2009; Sprinz, 2005, 2009, 2012, 2014; Sprinz & von Bünau, 2013) and, very prominently, in the context of democratic rule (Boston, 2017, 2021) and reforming pension systems (Jacobs, 2008, 2011). To keep it traceable, I have grouped a range of prominent policy proposals into the categories of

- Institutional design;
- Information;
- Dis/incentives; and
- Direct regulation and enforcement.

What follows is open to future refinement. I will briefly cover each of these four groups; the merits of their usefulness will most likely vary across specific LoPo challenges. I invite readers to think about an abstract menu from which to choose from, given the challenges at hand and the paucity of the literature on long-term policymaking.

Designing and redesigning institutions is the work of constitutionalists and politicians, the core of statecraft, but also the nexus between those who govern and those who choose who governs on their behalf. The old adage still holds that if a scandal occurs or a government does not know what to do, a commission is appointed. In nearly all cases, proposals are made to design new or strengthen existing formal institutions or, sometimes, to close institutions that have exceeded their (past) usefulness. Elections in select democracies more recently highlight the importance of environmental LoPo challenges as governmental priorities. It is not far-fetched to anticipate that future election results may be influenced more strongly by electoral sentiment on environmental LoPo issues. With a view to give more weight to those most effected by decisions, various proposals suggest strengthening the representation and influence of younger and future generations, including lowering the age for eligibility to vote, additional votes for parents (to be executed on behalf of their children), and the creation of particular youth councils. While each of these proposals ought to strengthen the rights of future generations and to ameliorate LoPo environmental challenges, more broadly based representation does not automatically imply more effective, long-term solutions.

Table 22.1 Policy options

Category	*Policy option*	*Example(s)*
Institutional Design	Create new or strengthen existing institutions	Hearings, commissions
Institutional Design	Electoral accountability and potentially delegate more weight to younger generations	Lower minimum age for eligibility to vote, additional votes for parents
Institutional Design	Nest intermediate goals within long-term goals	Short-, medium-, and long-term goals
Institutional Design	Delegation of authority	Carbon emissions and/or removals bank
Institutional Design and Information	Eliminate alternative options	Reduce choice set, construction of long-term infrastructures
Information	Transparency	Advisory councils, youth or intergenerational councils, labeling
Information	Rational ignorance: collecting no new information	Exit polluting sector, time limits for decisions
Dis-/incentives	Government fiscal policy	Subsidies and financial offlifting to taxpayers, taxes
Dis-/incentives	Compensation and sanctions	Compensation for past, present, and potentially future damages, penalties
Regulation and Enforcement	Government regulation and enforcement	Emission permits, prohibition of polluting activities

Two additional institutional design options appear helpful: nesting short- and medium-term performance benchmarks within long-term goals and delegating authority. Nesting goals across time allows the (s)electorate (Bueno de Mesquita et al., 2003) to more clearly check progress over time and diagnose overcompliance as well as undercompliance. Introducing climate-related emission reduction goals for 2030 as part of a long-term climate neutrality goal by, for example, the year 2050, is such a practical performance standard. Missing such intermediate goals is the purview of the non-compliance literature, both domestically and internationally. Perhaps most important is to realize that democratic and other institutions suffer from much "presentist bias" (Boston, 2017), and delegation to technocratic institutions is advised, much as originally formulated by Kydland and Prescott who suggested over four decades ago:

> The implication of our analysis is that policymakers should follow rules rather than have discretion. The reason that they should not have discretion is not that they are stupid or evil but, rather, that discretion implies selecting the decision which is best, given the current situation. … There could be institutional arrangements which make it a difficult and time-consuming process to change the policy rules in all but emergency situations.
>
> *(Kydland & Prescott, 1977, 477–487)*

Suggestions for an "energy agency" (Helm et al., 2003) or a carbon emissions (and/or removals) bank that handles carbon emissions (and/or removals) is one prominent option, given the failure of many (yet not all) governments to reduce absolute greenhouse gas emissions since 1992. Finally, one institutional design option may cross over to the informational category: the elimination of options. A prominent example is the debate about the phaseout of the coal-based electricity sector. Once concluded (i.e., coal-based power plants disassembled), this may be difficult to revert to due to the high infrastructural costs of a renewed buildup as well as the competitiveness of scaled-up renewable energy. Constitutional court decisions may serve the same function, assuming observance of the rule of law. Infrastructures built in one location are the strongest way to reduce choice sets: it is often extremely difficult to repurpose the same location for other policy options. More generally, high fixed costs make it difficult to initiate long-term strategies, yet also difficult to revert policy exits once decommissioning high fixed cost policies is under way. Eliminating endless searches for even better options, often a surrogate for procrastination in the absence of supermajorities in favor of policy resolution, brings us closer to the informational category.

Improving the informational basis for more suitable decisions[8] is the classical purview of advisory councils and information agencies, such as the Intergovernmental Panel on Climate Change (IPCC), the Intergovernmental Science-Policy Platform on Biodiversity and Ecosystem Services (IPBES), or, at the national level, the German Advisory Council on Global Change (WBGU), national councils on environmental quality, or the European Environment Agency (EEA). In addition, youth or intergenerational councils and the courts may lengthen the time horizon and solutions space to be presented to institutions that take final decisions. The ultimate challenge for such institutions is that words may be fruitful, yet "action speaks louder than words." Most of the time, informational institutions are separate from decision-making institutions, and the appetite for science-based policies, in the stricter LoPo sense, appears, as of now, to be limited. Yet without the strength of informational institutions, we should not expect decision-making bodies to be sufficiently informed about the LoPo aspects of environmental decisions. After "endless" searches it may, however, be rational to simply stop searching and to take a decision, for example, to exit a polluting sector, such as coal for electricity – unless an emergency arises before completely decommissioning these infrastructures.

Financial resources are often helpful, if not transformative, conditional on their availability and the appropriation to LoPo rather than short-term policy challenges. Restructuring the coal sector in several countries will be contingent on partial or complete buy-outs of constituents in these sectors. This "sugar daddy" solution, is, however, not available for all LoPo environmental challenges most of the time (Sprinz, 2008). Using taxes on pollutants belongs to the repertoire of economists. Contested at the international environmental level, yet standard at the domestic policy level, compensation for environmental damages is an option for challenges not mitigated and where adaptation proves insufficient. The same holds for penalties. Direct regulation and enforcement of the law are also possible, for example offering the right to migration for otherwise inundated countries as the result of sea-level rise or directly regulating a polluting industry (Sprinz & von Bünau, 2013; Verheyen & Roderick, 2008). Overall, this list of policy options is not exhaustive, yet should capture many important levers to limit environmental LoPo challenges.

22.5 Future Research

LoPo challenges are enduring challenges, and perhaps surprisingly a general (rather than single-issue) literature is in short supply. Selectively building on Sprinz (2009), I sketch a range of general challenges that academic research as well as research by practitioners should be devoted to.

First, democratic politics is fundamentally about (re-)elections and who gains office with which policy priorities. Given that LoPo challenges cannot be solved within one electoral period by definition (see Section 22.2), any LoPo policy undertaken in period 1 will not yet be able to demonstrate whether it is fruitful at the end of that period, thus no clear signal can be sent to the electorate in time for the election at the end of the first electoral period.

This begs the question of rewards for LoPo in the presence of competing short-term policy challenges and the challenge as to whether LoPo is continued in period 2 by the same or a newly formed government. This challenge applies to the environmental and non-environmental fields alike, especially given the longevity of and slow reaction to typical environmental LoPo challenges (e.g. the inertia of the climate system, recovery of overfished stocks, reversal of land degradation). More research should be directed at how long-term constituencies for LoPo (Lempert, 2007) can be created. The influence that Fridays for Future is trying to exert on politicians and other stakeholders is one prominent example thereof.

Second, can we predict environmental LoPo policymaking? As we have seen above (see Section 22.3), this should be feasible, yet we witness comparatively little such effort. By contrast, much normative argumentation of the need for environmental LoPo can be found, yet credible LoPo in action has been in short supply. The use of policy prediction models should be explored in depth, for example on the implementation of net zero greenhouse gas emission goals, their timing, intermediate goals, and the revisions thereof (see above), but also on whether and when overfished areas will (not) be reopened for harvest, and by which time stringent policies to build net carbon sinks will be pursued so as to add credibility to net zero climate emissions goals.

Third, assessing the effectiveness of LoPo, that is, the causal effect of policy decisions on LoPo, should be undertaken, bearing in mind that causal attribution over time will prove challenging as well as rewarding.

Fourth, the study of global and sub-global environmental policies has galvanized a lot of research capacity over the past decades, including research on multi-level environmental governance. Given Putnam's (1988) dictum of two-level games about the nexus of international and domestic (environmental) policies, it would be particularly fruitful to undertake *systematic* research on LoPo two-level challenges. This applies especially to the class of global environmental challenges (such as climate change, land degradation, biodiversity) but also to decision-making in supranational institutions, such as the EU.

Fifth, turning the time inconsistency challenge into a research question: Which policy tools prove most successful in avoiding delayed starts? While this partially overlaps with previous suggestions, here the lenses magnify less the "whether" than the "when" aspect. Much like the pursuit of robust decision-making (see above) focuses on the choice of short-term action with least regret about future outcomes, we know too little as to when and under which circumstances such policies commence.

Long-term environmental policy challenges are particularly difficult challenges, both in need of much more comparative as well as methodologically refined research. Unlike short-term environmental policies, there is no "end" in sight – as they are likely to be perennial and wicked challenges demanding our attention, research, wit, and dedication to keep them in check, if not to solve them.

Box 22.1 Chapter Summary

- By definition, LoPo challenges are difficult to solve.
- A broad range of social science methods is suitable for the study of LoPo.
- Policy options to cope with LoPo encompass institutional design, information, dis-/incentives, as well as regulation and enforcement.
- Predicting LoPo choices, coping with time inconsistency, and assessing the effectiveness of LoPo policies are among three challenges that merit dedicated future LoPo research.

Appreciation

I am grateful to Jonathan Boston, Helge Jörgens, and Christoph Knill for helpful comments on an earlier draft.

Notes

1 https://unfccc.int/news/cop26-pivotal-progress-made-on-sustainable-forest-management-and-conservation (last accessed: 23 April 2023).
2 https://forestdeclaration.org/about/ (last accessed: 23 April 2023).
3 https://www.globalforestwatch.org/blog/data-and-research/global-tree-cover-loss-data-2021/ (last accessed: 22 October 2022).
4 https://www.greatgreenwall.org/about-great-green-wall (last accessed: 23 April 2023). While ambitions for 2030 are clearly stated, the overall degree of *relative* progress to achievement has not been assessed for the period since inception.
5 https://www.greatgreenwall.org/results (last accessed: 22 October 2022).
6 Creating and maintaining such independent, rule-based institutions is itself a credible commitment challenge.
7 For a more detailed, select treatment of methods, see Lempert et al. (2009) and Sprinz & Wolinsky-Nahmias (2004).
8 Jacobs (2016) provides insightful reviews and especially suggestions as to how informational status might impinge on LoPo decisions. Here, I focus merely on institutional forms observable in the environmental policy field.

References

Ascher, W. (2009). *Bringing in the Future: Strategies for Farsightedness and Sustainability in Developing Countries*. University of Chicago Press.

Bennett, A. (2004). Case Study Methods: Design, Use, and Comparative Advantages. In D. F. Sprinz & Y. Wolinsky-Nahmias (Eds.), *Models, Numbers, and Cases: Methods for Studying International Relations* (pp. 19–55). University of Michigan Press.

Boston, J. (2017). *Governing for the Future: Designing Democratic Institutions for a Better Tomorrow* (First edition ed.). Emerald.

Boston, J. (2021). Assessing the Options for Combatting Democratic Myopia and Safeguarding Long-Term Interests. *Futures, 125*, 102668. https://doi.org/10.1016/j.futures.2020.102668.

Bueno de Mesquita, B. (2009a). *The Predictioneer's Game: Using the Logic of Brazen Self-Interest to See and Shape the Future* (1st ed.). Random House.

Bueno de Mesquita, B. (2009b). Recipe for Failure - Why Copenhagen Will Be a Bust, and Other Prophecies from the Foreign-Policy World's Leading Predictioneer. *Foreign Policy*. http://foreignpolicy.com/2009/10/19/recipe-for-failure/.

Bueno de Mesquita, B., Newman, D., & Rabushka, A. (1985). *Forecasting Political Events - The Future of Hong Kong*. Yale University Press.

Bueno de Mesquita, B., Smith, A., Siverson, R. M., & Morrow, J. D. (2003). *The Logic of Political Survival*. MIT Press.

Bueno de Mesquita, B. J., Newman, D., & Rabushka, A. (1996). *Red Flag Over Hong Kong*. Chatham House Publishers.

Dijkstra, J., Van Assen, M. A. L. M., & Stokman, F. N. (2008). Outcomes of Collective Decisions With Externalities Predicted. *Journal of Theoretical Politics, 20*(4), 415–441. https://doi.org/10.1177/0951629808093774.

Dimitrov, R., Hovi, J., Sprinz, D. F., Sælen, H., & Underdal, A. (2019). Institutional and Environmental Effectiveness: Will the Paris Agreement Work? *Wiley Interdisciplinary Reviews: Climate Change, 10*(4), e583. https://doi.org/10.1002/wcc.583.

Elster, J. (2000). *Ulysses Unbound: Studies in Rationality, Precommitment, and Constraints*. Cambridge University Press.

George, A. L., & Bennett, A. (2005). *Case Studies and Theory Development in the Social Sciences*. The MIT Press.

Georghiou, L., Harper, J. C., Keenan, M., Miles, I., & Popper, R. (Eds.). (2008). *The Handbook of Technology Foresight: Concepts and Practice*. Edward Elgar. https://www.e-elgar.com/shop/gbp/the-handbook-of-technology-foresight-9781848448100.html.

Gordon, T. J., & Helmer, O. (1964). *Report on a Long-Range Forecasting Study*. RAND Corporation. https://www.rand.org/pubs/papers/P2982.html.

Groves, D. G., & Lempert, R. J. (2007). A New Analytic Method for Finding Policy-Relevant Scenarios. *Global Environmental Change*, *17*, 73–85. https://www.rand.org/pubs/reprints/RP1244.html.

Helm, C., & Sprinz, D. F. (2000). Measuring the Effectiveness of International Environmental Regimes. *Journal of Conflict Resolution*, *44*(5), 630–652. https://doi.org/10.1177/0022002700044005004.

Helm, D., Hepburn, C., & Mash, R. (2003). Credible Carbon Policy. *Oxford Review of Economic Policy*, *19*(3), 438–450.

Hovi, J., & Areklett, I. (2004). Enforcing the Climate Regime: Game Theory and the Marrakesh Accords. *International Environmental Agreements: Politics, Law and Economics*, *4*(1), 1–26.

Hovi, J., Huseby, R., & Sprinz, D. F. (2005). When Do (Imposed) Economic Sanctions Work? *World Politics*, *57*(4), 479–499.

Hovi, J., Sælen, H., & Sprinz, D. F. (2020). Can the 2015 Paris Agreement on Climate Change Deliver 2°C? *Global Cooperation Research - A Quarterly Magazine*, *4/2020*(4), 3–7. https://www.gcr21.org/fileadmin/website/publications/Quarterly_Magazine/GCR21_Quarterly_Magazine_4-2020_December-online.pdf.

Hovi, J., Sprinz, D. F., Sælen, H., & Underdal, A. (2019). The Club Approach: A Gateway to Effective Climate Co-operation? *British Journal of Political Science*, *49*(3), 1071–1096. https://doi.org/10.1017/S0007123416000788.

Hovi, J., Sprinz, D. F., & Underdal, A. (2009). Implementing Long-Term Climate Policy: Time Inconsistency, Domestic Politics, International Anarchy. *Global Environmental Politics*, *9*(3), 20–39. https://doi.org/10.1162/glep.2009.9.3.20.

Jacobs, A. M. (2008). The Politics of When: Redistribution, Investment and Policy Making for the Long Term. *British Journal of Political Science*, *38*(2), 193–220. https://doi.org/10.1017/S0007123408000112.

Jacobs, A. M. (2011). *Governing for the Long Term: Democracy and the Politics of Investment*. Cambridge University Press. https://doi.org/10.1017/CBO9780511921766.

Jacobs, A. M. (2016). Policy Making for the Long Term in Advanced Democracies. *Annual Review of Political Science*, *19*(1), 433–454. https://doi.org/10.1146/annurev-polisci-110813-034103.

Kilgour, D. M., & Wolinsky-Nahmias, Y. (2004). Game Theory and International Environmental Policy. In D. F. Sprinz & Y. Wolinsky-Nahmias (Eds.), *Models, Numbers, and Cases: Methods for Studying International Relations* (pp. 317–343). University of Michigan Press.

Kydland, F. E., & Prescott, E. C. (1977). Rules Rather Than Discretion: The Inconsistency of Optimal Plans. *Journal of Political Economy*, *85*(3), 473–491.

Lempert, R., Scheffran, J., & Sprinz, D. F. (2009). Methods for Long-Term Environmental Policy Challenges. *Global Environmental Politics*, *9*(3), 106–133. https://doi.org/doi:10.1162/glep.2009.9.3.106.

Lempert, R. J. (2002). A New Decision Sciences for Complex Systems. *Proceedings of the National Academy of Sciences*, *99*(90003), 7309–7313. https://doi.org/10.1073/pnas.082081699.

Lempert, R. J. (2007). Creating Constituencies for Long-Term, Radical Change. John Brademas Center for the Study of Congress, New York University, Research Brief.

Lempert, R. J., Bankes, S. C., & Popper, S. W. (2003). *Shaping the Next One Hundred Years: New Methods for Quantitative, Long-Term Policy Analysis*. Rand Corporation. https://www.rand.org/pubs/monograph_reports/MR1626/index.html.

Levin, K., Cashore, B., Bernstein, S., & Auld, G. (2012). Overcoming the Tragedy of Super Wicked Problems: Constraining Our Future Selves to Ameliorate Global Climate Change. *Policy Sciences*, *45*(2), 123–152. https://doi.org/10.1007/s11077-012-9151-0.

Lindblom, C. E. (1959). The Science of "Muddling Through". *Public Administration Review*, *19*(2), 79–88. https://doi.org/10.2307/973677.

Olson, M. (1971). *The Logic of Collective Action - Public Goods and the Theory of Groups*. Harvard University Press.

Putnam, R. D. (1988). Diplomacy And Domestic Politics: The Logic of Two-Level Games. *International Organization*, *42*(3), 427–460.

Rittel, H. W. J., & Webber, M. M. (1973). Dilemmas in a General Theory of Planning. *Policy Sciences*, *4*(2), 155–169. https://doi.org/10.1007/BF01405730.

Sælen, H. (2020). Under What Conditions Will the Paris Process Produce a Cycle of Increasing Ambition Sufficient to Reach the 2°C Goal? *Global Environmental Politics, 20*(2), 83–104. https://doi.org/10.1162/glep_a_00548.

Sælen, H., Hovi, J., Sprinz, D. F., & Underdal, A. (2020). How US Withdrawal Might Influence Cooperation Under the Paris Climate Agreement. *Environmental Science and Policy, 108*, 121–132. https://doi.org/10.1016/j.envsci.2020.03.011.

Sprinz, D. F. (2004). Environment Meets Statistics: Quantitative Analysis of International Environmental Policy. In D. F. Sprinz & Y. Wolinsky-Nahmias (Eds.), *Models, Numbers, and Cases: Methods for Studying International Relations* (pp. 177–192). University of Michigan Press.

Sprinz, D. F. (2005). "Für das Klima haften?" [Liability for Climate Change?]. *Politische Ökologie* (December 2005/January 2006), 77.

Sprinz, D. F. (2008). Responding to Long-Term Policy Challenges: Sugar Daddies, Airbus Solution or Liability? *Ökologisches Wirtschaften* (2/2008), 16–19.

Sprinz, D. F. (2009). Long-Term Environmental Policy: Definition, Knowledge, Future Research. *Global Environmental Politics, 9*(3), 1–8. https://doi.org/doi:10.1162/glep.2009.9.3.1.

Sprinz, D. F. (2012). Long-Term Environmental Policy: Definition—Origin—Response Options. In P. Dauvergne (Ed.), *Handbook of Global Environmental Politics* (second ed., pp. 183–193). Edward Elgar.

Sprinz, D. F. (2014). Long-Term Policy Problems: Definition, Origins, and Responses. In F. W. Wayman, P. R. Williamson, S. Polachek, & B. Bueno de Mesquita (Eds.), *Predicting The Future In Science, Economics, And Politics* (pp. 126–143). Edward Elgar. https://doi.org/10.4337/9781783471874.00017.

Sprinz, D. F., Bueno de Mesquita, B., Kallbekken, S., Stokman, F., Sælen, H., & Thomson, R. (2016). Predicting Paris: Multi-Method Approaches to Forecast the Outcomes of Global Climate Negotiations. *Politics and Governance, 4*(3), 172–187. https://doi.org/10.17645/pag.v4i3.654.

Sprinz, D. F., & Helm, C. (1999). The Effect of Global Environmental Regimes: A Measurement Concept. *International Political Science Review, 20*(4), 359–369.

Sprinz, D. F., Sælen, H., Underdal, A., & Hovi, J. (2018). The Effectiveness of Climate Clubs under Donald Trump. *Climate Policy, 18*(7), 828–838. https://doi.org/10.1080/14693062.2017.1410090.

Sprinz, D. F., & von Bünau, S. (2013). The Compensation Fund for Climate Impacts. *Weather, Climate, and Society, 5*(3), 210–220. https://doi.org/10.1175/wcas-d-12-00010.1.

Sprinz, D. F., & Wolinsky-Nahmias, Y. (Eds.). (2004). *Models, Numbers, and Cases: Methods for Studying International Relations*. University of Michigan Press.

Stokman, F., & Van Oosten, R. (1994). The Exchange of Voting Positions: An Object-Oriented Model of Policy Networks. In B. Bueno de Mesquita & F. N. Stokman (Eds.), *European Community Decision-Making: Models, Applications, and Comparisons* (pp. 105–127). Yale University Press.

Verheyen, R., & Roderick, P. (2008). *Beyond Adaptation: The Legal Duty to Pay Compensation for Climate Change Damage*. WWF-UK.

23

CITIES AND URBAN TRANSFORMATIONS IN MULTI-LEVEL CLIMATE GOVERNANCE

Kristine Kern

23.1 Introduction

To the extent that national efforts and international negotiations have proven insufficient to successfully combat climate change, cities have become the focus of attention in research and practice. However, the institutionalization of environmental protection at the national level preceded similar initiatives at the local level. When the first local environmental departments were set up in the 1980s, environmental ministries and national agencies had already existed for around ten years. Many of them were founded shortly before the UN Conference on the Human Environment took place in Stockholm in 1972 (see Chapter 2).

Debates on sustainable development followed in the late 1980s, starting with the Brundtland report "Our Common Future" in 1987. This led to the Earth Summit in 1992 in Rio de Janeiro and an action plan called Agenda 21. Chapter 21 of this plan contained various provisions on local authorities' initiatives in support of Agenda 21. As many of the problems mentioned in this document are caused by local activities, Chapter 28 stated that the participation and cooperation of local authorities – as the level of governance closest to the people – would be a determining factor in fulfilling objectives of Agenda 21. Thus, it asked for the adoption of a "Local Agenda 21" by local authorities, cooperation between them, activities by city associations to increase cooperation and coordination with the goal of enhancing knowledge exchange and transfer between local authorities. This initiative fostered many Local Agenda 21 processes around the globe and became one of the most remarkable results of Agenda 21 (Lafferty, 1999; Kern et al., 2007). Moreover, many international organizations such as UN Habitat and the World Bank developed their own concepts of sustainable development (UN Habitat, 2022). The last major UN initiative in this area was Agenda 2030 (2015), which contains 17 Sustainable Development Goals. Most of them are relevant for cities, but SDG 11 in particular aims at "making cities and human settlements inclusive, safe, resilient and sustainable" (United Nations, 2015).

The first transnational municipal networks focusing on sustainable development and climate change emerged almost simultaneously with the Rio Summit in 1992. The International Council for Local Environmental Initiatives (ICLEI) was founded in 1990, followed by an

DOI: 10.4324/9781003043843-27

increasing number of similar networks. In this early phase of transnational city networking, these networks were set up "by leading cities, for leading cities" (Kern and Bulkeley, 2009). After a very dynamic start and a rapid diffusion of membership in networks such as ICLEI, these networks stabilized their membership and consolidated their internal organization and external relations. New generations of networks of cities and towns emerged, including the C40 Cities Climate Leadership Group founded in 2005, and 100 Resilient Cities set up in 2013 (Haupt et al., 2020).

In contrast to these networks of leading cities, the EU Covenant of Mayors (CoM), which was set up by the EU Commission in 2008 and is run by a consortium of various city networks (Energy Cities, Climate Alliance, EUROCITIES, ICLEI, etc.), attracted not only forerunner cities but also followers and latecomers, including many small and medium-sized cities and towns in Spain and Italy. In 2016, the EU CoM merged with the global Compact of Mayors and became the Global Covenant of Mayors. Today, this initiative counts almost 11,000 signatories with around 340 million inhabitants in more than 50 countries (Covenant of Mayors, 2022; Melica et al., 2022). New signatories need to adopt an integrated climate mitigation and adaptation strategy and pledge to reduce their CO_2 emissions by at least 55% by 2030. In 2021, the EU set up so-called EU Missions as part of Horizon Europe (2021–2027), including the EU Mission on 100 Climate-Neutral and Smart Cities. In contrast to the CoM, the majority of the cities that participate in this Mission are leading cities. Overall, this shows that international organizations and the EU have supported sustainable development and climate policies at the local level since the early 1990s and that these initiatives were accompanied by networking between (leading) cities and towns.

23.2 From Local Agenda 21 to Carbon-Neutral Cities

After the Rio Earth Summit in 1992, local authorities developed various concepts and visions on different aspects of sustainability and climate policy. Cities and towns set up strategies to become sustainable, low-carbon, resilient, smart, circular, and climate-neutral. Bibliometric analyses of the trajectory of these concepts show that they can be separated into two clusters around sustainable cities with an emphasis on eco-economic issues and smart cities focusing more on socio-economic issues, while other concepts, such as low-carbon cities or eco-cities, seem to be hybrid forms that help to enrich the traditional sustainable city concepts. Moreover, there is empirical evidence for differing regional patterns since the debates in Europe and America tend to focus mainly on sustainable and smart city concepts, while low-carbon and eco-city concepts are more popular in Asia, in particular in China (De Jong et al., 2015; Fu and Zhang, 2017; Schraven et al., 2021; Haupt et al., 2022).

De Jong et al. (2015) regard the sustainable city concept as a broad conceptual or "umbrella" category, while they see other concepts such as the smart city concept as more tightly defined categories. The smart city concept appears to be a rather distinct approach which emphasizes innovation, smart grids, smart meters, the Internet of Things, and so on (de Jong et al., 2015). Current developments suggest that both concepts may coexist and that the (wider) sustainable city concept will be reinvigorated and may, again, lead to the development of new (hybrid) concepts closely related to the sustainable city concept or may even stimulate the emergence of new concepts which may bridge the gap between sustainable and smart city concepts. The recent debates concentrate mainly on circular and carbon-neutral cities.

Many leading cities worldwide such as Sydney, New York, and Krakow have declared a climate emergency and decided to become carbon-neutral. However, a study of more than 300 European cities showed that these cities need to roughly double their ambitions to reach the 1.5 degree goal of the Paris Agreement. Only around 25% of the cities opted for climate neutrality, in particular cities that are members of city networks (Climate Alliance, Covenant of Mayors) (Salvia et al., 2021; Huovila et al., 2022). In the past, research on local climate action has focused mainly on mitigation policies (plans and strategies), on case studies of big forerunner cities in the Global North, and on transnational city networks such as ICLEI and C40. Although the number of studies on initiatives in the Global South (e.g. Fisher, 2014; Stehle et al., 2020) has increased, there is still a lack of studies on mid-sized and smaller cities (Hoppe et al., 2016; Otto et al., 2021), on "ordinary cities" (van der Heijden, 2019; Haupt et al., 2021), on national and subnational municipal networks (Karhinen et al., 2021), on the relationship between mitigation and adaptation (e.g. Göpfert et al., 2019), and on the implementation and evaluation of local climate policy. Although quantitative studies have become standard (e.g. Reckien et al., 2018), innovative approaches that combine qualitative and quantitative methods and can cope with the analysis of big data are still missing.

23.3 Can Cities Save the Planet?

Many cities worldwide have set up their climate policies earlier than the nation-states in which they are located, and they have also set more ambitious goals than their national governments. Thus, expectations that cities can manage the growing challenges of climate change have increased. Cities are the places where CO_2 emissions are generated, but they are also the places where creative solutions to mitigate and adapt to climate change are invented and tested (Kern and Alber, 2009; van der Heijden, 2022). However, in spite of many innovative experiments at the local level, it has become evident that a lack of national climate policy cannot be compensated by local initiatives because smaller cities and towns do not have sufficient capacities (Otto et al., 2021). This applies especially to cities in poor countries, but even in the Global North the majority of municipalities does not have the necessary capacities to become climate-neutral in the near future. Even if technical solutions exist, the majority of cities face a lack of capacities for starting and implementing successful climate mitigation and adaptation policies.

As the discussion on local climate action has focused on wealthy forerunner cities, it has neglected the constraints which "ordinary" cities face on a daily basis. Cities control only around 30% of the CO_2 emissions that are generated within their territorial borders. National laws in climate-relevant areas (built environment, transport, etc.) may hamper and prevent local climate action. Moreover, the majority of municipalities heavily depend on national funding, including general financial transfers as well as specific programs in the areas of climate mitigation and adaptation.

As it has become relatively unlikely that the 1.5 degree goal of the Paris Agreement can be reached (IPCC, 2021), cities shift their attention to climate adaptation but lack the means to fund big infrastructure projects, which are needed to make them climate resilient. On the local political agenda, climate change issues compete with other issues (such as the COVID-19 pandemic). They become more salient during and after specific events such as international climate conferences or extreme weather events (heat waves, flooding), though such windows of opportunity tend to close quickly without leading to transformational change (Berglund et al., 2022).

Climate policies may not be compatible with initiatives in other policy areas. Urbanization causes urban sprawl, a lack of affordable housing, and the need to construct new homes for an increasing population. However, the construction industry has become one of the main emitters of CO_2 emissions in cities, and climate adaptation requires green and blue infrastructure and the development of nature-based solutions as part of integrative and inclusive planning strategies (Sharifi, 2021; UN-Habitat, 2022).

The ongoing polycrisis makes the situation worse because increasing energy costs and inflation rates lead to growing inequalities between different groups of citizens and the neighborhoods where they live. Although the energy crisis offers chances for changing individual behavior and creating new pathways for energy transitions, it also leads to energy injustice and policies that prevent the decarbonization of the energy sector, maintain or even increase the dependency on fossil fuel, and do not lead to a reduction of CO_2 emissions.

23.4 Dimensions of Urban Transformations

23.4.1 Temporal Dimension of Urban Transformations

Transformational changes can be either incremental or, as a reaction to external shocks, radical and abrupt. Most often we see incremental transformation pathways and occasional punctuation (Berglund et al., 2022) that are characterized by slow changes, reactive behavior, institutional stability, and resistance to change. In contrast, radical and abrupt transformation pathways may be caused by disruptive events such as flooding and lead to rapid institutional change.

However, in leading cities we also find strategic transformation pathways characterized by proactive interventions that anticipate shocks and disruptions, develop integrative strategies, and stimulate institutional changes. Disruptive events such as COVID-19 and increasing risks and uncertainties require that cities engage in path creation and develop capacities for strategic, integrative, adaptive, and innovative action (UN Habitat, 2022; Kern et al., 2021, 2022).

In addition, organizational and governance arrangements do not always develop in a linear way. Reorganizations may follow political cycles (e.g. election of a new mayor). The main challenges are inconsistent timescales. Long-term climate strategies and emission reduction goals are not compatible with short-term political cycles (see Chapter 22). Interim and sectoral goals are needed, but they are difficult to institutionalize in such a way that they withstand local elections and a change of local leadership. This applies in particular to municipalities with increasing populist movements at the local level.

The causal relationships between place-based experiments and long-term urban sustainability transformations are still neglected. Most often local experiments are pilot or demonstration projects that work in specific places but not in others. If context conditions differ considerably, such experiments have only limited effects. Thus, there is a need for process learning that drives organizational change and enables organizations to experiment more effectively (Evans et al., 2021).

Inconsistent timescales also exist with respect to funding. Long-term climate mitigation and adaptation requires long-term funding of the local infrastructure. As most municipalities do not have the capacities to fund such long-term projects, they need to apply for funds and build capacities for the preparation of such applications. Even if funds are available at the national or regional level, funding schemes are restricted to a limited period of time (such as funding a climate manager for a fixed term). However, this "projectification of funding" prevents experiments leading to changes of organizational routines (Torrens and von Wirth, 2021), which impedes long-term strategic changes.

23.4.2 Spatial Dimension of Urban Transformations

There are spatial differences with respect to the preconditions and options for urban transformations. Cities in the Global South face different challenges than their peers in the Global North because rapid urbanization without structural transformation increases the vulnerability to future shocks in the Global South (Revi et al., 2014; IPCC, 2022; UN Habitat, 2022).

Moreover, varying national characteristics may have considerable effects on the chances and challenges to reduce CO_2 emissions. This is most evident for the national energy mix because some countries heavily depend on fossil fuel, which influences the local energy mix and limits the scope of action at the local level (Kern et al., 2022).

There are three dimensions of urban transformation pathways. In the first dimension, *within cities*, biophysical characteristics and vulnerabilities to climate risks, such as flooding or the capacity for the generation of renewable energy, need to be taken into account because they differ between cities but also between neighborhoods within the same city. Based on these preconditions, cities develop strategies, set up organizations, and execute projects. The success of climate policies depends on organizational arrangements and the specific combination of the influence of the lead unit on the one hand and the degree of (de)centralization on the other. Research on American cities shows that lead agency coordination is the best option, while decentralized networks often fail (Krause and Hawkins, 2021).

As most studies on local climate action focus on cities, a broader view *beyond cities* and their territorial borders is needed. There are pronounced differences between metropolitan and neighboring rural regions, for example with respect to car dependency. However, metropolitan regions need to develop integrated strategies because the capacities of a city to generate renewable energy or food on its own territory are rather limited. Even the most innovative cities cannot implement their ambitious plans and strategies without cooperation and coordination with the surrounding region. Cities depend on the sustainable use of natural resources in the region, regionally produced energy and food, and the regional coordination of transport networks. Making cities more resilient and better prepared for extreme weather events also requires regional cooperation. Therefore, cities need to recognize the interdependent relations between urban centers and the periphery, engage in regional planning, and develop new spatial arrangements (e.g. for the functional space of a river delta) (Dabrowski et al., 2021), which do not follow administrative borders. Governing such spaces may require institutional arrangements (such as specific districts) that may overlap and even compete with each other.

Finally, transformative changes *across cities* are also highly relevant because nationwide transformations require initiatives not only in big forerunner cities but also in mid-sized and small cities and towns with far less capacities. Comprehensive transformations do not take place on a voluntary basis because smaller municipalities do not have the capacity to follow the big forerunners. Urban transformations across cities are facilitated by city networking, which change over time. While transnational municipal networks are less important nowadays than they were in the 1990s, new territorial networks (e.g. within the metropolitan region) and functional networks (e.g. networks of climate managers) have emerged and help to transfer knowledge and support urban transformations across cities (Kern and Bulkeley, 2009; Kern, 2019; Haupt et al., 2020). Urban transformations within, beyond, and across cities require institutional and organizational changes but also governance arrangements that facilitate the scaling of place-based experiments.

23.5 The Challenge of Scaling

23.5.1 Concepts and Types of Scaling

Debates on scaling have gained momentum during recent years. Van der Heijden (2022) even asks for a "Science of Scaling". Scaling has been defined as "expanding, adapting and sustaining successful policies, programmes or projects in different places and over time to reach a greater number of people" (World Bank, 2005). The concept of scaling shares many characteristics with similar approaches such as policy diffusion, policy transfer, and policy mobility (see also Chapter 18). The main difference lies in the fact that scaling is not a voluntary but an intentional process. In contrast to voluntary policy diffusion, scaling requires a certain degree of state intervention, that is the involvement of higher levels of government. At least in the long run, scaling should lead to urban transformations.

Successful urban transformations depend on the degree of policy change and social learning. There is a need to switch from existing to new policy instruments, and to ultimately change the problem interpretation and the underlying policy paradigm (Hall, 1993). Thus, there are three orders of scaling. First-order scaling focuses on the substance of scaling, that is on specific policy innovations. This is most evident with respect to the discussions on best practice transfer. Second-order scaling refers to the processes that purposefully seek first-order scaling. This includes institutional innovations that influence scaling trajectories such as setting up transfer agencies. Finally, third-order scaling adds a reflexive perspective to scaling and refers to the principles of scaling (van Doren et al., 2018; van der Heijden, 2022).

Moreover, there are different types of scaling, ranging from horizontal relations and interactions (such as the coordination of a network of local climate managers) to vertical interdependent relations and interactions (such as guidelines and subsidy schemes) and hierarchical relations and interactions (such as local ordinances) (Kern, 2019). This means that "soft" governmental interventions may be sufficient for the scaling of local experiments in certain areas (such as the introduction of bus rapid transit), while "hard" governmental interventions may be needed in other areas (such as the management of parking spaces).

Therefore, the three types of scaling (horizontal, vertical, hierarchical) correspond to policy instruments, ranging from regulations and mandates and economic instruments to voluntary instruments such as certification schemes. As climate change policy is still a voluntary task in many local authorities worldwide, we find a "hardening" of soft instruments in local climate policy. Examples include Oslo's climate budget (Vedeld et al., 2021) and (binding) climate contracts between the national government and individual cities in Sweden (Viable Cities, 2022). This indicates that the causal mechanisms of scaling reach from emulation and learning to competition and coercion (van der Heijden, 2022).

Sustainability transformations require that local experiments spread within, beyond, and across cities. Scaling *within cities* refers to place-based experiments (pilot and demonstration projects) that may spread between neighborhoods in the same city (van Doren et al., 2020; van Winden and van der Buse, 2017) on a voluntary basis or become part of a city-wide strategy. Scaling *beyond cities* is often limited due to insufficient coordination and cooperation between city and region (e.g. bike infrastructure). Therefore, the scalability of local experiments often ends at city borders. Scaling *across cities* focuses on the spread of local experiments to cities and towns outside the metropolitan region. Scaling across cities is challenging due to contextual differences but also because internally and externally developed experiments need to be compatible (Wang and Bai, 2022).

23.5.2 City Types, Leadership, and Scaling

Furthermore, the potential for innovation and scaling differs between types of cities. There are clear disparities between forerunner, follower, and latecomer cities with respect to structural characteristics. This includes distinct types of forerunner cities, including pioneers with a predominantly internal orientation and externally oriented leaders striving for external reputation. Substantive leaders combine high internal with high external ambitions, while symbolic leaders cannot match their high external ambitions with high internal ambitions (Wurzel et al., 2019; see also Chapter 17).

Internal ambitions depend on structural preconditions for a more or less successful sustainability and climate policy: (1) socio-demographic characteristics (city size, growing/shrinking cities, average age of population, level of education); (2) economic structure and development (number of jobs in the service and green tech industry); (3) political and administrative characteristics (support for climate action, political influence of green parties); (4) ownership of public utilities and service companies (in particular energy, transport, and housing); (5) supporting research environments and city–university partnerships, in particular with research institutes that focus on climate issues; (6) civil society, especially environmental and climate action groups such as Fridays for Future (Zahran et al., 2008; Krause, 2011; Sharp et al., 2011; Bedsworth and Hanak, 2013; Fitzgerald & Lenhart, 2016; Hoppe et al., 2016; Homsy, 2018; Eckersley, 2018; Kern, 2019; Bery and Haddad, 2022; Kern et al., 2022).

External ambitions of cities are driven by efforts to become nationally and internationally known as a green, sustainable, and carbon-neutral city. Cities like Freiburg and Copenhagen have successfully managed to build up such a reputation. Indicators for such ambitions are the early engagement in city networks, frequent participation in national and international conferences, applications for awards such as the European Green Capital award, and fundraising activities (Eisenack and Roggero, 2022).

These internal and external factors are decisive for becoming a leader or laggard in sustainability and climate policy. Despite some overlap, in particular with respect to bigger cities, university cities show the best performance; historical cities face challenges but also enjoy advantages due to their urban form as compact cities; industrial cities are in a less favorable position, especially if they are still in an early stage of urban transformation (Haupt et al., 2022; Bery and Haddad, 2022).

There is still a research gap with respect to the different types of followers, latecomers, and laggards. It does not come as a surprise that studies have concentrated on forerunners because cities do not want to be labeled as latecomer or even laggards and are not willing to report on insufficient or even non-existent climate action. However, the relations between leaders and laggards may change over time because leadership may fade away and laggards may catch up with the forerunners. Here, six types of municipalities exist (Kern, 2022):

1. *Forerunners staying ahead*: active and internationally networked (larger) cities that were already active at an early stage, receive national and international attention, and start local experiments on a frequent basis;
2. *Followers catching up*: active cities seeking to catch up with the forerunners, adopt and adapt to the successful initiatives of forerunners, and become internationally active;
3. *Latecomers stepping in*: smaller cities and towns that have been passive, have little chance to catch up with the forerunners, but have launched their first initiatives due to the urgency of the issue;
4. *Stragglers falling behind*: cities that were active in the past but have fallen behind the forerunners, and only undertake incremental steps;
5. *Dropouts stepping out*: municipalities that began climate-policy initiatives which led to considerable local conflicts and ultimately to failure;
6. *Laggards staying behind*: the majority of smaller municipalities, in which climate mitigation and adaptation have never been on the political agenda, due to a lack of capacities.

Leadership requires agency within, beyond, and across cities. The agents of knowledge transfer include institutional actors (lead agencies within city administrations, regional planning agencies, city networks) and individual actors (mayors, climate managers, employees of regional planning agencies or city networks). Place-based leaders may become change agents by engaging in city-wide strategies, taking a new job in a regional planning agency or another city administration. Successful climate policies need appropriate leadership, which depends on the personality and professional orientation of leaders. Urban transformations require change agents, which have the ability to translate experiments into organizational changes, go beyond administrative routines and silo thinking, switch to strategic and design thinking instead, and engage in co-creational arrangements (Bery and Haddad 2022; Hofstad et al., 2022).

23.5.3 Matching Cities

Moreover, focusing on *matching cities* is a systematic approach to support scaling. It is a descriptive as well as a normative concept that starts with identifying the structural similarities (such as size, cultural heritage, national/regional context) of cities, for example mid-sized forerunner cities such as Turku (Finland), Groningen (the Netherlands), and Bern (Switzerland). However, this approach is not limited to analyzing forerunner cities but can also be used to study latecomer or laggard cities. It is based on the assumption that scaling works best among matching cities – in particular if the national context is identical. There are various dimensions to studying matching cities (Kern et al., 2022):

- *Comparison of matching cities.* Based on the structural characteristics of the selected cities and a comparative case-study approach, the strengths and weaknesses of their sustainability and climate policies are identified.
- *Assessment of the scaling potential of local experiments.* Since matching cities are not identical twins, they may differ with respect to their performance in specific policy areas (such as energy, transport, and green spaces). A city may lead in certain policy areas and lag behind in others. This enables them to learn from each other on a bilateral basis (first-order scaling).
- *Analysis of actual scaling between matching cities.* How matching cities exchange their knowledge and experiences, and adjust innovations to local conditions, is examined. This analysis is not limited to the transfer of existing best practice because matching cities have the potential to set up procedures that help them to co-design new solutions to emerging problems (second-order scaling).
- *Analysis of potential scaling between matching cities.* Emerging challenges that matching cities face in a similar way may require not only new policy instruments and procedures but also a change of the rules of the game (i.e., the underlying policy paradigm), which is a necessary precondition for successful urban transformations (third-order scaling).

23.6 Emerging Challenges for Urban Transformations

Although leading cities like Oslo have reduced their CO_2 emissions and are on a transformation pathway that may enable them to become climate neutral (Salvia et al., 2021; Melica et al., 2022), in many other cities CO_2 emissions have only marginally decreased or have even increased over time. Despite excellent climate strategies, changes on the ground follow incremental pathways when there is a lack of implementation due to insufficient local capacities (Otto et al., 2021; Irmisch et al., 2022). This is most evident in specific sectors such as transport and mobility where paradigmatic changes are urgently needed.

Cities and towns need indicators and to set up monitoring systems to measure their performance across all relevant sectors on a regular basis (Tyler et al., 2016; Barry and Hoyne, 2021). This applies not only to climate mitigation but also to climate adaptation policies. There is still a lack of indicators that can be used by different types of cities with varying vulnerabilities. Smaller cities and towns need harmonized monitoring tools that can be easily applied. Furthermore, there is a need for sharing information and knowledge within, beyond, and across cities. Digitalization offers chances for improving local climate policies, but it also leads to additional challenges due to the risk of a digital divide and the need to create digital platforms for creating, storing, and using big data. In many cities there is a lack of administrative reorganization which is needed to enable local governments to fulfil their tasks, especially in a situation of disruptive events.

Urban transformations require funding. Therefore, funding has become the main challenge for the implementation of climate strategies in many cities and towns (Deutsch-Französisches Zukunftswerk, 2022). Although wealthy cities may be able to build new infrastructure for becoming climate neutral and resilient, smaller cities do not have the necessary means to do so. This lack of funding limits the development and implementation of long-term climate strategies. It can be concluded that cities cannot save the planet on their own, they need support by national governments and international organizations.

23.7 Conclusion and Future Research

Since the first studies on local climate action appeared around 20 years ago, this topic has developed into a prolific research area. However, various research gaps can be identified. First, research tends to concentrate on leading cities, while "ordinary" cities are neglected. Thus, there is still a lack of knowledge on smaller cities and towns which heavily depend on national funding due to the lack of their own capacities. Second, the same applies to research on city networks that have drawn scholarly attention and stimulated many studies. Researchers prefer examining transnational city networks, such as ICLEI and C40, but overlook national and regional networks. Third, research on local sustainability and climate strategies needs to be complemented by studies on the actual implementation of such strategies on the ground. Fourth, there is still a lack of knowledge on the relationship between climate mitigation and adaptation at the local level, in particular with respect to strategies, organizational solutions, and project integration. Fifth, research on new governance arrangements and emerging climate policies, such as climate budgets and climate contracts between cities and national agencies, is still missing. Sixth, the relations between cities and their "hinterland" have hardly been studied, although sustainable, climate neutral, and resilient cities depend on their periphery. Finally, innovative methods which go beyond the state-of-the-art are needed because most research is based on established research methods (case studies, statistical analysis).

Box 23.1 Chapter Summary

- Since the Rio Conference in 1992 international organizations have supported sustainable development and climate policy at the local level.
- Leading cities have become players in global climate policy, but local climate action is not a panacea.
- As "ordinary cities" have less capacities than leading cities, they need support from national governments.
- Urban experiments may not lead to sustainability transformations but to "projectification".
- The decarbonization of cities and towns requires the scaling of urban experiments within, beyond, and across cities.

References

Barry, D. and S. Hoyne (2021): Sustainable Measurement Indicators to Assess impacts of Climate Change: Implications for the New Green Deal Era, *Current Opinion in Environmental Science and Health* 22: 100259. https://doi.org/10.1016/j.coesh.2021.100259

Bedsworth, L. and E. Hanak (2013): Climate Policy at the Local Level: Insights from California. *Global Environmental Change*, 23(3): 664–677. https://doi.org/10.1016/j.gloenvcha.2013.02.004

Berglund, O., Dunlop, C., Koebele, E. and C. Weible (2022): Transformational Change through Public Policy, *Policy and Politics* 50(3): 302–322. https://doi.org/10.1016/j.gloenvcha.2013.02.004

Bery, S. and M. Haddad (2022): Walking the Talk: Why Cities Adopt Ambitious Climate Action plans, Urban Affairs Review. https://doi.org/10.1177/10780874221098951

Covenant of Mayors (2022): *Covenant of Mayors – Europe.* https://eu-mayors.ec.europa.eu/en/home.

Dabrowski, M., Stead, D., He, J. and F. Yu (2021): Adaptive Capacity of the Pearl River Delta Cities in the Face of the Growing Flood Risk: Institutions, *Ideas and Interests, Urban Studies* 58(13): 2683–2702. https://doi.org/10.1177/0042098020951471

De Jong, M., S. Joss, D. Schraven, C. Zhan and M. Weijnen (2015): Sustainable–Smart–Resilient– low Carbon–Eco–Knowledge Cities: Making Sense of a Multitude of Concepts Promoting Sustainable Urbanization, *Journal of Cleaner Production* 109: 25–38. https://doi.org/10.1016/j.jclepro.2015.02.004

Deutsch-Französisches Zukunftswerk (2022): *Sozial-ökologische Transformation beschleunigen*, Deutsch-Französisches Zukunftswerk.

Eckersley, P. (2018): *Power and Capacity in Urban Climate Governance.* Peter Lang UK.

Eisenack, K. and M. Roggero (2022): Many Roads to Paris: Explaining urban climate action in 885 European cities, *Global Environmental Change* 72. https://doi.org/10.1016/j.gloenvcha.2021.102439

Evans, J., Vácha, T., Kok, H. and K. Watson (2021): How Cities Learn: From experimentation to transformation, *Urban Planning* 6(1): 171–182. https://doi.org/10.17645/up.v6i1.3545

Fisher, S. (2014): Exploring Nascent Climate Policies in Indian Cities: A Role for Policy Mobilities? *International Journal of Urban Sustainable Development* 6(2): 154–173. https://doi.org/10.1080/19463138.2014.892006

Fitzgerald, J. and J. Lenhart (2016): Eco-districts: can they accelerate urban climate planning? *Environment and Planning C: Government and Policy* 34(2): 364–380. https://doi.org/10.1177/0263774X15614666

Fu, Y. and X. Zhang (2017): Trajectory of Sustainability Concepts: a 35-year bibliometric analysis, *Cities* 60: 113–123. http://dx.doi.org/10.1016/j.cities.2016.08.003

Göpfert, C., Wamsler, C. and W. Lang (2019): A Framework for the Joint Institutionalization of Climate Change Mitigation and Adaptation in City Administrations, *Mitigation and Adaptation Strategies for Global Change* 24(1): 1–21. https://doi.org/10.1007/s11027-018-9789-9

Hall, P. (1993): Policy Paradigms, Social Learning, and the State: The Case of Economic Policymaking in Britain, *Comparative Politics* 25(3): 275–296. https://doi.org/10.2307/422246

Haupt, W., Chelleri, L., van Herk, S. and C. Zevenbergen (2020): City-to-city Learning within Climate City Networks: Definition, Significance, and Challenges from a Global Perspective. *International Journal of Urban Sustainable Development* 12(2): 143–159. https://doi.org/10.1080/19463138.2019.1691007

Haupt, W., Eckersley, P. and K. Kern (2021): How can "ordinary" cities become climate pioneers? In: Horwarth, C., Lane, M. and A. Slevin (eds.), *Addressing the Climate Crisis: Local Action in Theory and Practice* (Cham: Palgrave Macmillan), pp. 83–92. https://doi.org/10.1007/978-3-030-79739-3-8

Haupt, W., Eckersley, P., Irmisch, J. and K. Kern (2022): How Do Local Factors Shape Transformation Pathways towards Climate-neutral and Resilient Cities? *European Planning Studies*. https://doi.org/10.1080/09654313.2022.2147394

Hofstad, H., Vedeld, T., Agger, A., Hansen, G., Tonnesen, A. and S. Valencia (2022): Cites as Public Agents: A Typology of Co-Creational Leadership for Urban Climate Transformation, *Earth System Governance* 13. https://doi.org/10.1016/j.esg.2022.100146

Homsy, G. (2018): Unlikely Pioneers: Creative Climate Change Policymaking in Smaller U.S. Cities, *Journal of Environmental Studies and Sciences* 8(2): 121–131. https://doi.org/10.1007/s13412-018-0483-8

Hoppe, T., van der Vegt, A. and P. Stegmaier (2016): Presenting a Framework to Analyze Local Climate Policy and Action in Small and Medium-Sized Cities. *Sustainability* 8(9), 847. https://doi.org/10.3390/su8090847

Huovila, A., Siikavirta, H., Antuña Rozado, C., Rökman, J., Tuominen, P., Paiho, S., Hedman, Å. and P. Ylén (2022): Carbon-neutral Cities: Critical Review of Theory and Practice, *Journal of Cleaner Production* 341: 130912. https://doi.org/10.1016/j.jclepro.2022.130912

Intergovernmental Panel on Climate Change (IPCC) (2021): Climate Change 2021. The Physical Science Basis. Working Group I Contribution to the IPCC Sixth Assessment Report, Intergovernmental Panel on Climate Change. https://www.ipcc.ch/report/ar6/wg1/

Intergovernmental Panel on Climate Change (IPCC) (2022): Climate Change 2022. Impacts, Adaptation and Vulnerability. Summery for Policymakers. Working Group II Contribution to the IPCC Sixth Assessment Report, Intergovernmental Panel on Climate Change.

Irmisch, J., Haupt, W., Eckersley, P., Kern, K. and H. Müller (2022): *Klimapolitische Entwicklungspfade deutscher Groß- und Mittelstädte, IRS Dialog, No. 2/2022*, Leibniz-Institut für Raumbezogene Sozialforschung (IRS), Erkner.

Karhinen, S., Peltomaa, J., Riekkinen, V. and L. Saikku (2021): Impact of a Climate Network: The Role of Intermediaries in Local Level Climate Action, *Global Environmental Change* 67. https://doi.org/10.1016/j.gloenvcha.2021.102225

Kern, K. (2019): Cities as Leaders in EU multilevel Climate Governance: Embedded Upscaling of Local Experiments in Europe, *Environmental Politics* 28(1): 125–145. https://doi.org/10.1080/09644016.2019.1521979

Kern, K. (2022): *Cities in EU Multilevel Climate Policy: governance capacities, spatial approaches and upscaling local experiments*, In: Rayner, T., Szulecki, K., Jordan, A. and S. Oberthür, Handbook on European Union Climate Change Policy and Politics, Edward Elgar (forthcoming).

Kern, K. and G. Alber (2009): Governing Climate Change in Cities. Modes of Urban Climate Governance in Multi-level Systems, in: *Competitive Cities and Climate Change, OECD Conference Proceedings*, Milan, Italy, 9–10 October 2008, Chapter 8, Paris: OECD, pp. 171–196.

Kern, K. and H. Bulkeley (2009): Cities, Europeanization and Multi-level Governance: Governing Climate Change through Transnational Municipal Networks, *Journal of Common Market Studies* 47(2): 309–333. https://doi.org/10.1111/j.1468-5965.2009.00806.x

Kern, K., Grönholm, S., Haupt, W. and L. Hopman (2022): Matching Forerunner Cities: Climate policy in Turku, Groningen, Rostock, and Potsdam, *Review of Policy Research* forthcoming.

Kern, K., Haupt, W. and S. Niederhafner (2021): Entwicklungspfade städtischer Klimapolitik. Bedeutung von Schlüsselereignissen und Schlüsselakteur:innen für die Klimapolitik in Potsdam, Remscheid und Würzburg, *disP - The Planning Review* 57(4): 32–49. https://doi.org/10.1080/02513625.2021.2060576

Kern, K., Koll, C. and M. Schophaus (2007): The diffusion of Local Agenda 21 in Germany: Comparing the German Federal States, *Environmental Politics* 16(4): 604–624. https://doi.org/10.1080/0964401070 1419139

Krause, R. (2011): Policy Innovation, Intergovernmental Relations, and the Adoption of Climate Protection Initiatives by U.S. Cities, Journal of Urban Affairs, 33(1): 45–60. http://dx.doi.org/10.1111/j.1467-9906.2010.00510.x

Krause, R. and C. Hawkins (2021, June): Viewpoints: Improving Cities' Implementation of Sustainability Objectives, *Cities* 113, 32–43.

Lafferty, W. (ed.) (1999): *Implementing LA21 in Europe. New Initiatives for Sustainable Communities* (Oslo: ProSus).

Melica, G., Treville, A., De Los, Franco Rios, C., Baldi, M., Monforti-Ferrario, F., Palermo, V., Ulpiani, G., Ortega Hortelano, A., Lo Vullo, E., Barbosa, Marinho Ferreira, and P. Bertoldi (2022): *Covenant of Mayors: 2021 Assessment, Publications Office of the European Union*, Luxembourg, 2022. http://dx.doi.org/10.2760/58412

Otto, A., Kern, K., Haupt, W., Eckersley, P. and A. Thieken (2021): Ranking Local Climate Policy: Assessing the Mitigation and Adaptation Activities of 104 German cities, *Climatic Change* 167: 5. https://doi.org/10.1007/s10584-021-03142-9

Reckien, D. et al. (2018): How Are Cities Planning to Respond to Climate Change? Assessment of Local Climate Plans from 885 Cities in the EU-28, *Journal of Cleaner Production* 191: 207–219. https://doi.org/10.1016/j.jclepro.2018.03.220

Revi, A., Satterthwaite, D., Aragón-Durand, F., Corfee-Morlot, J. … W. Solecki (2014): Urban Areas. In: *Climate Change 2014: Impacts, Adaptation, and Vulnerability. Part A: Global and Sectoral Aspects. Contribution of Working Group II to the Fifth Assessment Report of the Intergovernmental Panel on Climate Change*, Cambridge University Press, Cambridge (UK) and New York (USA), pp. 535–612.

Salvia, M. et al. (2021): Will Climate Mitigation Ambitions Lead to Carbon Neutrality? An Analysis of the Local-Level Plans of 327 Cities in the EU, *Renewable and Sustainable Energy Reviews* 135. https://doi.org/10.1016/j.rser.2020.110253

Schraven, D., Joss, S. and M. de Jong (2021): Past, Present, Future: Engagement with Sustainable Urban Development Through 35 City Labels in the Scientific Literature 1990–2019, *Journal of Cleaner Production* 292. https://doi.org/10.1016/j.jclepro.2021.125924

Sharifi, A. (2021): Co-benefits and Synergies between Urban Climate Change Mitigation and Adaptation Measures: A literature review, *Science of the Total Environment* 750: 141642. http://dx.doi.org/10.1016/j.scitotenv.2020.141642

Sharp, E., Daley, D. and M. Lynch (2011): Understanding Local Adoption and Implementation of Climate Change Mitigation Policy, *Urban Affairs Review* 47(3): 433–457. https://doi.org/10.1177/1078087410392348

Stehle, F., Hickmann, T. Lederer, M. and C. Höhne (2020): Urban Climate Politics in Emerging Economies: A multi-level governance perspective, *Urbanizations*. https://doi.org/10.1177/2455747120913185

Tyler, S., Nugraha, E., Nguyen, H.K., Nguyen, N.V., Sari, A.D., Thinpanga, P., Tran, T. T. and S.S. Verma (2016): Indicators of Urban Climate Resilience: A contextual approach, *Environmental Science and Policy* 66: 420–426. https://doi.org/10.1016/J.ENVSCI.2016.08.004

Torrens, J. and T. von Wirth (2021): Experimentation or Projectification of Urban Change? A critical appraisal and three steps forward, *Urban Transformations* 3(8), 112–132. https://doi.org/10.1186/s42854-021-00025-1

United Nations (2015): Transforming our World: the 2030 Agenda for Sustainable Development, Resolution 70/1, adopted by the General Assembly on 25 September 2015.

UN-Habitat (2022): World Cities Report, United Nations Human Settlements Programme.

Van der Heijden, J. (2019): Studying Urban Climate Governance: Where to begin, what to look for, and how to make a meaningful contribution to scholarship and practice. Earth System Governance 1, 212–224. https://doi.org/10.1016/j.esg.2019.100005

Van der Heijden, J. (2022): Towards a Science of Scaling for Urban Climate Action and Governance, European Journal of Risk Regulation, 67–98. https://doi.org/10.1017/err.2022.13

Van Doren, D., Driessen, P. and M. Giezen (2018): Scaling-up low-carbon urban initiatives: Towards a better understanding, Urban Studies 55(1): 175–194. https://doi.org/10.1177/0042098016640456

Van Doren, D., Driessen, P., Runhaar, H. and M. Giezen (2020): Learning within Local Government to Promote the Scaling-up of Low-carbon Initiatives: A case study in the City of Copenhagen, Energy Policy 136, January 2020:111030. https://doi.org/10.1016/j.enpol.2019.111030

Van Winden, W. and D. van der Buse (2017): Smart City Pilot Projects: explaining the dimensions and conditions of scaling up, Journal of Urban Technology 24(4): 51–72. https://doi.org/10.1080/10630732.2017.1348884

Vedeld, T., Hofstad, H., Solli, H. and G. Hansen (2021): Polycentric Urban Climate Governance: creating synergies between integrative and interactive governance in Oslo, Environmental Policy and Governance. https://doi.org/10.1002/eet.1935

Viable Cities (2022): Together towards Climate-neutral Cities. https://en.viablecities.se/

Wang, S. and X. Bai (2022): Compatibility in Cross-city Innovation Transfer: Importance of Existing Local Experiments, Environmental Innovation and Social Transitions 45: 52–71. https://doi.org/10.1016/j.eist.2022.09.003

World Bank (2005): Reducing Poverty, Sustaining Growth: Scaling Up Poverty Reduction. Case Study Summaries, Conference in Shanghai, May 25–27, 2004.

Wurzel, R., Liefferink, D. and D. Torney (2019): Pioneers, Leaders and Followers in Multilevel and Polycentric Climate Governance, Environmental Politics 28(1): 1–21. https://doi.org/10.1080/09644016.2019.1522033

Zahran, S., Brody, S., Vedlitz, A., Grover, H. and C. Miller (2008): Vulnerability and Capacity: Explaining Local Commitment to Climate-Change Policy. Environment and Planning C: Government and Policy, 26(3): 544–562. https://doi.org/10.1068/c2g

24

POLICY MIXES FOR ADDRESSING ENVIRONMENTAL CHALLENGES

Conceptual Foundations, Empirical Operationalisation, and Policy Implications

Karoline S. Rogge and Qi Song

24.1 Introduction

The systemic nature of today's complex and often interconnected environmental challenges, such as the climate crisis or biodiversity loss, and their urgency require multi-faceted but often politically contested policy interventions that go beyond the realm of environmental policy and single policy instruments. This calls for the combination of policies addressing these environmental challenges by coordinating interventions across multiple policy fields, governance levels, and socio-ecological systems. For such a broad perspective on policy mixes – as a combination of policy strategies and instruments designed and implemented across multiple dimensions – we draw on the transitions literature, which has argued that effective policy mixes need to aim at the transformation of our existing unsustainable systems of production and consumption (Rogge and Reichardt, 2016). Such transformation is also referred to as "system innovation" and can be stimulated through policy mixes for "creative destruction" (Kivimaa and Kern, 2016): on the one hand, adopting policies supporting the creation of novel solutions – be they technological, social, or business model innovations, for example – and on the other hand further accelerating transformation processes by simultaneously implementing policies that phase out unsustainable fuels, technologies, or practices. For this, much can be learned from the literature on policy mixes for sustainability transitions which will be introduced in this chapter.

Over the past two decades increasing attention has been given to the role that policy mixes play in promoting environmental innovation and sustainability transitions. However, research initially has focused on policy mixes in a narrow sense, namely on the interaction between different policy instruments and their optimal combination in instrument mixes (Bouma et al.,

DOI: 10.4324/9781003043843-28

2019; Kern et al., 2019). In the last decade, increasing attention has been given to the role that policy mixes – also referred to as policy packages or policy portfolios – play in sustainability transitions in different sectors, such as in energy (Rogge et al., 2017), transport (Givoni et al., 2013), industry (Scordato et al., 2018), agri-food (Kalfagianni and Kuik, 2017), or forestry (Scullion et al., 2016). This newer line of interdisciplinary policy mix thinking for addressing environmental and sustainability challenges combines insights from various disciplines (Kern et al., 2019; Quitzow, 2015), in particular environmental economics and policy (Braathen, 2007; Lehmann, 2012), but also policy sciences (Capano and Howlett, 2020; Howlett et al., 2015) and innovation and transition studies (Flanagan et al., 2011; Rogge and Reichardt, 2016), with the latter two complementing the predominant analysis of instrument interactions with other important aspects of policy mixes. It is this emerging interdisciplinary literature on broader policy mixes for transitions towards more sustainable systems of production and consumption that is the focus of this chapter.

The formulation and implementation of policy mixes for addressing environmental and sustainability challenges can be justified by multiple policy rationales (Bouma et al., 2019; Weber and Rohracher, 2012). To be specific, the interdisciplinary policy mix literature stresses that *market failures*, such as the negative externalities of greenhouse gas emissions (Edenhofer et al., 2013) and other externalities (Lehmann et al., 2019), are an important but not the only justification for policy intervention (Jacobsson et al., 2017). Instead, the development of environmentally friendly solutions also requires an awareness of structural as well as transformational system failures (Weber and Rohracher, 2012).

That is, it is not seen as sufficient to internalise environmental externalities associated with environmental challenges. Instead, effective policy mixes also need to tackle *structural system failures* associated with environmental innovation and investment, such as failures in building up low-carbon infrastructure (e.g. aligning electricity grids and storage with the requirements of new low-carbon technology) and in adjusting existing institutions to sustainable solutions (e.g. reforming electricity market designs) (Bak et al., 2017; Patt and Lilliestam, 2018).

In addition, addressing environmental and sustainability challenges and thus reaching policy commitments tackling them, such as the Paris Agreement for limiting climate change, also requires that *transformational system failures* are considered when designing such policy mixes. These include, for example, the provision of a clear direction (e.g. through the elaboration of shared visions, unambiguous guidance for sustainable solutions, and coordination of actors involved in the transformation process). Transformational system failures, to give another example, also include the need to overcome policy silos through better coordination across policy fields (e.g. environmental policy and industrial policy) and governance levels (e.g. the national and regional level) (Nemet et al., 2017; Uyarra et al., 2016). Single policy instruments, such as carbon pricing or a pollution tax, are not able to address all of these failures, thereby justifying the implementation of broader policy mixes (Bouma et al., 2019; del Río, 2017; Tvinnereim and Mehling, 2018).

In this chapter, we first provide an overview of how policy mixes have been defined and how this varies across different disciplinary perspectives, followed by an introduction to an emerging interdisciplinary understanding of broader policy mixes (*conceptualising policy mixes*). We then introduce two methodological approaches for delineating policy mixes addressing environmental challenges *(delineating policy mixes)*. Thereafter, we provide a synthesis of empirical policy mix insights and their policy implications for the concrete example of climate policy mixes (*informing policy mix design*). Finally, we discuss areas that warrant more attention in future research on policy mixes addressing environmental challenges (*research outlook*).

24.2 Conceptualising Policy Mixes

In this section we first provide an overview of the different definitions of policy mixes in the three main disciplines that have significantly contributed to policy mix research. Based on this, we then introduce an interdisciplinary framework for a broader policy mix concept that we find particularly suitable for guiding future research on policy mixes addressing environmental challenges.

24.2.1 Defining Policy Mixes

The ambiguous meaning of the term "policy mix" has raised a series of challenges for researchers in identifying the scope and focus of their research. In this section, we therefore start by providing an overview of key policy mix definitions applied in different disciplines, namely environmental economics, policy studies, and innovation and transition studies (see Table 24.1).

This overview shows that policy mix definitions reflect the different research foci present in the different disciplines, but also that definitions vary within and across disciplines. First, research grounded in *environmental economics* has been largely focusing on instrument interactions and the design of optimal instrument mixes (Braathen, 2007; Lehmann, 2012). In that sense, this line of research would more accurately be better referred to as instrument mix research. A prime example of this are studies investigating the interaction effect of emissions trading and other policy instruments such as support for renewables (Sorrell and Sijm, 2003). Second, research in *policy studies* typically defines policy mixes as a combination of multiple instruments and goals (Howlett and Rayner, 2013; Kern and Howlett, 2009), and has been largely focusing on the sequencing and evolution of policy mixes over time (Howlett, 2009; Taeihagh et al., 2013). In doing so, policy scholars have often characterised the consistency and coherence of the resulting policy mixes, while, however, neglecting policy mix impacts.

Table 24.1 Three main disciplines addressing policy mixes with exemplary definitions

	Environmental economics	*Policy studies*	*Innovation and transition studies*
Example 1	Instrument mixes are defined as a situation in which "several – instead of one – policy instruments are used to address a particular environmental problem" (Braathen, 2007, p. 186).	"Policy mixes are complex arrangements of multiple goals and means which, in many cases, have developed incrementally over many years" (Kern and Howlett, 2009, p. 395).	"A policy mix is defined as: The combination of policy instruments, which interact to influence the quantity and quality of R&D investments in public and private sectors" (Cunningham et al., 2009, p. 3).
Example 2	"Polluting sources may be affected directly or indirectly by several policies addressing the same pollution problem. This is referred to as a policy mix" (Lehmann, 2012, p. 71).	"Policy mixes or portfolios feature the use of combinations of different kinds of policy tool – market based, hierarchical, network and others – whose exact configuration changes from location to location" (Rayner et al., 2017, p. 473).	"Policy mixes favourable to sustainability transitions need to involve both policies aiming for the 'creation' of new and for 'destroying' (or withdrawing support for) the old" (Kivimaa and Kern, 2016, p. 206).

Source: Own compilation of policy mix definitions using matrix logic from Rogge et al. (2017).

Finally, *innovation and transition studies* have focused initially on instrument mixes promoting technological change; but, particularly in studies investigating sustainability transitions, it has extended its scope to policy mixes promoting system innovation (Kivimaa and Kern, 2016). In addition, and regardless of the disciplinary grounding, policy mix studies have applied a variety of definitions of different aspects of policy mixes (Rogge and Reichardt, 2016), for example regarding the terms "consistency" or "coherence" of policy mixes. Such conceptual and terminological diversity complicates the synthesis of insights from policy mix research, but provides a rich foundation for an interdisciplinary conceptualisation of the term "policy mix" that combines the strengths of the respective approaches.

24.2.2 Interdisciplinary Conceptual Framework

Such an extended interdisciplinary framework for analysing complex policy mixes for environmental innovation and sustainability transitions, which builds on all three disciplinary traditions, has been proposed by Rogge and Reichardt (2016). Their broader conceptualisation not only offers a comprehensive and clearly defined policy mix concept but also provides resourceful guidance for developing future empirical studies. According to their work, a policy mix refers to "a combination of the three building blocks elements, processes and characteristics, which can be specified using different dimensions" (Rogge and Reichardt, 2016, p. 1622). Their conceptualisation has integrated most conceptual advances from the earlier policy mix literature: inspired by innovation studies it highlights long-term strategies by introducing policy strategy as one policy mix element; drawing on policy studies it captures associated policy processes and different types of policy mix characteristics; and building on environmental economics it also captures the importance of policy instrument design.

As Figure 24.1 shows, the first building block of this extended policy mix concept – policy mix *elements* – incorporates both policy strategies and instrument mixes articulated and introduced by associated governing entities in addressing specific policy challenges and fulfilling certain policy functions (Rogge and Reichardt, 2016). First, considering the role of the

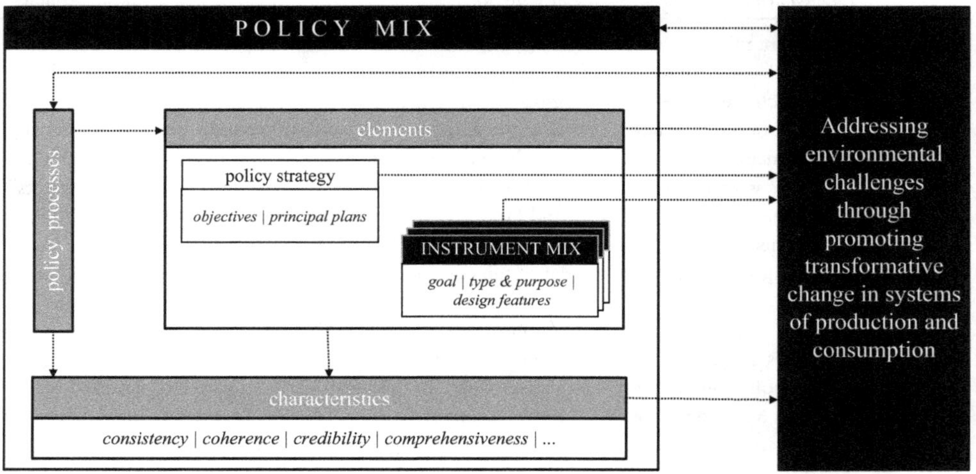

Figure 24.1 Implications of extended conceptual framework for investigating policy mixes for environmental challenges.

Source: Adapted from Rogge and Reichardt (2016) and Rogge (2019).

long-term horizon in designing real-world transformative policy solutions (Voß et al., 2009) underlines the importance assigned to policy strategies as the first part of this building block. Such policy strategies include a set of policy objectives – clarified by quantifiable targets – and the principal plans for materialising them. Second, various types of policy instruments, associated with specific goals and different design features, and their combinations in instrument mixes, are conceptualised as the second part within this building block. This implies that the conventional focus on instrument interactions is considered to be only one of several aspects relevant to the analysis of policy mix elements (Rogge and Reichardt, 2016). As a result, the extended policy mix framework offers researchers a more precise guidance on how to delineate the scope of complex policy mixes composed of both strategic and instrumental components. Therefore, the broader policy mix concept also explicitly distinguishes the two previously overlapping conceptual constructs of "policy mix" and "instrument mix".

Drawing on intellectual discussions within policy studies, the second building block – *policy processes* – suggests more dedicated attention to the associated policy processes that shape the concrete content of the policy mix elements (Rogge and Reichardt, 2016). Given the existence of controversial political dynamics (Meadowcroft, 2009) and complex power relationships (Stirling, 2014) observed in developing transformative policy solutions addressing environmental problems, the introduction of this building block enables researchers to critically examine the role of various actors with divergent beliefs, expectations, and interests in formulating and implementing policy strategies and instruments. The black-boxed decision-making mechanisms embedded in governing complex environmental issues, therefore, are invited to be unpacked by mobilising well-developed policy process theories and perspectives for interdisciplinary investigations of the politics and policies involved in governing sustainability transitions (Kern and Rogge, 2018). One example of such interdisciplinary policy mix analysis concerns the co-evolution of policy mix change and socio-technical change (Edmondson et al., 2019) that enable insights into the systemic dynamics and causal links across policy subsystems and the socio-technical systems.

Moreover, the third building block – policy mix *characteristics* – captures various attributes of the focal policy mix, specifying the nature of policy mix elements and the features of associated policy processes for informing. These characteristics are meant to describe and evaluate the design and impact of complex, real-world policy mixes (Rogge and Reichardt, 2016), and in the following we introduce four of them. First, the consistency of policy mix elements captures the degree of contradictions or synergies across three different but interconnected levels: within policy strategies, within the instrument mix, and between strategies and instruments. Second, recognising the role of policy coordination and policy integration in dealing with cross-sectoral policy challenges (Magro et al., 2014; Trein et al., 2021), the coherence of policy processes is also included in this third building block to underline the analytical importance of synergistic and systematic policy formulation and implementation processes for understanding the performance of real-world policy mixes (Rogge and Reichardt, 2016). Third, the credibility of policy mixes is also highlighted by the extended policy mix framework as another core characteristic which is considering how well target groups believe and trust in the policy mix – a critical factor for addressing environmentally related policy issues such as climate change (Nemet et al., 2017). Finally, a fourth policy mix characteristic that has received significant attention is the comprehensiveness of policy mixes. It captures the existence of different policy instrument types (e.g. economic, regulatory, and information tools, identified environmental policy researchers) and instrument purposes (e.g. technology-push, demand-pull and systemic concerns, highlighted by innovation scholars), as well as the extensiveness of actor involvements in associated policy processes (Rogge and Reichardt, 2016).

24.3 Delineating Policy Mixes

After having introduced the relevant terms and conceptual linkages of a broader policy mix concept, in this section we provide an overview of how to measure such real-world policy mixes. Indeed, despite the emergence of an increasing number of studies investigating real-world policy mixes (Kern et al., 2019), the absence of widely recognised standards for operationalising such mixes hampers the synthesis of the current empirical evidence. This section therefore introduces two methodological approaches – the top-down approach and the bottom-up approach – for delineating policy mixes, thereby providing consistent guidance for best practices in conducting empirical research programmes in this area (Ossenbrink et al., 2019).

24.3.1 The Top-Down Approach

Viewing the elements of focal policy mixes, implicitly or explicitly, as the outputs of associated policy processes navigated by a set of key governing entities at certain governance levels, for the top-down approach the identification of the overarching strategic intent is the key analytical clue to identifying relevant policy strategies and instruments (Ossenbrink et al., 2019). Following this approach and applying it to environmental challenges, the empirical investigation of a focal policy mix should start by scoping the stated environmental goals, for example in terms of reducing greenhouse gas emissions. Analysts then need to specify which governance level(s) and policy field(s) to include in their analysis of the focal policy mix, and for these identify the relevant governing entities. Depending on the research question this could be, for example, the environmental ministry at the national level, but could also be extended to include environmental ministries at other governance levels, or even might be extended to further policy fields if involved in pursuing the particular environmental goal. Analysts will then identify the corresponding policy strategies and supporting policy instruments implemented by the identified governing entities, typically by utilising publicly available data from secondary sources, such as agency and ministerial publications, program reports, and regulatory and legislative documents (Howlett et al., 2006). Although more and more online field-specific policy databases and observatories facilitate the mapping of relevant policy mix elements (Meissner and Kergroach, 2021), field-specific knowledge or professional expertise will still be important for such top-down policy mix mappings, for example in compiling the list of relevant governing entities, in formulating a string for keyword-based policy document searches, or in developing a theory-informed codebook for documentary analysis.

The availability of in-depth reports and strategic documents collected from these secondary sources can help researchers to obtain a quick overview of the focal policy mix. As Ossenbrink et al. (2019) note, some high-quality policy or industry reports may already have provided a detailed list of core public agencies, milestone policy documents, and key events in their focused policy issues, and researchers thus can use the insights from such summarising documents to lay the foundation for their own empirical analysis. In some cases, researchers can start systemic mapping work by analysing formalised strategy documents in the focal policy field(s), as these documents "not only articulate a set of policy objectives but typically also define a set of governmental actors" (Quitzow, 2015, p. 237). For instance, in their research concerning sustainable energy transitions in China, Li and Taeihagh (2020) capture the strategic elements of the focal policy mix (i.e. policy objectives and principal plans) by scoping all energy-relevant statements in China's Five-Year Plans – one of the most influential strategic document series at the national level, which also informs the subsequent keyword-based search for identifying specific policy instruments and their mixes supporting the aforementioned strategic elements.

In order to validate and complement these secondary data sources, expert interviews with competent officials and policy elites could be conducted to refine preliminary findings on the composition of the policy mix derived from the desktop analysis (Ossenbrink et al., 2019). Such expert interviews, in this case, can also help analysts to understand underlying policy formulation and implementation processes, or policy mix impacts. That is, such interviews can also go beyond the pure mapping of policy mix elements and already include further analytical steps, thereby enabling researchers to gain empirical evidence with respect to explaining complex political dynamics and causal mechanisms. In this line, as the example provided by Xu and Su (2016) shows, expert interviews with elite policy actors not only help researchers to obtain professional accounts of the rationale and dynamics of observed policy changes, but can also provide a valuable opportunity to access internal documents and confidential transcriptions in revealing invisible background information, which becomes particularly relevant for a comprehensive policy mix analysis which incorporates not only policy outputs but also the policy-making processes leading to them, as well as policy mix impacts.

24.3.2 The Bottom-Up Approach

In contrast, the bottom-up approach to delineating policy mixes pays more attention to the impact of a policy mix, which highlights the analytical importance of identifying the specific impact domain influenced by policy instruments and strategies (Ossenbrink et al., 2019). Here, it is thus key to start with the identification of a well-bounded focal impact domain, such as the diffusion of an environmentally friendly technology in a given country (e.g. electric vehicles as part of transitions to e-mobility). For the focal impact domain, scholars then need to identify relevant actors and all policy instruments that influence their activities throughout the bottom-up data collection and analysis process. That is, one can regard these actors – in the selected geographical scope – as the recipients of policy instruments from potentially various governance levels and policy fields, and thus use their accounts to sketch the big picture of the relevant focal policy mix (Ossenbrink et al., 2019). Perhaps most important for this bottom-up approach is that the role of some unintentional policy impacts of instruments from other policy fields or governance levels can be captured by researchers, shining a light on the layering structure of focal policy mixes and helping to identify otherwise potentially overlooked but relevant policies (Kern et al., 2017; Sovacool, 2009). However, as Ossenbrink et al. (2019) point out, the inductive nature of this bottom-up approach requires considerable research efforts in collecting and synthesising diverse insights from multiple relevant actors, so several research design issues, such as case selection rationales and the availability of research resources, need to be critically considered.

Defining the focal impact domain is a critical and challenging step when applying the bottom-up approach. Ideally, scholars should first identify the environmental, social, technological, and/or economic dimensions of interest. In this regard, the narrower these dimensions can be defined, the simpler data collection and analysis will be. As an illustrative case, Ossenbrink et al. (2019) define their impact domain as the economic dimension of energy storage in California's residential photovoltaic self-consumption, which is much narrower than the mentioned strategic intent of their top-down approach (i.e. the policy mix for energy storage in California). One key reason for formulating narrow impact domains is the limited resources that researchers can typically mobilise for conducting their projects. Indeed, if the definition of the focal impact domains is too broad, there would be very long lists of massive actors on the table, which are likely not possible for researchers to comprehensively cover in their data collection and analysis. In addition, the complexity of real-world policy mixes will be more

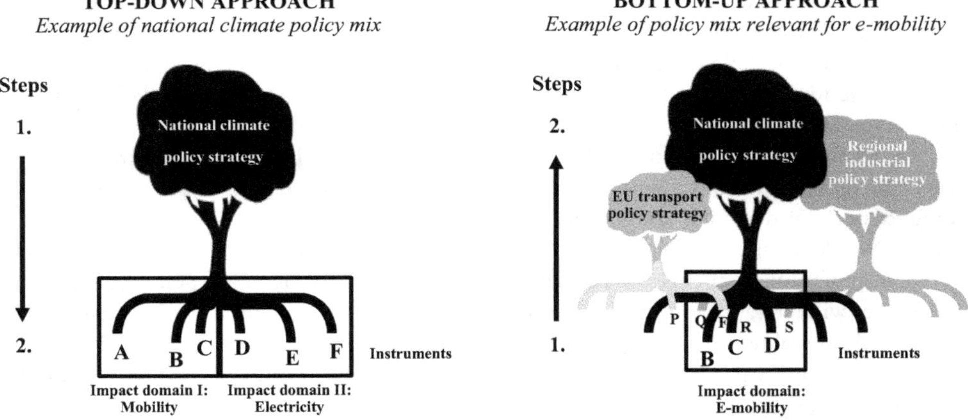

Figure 24.2 Illustration of top-down and bottom-up approach to delineating policy mixes.

Note: The figure shows that the relationship between policy mixes (tree, with the treetop = strategy, and roots = instruments) and impact domains (soil) is not one-to-one, which means as illustrated on the left, a given policy mix (e.g. the black one) may, intendedly or unintendedly, affect multiple impact domains (e.g. I and II), and as illustrated on the right, a given impact domain (e.g. e-mobility) can be affected by the elements of multiple policy mixes (e.g. the grey, black, and blue ones, though in reality these will typically be many more) – here the trees represent several policy fields (e.g. climate, industrial, and transport) and different governance levels (e.g. EU, national, regional) influencing the impact domain.

Source: Adapted from Ossenbrink et al. (2019).

challenging to handle, the larger the system boundaries are. However, it is also true that in times of increasingly interconnected systems researchers and policymakers alike might have to engage with multi-system policy mixes, and such consideration of complex policy mixes may enable the identification of synergies and conflicts of key relevance for policy mix effects. For example, and as visualised in Figure 24.2 (left-hand side), research investigating the decarbonisation of the transport sector through its electrification could take as a top-down starting point the climate policy strategy, but zoom in on the mobility and electricity system as main impact domains. Alternatively, following the bottom-up approach researchers could choose e-mobility as the impact domain of interest (or even more narrowly, electric vehicles), and then map all policy instruments that influence actors in this domain. As visualised in Figure 24.2 (right-hand side), this may lead to the identification of policy mix elements at different governance levels (EU, national, regional) and different policy fields (climate, industrial, and transport policy).

Regardless of approach, policy data collection and analysis should be seen as an iterative process in which it is beneficial to combine both secondary data (e.g. policy documents and academic literature) and primary data (e.g. interviews and participant observations). Considering the relevant actor networks in selected geographical scales, scoping reviews of the existing academic and grey literature can help researchers in terms of identifying representatives of each actor group. Snowballing techniques, moreover, should also be considered as a complementary method for finding relevant, but previously overlooked, actors. The collection and classification of policy instruments – which then can be traced back to relevant policy strategies at various governance levels and policy fields – could also be informed by existing theories at the initial stage (e.g. Kivimaa et al., 2017; Kivimaa and Kern, 2016).

Based on the guideline proposed by Ossenbrink et al. (2019), this section has introduced two archetypical analytical procedures for delineating policy mixes in line with the relevant research question. However, as Ossenbrink et al. (2019) also suggest, there is no best solution targeting all policy mix studies across different research contexts, and researchers need to find a way that fits their own research projects with their unique research questions. Moreover, the two approaches are complementary to each other, and when combined enable comprehensive insights. Yet such a combination does require significant time and effort and thus may not be possible in many instances, when it will be more important to choose the most appropriate delineation approach to proceed with for further empirical inquiries.

24.4 Informing Policy Mix Design: The Example of Climate Change

Having discussed terminology and delineation of policy mixes, we now turn to empirical insights gained on policy mix design and its impacts regarding addressing environmental challenges. As most policy mix studies have so far focused on tackling climate change, in this section we synthesise the main insights gained from the literature investigating broader climate policy mixes and their relevance for sustainability. In total, we derive seven findings with real-world policy relevance. Each of these insights are summarised in one sentence, and further elaborated in the following paragraph.

24.4.1 Coordinating the Design of Policy Mixes to Meet Multiple Policy Objectives Can Reduce the Overall Costs of Achieving Sustainability Objectives

Governments typically pursue multiple policy objectives beyond greenhouse gas mitigation, such as energy security, air quality, health, or energy access. The existence of such multiple policy objectives provides a rationale for coordination in policy mix design, as it allows policymakers to strive for synergies and to minimise trade-offs (Howlett and del Rio, 2015; Obersteiner et al., 2016). Integrated model studies suggest that well-designed transformative climate policy mixes paying attention to the co-benefits of climate mitigation for non-climate policy objectives can reduce the overall cost of achieving multiple sustainability objectives (von Stechow et al., 2015).

24.4.2 Climate Policy Mixes Need to Be Credible to Accelerate Low-Carbon Transitions

Long-term targets are an important element of climate policy mixes as they provide guidance to strategic investments and innovation (Schmidt et al., 2012). However, to be credible and effective they need to be backed up by stringent and consistent policy instruments (Rogge and Schleich, 2018). Given the outstanding importance of policy credibility for low-carbon investment and innovation shown in modelling studies (Bosetti and Victor, 2011; Faehn and Isaksen, 2016), several attributes have been identified in the literature to assess the extent to which policy mixes are believable and reliable (Jakob, 2017; Nemet et al., 2017): the design of rules (e.g. Are targets reviewed periodically?), transparency and trust (e.g. Does an independent authority oversee target achievement?), political economy and distribution (e.g. Are policies compensating losers of stringent climate policy?), and robustness (e.g. Are multiple policy instruments in place, potentially also at different governance levels?). Empirical evidence for Germany demonstrates that companies' perceptions of the credibility of the policy mix relevant for renewables can be linked not only to the existence of well-aligned instruments but also to the coherence of climate policy-making processes and the existence of ambitious phase-out policies for societally undesirable energy technologies (Rogge and Schleich, 2018). The literature thus suggests multiple avenues for enhancing climate policy credibility, which is key for accelerating low-carbon transitions.

24.4.3 Comprehensive, Balanced, and Consistent Instrument Mixes Can Help Drive Low-Carbon Transformative Change

The interdisciplinary literature on policy mixes points to the importance of evaluating policy mixes through their characteristics, such as comprehensiveness (capturing the extensiveness of policy mixes, e.g. in terms of whether a policy mix addresses all market and system failures), balance (capturing whether policy support is balanced between different instrument purposes), and consistency (capturing the alignment of policy instruments and the policy strategy) (Rogge, 2019). For example, for the case of energy efficiency policies in OECD countries it has been shown that a comprehensive instrument mix which balances technology push instruments supporting research and development (such as public R&D funding) and demand-pull instruments creating a demand for energy efficient products (such as through an energy tax) is beneficial for innovation in energy efficiency (Costantini et al., 2017). Similarly, comprehensive instrument mixes that include carbon pricing, policies supporting new low-carbon technologies, and a moratorium on coal-fired power plants may not only be politically more feasible than stringent carbon pricing, but may also limit efficiency losses and lower distributional impacts (Bertram et al., 2015). In addition, policy mix consistency has been identified as an important driver of low-carbon transformative change, particularly for renewable energy (Lieu et al., 2018; Rogge and Schleich, 2018).

24.4.4 Phasing Out Policies Supporting Carbon-Intensive Fuels, Technologies, or Practices Can Accelerate Low-Carbon Transitions

Climate policy mixes can be differentiated into policies supporting low-carbon niches (e.g. feed-in tariffs for renewable energy) and those destabilizing existing carbon-intensive regimes (e.g. reduction of subsidies for fossil fuels). If climate policy mixes contain both elements of creation and destruction – and thus do not only aim for the support of innovation but also its flipside of exnovation (capturing the termination of fossil-based technological trajectories in a deliberate fashion) – they stand a greater chance of accelerating low-carbon transitions (David, 2017; Kivimaa and Kern, 2016). Such destabilization policies include control policies (e.g. stringent carbon pricing), significant changes in regime rules (e.g. reform of the design of electricity markets), reduced support for dominant regime technologies (e.g. removing tax deductions for private motor transport), changes in social networks and replacement of actors (e.g. more balanced representation of incumbents and new entrants in policy advisory councils), and changes in organisational and institutional practices (e.g. enhanced coordination between governing entities from different policy fields) (Kivimaa and Kern, 2016; Kivimaa et al. 2017). Analysis has so far been done through the perspective of technological innovation systems and their functions, such as for Norway's transport and energy sector (Ćetković and Skjærseth, 2019), Sweden's pulp and paper industry (Scordato et al., 2018), and Finland's building sector (Kivimaa et al., 2017). In addition, computable general algorithm (CGE) modelling for China's fossil fuel subsidy reform found that integrating both creation and destabilization policies is able to reduce rebound effects and make the policy mix more effective (Li et al., 2017).

24.4.5 Transformative Climate Policy Mixes Have to Navigate Resistance from Vested Interests

Climate policy mixes create not only winners (e.g. low-carbon entrepreneurs, future generations) but also losers (e.g. incumbents with vested interests, neighbours of low-carbon infrastructure projects, coal miners at risk of job loss) (Geels, 2014; Rosenbloom, 2018). A broader understanding of climate policy mixes thus takes into consideration that low-carbon transitions are contested and deeply political processes (Kern and Rogge, 2018; Roberts et al., 2018). For example, it has been argued that such resistance justifies supplementing carbon pricing with other policies that are designed for limited impact on incumbents while supporting new entrants (Passey et al., 2012). Another option is the design of short-term policies which might help to provide later entry points for more ambitious climate policy (Kriegler et al., 2018). In this context the sequencing of policies has been discussed as a way to build coalitions for climate change, starting with green industrial policy (e.g. supporting renewable energies through feed-in tariffs) and introducing carbon pricing (or making it more stringent) when supportive coalitions of ambitious climate policy have been formed (Meckling et al., 2015). In addition, low-carbon technological innovation can play a key role in ratcheting up climate policy over time, for example through cost reductions and job creation (Schmidt and Sewerin, 2017). However, apart from such positive policy feedbacks there can also be cases of negative policy feedback, for example arising from ineffective policy instruments, competing policy objectives, exogenous factors (such as the financial crisis), and global dynamics (such as international competition). The resulting negative policy feedbacks may over time lead to a weakening of ambitious policy targets, as has been the case with the UK zero-carbon homes target introduced in 2006 but which was eventually scrapped in 2016 (Edmondson et al., 2019). This calls for dedicated attention to the co-evolution of policy mixes and socio-technical systems occurring through resource, interpretative, and institutional effects (e.g. increase of public R&D support for low-carbon solutions, information provision at climate policy conferences, expanding state capacities for policy evaluation, and/or enforcement), and their socio-political, administrative, and fiscal feedbacks (e.g. mobilisation of supporters vs opponents, avoiding or causing budgetary strains, and strengthening vs weakening of implementing agencies' reputations).

24.4.6 Accelerating Decarbonisation Calls for Enhancing Policy Coordination across Governance Levels and Policy Fields

Low-carbon transitions cannot only be slowed down through resistance from vested interests, but also through a lack of public acceptance (Bicket and Vanner, 2016). Therefore, several interdisciplinary studies have incorporated stakeholder views, for example by applying Q methodology in the case of building-integrated photovoltaics in Singapore (Chang et al., 2019) or transport backcasting scenarios with multi-criteria analysis in Spain (Soria-Lara and Banister, 2018). Similarly, the public acceptance of climate policy has been increasingly investigated, for example through choice experiments for sustainable passenger transport in China, Germany, and the USA (Wicki et al., 2019) or for climate change mitigation policies in the Czech Republic, Poland, and the UK (Ščasný et al., 2017). In addition, the emerging energy democracy literature argues for policy mixes that resist the dominant energy agenda (e.g. by ending subsidies for fossil fuels and supporting those dependent on jobs in fossil fuel industries), that reclaim the energy sector (e.g. by normalising public control of energy production and consumption), and that restructure the energy sector (e.g. by governing energy systems as a commons) (Burke and Stephens, 2017).

24.4.7 Systematic Mapping of the Policy Mix Is a Precondition for Policy Mix Analysis and Design

Accelerating low-carbon transitions can be supported by policy mixes spanning multiple governance levels (e.g. local, regional, national, supranational, and international) and policy fields (e.g. climate, energy, industry, economy, innovation, environment). Siloed rather than integrated policy mixes have been identified as bottleneck to low-carbon transitions, such as in the case of South Korea's renewable energy policy (Yoon and Sim, 2015). Policy coordination provides an avenue to manage trade-offs between different policy objectives and to seek policy synergies, although coordination is no panacea and may require institutional remedies, given the complexity, uncertainty, and cross-cutting character of transition processes (Gebara et al., 2019; Matti et al., 2017). An example includes urban planning where local authorities can use a variety of instruments that assist in implementing both planning and energy policy targets (Petersen and Heurkens, 2018) and where mainstreaming climate policy with urban planning can lead to win–win strategies (Viguié and Hallegatte, 2012). Another example includes power added to climate-relevant bureaucracies as a result of international and domestic climate policies, which can impact the direction and practical policy limits for climate change policy (Rahman and Giessen, 2017).

In conclusion, while these seven key insights of relevance for policymakers and others interested in effective climate policy mixes are not meant to represent a complete list and were specifically derived from the literature on climate policy mixes, we argue that many of these insights may be transferable to other environmental challenges. Future research on policy mixes addressing environmental challenges other than the climate crisis, such as the dramatic loss of biodiversity or plastic pollution in oceans, could help to clarify general and context-specific insights on effective policy mixes.

24.5 Research Outlook

In this chapter, we have outlined conceptual foundations and empirical approaches and advances of policy mix research and how these enable the provision of policy implications for tackling environmental challenges, such as climate change. In the final section, we now turn to what we consider to be three key research areas for further advancing much-needed insights on real-world policy mixes addressing environmental challenges. These three areas include capturing transformative change, incorporating multi-level governance settings, and rethinking policy regimes.

24.5.1 Capturing Transformative Change

Addressing environmental challenges for redirecting and accelerating socio-technical change towards sustainability calls for policy mixes which become transformative by targeting not only one but several system functions and/or policy intervention points (Kanger et al. 2020; Kivimaa and Kern, 2016).

Adopting an extended framework drawing from technological innovation systems, strategic niche management, and other innovation and transition studies insights (Hekkert et al. 2007; Schot and Geels, 2008), Kivimaa and Kern (2016) complement the seven well-established "creative" functions supporting green niches (e.g. plant-based meat substitutes), with four "destruction" functions aimed at destabilizing unsustainable regimes (e.g. stringent and compulsory sustainability standards for animal farming).[1] By doing so, this framework allows for the evaluation of the comprehensiveness and balance of policy mix elements that goes well beyond the separation of policies supporting the development of new technologies and practices and their adoption by consumers. As such, it probably could be applied to most environmental challenges.

In contrast, Kanger et al. (2020) propose a more deductive approach drawing on the multi-level perspective (Geels, 2004) to identify six key policy intervention points that policy mixes can target to influence transition processes – covering the environmentally innovative niche, the existing unsustainable regime, and external developments at the so-called landscape level. For example, to accelerate low-carbon energy transitions policies should not only stimulate the development of low-carbon niche technologies such as solar PV, but also need to create substantial market demand for these by reducing support for unsustainable regimes such as coal-based energy generation – which up to this point is in line with Kivimaa and Kern's approach. In addition, they also call for cushioning the repercussions of associated structural changes through dedicated policies, such as skills retraining and job creation for redundant coal miners. Another policy intervention point they identify are cross-sectoral trends and their coordination, such as greater electricity demand and thus expansion needs for renewable energies due to electrification of other sectors such as mobility and heat. Finally, given the global nature of many pressing environmental challenges, policymakers would also be well advised to attempt to tilt the general framework conditions, for example through binding international environmental agreements. Note that the combinations of policy efforts targeting various intervention points depend on the type of transition pathway of any given system and can change over time.

Despite drawing on different theoretical approaches to investigate the functional role of policy mixes in achieving transitions, similar insights can be derived from both for designing truly transformative policy mixes. For instance, the importance of public support concerning various niche technologies has been highlighted in both contributions. A particular strength of the "creative destruction" framework developed by Kivimaa and Kern (2016) is its provision of a toolbox to link policies to relevant socio-technical system functions. Meanwhile, Kanger et al. (2020) add a "global sense" to the analysis, especially regarding their systematic thinking about the role of policy mixes in different transition pathways. As they argued, the formulation and deployment of effective policy mixes will depend on the comprehensive mapping of all possible loci for accelerating socio-technical changes. In this sense, future research may benefit from drawing on and extending both approaches for deepening our understanding of the role of transformative policy mixes in accelerating desired system changes addressing environmental challenges (Kivimaa and Rogge, 2022).

24.5.2 Multi-Level Governance Settings

The role of multi-level governance settings, including both vertical and horizontal governance levels, has been highlighted by recent policy mix research (Howlett and del Rio, 2015) and will likely play an important but differentiated role in addressing the various current and future environmental challenges and crises. The recognition of this governance level dimension brings a series of future analytical challenges for which we would like to emphasise three aspects.

First, regarding the elements of multi-level policy mixes we expect the inclusion of policy strategies and instruments from different governance levels to significantly increase the challenge of delineating policy mixes, particularly those addressing global challenges such as the climate crisis. In this case, the environmental policy strategies of governing entities as well as the impact domains of transformative policy mixes addressing global or cross-sectoral environmental challenges may easily become too complex for scholars, policy analysts, and decision-makers alike for comprehensively mapping and harmonising all policy strategies and instrument mixes across different administrative levels and policy departments (Ossenbrink et al., 2019). Consequentially, there will be practical limits to seeking synergies and avoiding contradiction.

Second, considering policy processes associated with multi-level policy mixes, both formulation and implementation processes of a multi-level policy mix tend to be very different from a single-level policy mix (Howlett and del Rio, 2015; Weber and Rohracher, 2012). For example, Magro and Wilson (2019) noted that current place-based innovation policy mixes require new governance arrangements and processes across different governance levels to be able to properly consider the multi-level aspect of formulating and implementing policy mixes in real-world contexts. In addition, as found for the case of solar water heating in Shandong, China, there may be at least two types of vertically interactive patterns – bottom-up and top-down – in multi-level policy mixes (Huang, 2019). Such patterns could evolve from unidirectional types towards highly complex bidirectional modes with the co-evolutionary development between policy mixes and a focal industry.

Third, evaluating the impact of multi-level policy mixes further complicates the already substantial challenge of evaluating policy mixes rather than single policy instruments. This is true even for analysts tasked with assessing the impact of single policies when in fact their impact can only be properly understood when considering interactions with other policies in the focal policy mix, including the same from other governance levels than the focal one. In this regard, Rogge and Schleich (2018) have shown for the case of renewable energies in Germany that policy mix characteristics, especially consistency and credibility, can help in evaluating the impact of policy mixes rather than attempting to differentiate the effects of single policy instruments, at least when asking actors in the impact domain. Thereby policies originating from multiple governance levels can be captured in actors' overall perception of the policy mix, while actors also appear to be able to differentiate between policy mix credibility across different governance levels (Rogge and Dütschke, 2018). Similarly, Mavrot et al. (2019) point out the impacts of the perceived implementation environment on the reactions and subsequent behavioural changes of target groups, which suggests that the degree of policy acceptance should be particularly noticed in analysing multi-level policy mixes.

24.5.3 Rethinking Policy Regimes

Finally, taking seriously the complex formulation and implementation processes behind policy mixes implies the need for an institutional reconfiguration of different policy regimes (e.g. the environmental policy regime and the innovation policy regime) towards an integrated one. In this regard, and recognising complex overlaps and systemic interactions of innovation systems and policy processes, Foxon and Pearson (2008) underlined the importance of promoting the development of a sustainable innovation policy regime by bringing together separate innovation and environmental policy regimes that on their own cannot provide sustainable support for addressing long-term sustainability challenges. This point is particularly related to current observations of complex policy mixes. As Rogge and Reichardt (2016) pointed out, the consistency and coherence of policy mixes across policy fields is one of the core issues in future research, as possible conflicts between different policy fields might lead to ineffective policy mixes. Such a heightened role of joined-up policy mixes across different policy fields may be particularly significant for the formation and stimulation of sustainable industries (Gomel and Rogge, 2020; Magro and Wilson, 2019).

Current research, however, has mainly investigated policy mix elements consisting of instruments from single policy fields, while associated dynamic policy processes and coordination between different governing entities has not been well-studied. This is supported, for example, by Greco et al. (2022) who refer to a policy mix including environmental policy and innovation policy as a "cross-instrumental policy mix" and investigated the effect of this on eco-innovations by using datasets from the Mannheim Innovation Panel concerning German firms. Based on their findings they suggest that decision-makers in various public agencies should better cooperate and coordinate their separate policy efforts in formulation and implementation processes of both innovation and environmental policies. Similarly, for the case of resource efficiency, Wilts and O'Brien (2019) have argued that the formulation and implementation of policy mixes "must" go beyond the conventional environmental policy regime in terms of achieving transformative change of whole socio-technical systems. This implies that the strategic coordination between governing entities across multiple policy regimes will therefore be critical in employing effective cross-cutting policy mixes. However, existing governance arrangements do not yet ensure sufficient capacities in terms of achieving such coordination processes (Kivimaa and Sivonen, 2021; Wilts and O'Brien, 2019), suggesting future research and practice around policy mixes addressing environmental challenges should pay greater attention to such governance, coordination, and capacity challenges.

Box 24.1 Chapter Summary

- Presenting an extended conceptual framework for policy mix research.
- Highlighting the role of policy strategies and associated policy processes.
- Summarising two analytical approaches for delineating complex policy mixes.
- Providing an example of global climate change for informing policy mix design.
- Suggesting three key research areas for future investigations on policy mixes.

Acknowledgement

This chapter was written within the EMPOCI project which has received funding from the European Research Council under the European Union's Horizon 2020 research and innovation programme (grant agreement No 852730).

Note

1 These four were later extended by a fifth "destruction" function concerning institutional routine and policy coherence (Kivimaa et al., 2017).

References

Bak, C., Bhattacharya, A., Edenhofer, O., & Knopf, B. (2017). Towards a comprehensive approach to climate policy, sustainable infrastructure, and finance. *Economics*, *11*(1), 1–13. https://doi.org/10.5018/economics-ejournal.ja.2017-33

Bertram, C., Luderer, G., Pietzcker, R. C., Schmid, E., Kriegler, E., & Edenhofer, O. (2015). Complementing carbon prices with technology policies to keep climate targets within reach. *Nature Climate Change*, *5*(3), 235–239. https://doi.org/10.1038/nclimate2514

Bicket, M., & Vanner, R. (2016). Designing policy mixes for resource efficiency: The role of public acceptability. *Sustainability*, *8*(4), 366. https://doi.org/10.3390/su8040366

Bosetti, V., & Victor, D. G. (2011). Politics and economics of second-best regulation of greenhouse gases: The importance of regulatory credibility. *The Energy Journal*, *32*(1), 1–24. https://doi.org/10.5547/ISSN0195-6574-EJ-Vol32-No1-1

Bouma, J. A., Verbraak, M., Dietz, F., & Brouwer, R. (2019). Policy mix: Mess or merit? *Journal of Environmental Economics and Policy*, *8*(1), 32–47. https://doi.org/10.1080/21606544.2018.1494636

Braathen, N. A. (2007). Instrument mixes for environmental policy: How many stones should be used to kill a bird? *International Review of Environmental and Resource Economics*, *1*(2), 185–236. http://dx.doi.org/10.1561/101.00000005

Burke, M. J., & Stephens, J. C. (2017). Energy democracy: Goals and policy instruments for sociotechnical transitions. *Energy Research & Social Science*, *33*, 35–48. https://doi.org/10.1016/j.erss.2017.09.024

Capano, G., & Howlett, M. (2020). The knowns and unknowns of policy instrument analysis: Policy tools and the current research agenda on policy mixes. *SAGE Open*, *10*(1). https://doi.org/10.1177/2158244019900568

Ćetković, S., & Skjærseth, J. B. (2019). Creative and disruptive elements in Norway´s climate policy mix: The small-state perspective. *Environmental Politics*, *28*(6), 1039–1060. https://doi.org/10.1080/09644016.2019.1625145

Chang, R., Cao, Y., Lu, Y., & Shabunko, V. (2019). Should BIPV technologies be empowered by innovation policy mix to facilitate energy transitions? - Revealing stakeholders' different perspectives using Q methodology. *Energy Policy*, *129*, 307–318. https://doi.org/10.1016/j.enpol.2019.02.047

Costantini, V., Crespi, F., & Palma, A. (2017). Characterizing the policy mix and its impact on eco-innovation: A patent analysis of energy-efficient technologies. *Research Policy*, *46*(4), 799–819. https://doi.org/10.1016/j.respol.2017.02.004

Cunningham, P., Nauewelaers, C., Boekholt, P., Mostert, B., Guy, K., Hofer, R., & Rammer, C. (2009). *Policy Mixes for R&D in Europe*. United Nations University, Maastricht Economic and Social Research Institute on Innovation and Technology. https://www.research.manchester.ac.uk/portal/en/publications/policy-mixes-for-rd-in-europe(e83b3c90-625d-484e-9306-32a3ec71afe8).html

David, M. (2017). Moving beyond the heuristic of creative destruction: Targeting exnovation with policy mixes for energy transitions. *Energy Research & Social Science*, *33*, 138–146. https://doi.org/10.1016/j.erss.2017.09.023

del Río, P. (2017). Why does the combination of the European Union Emissions Trading Scheme and a renewable energy target makes economic sense? *Renewable and Sustainable Energy Reviews*, *74*, 824–834. https://doi.org/10.1016/j.rser.2017.01.122

Edenhofer, O., Hirth, L., Knopf, B., Pahle, M., Schlömer, S., Schmid, E., & Ueckerdt, F. (2013). On the economics of renewable energy sources. *Energy Economics*, *40*, S12–S23. https://doi.org/10.1016/j.eneco.2013.09.015

Edmondson, D. L., Kern, F., & Rogge, K. S. (2019). The co-evolution of policy mixes and socio-technical systems: Towards a conceptual framework of policy mix feedback in sustainability transitions. *Research Policy*, *48*(10), 103555. https://doi.org/10.1016/j.respol.2018.03.010

Faehn, T., & Isaksen, E. T. (2016). Diffusion of climate technologies in the presence of commitment problems. *The Energy Journal*, *37*(2), 155–180. https://doi.org/10.5547/01956574.37.2.tfae

Flanagan, K., Uyarra, E., & Laranja, M. (2011). Reconceptualising the 'policy mix' for innovation. *Research Policy*, *40*(5), 702–713. https://doi.org/10.1016/j.respol.2011.02.005

Foxon, T., & Pearson, P. (2008). Overcoming barriers to innovation and diffusion of cleaner technologies: Some features of a sustainable innovation policy regime. *Journal of Cleaner Production*, *16*(1, Supplement 1), S148–S161. https://doi.org/10.1016/j.jclepro.2007.10.011

Gebara, M. F., Sills, E., May, P., & Forsyth, T. (2019). Deconstructing the policyscape for reducing deforestation in the Eastern Amazon: Practical insights for a landscape approach. *Environmental Policy and Governance*, *29*(3), 185–197. https://doi.org/10.1002/eet.1846

Geels, F. W. (2004). From sectoral systems of innovation to socio-technical systems: Insights about dynamics and change from sociology and institutional theory. *Research Policy*, *33*(6), 897–920. https://doi.org/10.1016/j.respol.2004.01.015

Geels, F. W. (2014). Regime resistance against low-carbon transitions: Introducing politics and power into the Multi-Level Perspective. *Theory, Culture & Society*, *31*(5), 21–40. https://doi.org/10.1177/0263276414531627

Givoni, M., Macmillen, J., Banister, D., & Feitelson, E. (2013). From policy measures to policy packages. *Transport Reviews*, *33*(1), 1–20. https://doi.org/10.1080/01441647.2012.744779

Gomel, D., & Rogge, K. S. (2020). Mere deployment of renewables or industry formation, too? Exploring the role of advocacy communities for the Argentinean energy policy mix. *Environmental Innovation and Societal Transitions*, *36*, 345–371. https://doi.org/10.1016/j.eist.2020.02.003

Greco, M., Germani, F., Grimaldi, M., & Radicic, D. (2022). Policy mix or policy mess? Effects of cross-instrumental policy mix on eco-innovation in German firms. *Technovation*, *117*, 102194. https://doi.org/10.1016/j.technovation.2020.102194

Hekkert, M. P., Suurs, R. A. A., Negro, S. O., Kuhlmann, S., & Smits, R. E. H. M. (2007). Functions of innovation systems: A new approach for analysing technological change. *Technological Forecasting and Social Change*, *74*(4), 413–432. https://doi.org/10.1016/j.techfore.2006.03.002

Howlett, M. (2009). Process sequencing policy dynamics: Beyond homeostasis and path dependency. *Journal of Public Policy*, *29*(3), 241–262. https://doi.org/10.1017/S0143814X09990158

Howlett, M., & del Rio, P. (2015). The parameters of policy portfolios: Verticality and horizontality in design spaces and their consequences for policy mix formulation. *Environment and Planning C: Government and Policy*, *33*(5), 1233–1245. https://doi.org/10.1177/0263774X15610059

Howlett, M., Kim, J., & Weaver, P. (2006). Assessing instrument mixes through program- and agency-level data: Methodological issues in contemporary implementation research. *Review of Policy Research*, *23*(1), 129–151. https://doi.org/10.1111/j.1541-1338.2006.00189.x

Howlett, M., Mukherjee, I., & Woo, J. J. (2015). From tools to toolkits in policy design studies: The new design orientation towards policy formulation research. *Policy & Politics*, *43*(2), 291–311. https://doi.org/10.1332/147084414X13992869118596

Howlett, M., & Rayner, J. (2013). Patching vs packaging in policy formulation: Assessing policy portfolio design. *Politics and Governance*, *1*(2), 170–182. https://doi.org/10.17645/pag.v1i2.95

Huang, P. (2019). The verticality of policy mixes for sustainability transitions: A case study of solar water heating in China. *Research Policy*, 48(10), 103758. https://doi.org/10.1016/j.respol.2019.02.009

Jacobsson, S., Bergek, A., & Sandén, B. (2017). Improving the European Commission's analytical base for designing instrument mixes in the energy sector: Market failures versus system weaknesses. *Energy Research & Social Science*, *33*, 11–20. https://doi.org/10.1016/j.erss.2017.09.009

Jakob, M. (2017). Ecuador's climate targets: A credible entry point to a low-carbon economy? *Energy for Sustainable Development*, *39*, 91–100. https://doi.org/10.1016/j.esd.2017.04.005

Kalfagianni, A., & Kuik, O. (2017). Seeking optimality in climate change agri-food policies: Stakeholder perspectives from Western Europe. *Climate Policy*, *17*(sup1), S72–S92. https://doi.org/10.1080/14693062.2016.1244508

Kanger, L., Sovacool, B. K., & Noorkõiv, M. (2020). Six policy intervention points for sustainability transitions: A conceptual framework and a systematic literature review. *Research Policy*, *49*(7), 104072. https://doi.org/10.1016/j.respol.2020.104072

Kern, F., & Howlett, M. (2009). Implementing transition management as policy reforms: A case study of the Dutch energy sector. *Policy Sciences*, *42*(4), 391–408. https://doi.org/10.1007/s11077-009-9099-x

Kern, F., Kivimaa, P., & Martiskainen, M. (2017). Policy packaging or policy patching? The development of complex energy efficiency policy mixes. *Energy Research & Social Science*, *23*, 11–25. https://doi.org/10.1016/j.erss.2016.11.002

Kern, F., & Rogge, K. S. (2018). Harnessing theories of the policy process for analysing the politics of sustainability transitions: A critical survey. *Environmental Innovation and Societal Transitions*, *27*, 102–117. https://doi.org/10.1016/j.eist.2017.11.001

Kern, F., Rogge, K. S., & Howlett, M. (2019). Policy mixes for sustainability transitions: New approaches and insights through bridging innovation and policy studies. *Research Policy*, *48*(10), 103832. https://doi.org/10.1016/j.respol.2019.103832

Kivimaa, P., Kangas, H.-L., & Lazarevic, D. (2017). Client-oriented evaluation of 'creative destruction' in policy mixes: Finnish policies on building energy efficiency transition. *Energy Research & Social Science*, *33*, 115–127. https://doi.org/10.1016/j.erss.2017.09.002

Kivimaa, P., & Kern, F. (2016). Creative destruction or mere niche support? Innovation policy mixes for sustainability transitions. *Research Policy*, *45*(1), 205–217. https://doi.org/10.1016/j.respol.2015.09.008

Kivimaa, P., & Rogge, K. S. (2022). Interplay of policy experimentation and institutional change in sustainability transitions: The case of mobility as a service in Finland. *Research Policy*, *51*(1), 104412. https://doi.org/10.1016/j.respol.2021.104412

Kivimaa, P., & Sivonen, M. H. (2021). Interplay between low-carbon energy transitions and national security: An analysis of policy integration and coherence in Estonia, Finland and Scotland. *Energy Research & Social Science*, *75*, 102024. https://doi.org/10.1016/j.erss.2021.102024

Kriegler, E., Bertram, C., Kuramochi, T., Jakob, M., Pehl, M., Stevanović, M., Höhne, N., Luderer, G., Minx, J. C., Fekete, H., Hilaire, J., Luna, L., Popp, A., Steckel, J. C., Sterl, S., Yalew, A. W., Dietrich, J. P., & Edenhofer, O. (2018). Short term policies to keep the door open for Paris climate goals. *Environmental Research Letters*, *13*(7), 074022. https://doi.org/10.1088/1748-9326/aac4f1

Lehmann, P. (2012). Justifying a policy mix for pollution control: A review of economic literature. *Journal of Economic Surveys, 26*(1), 71–97. https://doi.org/10.1111/j.1467-6419.2010.00628.x

Lehmann, P., Sijm, J., Gawel, E., Strunz, S., Chewpreecha, U., Mercure, J.-F., & Pollitt, H. (2019). Addressing multiple externalities from electricity generation: A case for EU renewable energy policy beyond 2020? *Environmental Economics and Policy Studies, 21*(2), 255–283. https://doi.org/10.1007/s10018-018-0229-6

Li, H., Bao, Q., Ren, X., Xie, Y., Ren, J., & Yang, Y. (2017). Reducing rebound effect through fossil subsidies reform: A comprehensive evaluation in China. *Journal of Cleaner Production, 141*, 305–314. https://doi.org/10.1016/j.jclepro.2016.09.108

Li, L., & Taeihagh, A. (2020). An in-depth analysis of the evolution of the policy mix for the sustainable energy transition in China from 1981 to 2020. *Applied Energy, 263*, 114611. https://doi.org/10.1016/j.apenergy.2020.114611

Lieu, J., Spyridaki, N. A., Alvarez-Tinoco, R., Van der Gaast, W., Tuerk, A., & Van Vliet, O. (2018). Evaluating consistency in environmental policy mixes through policy, stakeholder, and contextual interactions. *Sustainability, 10*(6), 1896. https://doi.org/10.3390/su10061896

Magro, E., Navarro, M., & Zabala-Iturriagagoitia, J. M. (2014). Coordination-mix: The hidden face of STI policy. *Review of Policy Research, 31*(5), 367–389. https://doi.org/10.1111/ropr.12090

Magro, E., & Wilson, J. R. (2019). Policy-mix evaluation: Governance challenges from new place-based innovation policies. *Research Policy, 48*(10), 103612. https://doi.org/10.1016/j.respol.2018.06.010

Matti, C., Consoli, D., & Uyarra, E. (2017). Multi level policy mixes and industry emergence: The case of wind energy in Spain. *Environment and Planning C: Politics and Space, 35*(4), 661–683. https://doi.org/10.1177/0263774X16663933

Mavrot, C., Hadorn, S., & Sager, F. (2019). Mapping the mix: Linking instruments, settings and target groups in the study of policy mixes. *Research Policy, 48*(10), 103614. https://doi.org/10.1016/j.respol.2018.06.012

Meadowcroft, J. (2009). What about the politics? Sustainable development, transition management, and long term energy transitions. *Policy Sciences, 42*(4), 323–340. https://doi.org/10.1007/s11077-009-9097-z

Meckling, J., Kelsey, N., Biber, E., & Zysman, J. (2015). Winning coalitions for climate policy. *Science, 349*(6253), 1170–1171. https://doi.org/10.1126/science.aab1336

Meissner, D., & Kergroach, S. (2021). Innovation policy mix: Mapping and measurement. *The Journal of Technology Transfer, 46*(1), 197–222. https://doi.org/10.1007/s10961-019-09767-4

Nemet, G. F., Jakob, M., Steckel, J. C., & Edenhofer, O. (2017). Addressing policy credibility problems for low-carbon investment. *Global Environmental Change, 42*, 47–57. https://doi.org/10.1016/j.gloenvcha.2016.12.004

Obersteiner, M., Walsh, B., Frank, S., Havlík, P., Cantele, M., Liu, J., Palazzo, A., Herrero, M., Lu, Y., Mosnier, A., Valin, H., Riahi, K., Kraxner, F., Fritz, S., & van Vuuren, D. (2016). Assessing the land resource–food price nexus of the Sustainable Development Goals. *Science Advances, 2*(9), e1501499. https://doi.org/10.1126/sciadv.1501499

Ossenbrink, J., Finnsson, S., Bening, C. R., & Hoffmann, V. H. (2019). Delineating policy mixes: Contrasting top-down and bottom-up approaches to the case of energy-storage policy in California. *Research Policy, 48*(10), 103582. https://doi.org/10.1016/j.respol.2018.04.014

Passey, R., Bailey, I., Twomey, P., & MacGill, I. (2012). The inevitability of 'flotilla policies' as complements or alternatives to flagship emissions trading schemes. *Energy Policy, 48*, 551–561. https://doi.org/10.1016/j.enpol.2012.05.059

Patt, A., & Lilliestam, J. (2018). The case against carbon prices *Joule, 2*(12), 2494–2498. https://doi.org/10.1016/j.joule.2018.11.018

Petersen, J.-P., & Heurkens, E. (2018). Implementing energy policies in urban development projects: The role of public planning authorities in Denmark, Germany and the Netherlands. *Land Use Policy, 76*, 275–289. https://doi.org/10.1016/j.landusepol.2018.05.004

Quitzow, R. (2015). Assessing policy strategies for the promotion of environmental technologies: A review of India's National Solar Mission. *Research Policy, 44*(1), 233–243. https://doi.org/10.1016/j.respol.2014.09.003

Rahman, M. S., & Giessen, L. (2017). The power of public bureaucracies: Forest-related climate change policies in Bangladesh (1992–2014). *Climate Policy, 17*(7), 915–935. https://doi.org/10.1080/14693062.2016.1197093

Rayner, J., Howlett, M., & Wellstead, A. (2017). Policy mixes and their alignment over time: Patching and stretching in the oil sands reclamation regime in Alberta, Canada. *Environmental Policy and Governance, 27*(5), 472–483. https://doi.org/10.1002/eet.1773

Roberts, C., Geels, F. W., Lockwood, M., Newell, P., Schmitz, H., Turnheim, B., & Jordan, A. (2018). The politics of accelerating low-carbon transitions: Towards a new research agenda. *Energy Research & Social Science*, *44*, 304–311. https://doi.org/10.1016/j.erss.2018.06.001

Rogge, K. S. (2019). Policy mixes for sustainable innovation: Conceptual considerations and empirical insights. In *Handbook of Sustainable Innovation* (pp. 165–185). Edward Elgar Publishing. https://www.elgaronline.com/view/edcoll/9781788112567/9781788112567.00016.xml

Rogge, K. S., & Dütschke, E. (2018). What makes them believe in the low-carbon energy transition? Exploring corporate perceptions of the credibility of climate policy mixes. *Environmental Science & Policy*, *87*, 74–84. https://doi.org/10.1016/j.envsci.2018.05.009

Rogge, K. S., Kern, F., & Howlett, M. (2017). Conceptual and empirical advances in analysing policy mixes for energy transitions. *Energy Research & Social Science*, *33*, 1–10. https://doi.org/10.1016/j.erss.2017.09.025

Rogge, K. S., & Reichardt, K. (2016). Policy mixes for sustainability transitions: An extended concept and framework for analysis. *Research Policy*, *45*(8), 1620–1635. https://doi.org/10.1016/j.respol.2016.04.004

Rogge, K. S., & Schleich, J. (2018). Do policy mix characteristics matter for low-carbon innovation? A survey-based exploration of renewable power generation technologies in Germany. *Research Policy*, *47*(9), 1639–1654. https://doi.org/10.1016/j.respol.2018.05.011

Rosenbloom, D. (2018). Framing low-carbon pathways: A discursive analysis of contending storylines surrounding the phase-out of coal-fired power in Ontario. *Environmental Innovation and Societal Transitions*, *27*, 129–145. https://doi.org/10.1016/j.eist.2017.11.003

Ščasný, M., Zvěřinová, I., Czajkowski, M., Kyselá, E., & Zagórska, K. (2017). Public acceptability of climate change mitigation policies: A discrete choice experiment. *Climate Policy*, *17*(sup1), S111–S130. https://doi.org/10.1080/14693062.2016.1248888

Schmidt, T. S., Schneider, M., Rogge, K. S., Schuetz, M. J. A., & Hoffmann, V. H. (2012). The effects of climate policy on the rate and direction of innovation: A survey of the EU ETS and the electricity sector. *Environmental Innovation and Societal Transitions*, *2*, 23–48. https://doi.org/10.1016/j.eist.2011.12.002

Schmidt, T. S., & Sewerin, S. (2017). Technology as a driver of climate and energy politics. *Nature Energy*, *2*(6), 1–3. https://doi.org/10.1038/nenergy.2017.84

Schot, J., & Geels, F. W. (2008). Strategic niche management and sustainable innovation journeys: Theory, findings, research agenda, and policy. *Technology Analysis & Strategic Management*, *20*(5), 537–554. https://doi.org/10.1080/09537320802292651

Scordato, L., Klitkou, A., Tartiu, V. E., & Coenen, L. (2018). Policy mixes for the sustainability transition of the pulp and paper industry in Sweden. *Journal of Cleaner Production*, *183*, 1216–1227. https://doi.org/10.1016/j.jclepro.2018.02.212

Scullion, J. J., Vogt, K. A., Winkler-Schor, S., Sienkiewicz, A., Peña, C., & Hajek, F. (2016). Designing conservation-development policies for the forest frontier. *Sustainability Science*, *11*(2), 295–306. https://doi.org/10.1007/s11625-015-0315-7

Soria-Lara, J. A., & Banister, D. (2018). Evaluating the impacts of transport backcasting scenarios with multi-criteria analysis. *Transportation Research Part A: Policy and Practice*, *110*, 26–37. https://doi.org/10.1016/j.tra.2018.02.004

Sorrell, S., & Sijm, J. (2003). Carbon trading in the policy mix. *Oxford Review of Economic Policy*, *19*(3), 420–437. https://doi.org/10.1093/oxrep/19.3.420

Sovacool, B. K. (2009). The importance of comprehensiveness in renewable electricity and energy-efficiency policy. *Energy Policy*, *37*(4), 1529–1541. https://doi.org/10.1016/j.enpol.2008.12.016

Stirling, A. (2014). Transforming power: Social science and the politics of energy choices. *Energy Research & Social Science*, *1*, 83–95. https://doi.org/10.1016/j.erss.2014.02.001

Taeihagh, A., Givoni, M., & Bañares-Alcántara, R. (2013). Which policy first? A network-centric approach for the analysis and ranking of policy measures. *Environment and Planning B: Planning and Design*, *40*(4), 595–616. https://doi.org/10.1068/b38058

Trein, P., Biesbroek, R., Bolognesi, T., Cejudo, G. M., Duffy, R., Hustedt, T., & Meyer, I. (2021). Policy coordination and integration: A research agenda. *Public Administration Review*, *81*(5), 973–977. https://doi.org/10.1111/puar.13180

Tvinnereim, E., & Mehling, M. (2018). Carbon pricing and deep decarbonisation. *Energy Policy*, *121*, 185–189. https://doi.org/10.1016/j.enpol.2018.06.020

Uyarra, E., Shapira, P., & Harding, A. (2016). Low carbon innovation and enterprise growth in the UK: Challenges of a place-blind policy mix. *Technological Forecasting and Social Change*, *103*, 264–272. https://doi.org/10.1016/j.techfore.2015.10.008

Viguié, V., & Hallegatte, S. (2012). Trade-offs and synergies in urban climate policies. *Nature Climate Change*, *2*(5), 334–337. https://doi.org/10.1038/nclimate1434

von Stechow, C., McCollum, D., Riahi, K., Minx, J. C., Kriegler, E., van Vuuren, D. P., Jewell, J., Robledo-Abad, C., Hertwich, E., Tavoni, M., Mirasgedis, S., Lah, O., Roy, J., Mulugetta, Y., Dubash, N. K., Bollen, J., Ürge-Vorsatz, D., & Edenhofer, O. (2015). Integrating global climate change mitigation goals with other sustainability objectives: A synthesis. *Annual Review of Environment and Resources*, *40*(1), 363–394. https://doi.org/10.1146/annurev-environ-021113-095626

Voß, J.-P., Smith, A., & Grin, J. (2009). Designing long-term policy: Rethinking transition management. *Policy Sciences*, *42*(4), 275–302. https://doi.org/10.1007/s11077-009-9103-5

Weber, K. M., & Rohracher, H. (2012). Legitimizing research, technology and innovation policies for transformative change: Combining insights from innovation systems and multi-level perspective in a comprehensive 'failures' framework. *Research Policy*, *41*(6), 1037–1047. https://doi.org/10.1016/j.respol.2011.10.015

Wicki, M., Fesenfeld, L., & Bernauer, T. (2019). In search of politically feasible policy-packages for sustainable passenger transport: Insights from choice experiments in China, Germany, and the USA. *Environmental Research Letters*, *14*(8), 084048. https://doi.org/10.1088/1748-9326/ab30a2

Wilts, H., & O'Brien, M. (2019). A policy mix for resource efficiency in the EU: Key instruments, challenges and research needs. *Ecological Economics*, *155*, 59–69. https://doi.org/10.1016/j.ecolecon.2018.05.004

Xu, L., & Su, J. (2016). From government to market and from producer to consumer: Transition of policy mix towards clean mobility in China. *Energy Policy*, *96*, 328–340. https://doi.org/10.1016/j.enpol.2016.05.038

Yoon, J.-H., & Sim, K. (2015). Why is South Korea's renewable energy policy failing? A qualitative evaluation. *Energy Policy*, *86*, 369–379. https://doi.org/10.1016/j.enpol.2015.07.020

25

FIFTY SHADES OF SUFFICIENCY

Semantic Confusion and No Policy

Doris Fuchs, Sylvia Lorek, Pia Mamut, and Anica Rossmoeller

25.1 Introduction

Sufficiency is a concept that receives increasing attention in the environmental policy discourse. An increasing share of environmental policy research focuses on the need to pursue an absolute reduction in global resource consumption and on the role of limits and the need for a paradigm shift towards sufficiency norms. Similarly, policy actors like the Deutsche Städtetag (Association of German Cities) have started to consider sufficiency as an important goal in their environmental governance strategies (e.g. see Deutscher Städtetag, 2021). For a long time, a focus on improving efficiency, that is reducing the resource input for a given level of output with the help of technical innovation and process optimisation) had dominated environmental governance.[1] Today, however, it appears to have become more and more impossible to ignore the need for an accompanying focus on sufficiency measures. This is especially the case as sustainability research has documented the failure of efficiency measures as leading to an actual reduction in resource use (Wiedmann et al., 2020). Rebound effects, in particular, have led to the failure of decades of environmental governance to achieve the decoupling between economic growth and environmental resource consumption that would be needed for a sustainability transition. In consequence, many sustainability scholars and activists today see sufficiency as a necessary condition for a shift to one-world living.

Thus, sufficiency clearly is an important topic when it comes to environmental governance. But what is sufficiency, and how can we assess its political pursuit implemented in the form of policies? These are the questions addressed in this chapter. In pursuit of our objectives, we show the wide variance in the definitions of "sufficiency" used. In addition, we delineate the difficulties involved in translating these definitions into a toolbox for evaluating policies with respect to their sufficiency focus. At the end, we show that "true" sufficiency policies are scarce still, notwithstanding the wide range of valuable policies that may contribute to some extent to its pursuit.

The structure of the chapter is as follows. The next section delineates the range of conceptualizations of sufficiency employed in the literature and political debate, and serves to identify the one that we find most convincing. Subsequently, we explore assessing policies with respect to their pursuit of sufficiency as an aim. The following section then illustrates these difficulties

DOI: 10.4324/9781003043843-29

for the consumption sectors of mobility, food, and housing. Finally, the concluding section summarizes our inquiry and delineates its implications for research and practice.

25.2 Sufficiency: What Does It Mean?

Both scholars and practitioners use the term *sufficiency* in a large variety of ways (Mamut, forthcoming). The term derives from the Latin verb *sufficere*, corresponding to the contemporary meaning of being enough or "being content with what is enough" (Scherhorn, 2002: 24). In this sense, sufficiency may be seen as a guiding principle in the art of living (Linz, 2013: 24) and relates to ideas of human well-being and quality of life (Schneidewind & Zahrnt, 2014). Indeed, some scholars consider "a good life" to be the central vantage point of sufficiency: to enable individuals as well as communities to strive for a good life for all and forever (Alexander, 2015: 12, Muller, 2009: 86, Samadi et al., 2017: 127). However, sufficiency may also be linked to rather negative connotations, like scarcity and deficit. Such connotations have their roots in the use of the term in association with (a new) asceticism, survivalism, or sacrifice (Linz, 2002: 9). Thus, rather than raising hopes for and visions of a good life, sufficiency as a concept may also foster scepticism if not fear of losing one's standard of living. These two partly adverse considerations of the concept already signal the semantic diversity that sufficiency offers.

A conventional and popular approach to sufficiency is to distinguish it from its sibling principles efficiency and consistency (e.g. Samadi et al., 2017). Within this approach, efficiency refers to "doing things better" in terms of the relationship of resource input to output. Consistency calls for "doing things differently", for example increasing the ecological sustainability of the output via the "foresightful and encompassing design of life-cycles", as in the sense of a circular economy to cradle perspectives (Metzner-Szigeth, 2019).[2] Finally, sufficiency focuses on the potential of "doing less", that means it focuses on the level of output sought. Sufficiency strategies ask "How much is enough?", which, in the context of the global consumer class, translates into calls for a reduction in output (with its resulting effect on resource input).[3] A sufficiency focus has far-reaching implications for societal infrastructures and institutions, as well as individuals' everyday life practices, across consumption sectors such as transport, food, housing, and the use of consumer goods and services (Creutzig et al., 2018, van den Berg et al., 2019). Inevitably, implementing sufficiency as a guiding principle for societies would imply fundamental changes in the way people live and societies organize their economic activities. Efficiency and consistency, on the other hand, generally imply technological improvement and substitution, dynamics which are very common in economies as well as in households.

Importantly, the argument for the need of sufficiency measures does not imply that efficiency and consistency measures will no longer be needed. Rather, scholars generally agree that sustainability transitions need to be equally built on the principles of efficiency, consistency, *and* sufficiency (Hayden, 2020: 160; Spengler, 2018: 921; van den Berg et al., 2019; Wissen & Brand, 2016). They also stress that sufficiency is the still-missing element in applied policies, however, and the "sufficient" condition for sustainability to materialize (Stengel, 2011: 294). Accordingly, an increasing number of scholars and activists call for sufficiency to become elevated to the role of a societal organizing principle (Hayden, 2020: 151; Lorek & Fuchs, 2013; Princen, 2003: 44).

Despite this growing consensus that a successful transformation requires embracing the principle of sufficiency in one way or another, a serious effort at the political level to pursue sufficiency is difficult to recognize at best. Recent reviews of policy practice show that sufficiency is sidelined or even entirely neglected in climate and energy scenarios at the global and national policy levels (Samadi et al., 2017: 130; Toulouse et al., 2019: 332–333). An analysis of European National Energy and Climate Plans revealed that sufficiency is still largely perceived as a micro-level individual behaviour change or as necessary exogenous trends that will need to take place but are not treated yet as a genuine field of policy action to provide the necessary framework for enabling societal change (Zell-Ziegler et al., 2021). Even at local policy levels, which are often considered innovative niches for transformation, sufficiency has failed to emerge as a strong principle. For instance, "sustainable" energy initiatives, rising in the aftermath of the 1992 Earth Summit, despite all the good intentions and noteworthy achievements, persistently and systematically avoided addressing "questions about how much [energy] is enough" (Fuchs et al., 2016: 299). In a similar vein, Vadovics and Živčič (2019: 165) concluded that "no, we are not yet ready" for sufficiency, when surveying European energy projects and their degree of sufficiency incorporation.

The observation that the current political-economic system appears not ready for sufficiency, combined with the conviction that sufficiency is a requirement for real sustainability improvements, poses the question as to what it is about the nature of sufficiency that creates these road blocks. One core reason for sufficiency's lack of practical incorporation is likely the challenge that the idea of "enoughness" involves for the capitalist political economy, with its inbuilt dynamics of constant accumulation and growth (Hayden, 2014). Indeed, institutional lock-ins in the politico-economic system prevent sufficiency from becoming a political objective, societal organizing principle, or dominant individual norm. The challenge becomes visible and is at least partially re-emphasized via different interpretations or discursive framings of sufficiency[4] (Mamut, forthcoming), four of which we sketch below.

25.2.1 Sufficiency as a Sacrifice

Some discursive uses of sufficiency link it to personal and societal sacrifice and paralysis. Moreover, they suggest that sufficiency is incompatible with dominant liberal conceptions of society (Muller and Huppenbauer, 2016). Specifically, they claim that sufficiency-oriented policies would invade individual freedoms by prescribing specific lifestyles or shaming and prohibiting others (Grunwald, 2010: 180), and that they would prioritize ecological concerns to an unacceptable degree (Kanschik, 2016). Thereby such interpretations and framings convey and reproduce political and societal resistance. Due to such interpretations of sufficiency, critical sustainability scholars stress the need to de-demonize the principle to make it at all accessible for wider policy application (Toulouse et al., 2019: 333).

25.2.2 Sufficiency as an Industrialized Country Problem

A second and yet related source of opposition to sufficiency as an objective of environmental governance results from specific interpretations linking it to global questions. Though the research on sufficiency is predominantly Eurocentric (Toulouse et al., 2019: 332), the concept itself aims to be globally applicable. Countering such global aspirations, some scholars and practitioners claim that sufficiency as a societal norm and objective stands against the current paradigm of development with its focus on growth and efficiency (Malghan, 2019: 51). They argue that the Global South has strongly and justifiably incorporated the wish for further economic development and the harvesting of the so-called "low-hanging fruits" of efficiency. In consequence, they conclude that sufficiency is not a relevant approach for the Global South but rather is an industrialized country "problem". Importantly, however, such a claim misses the fact that the common development frame itself is not a consensus in the Global South. Critical discourses exist there as well that formulate demands of enoughness (Kothari et al.,

2019: xxix), though termed differently. Latin American critique on development, including the well-known *Buen Vivir* movement, has often bemoaned the limitless consumerist way of life since at least the 1970s (Svampa, 2019: 18–19). In India, the Gandhian concept of *swaraj* similarly included a critical assessment of consumerist desires (Shrivastava, 2019: 285). Indeed, several alternative concepts, best subsumed under the heading of post-development, take up the idea of sufficiency in one way or the other (Kothari et al., 2019: xxviii). Therefore, sufficiency, though being at the moment a Eurocentric research stream, has the potential to join research as well as activism on sustainable transformations in the Global North as well as the Global South.

25.2.3 Sufficiency as Efficiency

While interpretations of sufficiency as sacrifice or as a Northern problem thus emphasize challenges to the economic system and capitalist development involved, an alternative interpretation of sufficiency exists that is rather well adapted to the current capitalist political economy and applies the principle in a way that does not necessitate absolute reductions and deeper structural shifts. This interpretation propagates sufficiency measures in pursuit of sustainable lifestyles, ranging from switching conventional light bulbs to energy-efficient ones, to using a car-sharing app, to voluntary simplicity. This use of sufficiency can be termed "eco-modern sufficiency" (Mamut, forthcoming). Conceptually it is problematic that this kind of sufficiency can mean anything and becomes almost indistinguishable from efficiency and consistency (Hayden, 2014). While the range of activities noted above may entirely embody valuable contributions to sustainability, they do not all relate to a sense of enoughness and an understanding of the need for absolute reductions. On the contrary, critical scholars argue that they contribute in sum to the impression that a vast range of small and large behavioural changes on the part of individual consumers in the form of greener consumption will magically lead to the desired outcome of sustainability, something sustainable consumption research has revealed to be a myth, for decades (Fuchs et al., 2021; Maniates, 2001: 34). Thus, while this view of sufficiency fits more easily into the existing political economy by proposing technological fixes to reconcile economic growth with planetary boundaries, critical scholars point out that this mainstreaming of the concept of sufficiency avoids taking serious the issue of ecological and social needs and limits. Needless to say, it is easier to agree upon the "limits to (in)efficiency" than on the absolute limits of consumption (Fawcett & Darby, 2019: 10; Vadovics & Živčič, 2019). Likewise, it is easier to ignore the systemic injustice inherent in today's consumption patterns than to face the conflicts associated with the fundamental changes in economic activities that sufficiency implies (Maniates & Meyer, 2010: 9).

25.2.4 Sufficiency as a Condition for a Good Life for All

Last but not least, a fourth interpretation of sufficiency emphasizes the idea of enoughness as its core, considers it a requirement on the way to a good life for all, and links it to consumption (and production) minima and maxima (Spengler, 2016). Applied to the context of the Global South, this idea of sufficiency implies more consumption for a substantial share of the population; applied to the Global North as well as to the economic elites and a rising middle class in the Global South, however, it implies less consumption for most. This conceptualization of sufficiency maintains a "radical" edge and core that challenges the prevailing growth compulsions of the capitalist political economy, while still being compatible with a liberal society and democracy, because it presupposes societal dialogue about what is "enough".

Conceptualized in this form, sufficiency can indeed become a core organizing principle for society that allows for the pursuit of justice and well-being in a world of limits and thus actually makes it a necessary and sufficient condition for sustainability transformation. Contrary to the overwhelmingly Eurocentric or Western situatedness of conceptions of sufficiency, moreover, this view of sufficiency has the ability to relate to other contexts, societies, and normative debates in the Global South (consider Mongsawad, 2012, on faiths and traditions). It even allows sufficiency to be considered as a right: for one thing, the right that "no one should always have to want more" and, respectively, the right to "be able to have something left for the future" (von Winterfeld, 2007: 53). This interpretation of sufficiency, then, allows envisioning a sustainable consumption corridor, that is the potential for a good life for all, living now and in the future on the basis of societal agreement on consumption minima and maxima (Di Giulio & Fuchs, 2014, Fuchs et al., 2021). It presupposes societal frameworks for sustainable consumption between consumption minima, that is a sufficient level of access to resources to allow individuals to satisfy their needs, and consumption maxima, the transgression of which destroys the possibility of sufficient resources for all. It thereby shows that sufficiency as a societal norm unfolding the corridor between these upper and lower limits "is not about how to implement a specific consumption level with each individual, but how much of a specific individual consumption level is possible in relation to the consumption of all other citizens and the corresponding aggregate" (Muller & Huppenbauer, 2016: 106). As pointed out above, sufficiency in the sense of enoughness requires democratic deliberation, it requires negotiating and agreeing on the "two types of enough" (Fuchs et al., 2021, Hayden, 2020; Muller & Huppenbauer, 2016; Spengler, 2016). Accordingly, it is fundamentally a political project (Mamut, forthcoming).

In sum, even if one approaches sufficiency as a companion to efficiency and consistency, a range of different interpretations and framings remains. Sufficiency may be connoted as sacrifice and loss or as a concept that is relevant only for the Global North. It may even be employed in a way that the difference to efficiency and consistency becomes difficult to detect. Finally, it may be interpreted as fundamentally and exclusively centering on the idea of enoughness. Only the latter conceptualization has the potential to foster a sustainability transformation. While interpretations and framings of sufficiency as sacrifice or as a Northern problem lead to its political death or marginalization, an eco-modern notion of sufficiency prolongs the failure of environmental governance in terms of achieving actual reductions in resource use that we have witnessed in recent decades. Sufficiency in the double sense of enoughness, enough for me and enough for others, however, has the potential to allow the combined pursuit of well-being and justice within planetary boundaries.

25.3 Sufficiency in Policy: Ready for Practice?

Given the breadth of interpretations of sufficiency, evaluating policies with respect to their embodiment of sufficiency objectives clearly is difficult. Some studies link sufficiency to policy-imposed bans and limits, the modification of relative prices, or altering individual preferences and behaviours (Samadi et al., 2017; see also Alcott, 2008 for price caps). Sachs (1993) envisioned it to involve reducing speed, distance, trade, and ownership. More recently, some scholars suggest that such assessments can rely on the avoid-shift-improve framework (Creutzig et al., 262), relating improvement to efficiency measures, shift to consistency measures, and avoid to sufficiency measures. Yet, we can easily identify complexities that render judgements difficult even if we use this framework. The following short discussions of relevant policies in the consumption fields of mobility, food, and housing illustrate this problem.

25.3.1 Mobility

A policy measure in the consumption realm of mobility that may appear to represent an exemplary sufficiency policy is providing financial incentives for discarding old cars under the condition that no new ones are bought (Spengler, 2018). This measure fosters "avoidance" or non-consumption well, as the targeted individual car users receive financial compensation under the condition that they dispose of their car. This condition also covers, according to our assumption, a substitution by a supposedly cleaner electric motorized vehicle.

Yet, it is important to note that this policy measure does not consider individual mobility needs or the alternatives regarding their satisfaction per se. After all, the individual former private car owner may "shift" to walking, biking, the use of public transport, or car-pooling and sharing, for example. One even has to wonder if such a shift, especially in the latter cases, does not merely constitute an improvement in the resource efficiency of travel. Clearly, the actual mobility infrastructure will have an influence, whatever form the shift may take. Similarly, the needs originally satisfied by the car-use play a role. If the former private car was only used for weekend or holiday trips that involved long-distance travel, for example, and those are conducted by plane after disposing with the car, the original intention of the policy measure would be perverted.

In sum, the suggested policy measure cannot be defined as a clear example of sufficiency in the sense of avoidance. It is more likely to foster a modal shift that may or may not lead to a reduction of ecological impacts. Clearly, the policy is well intended and may lead to the desired improvement in the ecological footprint of some individuals. For an absolute reduction in resource use to be ensured, however, such a policy measure will at least have to take into account and try to influence the choice of alternative modes of mobility. Better still, it would be accompanied by measures fostering a reduction in the need for mobility, for instance via an increased use of teleconferencing or working at home.[5] Even then, however, the question of what is "enough" has not been touched on.

25.3.2 Food

Our current consumerist lifestyle with its excesses of food waste and means of production that constantly breaches planetary boundaries offers many opportunities to apply a sufficiency perspective when it comes to food. We will thus have a closer look at two policies that may have the potential to foster sufficiency: one penalizing supermarkets for food waste and one taxing conventional meat and dairy products (i.e., not those from organic farming) (Spengler, 2018).

Policies penalizing supermarkets for food waste tend to aim at "avoiding" wasteful resource use. Whether the intended reduction in resource use actually happens is, however, a function of retailers' reactions to such a policy. They could decide to reduce the price of products due to expire. If consumers then simply decide to buy this product rather than one with a later due date, this may indeed lead to an avoidance of resource waste. If, however, consumers decide to purchase more of a product due to the cheaper price (and not reduce the purchase of other products in a corresponding manner), a reduction in resource consumption may well not occur. Indeed, there is a risk that consumers overstock and that the waste then simply occurs in households rather than supermarkets. An alternative potential response to the identified policy option by retailers could be to reduce products in stock. Such a measure would, in fact, likely lead to an "avoidance" of food waste and therefore resources use, especially if taken by supermarkets across the board. The latter condition already signals that we cannot trust supermarkets to react in this way. In an often highly competitive market, the view pertains that consumers expect full shelves at all times, and that empty shelves will drive them to competing markets. Thus, surplus production is actually a structural default built into the system (Gumbert, 2022). Accordingly, it is by no means certain that the identified policy measure will really lead to an "avoidance" of resource consumption. To raise the potential of achieving this goal, at a minimum, dialogue with food retailers regarding stocking practices and consumers regarding the avoidance of surplus consumption would be needed.

A policy taxing meat and dairy products from conventional farming aims to reduce the consumption of products that are particularly harmful to the environment. The way to achieve this is by financially incentivizing consumers to buy organic products, hence avoid foods with bigger CO_2 footprints. However, such a policy will in many cases only lead to a "shift" in consumption, which may or may not imply an improvement in the ecological footprint of one's diet at the end of the day. Typically, a shift to an organic, plant-based diet is expected to foster such an improvement. However, it is not guaranteed. An organic avocado adds more to global CO_2 emissions than a local, non-organic apple. Accordingly, food policies decidedly aiming at reducing the ecological footprint of diets have to balance delicately the complexities of production forms and distances travelled by products.

In conclusion, the two examples of policies aiming to foster an avoidance of food waste or the consumption of CO_2-intensive foods show the difficulties associated with achieving a real reduction in resource consumption. Again, in many cases the reaction will likely involve a shift, and attempts to raise the reduction potential of such a shift will require additional considerations. Again, this does not mean that the two policies are not well intended and hold potential for an absolute reduction in resource use. It simply shows the complex dynamics to be considered and the potential need for accompanying measures. Even in the best case, however, they are far from fostering sufficiency in the sense of "enoughness" and the necessary balancing of individual needs with existing planetary boundaries.

25.3.3 Housing

One of the most promising sufficiency approaches in housing is to reduce the per capita living area – which to date is still on a constantly increasing path. Not only would this remove the burden from further soil sealing, but so would every square metre that does not need to be heated or cooled and is not equipped with additional stuff. Policy support could approach it from two directions: providing benefits for those who like to reduce their individual space and increasing burdens for those living on a larger sized area. Policies for the former can include a broad range of instruments including a reduced price for buying community/municipality ground if it is for example a more generational or community initiative. National policies could set reduced taxes/fees for owners' associations proving that they build – or better refurbish existing – houses based on a low individual per capita living area. All such instruments could enable individuals and households to move to a place where some rooms and facilities are shared and thus less individual living space is needed.

Progressive property taxation (Cohen, 2020) comes in at the other end as it increases the financial burden on buying large houses or apartments. As a side effect it contributes to better equity as it especially targets wealthier parts of society. The progressive taxation would, nevertheless, provide incentives for smaller flats and houses, perhaps even in the construction phase.

While all mentioned examples could be classified as avoidance strategies there are, nevertheless, arguments that do not perceive it this way. Taxes, after all, are an example of a classical steering instrument. Also, the steepness of progression would make a difference and in sum the instrument runs the risk of mainly being perceived as a burden and not as something which helps us to reconsider our needs. For the positive incentives it is interesting to recognize that even sufficiency proponents tend to sell living space reduction as an efficiency. Thus, floor space reduction could be defined as an improvement as well. This depends on whether we argue via a need for a certain amount of space to fulfil individual needs for shelter and comfort, or whether we argue via the efficiency improvements realized through shared living space.

25.4 Conclusion and Future Research

In this chapter, we have shown the breadth of the interpretations and framing of sufficiency as well as the difficulties associated with identifying a sufficiency focus and potential in policy measures. With respect to the former, we have argued that a notion of sufficiency that (re) focuses on the double sense of "enoughness" at its core is the only one that will allow the pursuit of a sustainability transformation. With respect to the latter, specifically, we have demonstrated that using the "improve-shift-avoid" framework suggested by some as an approach to differentiating between different policy types or objectives does not provide the clear answers one might expect, using policy examples from the consumption sector, mobility, food, and housing. In particular, policies potentially aiming at "avoiding" certain resource uses will always have to consider the range and likelihood of alternative ways of satisfying a given need. In many if not most cases, policies aiming at "avoid" at first glance will actually lead to "shift" if not (merely) "improve", and may even increase resource use via alternative forms of need satisfaction. In the end, the only way to identify clearly sufficiency policies is to look for a definition of "enoughness". After all, if we make that the standard, we end up with little evidence of sufficiency policies being implemented or placed on the policy agenda. Of course, this finding should not be understood as a general discrediting of policies aiming at absolute reductions in resource use that exist or are being discussed. All discussed policies are valuable in their aim to reduce our resource use. Still, our results spotlight the ongoing lack of consideration of sufficiency policies in practice.

Given the crucial and short-term need for a paradigm shift towards sufficiency, our analysis should not be understood as questioning the need and potential for sufficiency-oriented policies. On the contrary, our findings only serve to underline that the development of methodologies for designing and evaluating sufficiency policies remains a pivotal task for scholars and practitioners alike. Moreover, it will be important to accompany relevant processes aiming to turn recent decisions by policy actors, such as the Deutsche Städtetag (Association of German Cities), to integrate sufficiency in their sustainability strategies into practice. In the end, most fundamentally, we will need a broad societal dialogue about the question of "what is enough".

Box 25.1 Chapter Summary

- Exploring the meaning of sufficiency in environmental policy discourse.
- Delineating the variance in definitions used and the associated effects for evaluating policies with respect to their sufficiency focus.
- Illustrating these difficulties with policy examples in the consumption areas of mobility, food, and housing.
- Highlighting the current lack of real sufficiency-oriented policies.

Notes

1 Often efficiency is paired with the principle of effectiveness, serving to evaluate governance or business strategies and policies. In this chapter, however, we are referring to efficiency as one of the three complementary sustainability principles: efficiency, consistency, and sufficiency.

2 Consistency is the most controversial of the three strategies, at the conceptual level, as much of the changes sought via a consistency strategy will lead to reductions in resource use for a given level of output and thus can also be viewed through an efficiency lens.

3 In this sense, a sufficiency lens is closely related to strong sustainable consumption governance (Lorek & Fuchs, 2013), consumption corridors (Fuchs et al., 2021; https://www.uni-muenster.de/Fuchs/en/forschung/projekte/konsumkorridore.html), and 1.5° lifestyle (onepointfivelifestyles.eu/) perspectives.

4 The difference between "interpretations" and "discursive framings" is the ascription of intentions to influence the connotations associated with a concept such as sufficiency on the part of recipients of a communication. As we do not pursue a research design here that would allow the ascription of such intentions, we have to remain agnostic to them and use the terms "interpretation" and "framing" interchangeably.

5 Teleconferencing and the home office, however, should not be seen as the panacea for mobility reduction concerning work as we are aware that mobility needs need to be balanced with social needs, like team exchanges in person and an adequate office environment.

References

Alcott, Blake. 2008. The sufficiency strategy: Would rich-world frugality lower environmental impact? *Ecological Economics* 64 (4): 770–786.

Alexander, Samuel. 2015. *Sufficiency Economy: Enough, For Everyone, Forever.* Melbourne: Simplicity Institute.

van den Berg, Nicole; Hof, Andries; Akenji, Lewis; Edelenbosch, Oreane; van Sluisveld, Mariesse; Timmer, Vanessa; von Vuuren, Detlef. 2019. Improved modelling of lifestyle changes in Integrated Assessment Models: Cross-disciplinary insights from methodologies and theories. *Energy Strategy Reviews* 26: 100420.

CreuCreutzig, Felix; Roy, Joyashree; Lamb, William F.; Azevedo, Inês M. L.; Bruine de Bruin, Wändi; Dalkmann, Holger; Edelenbosch, Oreane Y.; Geels, Frank W.; Grubler, Arnulf; Hepburn, Cameron; Hertwich, Edgar G.; Khosla, Radhika; Mattauch, Linus; Minx, Jan C.; Ramakrishnan, Anjali; Rao, Narasimha D.; Steinberger, Julia K.; Tavoni, Massimo; Ürge-Vorsatz, Diana; & Weber, Elke U. 2018. Towards demand-side solutions for mitigating climate change. *Nature Climate Change* 8 (4): 260–263.0–263.

Cohen, Maurie. 2020. New conceptions of sufficient home size in high-income countries: Are we approaching a sustainable consumption transition? *Housing, Theory and Society* 38 (2): 1–31.

Deutscher Städtetag. 2021. Nachhaltiges und suffizientes Bauen in den Städten. 2wegewerk GmbH (Eds). Retrieved 06.09.2021. from: https://www.staedtetag.de/publikationen/weitere-publikationen/2021/handreichung-nachhaltiges-und-suffizientes-bauen.

Di Giulio, Antonietta; Fuchs, Doris. 2014. Sustainable consumption corridors: concept, objections, and responses. *GAIA* 23 (1): 184–192.

Fawcett, Tina; Darby, Sarah. 2019. Energy sufficiency in policy and practice: the question of needs and wants. *Proceedings of European Council for an Energy Efficient Economy, Summer Study*, France, Belambra Presqu'île de Giens: 361–370.

Fuchs, Doris; Di Giulio, Antonietta; Glaab, Katharina; Lorek, Sylvia; Maniates, Michael; Princen, Tom; Ropke, Inge. 2016. Power: The missing element in sustainable consumption and absolute reductions research and action. *Journal of Cleaner Production* 132: 298–307.

Fuchs, Doris; Sahakian, Marlyne; Gumbert, Tobias; Di Giulio, Antonietta; Maniates, Michael; Lorek, Sylvia; Graf, Antonia. 2021. *Consumption Corridors: Living Well within Sustainable Limits.* London: Routledge.

Gumbert, Tobias. 2022. *Responsibility in Environmental Governance. Unwrapping the Global Food Waste Dilemma.* Basingstoke: Palgrave Macmillan.

Grunwald, Armin. 2010. Wider die Privatisierung der Nachhaltigkeit – Warum ökologisch korrekter Konsum die Umwelt nicht retten kann. *GAIA* 19 (3): 178–182.

Hayden, Anders. 2014. *When Green Growth Is Not Enough: Climate Change, Ecological Modernization, and Sufficiency.* Montreal: MQUP.

Hayden, Anders. 2020. Sufficiency. In Kalfagianni, Fuchs, Hayden, Anders (Eds.) *Routledge Handbook of Global Sustainability Governance*. London: Routledge, 151–163.

Kanschik, Philipp. 2016. Eco-Sufficiency and distributive sufficientarianism-friends or foes? *Environmental Values* 25 (5): 553–571.

Kothari, Ashish; Salleh, Ariel; Escobar, Arturo; Demaria, Federico; Acosta, Alberto 2019. Introduction: Finding Pluriversal Paths. In Kothari, Ashish; Salleh, Ariel; Escobar, Arturo; Demaria, Federico; Acosta, Alberto (Eds.) *Pluriverse. A Post-Development Dictionary*. New Delhi: Tulika Books, xxi–xl.

Linz, Manfred. 2002. Warum Suffizienz unentbehrlich ist. In Linz, Bartelmus, Hennicke, Jungkeit, Sachs, Scherhorn, Wolke, von Winterfeld (Eds.) Von nichts zuviel. *Wuppertal Papers* 125: 7–14.

Linz, Manfred. 2013. Ohne sie reicht es nicht: Zur Notwendigkeit von Suffizienz. *Politische Ökologie* 135: 24–32.

Lorek, Sylvia; Fuchs, Doris. 2013. Strong sustainable consumption governance–precondition for a degrowth path? *Journal of Cleaner Production* 38: 36–43.

Malghan, Deepak. 2019. Efficiency. In Kothari, Ashish et al. (Eds.) *Pluriverse. A Post-Development Dictionary*. New Delhi: Tulika Books, 50–52.

Mamut, Pia (forthcoming). *Sufficiency – An Emerging Discourse?* [Sustainable Development in the 21st Century]. Baden-Baden: Nomos.

Maniates, Michael. 2001. Individualization: Plant a Tree, Buy a Bike, Save the World? *Global Environmental Politics* 1 (3): 31–52.

Maniates, Michael; Meyer, John. 2010. Asking the Right Questions. In Maniates, Meyer (Eds.) *The environmental politics of sacrifice*. Cambridge: MIT Press, 9–12.

Metzner-Szigeth, Andreas. 2019. Strategies for Eco-Social Transformation: Comparing Efficiency, Sufficiency and Consistency. In Ambrosio, Vezzoli (Eds.) *Designing Sustainability for All*. Proceedings of the 3rd LeNS World Distributed Conference, 649–654.

Mongsawad, Prasopchoke. 2012. The philosophy of the sufficiency economy: a contribution to the theory of development. *Asia-Pacific Development Journal* 17 (1): 123–143.

Muller, Adrian. 2009. Sufficiency – does energy consumption become a moral issue? Proceedings of European Council for an energy efficient economy, 2007 Summer Study, 83–90.

Muller, Adrian; Markus Huppenbauer. 2016. Sufficiency, liberal societies and environmental policy in the face of planetary boundaries. *GAIA* 25 (2): 105–109.

Princen, Thomas. 2003. Principles for sustainability: From cooperation and efficiency to sufficiency. *Global Environmental Politics* 3 (1): 33–50.

Sachs, Wolfgang. 1993. Die vier E's -Merkposten für einen maß-vollen Wirtschaftsstil. *Politische Ökologie* 33: 69–72.

Samadi, Sascha; Gröne, Marie-Christine; Schneidewind, Uwe; Luhmann, Hans-Jochen; Venjakob, Johannes; Best, Benjamin. 2017. Sufficiency in energy scenario studies: Taking the potential benefits of lifestyle changes into account. *Technological Forecasting and Social Change* 124: 126–134.

Scherhorn, Gerhard. 2002. Die Logik der Suffizienz. In Linz, Bartelmus, Hennicke, Jungkeit, Sachs, Scherhorn, Wolke, von Winterfeld (Eds.) Von nichts zuviel. *Wuppertal Papers* 125. Wuppertal: Wuppertal Institut 15–26.

Schneidewind, Uwe; Zahrnt, Angelika. 2014. *The Politics of Sufficiency. Making it easier to live the good life.* Munich: Oekom.

Shrivastava, Aseem. 2019. Prakritik Swaraj. In Ashish Kothari, Ariel Salleh, Arturo Escobar, Federico Demaria, and Alberto Acosta (Eds.) *Pluriverse. A Post-Development Dictionary*. New Delhi: Tulika Books, 283–286.

Spengler, Laura. 2016. Two types of 'enough': sufficiency as minimum and maximum. Environmental Politics 25 (5): 921–940.

Spengler, Laura. 2018. *Sufficiency as Policy*. Baden-Baden: Nomos.

Stengel, Oliver. 2011. *Suffizienz: Die Konsumgesellschaft in der ökologischen Krise*. Munich: Oekom.

Svampa, Maristella. 2019. The Latin American Critique of Development. Kothari, Ashish et al. (Eds.) *Pluriverse. A Post-Develeopment Dictionary*. New Delhi: Tulika Books, 18–21.

Toulouse, Edouard; Sahakian, Marlyne; Lorek, Sylvia; Bohnenberger, Katharina; Bierwirth, Anja; Leuser, Leon. 2019. Energy sufficiency: How can research better help and inform policy-making? *Proceedings of European Council for an Energy Efficient Economy*, Summer Study, France, Belambra Presqu'île de Giens: 331–339.

Vadovics, Edina; Živčič, Lidija. 2019. Energy sufficiency: are we ready for it? *Proceedings of European Council for an Energy Efficient Economy, Summer Study*, France, Belambra Presqu'île de Giens. 159–168.

Wiedmann, Thomas; Lenzen, Manfred; Keyßler, Lorenz; Steinberger, Julia. 2020. Scientists' warning on affluence. *Nature Communications* 11: 3107.

von Winterfeld, Uta. 2007. Keine Nachhaltigkeit ohne Suffizienz: Fünf Thesen und Folgerungen. *Vorgänge* 176 (3): 46–54.

Wissen, Markus; Brand, Ulrich. (2016): Imperiale Lebensweise und die politische Ökonomie natürlicher Ressourcen. In Fischer, Karin, Jäger, Johannes, Schmidt, Lukas (Eds.): *Rohstoffe und Entwicklung. Aktuelle Auseinandersetzungen im historischen Kontext.* Wien: New Academic Press, 235–248.

Zell-Ziegler, Carina; Thema, Johannes; Best, Benjamin; Wiese, Frauke; Lange, Jonas; Schmidt, Annika; Toulouse, Edouard; Stagl, Sigrid. 2021. Enough? The role of sufficiency in European energy and climate plans. *Energy Policy* 157: 112483.

26

THE NEW CLIMATE MOVEMENT

Organization, Strategy, and Consequences

Aron Buzogány and Patrick Scherhaufer

26.1 Introduction

Civil society organizations are widely recognized as essential at all environmental and climate governance levels. Environmental civil society organizations, such as Greenpeace, Friends of the Earth, or the World Wildlife Fund for Nature, have become respected actors in national and global environmental policy-making (Berny and Rootes, 2018). In the fields of energy and climate policy, civil society organizations are often seen as promoting the diffusion of new energy technologies (Vasi, 2011), opposing new energy projects (McAdam and Boudet, 2012), or experimenting with new energy ownership models (Bauwens et al., 2016; Moss et al., 2015). Such policy-oriented studies highlight the role of actor coalitions and public mobilization in policy changes affecting the energy sector (Hess, 2018). The emphasis is on interest intermediation; civil society is recognized as one actor among many seeking to maximize its policy goals (Brulle, 2010; David, 2018; North, 2011). The public policy literature tends to hold functionalist, goal-oriented views on the role of public participation and civil society involvement in the policy process, emphasizing the role of civil society as a bottom-up force for policy innovation and societal democratization, the importance of open planning cultures, and the inclusion of key stakeholders and the lay public in achieving policy goals, such as more progressive climate policies, higher levels of renewable energy deployment, or changes in the means and sources of energy production.

What policy-oriented perspectives in environmental and climate governance often fail to see are new developments in social mobilization related to environmental and climate policy that have become particularly evident in the last few of years. The new climate movement is the latest example of a movement that can justifiably claim to have the transformative power needed for large-scale historical change (Hadden, 2014; Perez et al., 2015; Stuart et al., 2020; Fisher and Nasrin, 2021). Many observers argue that if there is hope that the future will lead to sustainable transformations that avert climate change, biodiversity loss, and fossil fuel dependency, this can only be achieved through large-scale bottom-up societal mobilization (Hess, 2018). With a world on the brink of climate tipping points, social tipping might be the only hope left (Milkoreit et al., 2018; Otto et al., 2020; Winkelmann et al., 2022).

Climate change and its consequences are among the issues with the highest mobilization potential across Europe (Wahlström et al., 2019). A new climate movement has emerged and

DOI: 10.4324/9781003043843-30

has succeeded in stirring the masses and bringing protests and civil disobedience as relevant forms of resistance back into the public space (Ruser, 2020; Scherhaufer et al., 2021; Buzogány & Scherhaufer, 2022). In 2019, in probably the largest protest event ever, over seven million people took to the streets simultaneously in 185 countries to protest for better climate policies. But the new climate movement cannot be reduced to large-scale and colourful mass protest: more disruptive actions including blocking city highways or throwing tomato soup on Vincent van Gogh's Sunflowers by protest groups, such as Just Stop Oil or Last Generation, are also part of the same trend. This new wave of protest takes place in a context of increased emergency and shows that conflicts over sustainable transformations are not only about the environmental, economic, and social consequences of climate change (Markard, 2018; Radtke and Scherhaufer, 2022) but also about democracy and the future of democratic decision-making (Burnell, 2012; Fischer, 2017).

Against this background, this chapter maps the existing scholarship and synthesizes what we know about the new climate movement. We will mainly focus on three emblematic groups – Fridays for Future (FFF), Extinction Rebellion (XR), and Ende Gelände – by looking at their 'knowledge repertoires' (Della Porta & Pavan, 2017) and their 'repertoires of action' (Tilly, 1977), that is the various activities used to promote climate and the consequences these actions have.

In so doing, we draw on different disciplines within the social sciences. The main area we review is social movement studies, which have developed a strong interest in climate mobilization (McAdam, 2017) and which can build on the rich research tradition of change-oriented social movements, such as the environmental movement. Methods and findings from social movement studies are essential for public policy scholarship interested in how policies are made and to what effect. As suggested by the nascent conversation between science and technology studies (STS) and social movement studies (Hess, 2015), we propose that these two subfields must cooperate more closely to understand the new climate movement. At the same time, we should not forget other fields, such as political theory (Mattheis, 2022; Scheuerman, 2022), media and communication studies, anthropology (Von Storch et al., 2021; Posmek, 2022), or energy social science, as new fields of research (Krüger et al., 2022; Sovacool, 2022) that offer their own critical perspectives on the climate movement and that have been included in this comprehensive review.

Section 26.2 reviews the literature on the climate movement. It provides an overview of scholarly research focused on the composition of protest participants and the structures of organizations involved in social activities. Section 26.3 dwells more profoundly on the strategies and tactics of these movements, highlighting legal as well as illegal repertoires of action. Section 26.4 provides an overview of the various direct and indirect consequences of the new climate movement. We conclude with suggestions for future research.

26.2 The New Climate Movement: Emergence, Narratives, Organization

In this section, we review the literature on the emergence, organizational forms, and narratives of the new climate and environmental movement from the end of the 2010s. While we draw on various examples, our empirical focus is on three emblematic groups representing different aspects of the new climate movement: Fridays for Future, Extinction Rebellion, and Ende Gelände.

Fridays for Future (FFF) grew out of the mobilization inspired by Greta Thunberg's school strike. FFF is predominantly run by teenagers and young adults and as an organization is committed to non-violence and non-partisanship (Wahlström et al., 2019; De Moor et al., 2020). FFF is a decentralized network of local and regional groups, as well as subgroups like Scientists for Future and Parents for Future (Capstick et al., 2022). FFF initially began with smaller school strikes that took place on Fridays, but also became involved in mobilizing for the Global Earth Strikes, which led to a change in the age structure of protest participants that now includes students, academics, and other groups (Sommer et al., 2019; Sommer and Haunss, 2020).

Extinction Rebellion (XR) emerged in 2018 from the RiseUp! Movement and was founded in Great Britain by Roger Hallam. While FFF is characterized by its large, worldwide demonstrations, XR focuses on smaller, more disruptive protests involving non-violent civil disobedience, including street blockades, parliamentary occupations, disruptions of commuter traffic, or activists sticking themselves to protest sites. XR's primary concern is climate change and species extinction, and it makes three central demands to politicians: "Tell the truth!", "Act now!", and "Live politics anew!" (Slaven and Heydon, 2020). A distinctive feature of XR's protests is an hourglass, which symbolizes the urgency to act. In 2022, XR counted 1,178 local groups that were active in 85 countries.[1]

Ende Gelände was born out of an insight within the climate movement that the broad, colourful, and large-scale protests against global climate summitry were utterly irrelevant at making an impact. Ende Gelände has been variously described as an 'alliance' (Bosse, 2017), as 'direct action' (Brock and Dunlap, 2018), as a 'movement' (Temper et al., 2015), as 'action alliance' (Häfner et al., 2016), as a 'network' (Rucht, 2017), or a 'campaign' (Toewe, 2017). In essence Ende Gelände is a broad network of anti-nuclear and anti-coal organizations in Europe, which share the belief that civil disobedience is needed to stop climate change (Kaufer and Lein, 2018; Temper, 2019; Kalt, 2021; Oßenbrügge, 2021).

26.2.1 Emergence

Why has the new wave of climate-focused mobilization emerged? Social movement scholars use various theories to explain collective action (McAdam et al., 1996). For deprivation theorists, social mobilization arises when people feel deprived of a particular resource. Resource mobilization theorists argue that deprivation alone is not enough to mobilize but emphasize the need of resources for mobilization. In contrast, theories of political opportunity structures underscore the importance of features of the political system and movement allies within the political system in explaining mobilization outcomes. Theories of social movements have been used to explain the emergence of 'old' social movements, such as the labour movement or 'new' postmaterialist movements such as the women's movement or environmentalism. Can these theories explain the emergence of the new climate movement?

Macro-comparative work uses these social movement theories to explain the uneven emergence of climate activism in Western democracies using multiple equifinal combinations of causal mechanisms, including trust in environmental movements, availability of resources through international non-governmental organizations, availability of information and communication technologies, and resonance of values related to environmental protection (Laux, 2021). However, most existing scholarship employs the toolbox of social movement studies either as a heuristic device or as a methodological toolbox rather than as an explanatory theory. Many explanations put forward are, therefore, somewhat ad hoc. Rucht and Sommer (2019) mention five main factors to explain the rise of the new climate movement: (1) Greta Thunberg as a role model and the effect that researchers describe as the 'Greta effect' (Hayes and O'Neill, 2021), (2) the attractiveness of the protest as a matter of "'truancy' (3), the appeal of simple demands to a political establishment that is difficult to understand in its complexity, (4) the mobilization effects of social networks based on digital communication, and (5) the role of climate activists who supposedly manipulate the young and seemingly clueless protesters. Coincidental events may also have played a role. The heatwave of 2018, for example, was a significant factor in the run-up to the Global Climate Action 2019.

Other explanations see long-term structural developments at work. Historical materialist and Polányian perspectives emphasize the role of capitalism and its inherent conflicts behind the emergence of the climate movement (Sander, 2016; Brechin and Fenner, 2017). Frustrations caused by failures to deal with climate change nationally and globally are also part of this explanation. At the same time, part of the frustration that has led to the emergence of the new movement is due to the failure of the 'old' environmental movement, which is seen as part of the eco-modernist governance system that it initially criticized (Mol, 2000). Discussions about the failure to civilize the global climate regime (Brunnengräber, 2011; Dietz and Garrelts, 2013, 2014) were instrumental in the rupture between reformist and radical groups within the climate movement (Della Porta and Parks, 2014). The reformist current of 'climate action' has emphasized the role of experimentation and bottom-up climate activism and included more moderate groups with a "focus on achieving technical goals (such as certain percentages of carbon emissions cuts or limitations to a certain number of degrees warming of the global climate)" (della Porta and Parks, 2014: 24).

In contrast, the 'climate justice' part used a more radical discourse that doubted the global capitalist political economy. The latter group manifested itself in climate camps and local struggles, including direct action at the local level (Bosse, 2017; della Porta and Parks, 2014; Gach, 2019). While these precursors are important, the metamorphosis and reconfiguration of the climate movement that began in 2018 was nonetheless unexpected (Boucher et al., 2021).

26.2.2 Profiles of Climate Protest Participants and Activists

Micro-level studies of climate mobilization ask who those are who take to the streets, what their motivations are, and how they became mobilized. We can distinguish between studies that profile participants in climate demonstrations and research focusing explicitly on climate activists. The first group typically uses protest surveys, many of which are carried out as a collaborative effort across multiple countries and are usually based on established methods of quantitative protest analysis. In contrast, the second group of studies is more qualitatively oriented.

Protest surveys show that at the individual level climate mobilization can be explained by personal networks and the motivations of participants (Van Laer, 2017). Surveys carried out in particular since 2019 allow for a very detailed study of FFF mobilization and have led to a growing number of publications on the sociological characteristics of climate protestors (Wahlström et al., 2019; de Moor et al., 2020; de Moor et al., 2021). Aside from its ability to mobilize young people to participate, FFF differs from other protests in its high proportion of women. Focusing on age cohort differences among FFF participants, studies show that participation levels and pro-environmental attitudes differ across generations (Cologna et al., 2021; Lorenzini et al., 2021; Zamponi et al., 2022). Multiple surveys of climate protestors illustrate the importance of personal and social factors, such as the participation of friends in the protest (Van Laer, 2017; Wallis and Loy, 2021), while Parth et al. (2020) show that participation in FFF strikes has a strong influence on political attitudes. Research also shows that while FFF protesters are disappointed, they are not resigned or disenchanted with politics (Daniel et al., 2020). They have a strong conviction that representative democracy and more aspiring political decision-making can limit climate change (Sommer et al., 2019). Cross-class alliances within the climate movement are emphasized by Della Porta and Portos (2021), who highlight the social heterogeneity of climate protestors but also find that the social backgrounds of protesters influence their repertoires of action. Experimental studies show significant differences among protest participants in their choice of policy instruments for addressing climate change (Soliev et al., 2021). Swedish FFF participants are shown to prioritize environmental framings over economic ones (Emilsson et al., 2020). Studies also confirm young people with more left-leaning values tend to be more concerned about climate change and the environment compared to those experiencing economic problems (Uba et al., 2022). Participants who trust business actors are less likely to participate in these demonstrations (Noth and Tonzer, 2022). Younger protestors, on the other hand, appear to hold more pro-business positions (Daniel et al., 2020). Using survey data collected during the US climate movement protests, Beer (2022) finds support for a fundamental systemic shift away from capitalism to addressing climate change.

Studies focusing on climate activists typically use qualitative methods and ethnographic observations to understand their worldviews and strategies. Jasny and Fisher (2022) find that climate justice and equality unite climate activists around coherent motivational frames. Based on interviews with activists in several European countries, de Moor (2021) finds that post-apocalyptic narratives are common among activists, but this does not affect their focus on climate mitigation. Doherty et al. (2020) find that XR's activists tend to be well-educated, middle-class, left-leaning, and engage in pro-environmental behaviour. A study of climate activists in 66 countries finds that protesters in the Western world are predominantly female and are committed to reducing greenhouse gases Boucher et al. (2021). Still, the actions they envisage differ by region. Similarly, Martiskainen et al. (2020) find variation among climate strikers regarding knowledge, emotions, motivations, and activities.

The turn to emotions as drivers of protest is a key aspect in the study of the new climate movement, analysed at the individual level through experimental and psychological studies and at the collective level by looking at the frames the climate movement uses for mobilization. Anger, worry, frustration, followed by anxiety and fear are the most dominant emotions for XR (Doherty et al., 2020). Surveys of activists show that collective mobilization is based on moral outrage, anger, global identification, or participative efficacy (Furlong and Vignoles, 2021). Another study shows that politicized social identity and group efficacy are positively related to the intention to participate in climate protests, while collective guilt and perceptions of environmental threats are indirectly associated with protest intentions (Haugestad et al., 2021). Focusing on the US climate movement, Nijjar (2022) shows how intergenerational differences affect how activists deal with climate change anxiety.

26.2.3 Mobilizing Frames

Beyond focusing on protest participants and climate activists individually, another large group of studies focuses on collective meaning-making and repertories of knowledge-production related to climate action (Della Porta and Pavan, 2017; Machin, 2022b). Building on various traditions within social movements studies, policy analysis, science and technology studies, and the analysis of frames – or discourses, narratives, and imaginaries – produced by climate activists, these studies use official documents and speeches, often complemented by qualitative interviews with activists and experts involved in the movement. While most studies focus on single-case studies and identify different issues of contestation (Daniel et al., 2020; Reichel et al., 2022; Spaiser et al., 2022), a growing number of comparative studies show important differences within the new climate movement (Melchior and Rivera, 2021; Buzogány and Scherhaufer, 2022; Friberg, 2022; Krüger et al., 2022).

While the 'climate justice' frame plays a central role in the meaning-making of the new climate movement, the use of apocalyptic and 'post-apocalyptic' imagery is prevalent (Cassegård and Thörn, 2018), and there are significant differences that also affect the activities of different parts of the climate movement. According to Buzogány and Scherhaufer (2022), FFF predominantly uses apocalyptic imaginaries of climate change. The responsibility for the success of sustainability transitions is associated with political decision-makers that act quickly and courageously. In contrast, XR mainly uses post-apocalyptic framings and evokes an ongoing climate catastrophe and mass extinction. Hence, more radical societal changes and political actions are necessary to save the planet and the survival of humankind.

A defining feature in the narratives of the climate movement is the *contestation of the existing political and economic system* – which is perceived as unjust, undemocratic, and capitalist (Machin, 2022b). But there are differences here as well. Regarding political order, FFF supports liberal democracy where elected representatives hold the authority, legitimacy, and capacity to deal with climate change. Still, mass protests are needed to get them to assume responsibility. XR, in turn, favours participatory and deliberative elements of democracy where citizens can make binding decisions themselves, often referring also to the anarchist tradition (Berglund and Schmidt, 2020). In general, XR expresses more disillusionment with traditional political governance capacities and calls for more grassroots deliberation and opportunities for democratic voice and participation.

Regarding the economic policy framework, FFF and XR criticize the existing power relations that hinder climate action. However, FFF's position oscillates between green pragmatism based on the paradigm of ecological modernization, which involves significant path dependency, and more radical perspectives related to degrowth. While sharing some of these positions, XR's communication in this field is less explicit and more geared towards emphasizing individual change and self-awareness. Both FFF and XR frequently refer to 'fossil capitalism' and the interests of established 'old industry', such as gas infrastructure and coal mining, rather than pointing fingers at 'capitalism' as such. Ende Gelände uses strongly anti-capitalist discourses (Bosse, 2017; Sander, 2017; Sander, 2016) and criticizes "the carbon economy as a symptom of larger inequalities created and exploited by the global capital" (Schlosberg and Collins, 2014, p. 364). The framing of 'capitalism' is closely intertwined with carbon-based modes of production, colonialism, and racism, often in conjunction with the Black Lives Matter movement. At the same time, FFF's heterogeneity, which includes platforms such as Entrepreneurs for Future, makes sure that the problem framing of FFF is well-balanced between moderate approaches that rely on the narrative of 'ecological modernization' and more radical ones which can be linked to the degrowth perspective (Marquardt, 2020). Interestingly, these debates are less present within more radical XR, which extends its general 'no blame' approach to polluting companies but also provides space for scientists linked to post-developmental views.

Due to these rather heterogeneous positions, the framing related to reforming the economic order remains ambiguous. FFF refers to the degrowth perspective and the typical strategies geared towards green growth, mostly implicitly relying on market-based solutions. FFF has frequently taken positions on concrete policy measures, such as the carbon tax. These positions are rarely radical and include standard textbook cases of neoclassic environmental economics. For instance, it is argued that most of the incumbent energy industry (coal, gas, nuclear) is not economically viable (Hochmann, 2020). As a solution, the strengthening of the market principles by creating a level playing field and removing subsidies with negative environmental impacts are proposed. While XR also takes a stand on specific policy issues and formulates prognostic positions, the emphasis here is on radical changes in everyday lifestyles and consumption behaviour.

One of the central characteristics of the new climate movement is the acceptance of (climate) science as the ultimate solution to climate change (Evensen, 2019). This includes the frequent invocation of calls such as 'Follow the Science' (FFF), 'Listen to the Science' (XR), 'Tell the Truth' (XR), coupled with a strong emphasis on the clarity of scientific findings on climate change. XR and FFF use much of their political communications to emphasize the role of science, blame climate sceptics, or disseminate scientific evidence and research on climate change. In the case of FFF, the direct proximity to science is also reflected in the involvement of the Scientists for Future movement, which brings together scientists who directly support FFF's demands. Using scientific evidence adds legitimacy to the climate movement's goals, increases the sense of urgency, and helps put pressure on the public debate, political parties, and individual politicians. Using climate science and acting as its ambassador creates a discursive alliance between the climate movement and climate science that is difficult to challenge. This is also related to the call for 'science-based' policies advocated by FFF and XR (Spaiser et al., 2022). While FFF often regards itself as an actor that aims to directly influence politics by amplifying the 'voice of science' and thus joins the chorus of interest groups active in this field, XR's focus is more on disseminating knowledge and bringing science closer to people, for instance by organizing teach-ins or workshops with established scientists. The emphasis is less on concrete policies and more on the participants' educational and discursive interaction and self-empowerment (Zantvoort, 2021).

Using a pro-science framing and capitalizing on its close connections with prominent scientists ultimately enabled FFF and XR to mobilize broad and sustained support for the movement and make climate change a central topic in public discourse (Rödder and Pavenstädt, 2022; Soßdorf and Burgi, 2022). While large segments of the scientific establishment have welcomed the attention given to research, both groups have been criticized for confirmation bias (i.e., selective use of scientific information) and for taking a rather simplistic and essentialist view of scientific production (Evensen, 2019; Bowman, 2020, Matthews, 2020). In contrast to the claim of a rationalist-technocratic focus of the new climate movement, there is also a discussion in the literature as to whether the movement uses – or indeed should use – a populist discourse, because the opposition to more progressive climate change policies is eminently political and can only be overcome politically (Zulianello and Ceccobelli, 2020; Buzogány and Mohamad-Klotzbach, 2022; Mouffe, 2022; Nordensvard and Ketola, 2022; Swyngedouw, 2022).

26.2.4 Organizational Structures

Whereas many studies on the new climate movement have focused on the knowledge repertoires used, little attention has been paid to the organizational structures that hold these groups together and make them functional. The distinction between the 'frontstage' and the 'backstage' (Rucht, 2017) of movements, the latter understood as the internal structures of groups, offers essential insights into the organizational cultures that have developed and the decision-making structures that characterize the new climate movement. From an organizational theory perspective, the main question is how the externally communicated norms of climate mobilization, such as democracy, decentral action, self-awareness, and mindfulness, correlate with internal decision-making structures (de Bakker et al., 2017). Due to potentially sensitive information, these questions are rarely addressed in research, and if so, the empirical basis is mostly case studies of individual local groups or campaigns.

The few existing studies highlight that, despite decentral or even anarchic organizational structures, the climate movement is impressively successful in maintaining its global reach and securing the authenticity of its 'brand' (Etchanchu et al., 2021). Important drivers for the global rise of new organizational chapters are landmark protest events (Gardner et al., 2022) and the relative vagueness and adaptability of their frames about how exactly to confront climate change (Kinniburgh, 2020). Some attention was paid to the management concept called 'holacracy', which XR adopted to spread power between the leading group and its many chapters acting autonomously and decentrally worldwide (Smiles and Edwards 2021). Studies of XR and Ende Gelände also emphasize protest camps and non-violence training as important venues of socialization for activists (Costa and Wittmann, 2021; Scherhaufer et al., 2021; Abajo-Sanchez, 2022). Climate camps and protest actions were crucial for prefigurative action, that is, spaces where movement activists develop (and live) their visions of a different society and economic system (Yates, 2021). Studies on prefigurative action draw attention to features of democratic quality in the climate movement, such as inclusion, leaderless organizational structures, transparency, or mindfulness. However, various studies show that many of these concepts are internally contested, and tensions emerge between global and local (Smiles and Edwards 2021) as well as individual and collective goals (Stuart, 2022). The decentralized structure of XR is crucial for the high frequency of actions, but it also makes it difficult to avoid controversial activities. The concept of a 'leaderless autonomous organization' has been questioned in interviews by XR members (Fotaki and Foroughi, 2022). The concept of 'mindfulness' also plays an important yet controversial role as it is often seen to stand for radical individualization and depoliticization; however, as Sauerborn (2022) argues, XR's usage offers a politicized version of the concept (see also Schweinschwaller, 2020; Westwell and Bunting, 2020).

26.3 Strategies

This section deals with repertoires of action – understood as various forms of activities – used by groups belonging to the new climate movement (Parks et al., 2022). Action repertoires related to energy and climate change occupy a continuum that ranges from non-violent legal actions to violent illegal forms of action (e.g., breaking into a nuclear power plant). In an overview article focusing on radical tactics in the US environmentalist movement, Brown (2021) distinguishes between legal activism, which includes lobbying, campaigns, litigation, boycotts, or participation in lawful demonstrations, and confrontational strategies, such as civil disobedience, sabotage, and violence. For Goodman (2022), social movements play three different roles: politicizing the impacts of climate change, contesting the causes, and advancing solutions. Nulman (2016) differentiates between five mechanisms of change which are based on different action repertoires: disruption, public preference, political access, judicial mechanisms, and international politics. Choosing action types from the available repertoire includes a strategic and tactical component, which effectively co-determines movement success (Doherty and Hayes, 2018). Tactical diversity means that movements are not limited to one strategy: the same movement can be involved in civil disobedience actions and litigation (Brown, 2021).

While absent from established repertoires of action, prefigurative action is increasingly recognized as a movement strategy of its own (Yates, 2021). Much of the literature confirms that prefiguration plays a central role for the new climate movement which builds on a long tradition of climate camps. Climate camps emerged in the 1990s as spaces of alternative heterotopias, mainly in the UK and Germany (Kaufer and Lein, 2018). Often underpinned by anarchist thought, climate camps provided spaces for radically reimagining and creating a new society that would break with the outside world and offer a utopian setting free of police violence and the constraints of a fossil lifestyle. While in practice, this utopia was confronted with internal rifts between reformers and radicals (Saunders and Price, 2009; Saunders, 2012), these camps became important sites for self-organized strategy-making, learning, and even specialized policy expertise for a new generation of activists (Frenzel, 2014; Hübinger, 2020; Corry and Reiner, 2021).

Organizing demonstrations and strikes, a second strategy, has become the signature of the new climate movement (Fisher and Nasrin, 2021). Climate strikes, such as the climate school strike, have attracted millions worldwide, and these large numbers of participants, coupled with the ongoing mobilization of young protestors, have had a powerful impact on societies. Major protest events successfully helped the movement spread worldwide (Gardner et al., 2022). Although COVID-19 was a threat to all forms of protests (della Porta, 2021), it can also be interpreted as an opportunity

> to experiment with new forms of actions like the 'rebellion of one' introduced by XR or the manifold livestream activities like concerts, interviews with celebrity guests and different formats for people to get involved within the framework of the Global Earth Strike in spring 2020 by FFF.
>
> *(Buzogány & Scherhaufer, 2022: 5)*

While the joyful protest events captured attention, parts of the new climate movement also used less visible strategies such as lobbying and litigation. Following these established strategies of environmentalist groups, the climate movement has targeted governments, international bodies such as the EU or the UN, but also business actors, fossil-fuel producers, investment funds, or universities with policy-related demands, comments, and criticism (Fisher and Nasrin, 2021). For instance, FFF Germany has commissioned a study from a well-known environmental think tank to assess the feasibility of achieving carbon neutrality by 2035 (Wuppertal Institute, 2020) and the results of this study were directly adopted as FFF's policy demands. FFF's list of demands to the EU includes profound reforms of agricultural subsidies or the inclusion of climate-related issues into trade negotiations[2] in a way that is not significantly different nor more radical the demands of other green organizations. These policy-related positions are often framed in terms of climate justice arguments. EU-targeted campaigns[3] and open letters are another sign of bitter criticism, but also of the acceptance of the need for common, EU, or global level action in the field of climate policy.[4]

Together with climate strikes, acts of civil disobedience can be regarded as one central strategy of the new climate movement. The predominant understanding of civil disobedience is non-violent, civil, conscientious, and public lawbreaking (Scheuerman, 2022). Civil disobedience occurs whenever the formal and informal channels of democratic institutions provide no space for manoeuvring, when policy change is unlikely or when "the urgency of problems, in the eyes of protesters, demands swift reaction" (Lauth, 2000: 38). The overall assumption of citizens relying on civil disobedience is that other legal ways of protest (such as demonstrations, boycotts, and petitions) to stop harmful activities are insufficient and therefore have to be expanded by non-legal actions. This understanding of civil disobedience is associated with a perception of justification, where non-legal actions are interpreted as 'just' as long as they help overcome 'unjust' laws (Dow, 2018; Rawls, 1985). Using references to Martin Luther King or Gandhi, non-violent civil disobedience has become an important theoretical and practical device for climate activists. For instance, the 'civil resistance model' (Hallam, 2019) of the founder of XR, Roger Hallam, uses political science research by Chenoweth and Stephan (2011) to show that the "only way to overcome entrenched political power is through extensive campaigns of large-scale nonviolent direct action" (Hallam, 2019). Climate school strikes were also often framed as civil disobedience, as school attendance is legally mandatory. At the same time, non-violent civil disobedience is increasingly embraced also by established environmental groups. This is undoubtedly true for Greenpeace, which was historically among the first environmentalist groups to use this strategy, but also for the lobbying-oriented group Sierra Club in the USA, or the Sunrise Movement (Brown, 2021, 7). As Sovacool and Dunlap (2022) highlight, non-violent civil disobedience includes many strategies, including demonstrations, mass arrests, occupations, sit-ins, boycotts, hacktivism, strikes, and hunger strikes. Examples of occupations of energy infrastructure include the anti-coal protests organized by Ende Gelände in Europe or the protests in the USA against the Dakota Access Pipeline. Such protests, termed by Naomi Klein (2014) 'Blockadia', are among the more radical forms of climate activism (Temper, 2019). They mark two significant developments: the turn towards civil disobedience and the increasing importance of the spatial dimension of contention, meaning that protests are taking place directly at the sites of conflict, such as forests, mines, or pits.

Anti-authoritarian strategies of resistance, together with guerrilla warfare perspectives, provide more radical and more violent strategies of action, which are increasingly reflected in the literature (Sovacool and Dunlap, 2022). The starting point is the general debate about the effectiveness of non-violent direct action (Gelderloos, 2007; Chenoweth and Stephan, 2011), which, despite essential differences in the original research focusing mainly on non-democracies, is increasingly being conducted around climate action (Delina, 2019; Matthews, 2020; Delina, 2022; Gelderloos, 2022). Andreas Malm's influential "How to Blow Up a Pipeline: Learning to Fight in a World on Fire" (Malm, 2021) argues that the violent flanks of historical social movements e.g. (anti-slavery, anti-apartheid) had an important role to play in those movements' long-term success. A comprehensive inventory put forward by Sovacool and Dunlap (2022) distinguishes between 20 direct action tactics, including unpermitted demonstrations, blockading, eco-sabotage, ecotage (or 'monkey wrenching'), or 'climatage' (climate action + sabotage), where protestors willingly destroy or blockade infrastructure. This can include disrupting traffic, occupying forests, deflating SUVs in affluent communities, or interfering with constructions. All this might include violence against things and, in some cases, also against persons, as the last resort in the strategic repertoire of social movements. While violence has been rarely used in general by environmentalists or the climate movement, some historical examples can be considered eco-terrorism (da Silva, 2020), such as Earth First!, the Animal Liberation Front (ALF), or the Earth Liberation Front (ELF). More recent examples include the El Paso and Christchurch killings, which used eco-fascist rhetoric that can be traced back to the racist 'blood and soil' doctrine of the German Nazi party and blames immigration for environmental destruction (Malm and The Zetkin Collective, 2021; Macklin, 2022; Silke and Morrison, 2022).

Little research has addressed why specific strategies were chosen from the available repertoire, a question that is important for the 'tactical diversity' argument. Hadden and Jasny (2019) suggest that peer pressure might play a role: protest tactics are often copied from adjacent organizations. At the same time, the characteristics of the political context and the internal structure of movement organizations might also play a role (Wong, 2012).

26.4 Consequences

What are the consequences of climate action, and how do the various strategies influence (policy) change? Many large-scale historical changes were triggered by social movements, including the labour, women, and environmental movements (Giugni, 2004; Bosi et al., 2016, Sovacool, 2022). While the short-term impacts of social movements on policies often remain unclear (Rochon and Mazmanian, 1993; Giugni et al., 1999; Amenta et al., 2010), both historical and contemporary evidence suggests that social movements have successfully contested entrenched power structures and led in the long term to social change. However, it is difficult to explicitly measure the success of movements because it is usually hard to attribute the results of activism to movement strategies (Burstein, 1999; Amenta et al., 2010). Therefore, defining success is a highly political issue for movements and societies. While most research concerns policy change due to activism and thus focuses on state institutions and policy output (Van Dyke et al., 2005), it is helpful to look also at different domains where change might happen. This includes attention to the consequences of climate action along four dimensions: people, discourses, parties, and policies.

People, the first dimension, focuses on movement *activists* themselves. In line with the literature on the biographical consequences of social movement activism (Giugni, 2008), prefiguratively performing new social practices or envisioning 'real life utopias' (Wright, 2010) has a personal impact on activists. While climate activists' demands are highly abstract and usually difficult to meet, success at the movement level is often measured more subjectively, for example by the size of protests, the growth of the number of activists, or media attention (Etchanchu et al., 2021).

This relates to *societal discourses* as a second dimension, focusing on social movements' role in societal debates. Social movements are essential for opening up new deliberative spaces and changing the terms of the debate through discursive shifts. They represent actors, locations, or practices in ways that take structural inequalities into account more than the established institutions of representative democracy normally do (della Porta and Rucht 2013), and they provide spaces for the production of new kinds of knowledge and innovation (Della Porta and Pavan, 2017). Therefore, some literature states that progressive social movements can evoke both the promotion of policy reforms and have a positive impact on the functioning of democracy (Della Porta, 2020). Frames, narratives, and imaginaries produced by the movement are essential to understand a movement's claims. As the literature on digital activism shows, social media provides important channels of dissemination for social movements and is also a low-threshold entry point for mobilization (Van Laer and Van Aelst, 2010). Several studies show that, given the existence of this digital backbone, the climate movement proved relatively resilient to the challenges posed by mobilization during the coronavirus pandemic (Mucha et al., 2020; Hunger and Hutter, 2021; Reif et al., 2021; Sorce and Dumitrica, 2021; Chen et al., 2022; Fernandez-Zubieta et al., 2022).

At the same time, the climate movement has been successful in bringing climate change onto the political agenda (Marquardt, 2020: 14). Although many have criticized the demands of the climate movement as unrealistic and the disruptive tactics it has used, climate action has changed the discourse from whether to how. Previously impossible options – such as the coal exit – became suddenly possible. Members of Fridays for Future and other groups now sit on talk shows, work with science to propose solutions, talk to politicians, start petitions, organize or accompany climate protection projects in communities, and much more.

Beyond the climate movement's production of knowledge about its actions, goals, and values, media and communication research show the external framing power of traditional and new social media concerning the climate movement. In contrast to own narratives produced by the climate movement, external media reporting has an obvious filtering function (Bergmann and Ossewaarde, 2020; Goldenbaum and Thompson, 2020; Huttunen and Albrecht, 2021; Von Zabern and Tulloch, 2021; Wozniak et al., 2021). Media studies highlight that reports about the climate movement, and climate change in general, have increased manifold. At the same time, press coverage has remained one-sidedly focused on events (such as demonstrations), personification (especially the iconic figure of Greta Thunberg), or political controversies (such as Roger Hallam's comparison of the Holocaust to the genocide in the Congo). In a study of online media coverage in Germany, Von Zabern and Tulloch (2021) found that while protests are given space and climate change is framed as conflicts of intergenerational justice, existing power structures are reproduced by marginalizing and depoliticizing the protests. A study of Australian mainstream media shows that school strikes were framed using anticipatory narratives emphasizing that pupils should be punished for staying away from school and protectionist narratives suggesting that the striking students were indoctrinated (Alexander et al., 2022).

Changes in the societal discourse can also be observed by looking at public opinion. The long-term change in climate awareness can be expressed in figures: in 2021, the Eurobarometer survey showed that 93 percent of respondents considered climate change a serious problem. Beyond this overall acceptance, there is much debate in the literature about how movements and their actions are perceived by the general public. On the one hand, several studies argue that radicalization can increase support for social movements. The controversy revolves around the so-called 'radical flank effect' (Haines, 2013). This suggests that radical flanks or fractions of social movements can increase support for moderate factions. An early study of Just Stop Oil protests shows no loss of support for climate policies despite disruptive protests in the UK (Ozden and Glover, 2022). Experimental studies of the climate and animal advocacy movement *suggest that radical strategies do not automatically reduce support for the movement* as a whole (Simpson et al., 2022). Similarly, Shuman et al. (2021) show that non-violent protest using street blockades or sit-ins can effectively generate support. Again, Bugden (2020) indicates that peaceful marches increase support for the movement among independents and Democrats in the USA, and that civil disobedience positively affects Democrats, while no backlash effects are found for any group. At the same time, radical flank effects can also lead to discursive changes of mainstream organizations (Schifeling and Hoffman, 2017).

In contrast, another group of studies points to the dilemma of activism, showing that peaceful and legal protests gain more support while radical protests lead to a decline in support. All this comes, however, at the price of a decrease in social identification with the movement (Feinberg et al., 2020). Following one year of intensive XR disruptive action in London, climate action is less rejected than before, but those who had experienced some disruption are more hostile towards climate protection (Kountouris and Williams, 2022). A meta-analysis of the effects of non-violent action on third-party groups by Orazani et al. (2021) shows a small-to-moderate positive impact on support, while violent strategies are found not to increase support. Ünal et al. (2022) argue that people's preferences to support violent or non-violent strategies depend on how they are affected by the conflict.

Public opinion shifts are relevant for political *parties* and their members of parliament (MPs), who are among the main targets of social movement action. Legislative scholars assume that parties and MPs respond to signals from different sources: from government, lobby or interests groups, constituencies, public opinion, or the media (Miler, 2010; Farrer, 2017). These studies also show that MPs respond to protest activities and that the size of a protest event and the unity of the protestors are the most important factors shaping their attitudes (Wouters and Walgrave, 2017). This reinforces the classic 'WUNC' argument made by Charles Tilly, who claimed that to be successful, a protest needs to demonstrate worthiness, unity, numerical strength, and commitment to becoming widely accepted (Tilly, 2006). In a later study, Wouters (2018) complements the argument to 'dWUNC' by showing that more *diverse* protest participants gather more support.

Social movement activities are also found to influence electoral results. Of particular relevance for the climate movement are the results of Green parties. In the USA, violent environmental action tends to lower the results of the niche Green Party (Farrer and Klein, 2022). In Germany, party reactions to the emergence of the new climate movement show that party responses are determined by traditional left–right positioning and that support on the right wing remains low (Berker and Pollex, 2021; Berker and Pollex, 2022).

At the same time, there is evidence that more climate protest in electoral districts increases attention of politicians to these issues (Schürmann 2023).

The final dimension concerns the consequences of social movements on *policies*. This is an area where we still need more evidence. Most research in this field is conceptual, speculative, and based on single case studies (Lorenzen, 2020). Piggot (2018), for instance, argues that social mobilization can lead to decarbonization policies if there is a window of opportunity and bottom-up support by citizens mobilized by 'compelling framings'. Environmental movement organizations are important for policy-making, but this varies greatly, depending on the political context and the stage of the policy cycle. In the problem definition phase, protests and strikes often play an essential role in initiating legislative processes (Longhofer et al., 2016).

Regarding agenda-setting, US environmental organizations have suitable access and testify in significant numbers in legislative hearings due to their unique expertise and perceived legitimacy (Ganz and Soule, 2019). Unsurprisingly, the mobilization of climate policy supporters is found to have a positive effect on enacting environmental policies (Olzak et al., 2016, Böhler et al., 2022). Nash and Steurer (2021) argue that the extent and intensity of the climate discourse have a more significant influence on a country's climate legislation than the government's political orientation. Even if the causal link is difficult to draw, global climate events, different types of protests, climate litigation, climate citizen assemblies, media reporting, social media, or parliamentary debates all contribute to the discourse and to whether the discourse finally becomes transmitted into government action. Less is known about the later phases of the policy cycle, such as implementation, where social movement organizations use naming and shaming strategies to push states into compliance (Börzel and Buzogány, 2010). The same is true for studies that go beyond the analysis of policy output and focus on environmental or climate policy performance. One key finding is that a strong environmental movement positively affects the green position of parties, which is beneficial for good environmental performance (Jahn, 2017). At the same time, in the USA, Muñoz et al. (2018) found declining emission levels in states with more pro-environmental protest.

26.5 Conclusion and Future Research

Public awareness of the importance of the environmental crisis has reached new heights following the difficulties in addressing these issues in international negotiations and the accumulation of extreme weather events, eventually leading to the worldwide diffusion of a new type of climate protest. In this chapter, we have attempted to synthesize the existing knowledge about the new climate movement by combining scholarship from research fields such as social movement studies, policy analysis, political theory, communication studies, and anthropology. We have first provided an overview of the literature dealing with the emergence, participant profiles, mobilizing frames, and organizational structures of groups belonging to the new climate movement. We then catalogued the strategies and tactics used by the movement, distinguishing between legal and illegal strategies. The third focus was on the consequences that the climate movement might trigger along four dimensions: people, discourses, parties, and policies.

Based on this state-of-the-art overview, the first observation is that the literature on the three main aspects we have highlighted here (movement characteristics, strategies, and consequences) hardly talk to each other. While we know much about the characteristics of the movements from protest surveys or analyses of their narratives, this is rarely taken up by those interested in strategy choices. The same is true for the missing interactions between the literature on strategies and the political or societal consequences of the climate movement. This observation could be described as 'disciplinary nationalism'. For example, while political sociology and social movement studies take the lead in analysing the new climate movement, their findings stay siloed due to content-focused or methodological concerns. Overstepping these boundaries will help us to gain a more comprehensive perspective on the climate movement.

A second observation is about biases both in research and within the movements themselves. Similar to the academic silos just mentioned, from what we read, the new climate movement seems to be a Western and Northern phenomenon, and little is talked about the Global South in this context. The same is true for the question of class in social movements, which is often politely forgotten or subsumed as 'middle class'. Addressing these issues calls for a focus on the movement's internal power structures, which exist despite the averments to the contrary. This also holds implications for the potential to form a truly global and cross-class coalition supporting post-fossil transformation. In this context, for example, Andreas Malm's work on militancy has been criticized within the movement for downplaying indigenous resistance in the field and ignoring other, often less spectacular, forms of resistance in a part of the world where the climate crisis is not a distant imaginary of the future but bitter everyday reality (Hansen Rübner, 2021). XR's disruptive tactics in the Western hemisphere holds arguably more serious implications to marginalized Black, Asian, and Minority Ethnic (BAME) activists or other 'working-class environmentalists' (Bell, 2021; Bell and Bevan, 2021; Kalt, 2021), who often come from communities that are more affected by the climate crisis because of socioeconomic disparities but are disproportionally targeted by police violence, while their narratives are little heard. To gather and maintain strength and be politically successful, a global new climate movement needs to take into account the 'diversity' criterion mentioned above in the work of Wouters (2018).

Beyond the question of diversity in movement, we see three additional important strands of literature about the new climate movement as parts of an emerging research agenda with theoretical and empirical implications. First, the study of party-movement relations (Hutter et al., 2018) and the influence of social movements on different forms of political participation, including voting (Rhodes, 2021), have already provided social movement studies with a solid embedding into more mainstream comparative politics (Wouters, 2018). Studying protest agendas, public opinion, and policy agendas together, including 'movement–party' and 'movement–voter' interactions, will undoubtedly become an essential field of research concerning climate change, which is addressed and contested increasingly also in party-political terms (Farstad, 2018; Buzogány and Mohamad-Klotzbach, 2022; Lüth and Schaffer, 2022; Otteni and Weisskircher, 2022).

The second line of investigation that is promising focuses on knowledge production, meaning-making, and epistemologies of social movements and is strongly represented in sociology, science-technology studies, communication studies, and political theory. How the climate movement can use discursive opportunity structures and produce new imaginaries by coupling emotions and highly complex, and thus insecure, scientific knowledge is a question that will also co-determine its societal resonance and relevance (Zulianello and Ceccobelli, 2020; Kern and Opitz, 2021; Machin, 2022a, 2022b).

Third and finally, the rich literature on democratic innovations provides an obvious link for scholars of the new climate movement. The questions addressed here have important implications beyond the narrow focus on climate policy (Willis et al., 2022). They are, and essentially so, also about democracy and the question of how (representative) democracy can face crises and reform itself through experimenting (Della Porta, 2020). One of the most direct implications of the new climate movement has been the diffusion of citizen climate assemblies worldwide – which was one of the three central demands of XR. What the consequences of such deliberative bodies will be in terms of policy or polity effects is far from clear (Boswell et al., 2022), and some voices are warning the climate movement from entering the swampy territory of institutionalized deliberation (Krüger, 2021; Ufel, 2021; Mouffe, 2022). At the same time, researchers must be aware that the new climate movement has entered a new phase. The initial aim was to increase political pressure with large-scale demonstrations. What is important now is to remain visible and influential beyond occasional strikes and demonstrations. This will inevitably include also institutional politics, bargains, and conflicts. The awareness is there; now is the time to act.

Box 26.1 Chapter Summary

- The new climate movement has succeeded in mobilizing the masses and bringing protests and civil disobedience as relevant forms of resistance back into the public space.
- There are significant distinctions in the profiles, narratives, and organizational forms of different groups of the new climate movement.
- The new climate movement uses both legal and illegal forms of protest.
- The activities of the new climate movement have consequences for public discourses, strategies of political parties, policies, and the activists themselves.
- Studying the new climate movement is essential for public policy scholarship interested in how climate and energy policies emerge and to what effect.

Acknowledgements

This research was funded in part by the Austrian Science Fund (FWF) [Grant number I 5786] and the Jean Monnet Network Green Deal-NET.

Notes

1 https://rebellion.global/groups/#countries
2 https://fridaysforfuture.de/wp-content/uploads/2020/10/Forderungen-EU.pdf
3 https://www.deutschlandfunk.de/sophia-marie-pott-fridays-for-future-keine-unterstuetzung-100.html
4 https://www.carbonbrief.org/climate-strikers-open-letter-to-eu-leaders-on-why-their-new-climate-law-is-surrender/

References

Abajo-Sanchez, C., 2022. Devenir activiste pour le climat: Formation à la désobéissance civile comme processus de socialisation chez des jeunes militants d'Extinction Rebellion à Paris. *Educação, Sociedade & Culturas*, 62. doi: 10.24840/esc.vi62.359.

Alexander, N., Petray, T. & McDowall, A., 2022. More learning, less activism: Narratives of childhood in Australian media representations of the School Strike for Climate. *Australian Journal of Environmental Education*, 38 (1), 96–111. doi: 10.1017/aee.2021.28.

Amenta, E., Caren, N., Chiarello, E. & Su, Y., 2010. The political consequences of social movements. *Annual Review of Sociology*, 36, 287–307.

Bauwens, T., Gotchev, B., & Holstenkamp, L. (2016). What drives the development of community energy in Europe? The case of wind power cooperatives. *Energy Research & Social Science*, 13, 136–147.

Beer, C.T., 2022. "Systems Change Not Climate Change": Support for a radical shift away from capitalism at mainstream U.S. climate change protest events. *The Sociological Quarterly*, 63 (1), 175–198. doi: 10.1080/00380253.2020.1842141.

Bell, K., 2021. Working-class people, Extinction Rebellion and the environmental movements of the Global North. *Diversity and Inclusion in Environmentalism*. Routledge, 63–81.

Bell, K. & Bevan, G., 2021. Beyond inclusion? Perceptions of the extent to which Extinction Rebellion speaks to, and for, Black, Asian and Minority Ethnic (BAME) and working-class communities. *Local Environment*, 26 (10), 1205–1220.

Berglund, O. & Schmidt, D., 2020. *Extinction Rebellion and climate change activism: Breaking the law to change the world*: Springer.

Bergmann, Z. & Ossewaarde, R., 2020. Youth climate activists meet environmental governance: ageist depictions of the FFF movement and Greta Thunberg in German newspaper coverage. *Journal of Multicultural Discourses*, 15 (3), 267–290. doi: 10.1080/17447143.2020.1745211.

Berker, L.E. & Pollex, J., 2021. Friend or foe?—comparing party reactions to Fridays for Future in a party system polarised between AfD and Green Party. *Zeitschrift für Vergleichende Politikwissenschaft*, 15 (2), 1–19.

Berker, L.E. & Pollex, J., 2022. Explaining differences in party reactions to the Fridays for Future-movement – a qualitative comparative analysis (QCA) of parties in three European countries. *Environmental Politics*, 1–22. doi: 10.1080/09644016.2022.2127536.

Berny, N. & Rootes, C., 2018. Environmental NGOs at a crossroads? *Environmental Politics*, 27 (6), 947–972. doi: 10.1080/09644016.2018.1536293

Böhler, H., Hanegraaff, M. & Schulze, K., 2022. Does climate advocacy matter? The importance of competing interest groups for national climate policies. *Climate Policy*, 22 (8), 961–975. doi: 10.1080/14693062.2022.2036089.

Börzel, T. & Buzogány, A., 2010. Environmental organisations and the Europeanisation of public policy in Central and Eastern Europe: the case of biodiversity governance. *Environmental Politics*, 19 (5), 708– 735.

Bosi, L., Giugni, M. & Uba, K., 2016. *The consequences of social movements*: Cambridge University Press.

Bosse, J. (2017). Analyse: "Zurück in der Grube. Ende Gelände 2" – Die Anti-Kohlebewegung in der Lausitz im Mai 2016. *Forschungsjournal Soziale Bewegungen*, 30(1), 88–92. doi: 10.1515/fjsb-2017-0011.

Boswell, J., Dean, R. & Smith, G., 2022. Integrating citizen deliberation into climate governance: Lessons on robust design from six climate assemblies. *Public Administration*.

Boucher, J.L., Kwan, G.T., Ottoboni, G.R. & McCaffrey, M.S., 2021. From the suites to the streets: Examining the range of behaviors and attitudes of international climate activists. *Energy Research & Social Science*, 72, 101866. doi: 10.1016/j.erss.2020.101866.

Bowman, B., 2020. 'They don't quite understand the importance of what we're doing today': the young people's climate strikes as subaltern activism. *Sustainable Earth*, 3 (1), 1–13.

Brechin, S.R. & Fenner, W.H., 2017. Karl Polanyi's environmental sociology: a primer. *Environmental Sociology*, 3 (4), 404–413. doi: 10.1080/23251042.2017.1355723.

Brock, A., & Dunlap, A. (2018). Normalising corporate counterinsurgency: Engineering consent, managing resistance and greening destruction around the Hambach coal mine and beyond. *Political Geography*, 62, 33–47.

Brown, J.M., 2021. Civil disobedience, sabotage, and violence in US environmental activism. *The Oxford handbook of comparative environmental politics*.

Brulle, R. J. (2010). From Environmental Campaigns to Advancing the Public Dialog: Environmental Communication for Civic Engagement. *Environmental Communication*, *4*(1), 82–98. doi: 10.1080/17524030903522397.

Brunnengräber, A., 2011. *Zivilisierung des Klimaregimes: NGOs und soziale Bewegungen in der nationalen, europäischen und internationalen Klimapolitik*: Springer-Verlag.

Bugden, D., 2020. Does climate protest work? Partisanship, protest, and sentiment pools. *Socius*, 6, 2378023120925949. doi: 10.1177/2378023120925949.

Burnell, P., 2012. Democracy, democratization and climate change: complex relationships. *Democratization*, 19 (5), 813–842.

Burstein, P., 1999. Social movements and public policy. *How Social Movements Matter*, 3, 7–8.

Buzogány, A. & Mohamad-Klotzbach, C., 2022. Environmental populism. *In* Oswald, M. ed. *The Palgrave handbook of populism*. London: Palgrave, 321–340.

Buzogány, A. & Scherhaufer, P., 2022. Framing different energy futures? Comparing fridays for future and Extinction Rebellion in Germany. *Futures*, 137, 102904.

Capstick, S., Thierry, A., Cox, E., Berglund, O., Westlake, S. & Steinberger, J.K., 2022. Civil disobedience by scientists helps press for urgent climate action. *Nature Climate Change*, 1–2.

Cassegård, C. & Thörn, H., 2018. Toward a postapocalyptic environmentalism? Responses to loss and visions of the future in climate activism. *Environment and Planning E: Nature and Space*, 1 (4), 561–578.

Chen, K., Molder, A.L., Duan, Z., Boulianne, S., Eckart, C., Mallari, P. & Yang, D., 2022. How climate movement actors and news media frame climate change and strike: Evidence from analyzing twitter and news media discourse from 2018 to 2021. *The International Journal of Press/Politics*. 19401612221106405.

Chenoweth, E. & Stephan, M., 2011. *Why civil resistance works: The strategic logic of nonviolent conflict*: Columbia University Press.

Cologna, V., Hoogendoorn, G. & Brick, C., 2021. To strike or not to strike? an investigation of the determinants of strike participation at the Fridays for Future climate strikes in Switzerland. *PLOS ONE*, 16 (10), e0257296. doi: 10.1371/journal.pone.0257296.

Corry, O. & Reiner, D., 2021. Protests and policies: How radical social movement activists engage with climate policy dilemmas. *Sociology*, 55 (1), 197–217.

Costa, J. & Wittmann, E., 2021. Fridays for Future als Lern-und Erfahrungsraum: Befunde zu den Beteiligungsformaten, den Motiven und der Selbstwirksamkeitserwartung von Engagierten. *ZEP: Zeitschrift für internationale Bildungsforschung und Entwicklungspädagogik*, 44 (3), 10–15.

Da Silva, J.R., 2020. The eco-terrorist wave. *Behavioral Sciences of Terrorism and Political Aggression*, 12 (3), 203–216.

Daniel, A., Deutschmann, A., Buzogány, A. & Scherhaufer, P., 2020. Die Klimakrise deuten und Veränderungen einfordern: Eine Framing-Analyse der Fridays for Future. *SWS-Rundschau*, 60 (4), 365–384.

David, M. (2018). The role of organized publics in articulating the exnovation of fossil-fuel technologies for intra-and intergenerational energy justice in energy transitions. *Applied Energy*, 228, 339–350.

De Bakker, F.G.A., Den Hond, F. & Laamanen, M., 2017. Social Movements: Organizations and Organizing. In Roggeband, C. & Klandermans, B. eds. *Handbook of social movements across disciplines*. Cham: Springer International Publishing, 203–231.

De Moor, J., 2021. Postapocalyptic narratives in climate activism: their place and impact in five European cities. *Environmental Politics*, 1–22. doi: 10.1080/09644016.2021.1959123.

De Moor, J., De Vydt, M., Uba, K. & Wahlström, M., 2021. New kids on the block: Taking stock of the recent cycle of climate activism. *Social Movement Studies*, 20 (5), 619–625.

De Moor, J., Uba, K., Wahlström, M., Wennerhag, M. & De Vydt, M., 2020. Protest for a future II: Composition, mobilization and motives of the participants in Fridays For Future climate protests on 20–27 September, 2019, in 19 cities around the world.

Delina, L.L., 2019. *Emancipatory climate actions: Strategies from histories*. Springer.

Delina, L.L., 2022. Moving people from the balcony to the trenches: Time to adopt "climatage" in climate activism? *Energy Research & Social Science*, 90, 102586.

Della Porta, D., 2020. *How social movements can save democracy: Democratic innovations from below*: John Wiley & Sons.

Della Porta, D., 2021. Progressive Social Movements, Democracy and the Pandemic. In Delanty, G. ed. *Pandemics, politics, and society*. Berlin: De Gruyter, 209–226.

Della Porta, D. & Parks, L., 2014. Framing processes in the climate movement: From climate change to climate justice. *Routledge Handbook of the Climate Change Movement*, 19–30.

Della Porta, D. & Pavan, E., 2017. Repertoires of knowledge practices: Social movements in times of crisis. *Qualitative Research in Organizations and Management: An International Journal*, 12 (4), 297–314, https://doi.org/10.1108/QROM-01-2017-1483.

Della Porta, D. & Portos, M., 2021. Rich kids of Europe? Social basis and strategic choices in the climate activism of Fridays for Future. *Italian Political Science Review/Rivista Italiana di Scienza Politica*, 1–26.

Della Porta, D., & Rucht, D. (2013). *Meeting democracy: power and deliberation in global justice movements*. Cambridge University Press.

Dietz, M. & Garrelts, H., 2013. *Die internationale Klimabewegung: ein Handbuch*. Springer-Verlag.

Dietz, M. & Garrelts, H., 2014. *Routledge handbook of the climate change movement*. Routledge.

Doherty, B. & Hayes, G., 2018. Tactics and strategic action. *The Wiley Blackwell Companion to Social Movements*, 269–288.

Doherty, B., Saunders, C. & Hayes, G., 2020. A New Climate Movement? Extinction Rebellion's Activists in Profile. CUSP Working Paper No 25. *CUSP Working Paper Series| No 25*, 25, 1–39.

Dow, J. M. (2018). Environmental Civil Disobedience. In D. Boonin (Ed.), *The Palgrave Handbook of Philosophy and Public Policy* (pp. 795–807): Palgrave Macmillan.

Emilsson, K., Johansson, H. & Wennerhag, M., 2020. Frame disputes or frame consensus?"Environment" or "welfare" first amongst climate strike protesters. *Sustainability*, 12 (3), 882.

Etchanchu, H., De Bakker, F.G. & Delmestri, G., 2021. Social movement organizations agency for sustainable organizing. *Research handbook of sustainability agency*. Edward Elgar Publishing.

Evensen, D., 2019. The rhetorical limitations of the# FridaysForFuture movement. *Nature Climate Change*, 9 (6), 428–430.

Farrer, B., 2017. *Organizing for policy influence: Comparing parties, interest groups, and direct action*: Routledge.

Farrer, B. & Klein, G.R., 2022. How radical environmental sabotage impacts US elections. *Terrorism and Political Violence*, 34 (2), 218–239.

Farstad, F.M., 2018. What explains variation in parties' climate change salience? *Party Politics*, 24 (6), 698–707. doi: 10.1177/1354068817693473

Feinberg, M., Willer, R. & Kovacheff, C., 2020. The activist's dilemma: Extreme protest actions reduce popular support for social movements. *Journal of Personality and Social Psychology*, 119 (5), 1086.

Fernandez-Zubieta, A., Guevara Gil, J.A., Caballero Roldan, R. & Robles Morales, J.M., 2022. Digital Activism Masked. The Fridays for Future movement and the "Global day of climate action": testing social function and framing typologies of claims on Twitter. *The Fridays for Future movement and the "Global day of climate action": testing social function and framing typologies of claims on Twitter (April 12, 2022).*

Fischer, F., 2017. *Climate crisis and the democratic prospect: Participatory governance in sustainable communities*: Oxford University Press.

Fisher, D.R. & Nasrin, S., 2021. Climate activism and its effects. *Wiley Interdisciplinary Reviews: Climate Change*, 12 (1), e683.

Fotaki, M. & Foroughi, H., 2022. Extinction rebellion: Green activism and the fantasy of leaderlessness in a decentralized movement. *Leadership*, 18 (2), 224–246. doi: 10.1177/17427150211005578.

Frenzel, F., 2014. Exit the system? Anarchist organisation in the British climate camps. *Ephemera: Theory & Politics in Organization*, 14 (4).

Friberg, A., 2022. Disrupting the present and opening the future: Extinction rebellion, fridays for future, and the disruptive Utopian method. *Utopian Studies*, 33 (1), 1–17. doi: 10.5325/utopianstudies.33.1.0001.

Furlong, C. & Vignoles, V.L., 2021. Social identification in collective climate activism: Predicting participation in the environmental movement, Extinction Rebellion. *Identity*, 21 (1), 20–35. doi: 10.1080/15283488.2020.1856664

Gach, E. (2019). Normative shifts in the global conception of climate change: The growth of climate justice. *Social Sciences*, *8*(1), 24.

Ganz, S.C. & Soule, S.A., 2019. Greening the Congressional record: Environmental social movements and expertise-based access to the policy process. *Environmental Politics*, 28 (4), 685–706.

Gardner, P., Carvalho, T. & Valenstain, M., 2022. Spreading rebellion?: The rise of extinction rebellion chapters across the world. *Environmental Sociology*, 1–12. doi: 10.1080/23251042.2022.2094995.

Gelderloos, P., 2007. *How nonviolence protects the state*. Cambridge, MA: South End Press.

Gelderloos, P., 2022. The solutions are already here: tactics for ecological revolution from below. London: Pluto Press.

Giugni, M., 2004. *Social protest and policy change: Ecology, antinuclear, and peace movements in comparative perspective*. Rowman & Littlefield.

Giugni, M., 2008. Political, biographical, and cultural consequences of social movements. *Sociology Compass*, 2 (5), 1582–1600.

Giugni, M., McAdam, D. & Tilly, C., 1999. *How social movements matter*. University of Minnesota Press.

Goldenbaum, M. & Thompson, C., 2020. Fridays for Future im Spiegel der Medienöffentlichkeit. In Sommer, M. & Haunss, S. eds. *Fridays for Future-Die Jugend gegen den Klimawandel*. Bielefeld: transcript-Verlag, 181–204.

Goodman, J., 2022. Social movements and climate change: "Climatizing" society from within. Oxford: Oxford University Press.

Hadden, J., 2014. Explaining variation in transnational climate change activism: The role of inter-movement spillover. *Global Environmental Politics*, 14 (2), 7–25.

Hadden, J. & Jasny, L., 2019. The power of peers: How transnational advocacy networks shape NGO strategies on climate change. *British Journal of Political Science*, 49 (2), 637–659. doi: 10.1017/S0007123416000582.

Häfner, D., Schmidtke, D., & Scholl, F. (2016). Pro Lausitzer Braunkohle vs. Ende Gelände / Eine erneute Annäherung an gesteuerte Bürgerinitiativen. *Forschungsjournal Soziale Bewegungen*, *29*(3), 237–241. doi: 10.1515/fjsb-2016-0244.

Haines, H.H., 2013. Radical Flank Effects. *The Wiley-Blackwell encyclopedia of social and political movements*.

Hallam, R., 2019. Now we know: Conventional campaigning won't prevent our extinction. *The Guardian*, 1. https://www.theguardian.com/commentisfree/2019/may/01/extinction-rebellion-non-violent-civil-disobedience

Hansen Rübner, B., 2021. The kaleidoscope of catastrophe: On the clarities and blind spots of Andreas Malm. *Viewpoint Magazine*, https://viewpointmag.com/2021/04/14/the-kaleidoscope-of-catastrophe-on-the-clarities-and-blind-spots-of-andreas-malm/.

Haugestad, C.A.P., Skauge, A.D., Kunst, J.R. & Power, S.A., 2021. Why do youth participate in climate activism? A mixed-methods investigation of the #FridaysForFuture climate protests. *Journal of Environmental Psychology*, 76, 101647. doi: 10.1016/j.jenvp.2021.101647.

Hayes, S. & O'Neill, S., 2021. The Greta effect: Visualising climate protest in UK media and the Getty images collections. *Global Environmental Change*, 71, 102392.

Hess, D.J., 2015. Publics as threats? Integrating science and technology studies and social movement studies. *Science as Culture*, 24 (1), 69–82.

Hess, D.J., 2018. Social movements and energy democracy: Types and processes of mobilization. *Frontiers in Energy Research*, 6, 135.

Hochmann, L., 2020. *economists4future: Verantwortung übernehmen für eine bessere Welt.* Murmann Publishers GmbH.

Hübinger, J., 2020. Yes, We Camp for Climate Action. Die Klimacamps im Rheinland als Orte des Protests und transformative Möglichkeitsräume. *Nachhaltigkeit, Postwachstum, Transformation.* Springer, 451–485.

Hunger, S. & Hutter, S., 2021. Fridays for Future in der Corona-Krise. *Forschungsjournal Soziale Bewegungen*, 34 (2), 218–234.

Hutter, S., Kriesi, H. & Lorenzini, J., 2018. Social movements in interaction with political parties. *The Wiley Blackwell companion to social movements*, 322–337.

Huttunen, J. & Albrecht, E., 2021. The framing of environmental citizenship and youth participation in the Fridays for Future Movement in Finland. *Fennia*, 199 (1).

Jahn, D., 2017. New Internal Politics in Western Democracies: The Impact of the Environmental Movement in Highly Industrialized Democracies. *In* Harfst, P., Kubbe, I. & Poguntke, T. eds. *Parties, governments and elites.* Wiesbaden: VS Verlag, 125–150.

Jasny, L. & Fisher, D.R., 2022. How networks of social movement issues motivate climate resistance. *Social Networks.* doi: 10.1016/j.socnet.2022.02.002.

Kalt, T., 2021. Jobs vs. climate justice? Contentious narratives of labor and climate movements in the coal transition in Germany. *Environmental Politics*, 30 (7), 1135–1154. doi: 10.1080/09644016.2021.1892979.

Kaufer, R. & Lein, P., 2018. Widerstand im Hambacher Forst: Analyse einer anarchistischen Waldbesetzung. *Forschungsjournal soziale Bewegungen*, 31 (4).

Kern, T. & Opitz, D., 2021. "Trust Science!" institutional conditions of frame resonance in the United States and Germany: The case of fridays for future. *International Journal of Sociology*, 51 (3), 249–256.

Kinniburgh, C., 2020. Can extinction rebellion survive? *Dissent*, 67 (1), 125–133.

Klein, N. (2014). *This Changes Everything: Capitalism vs. the Climate.* New York: Simon & Schuster.

Kountouris, Y. & Williams, E., 2022. Do protests influence environmental attitudes? Evidence from Extinction Rebellion. *Environmental Research Communications* 5, 011003, doi: 10.1088/2515-7620/ac9aeb.

Krüger, T., 2021. Die energiewende im kontext von klima-und demokratiekrise: Die grenzen der deliberation und radikaldemokratische alternativen. *GAIA-Ecological Perspectives for Science and Society*, 30 (3), 181–188.

Krüger, T., Eichenauer, E., & Gailing, L. (2022). Whose future is it anyway? Struggles for just energy futures. *Futures*, *142*, 103018. https://doi.org/10.1016/j.futures.2022.103018

Lauth, H. J. (2000). Informal Institutions and Democracy. *Democratization*, 7(4), 21–50. doi: 10.1080/13510340008403683.

Laux, T., 2021. What makes a global movement? Analyzing the conditions for strong participation in the climate strike. *Social Science Information*, 60 (3), 413–435.

Longhofer, W., Schofer, E., Miric, N. & Frank, D.J., 2016. NGOs, INGOs, and environmental policy reform, 1970–2010. *Social Forces*, 94 (4), 1743–1768.

Lorenzen, J.A., 2020. 'For us climate action never dies': a legislative process analysis of environmental movement tactics in Oregon. *Environmental Sociology*, 6 (4), 375–389. doi: 10.1080/23251042.2020.1792265.

Lorenzini, J., Monsch, G.-A. & Rosset, J., 2021. Challenging climate strikers' youthfulness: The evolution of the generational gap in environmental attitudes since 1999. *Frontiers in Political Science*, 3, 633563.

Lüth, M. & Schaffer, L.M., 2022. The electoral importance and evolution of climate-related energy policy: evidence from Switzerland. *Swiss Political Science Review*, 28 (2), 169–189. doi: 10.1111/spsr.12520.

Machin, A., 2022a. Climates of democracy: Skeptical, rational, and radical imaginaries. *Wiley Interdisciplinary Reviews: Climate Change*, e774.

Machin, A., 2022b. Green democracy: Political imaginaries of environmental movements. *The Routledge handbook of environmental movements*. Routledge, 552–563.

Macklin, G., 2022. The extreme right, climate change and terrorism. *Terrorism and Political Violence*, 34 (5), 979–996. doi: 10.1080/09546553.2022.2069928.

Malm, Andreas 2021. *How to blow up a pipeline: Learning to fight in a world on fire*. London: Verso.

Malm, A. & The Zetkin Collective, 2021. *White skin, black fuel: on the danger of fossil fascism*: Verso Books.

Markard, J., 2018. The next phase of the energy transition and its implications for research and policy. *Nature Energy*, 3 (8), 628.

Marquardt, J., 2020. Fridays for Future's disruptive potential: An inconvenient youth between moderate and radical ideas. *Frontiers in Communication 5*, 48. doi: 10.3389/fcomm.

Martiskainen, M., Axon, S., Sovacool, B.K., Sareen, S., Furszyfer Del Rio, D. & Axon, K., 2020. Contextualizing climate justice activism: Knowledge, emotions, motivations, and actions among climate strikers in six cities. *Global Environmental Change*, 65, 102180. doi: 10.1016/j.gloenvcha.2020.102180.

Mattheis, N., 2022. Unruly kids? Conceptualizing and defending youth disobedience. *European Journal of Political Theory*, 21 (3), 466–490.

Matthews, K., 2020. Social movements and the (mis) use of research: Extinction Rebellion and the 3.5% Rule. *Interface: A Journal for and About Social Movements*, 12 (1), 591–615.

McAdam, D., 2017. Social movement theory and the prospects for climate change activism in the United States. *Annual Review of Political Science*, 20 (1), 189–208.

McAdam, D., & Boudet, H. (2012). *Putting social movements in their place: Explaining opposition to energy projects in the United States, 2000–2005*. Cambridge University Press.

McAdam, D., McCarthy, J.D. & Zald, M.N., 1996. *Comparative perspectives on social movements: political opportunities, mobilizing structures, and cultural framings* Cambridge [England]. New York: Cambridge University Press.

Melchior, M. & Rivera, M., 2021. *Klimagerechtigkeit erzählen. Narrative, Wertbezüge und Frames bei Extinction Rebellion, Ende Gelände und Fridays for Future* Potsdam: IASS.

Miler, K.C., 2010. *Constituency representation in congress: The view from capitol hill* Cambridge: Cambridge University Press.

Milkoreit, M., Hodbod, J., Baggio, J., Benessaiah, K., Calderón-Contreras, R., Donges, J.F., Mathias, J.-D., Rocha, J.C., Schoon, M. & Werners, S.E., 2018. Defining tipping points for social-ecological systems scholarship—An interdisciplinary literature review. *Environmental Research Letters*, 13 (3), 033005. doi: 10.1088/1748-9326/aaaa75.

Mol, A.P., 2000. The environmental movement in an era of ecological modernisation. *Geoforum*, 31 (1), 45–56.

Moss, T., Becker, S., & Naumann, M. (2015). Whose energy transition is it, anyway? Organisation and ownership of the Energiewende in villages, cities and regions. *Local Environment*, 20(12), 1547–1563.

Mouffe, C., 2022. *Towards a green democratic revolution: Left populism and the power of affects*. Verso Books.

Mucha, W., Soßdorf, A., Ferschinger, L. & Burgi, V., 2020. Fridays for future meets citizen science. Resilience and digital protests in times of covid-19. *Voluntaris*, 8 (2), 261–277.

Muñoz, J., Olzak, S. & Soule, S.A., 2018. Going green: Environmental protest, policy, and CO_2 emissions in US states, 1990–2007. *Sociological Forum*, 33 (2), 403–421.

Nash, S.L. & Steurer, R., 2021. Climate change acts in Scotland, Austria, Denmark and Sweden: The role of discourse and deliberation. *Climate Policy*, 21 (9), 1120–1131.

Nijjar, M.S., 2022. *Frames, fear, and identity in the american environmental movement*. Fairfax: George Mason University.

Nordensvard, J. & Ketola, M., 2022. Populism as an act of storytelling: analyzing the climate change narratives of Donald Trump and Greta Thunberg as populist truth-tellers. *Environmental Politics*, 31 (5), 861–882.

North, P. (2011). The politics of climate activism in the UK: a social movement analysis. *Environment and Planning A*, 43(7), 1581–1598.

Noth, F. & Tonzer, L., 2022. Understanding climate activism: Who participates in climate marches such as "Fridays for Future" and what can we learn from it? *Energy Research & Social Science*, 84, 102360.

Nulman, E., 2016. *Climate change and social movements: civil society and the development of national climate change policy*: Springer.

Olzak, S., Soule, S.A., Coddou, M. & Muñoz, J., 2016. Friends or foes? How social movement allies affect the passage of legislation in the US congress. *Mobilization: An International Quarterly*, 21 (2), 213–230.

Orazani, N., Tabri, N., Wohl, M.J.A. & Leidner, B., 2021. Social movement strategy (nonviolent vs. violent) and the garnering of third-party support: A meta-analysis. *European Journal of Social Psychology*, 51 (4–5), 645–658. doi: 10.1002/ejsp.2722.

Oßenbrügge, J. (2021). Von der Anti-AKW-Bewegung bis Ende Gelände: Soziale Bewegungen in der deutschen Energiepolitik. In S. Becker, B. Klagge, & M. Naumann (Eds.), *Energiegeographie* (pp. 135–146). Stuttgart: Ulmer.

Otteni, C. & Weisskircher, M., 2022. AfD gegen die Grünen? Rechtspopulismus und klimapolitische Polarisierung in Deutschland. *Forschungsjournal Soziale Bewegungen*, 35 (2), 317–335.

Otto, I.M., Donges, J.F., Cremades, R., Bhowmik, A., Hewitt, R.J., Lucht, W., Rockström, J., Allerberger, F., McCaffrey, M., Doe, S.S.P., Lenferna, A., Morán, N., Van Vuuren, D.P. & Schellnhuber, H.J., 2020. Social tipping dynamics for stabilizing Earth's climate by 2050. *Proceedings of the National Academy of Sciences*, 117 (5), 2354–2365. doi: 10.1073/pnas.1900577117.

Ozden, J. & Glover, S., 2022. *Disruptive climate protests in the UK didn't lead to a loss of public support for climate policies*. Social Change Lab.

Parks, L., Della Porta, D. & Portos, M., 2022. Environmental and Climate Activism and Advocacy in Europe In Oberthür, S. ed. *Elgar handbook on European union climate change policy and politics*.

Parth, A.-M., Weiss, J., Firat, R. & Eberhardt, M., 2020. "How dare you!"—the influence of Fridays for Future on the political attitudes of young adults. *Frontiers in Political Science*, 2, 611139.

Perez, A.C., Grafton, B., Mohai, P., Hardin, R., Hintzen, K. & Orvis, S., 2015. Evolution of the environmental justice movement: activism, formalization and differentiation. *Environmental Research Letters*, 10 (10), 105002.

Piggot, G., 2018. The influence of social movements on policies that constrain fossil fuel supply. *Climate Policy*, 18 (7), 942–954. doi: 10.1080/14693062.2017.1394255.

Posmek, J., 2022. Die fridays for future-bewegung Deutschland – Skizzierung eines ethnographischen Forschungsprogramms. *Soziale Passagen*. doi: 10.1007/s12592-022-00406-8.

Radtke, J. & Scherhaufer, P., 2022. A social science perspective on conflicts in the energy transition: An introduction to the special issue. *Utilities Policy*, 78, 101396. doi: 10.1016/j.jup.2022.101396.

Rawls, J. (1985). Justice as Fairness: Political not Metaphysical. *Philosophy & Public Affairs*, *14*(3), 223–251. http://www.jstor.org/stable/2265349

Reichel, C., Plüschke-Altof, B. & Plaan, J., 2022. Speaking of a 'climate crisis': Fridays for Future's attempts to reframe climate change. *Innovation: The European Journal of Social Science Research*, 1–19. doi: 10.1080/13511610.2022.2108006.

Reif, A., Peter, E., Gillner, T., Hortig, L.-M., Joost, A. & Taddicken, M., 2021. Vom Bildschirm auf die Straße? *Medien & Kommunikationswissenschaft*, 69 (4), 578–597.

Rhodes, A., 2021. Social Movements in Elections: UK Anti-Austerity and Environmental Campaigning 2015-19. Cham: Springer.

Rochon, T.R. & Mazmanian, D.A., 1993. Social movements and the policy process. *The Annals of the American Academy of Political and Social Science*, 528 (1), 75–87.

Rödder, S., & Pavenstädt, C. N. (2022). 'Unite behind the Science!' Climate movements' use of scientific evidence in narratives on socio-ecological futures. *Science and Public Policy*, *50*(1), 30–41. doi:10.1093/scipol/scac046

Rucht, D. (2017). Exploring the Backstage: Preparation and Implementation of Mass Protests in Germany. *American Behavioral Scientist*, 61(13), 1678–1702. doi: 10.1177/0002764217744135.

Rucht, D. & Sommer, M., 2019. Fridays for future. Vom Phänomen Greta Thunberg, medialer Verkürzung und geschickter Mobilisierung: Zwischenbilanz eines Höhenflugs. *Internationale Politik*, 74 (4), 121–125.

Ruser, A., 2020. Radikale Konformität und konforme Radikalität? Fridays for Future und Ende Gelände. *Forschungsjournal Soziale Bewegungen*, 33 (4), 801–814.

Sander, H., 2016. Die Bewegung für Klimagerechtigkeit und Energiedemokratie in Deutschland.: Eine historisch-materialistische Bewegungsanalyse. *PROKLA. Zeitschrift für kritische Sozialwissenschaft*, 46 (184), 403–421.

Sander, H. (2017). Ende Gelände: Anti-Kohle-Proteste in Deutschland. *Forschungsjournal Soziale Bewegungen*, *30*(1), 26–36.

Sauerborn, E., 2022. The politicisation of secular mindfulness – Extinction Rebellion's emotive protest practices. *European Journal of Cultural and Political Sociology*, 9 (4), 451–474. doi: 10.1080/23254823.2022.2086596.

Saunders, C., 2012. Reformism and radicalism in the Climate Camp in Britain: Benign coexistence, tensions and prospects for bridging. *Environmental Politics*, 21 (5), 829–846. doi: 10.1080/09644016.2012.692937.

Saunders, C. & Price, S., 2009. One person's eu-topia, another's hell: Climate Camp as a heterotopia. *Environmental Politics*, 18 (1), 117–122. doi: 10.1080/09644010802624850.

Scherhaufer, P., Klittich, P. & Buzogány, A., 2021. Between illegal protests and legitimate resistance. Civil disobedience against energy infrastructures. *Utilities Policy*, 72, 101249.

Scheuerman, W.E., 2022. Political disobedience and the climate emergency. *Philosophy & Social Criticism*, 48 (6), 791–812.

Schifeling, T. & Hoffman, A.J., 2017. Bill McKibben's Influence on U.S. climate change discourse: Shifting field-level debates through radical flank effects. *Organization & Environment*, 32 (3), 213–233. doi: 10.1177/1086026617744278.

Schürmann, L., 2023. The impact of local protests on political elite communication: evidence from Fridays for Future in Germany. *Journal of Elections, Public Opinion and Parties*, 1–21. doi:10.1080/17457289.2023.2189729.

Schweinschwaller, T., 2020. Extinction Rebellion - Zwischen Achtsamkeit und Aktion Eine Interviewstudie. *SWS-Rundschau*, 4, 385–407.

Shuman, E., Saguy, T., Van Zomeren, M. & Halperin, E., 2021. Disrupting the system constructively: Testing the effectiveness of nonnormative nonviolent collective action. *Journal of Personality and Social Psychology*, 121, 819–841. doi: 10.1037/pspi0000333.

Silke, A. & Morrison, J., 2022. Gathering storm: An introduction to the special issue on climate change and terrorism. *Terrorism and Political Violence*, 34 (5), 883–893. doi: 10.1080/09546553.2022.2069444.

Simpson, B., Willer, R. & Feinberg, M., 2022. Radical flanks of social movements can increase support for moderate factions. *PNAS Nexus*, 1 (3). doi: 10.1093/pnasnexus/pgac110.

Slaven, M. & Heydon, J., 2020. Crisis, deliberation, and Extinction Rebellion. *Critical Studies on Security*, 8 (1), 59–62. doi: 10.1080/21624887.2020.1735831.

Smiles, T. & Edwards, G.A.S., 2021. How does Extinction Rebellion engage with climate justice? A case study of XR Norwich. *Local Environment*, 26 (12), 1445–1460. doi: 10.1080/13549839.2021.1974367.

Soliev, I., Janssen, M.A., Theesfeld, I., Pritchard, C., Pirscher, F. & Lee, A., 2021. Channeling environmentalism into climate policy: An experimental study of Fridays for Future participants from Germany. *Environmental Research Letters*, 16 (11), 114035.

Sommer, M. & Haunss, S., 2020. *Fridays for Future-Die Jugend gegen den Klimawandel: Konturen der weltweiten Protestbewegung (Edition 1)*. transcript Verlag.

Sommer, M., Rucht, D., Haunss, S. & Zajak, S., 2019. Fridays for future: Profil, Entstehung und Perspektiven der Protestbewegung in Deutschland.

Sorce, G. & Dumitrica, D., 2021. #fighteverycrisis: Pandemic shifts in fridays for future's protest communication frames. *Environmental Communication*, 1–13. doi: 10.1080/17524032.2021.1948435.

Soßdorf, A. & Burgi, V., 2022. "Listen to the science!"—The role of scientific knowledge for the Fridays for Future movement. *Frontiers in Communication*, 7. doi: 10.3389/fcomm.2022.983929.

Sovacool, B.K., 2022. Beyond science and policy: Typologizing and harnessing social movements for transformational social change. *Energy Research & Social Science*, 94, 102857. doi: 10.1016/j.erss.2022.102857.

Sovacool, B.K. & Dunlap, A., 2022. Anarchy, war, or revolt? Radical perspectives for climate protection, insurgency and civil disobedience in a low-carbon era. *Energy Research & Social Science*, 86, 102416.

Spaiser, V., Nisbett, N. & Stefan, C.G., 2022. "How dare you?"—The normative challenge posed by Fridays for Future. *PLOS Climate*, 1 (10), e0000053. doi: 10.1371/journal.pclm.0000053.

Stuart, D., 2022. Tensions between individual and system change in the climate movement: an analysis of Extinction Rebellion. *New Political Economy*, 1–14.

Stuart, D., Gunderson, R. & Petersen, B., 2020. The climate crisis as a catalyst for emancipatory transformation: An examination of the possible. *International Sociology*, 35 (4), 433–456.

Swyngedouw, E., 2022. The unbearable lightness of climate populism. *Environmental Politics*, 31 (5), 904–925.

Temper, L., 2019. Radical Climate Politics: From Ogoniland to Ende Gelände. *Routledge Handbook of Radical Politics*. Routledge, 97–106.

Temper, L., del Bene, D., & Martinez-Alier, J. (2015). Mapping the frontiers and front lines of global environmental justice: the EJAtlas. *Journal of Political Ecology*, 22, 255–275. doi: 10.2458/v22i1.21108.

Tilly, C. (1977). *From mobilization to revolution*. New York: Random House.

Tilly, C., 2006. WUNC. In Schnapp, J. & Tiews, M. eds. *Crowds*. Stanford: Stanford University Press, 289–306.

Toewe, S. (2017). Ende Gelände! hat es geschafft, handlungsfähige Akteure in einem internationalen Prozess zusammenzubinden. Interview mit Insa Vries, Sprecherin von Ende Gelände! *Forschungsjournal Soziale Bewegungen* 1, 92–96.

Uba, K., Lavizzari, A. & Portos, M., 2022. Experience of economic hardship and right-wing political orientation hinder climate concern among European young people. *Journal of Contemporary European Studies*. doi: 10.1080/14782804.2022.2061433.

Ufel, W., 2021. Are citizens' assemblies a good strategy for climate activists? *Interfere*, 2, 124–139.

Ünal, H., Adelman, L. & Leidner, B., 2022. The sword or the plowshare: Conflict and third-party groups' reaction to violent versus nonviolent resistance. *Journal of Social Issues*. doi: 10.1111/josi.12522.

Van Dyke, N., Soule, S. & Taylor, V., 2005. The targets of social movements: Beyond a focus on the state. *Research in Social Movements, Conflicts and Change*, 25, 27–51. doi: 10.1016/S0163-786X(04)25002-9.

Van Laer, J., 2017. The mobilization dropout race: Interpersonal networks and motivations predicting differential recruitment in a national climate change demonstration. *Mobilization*, 22 (3), 311–329.

Van Laer, J. & Van Aelst, P., 2010. Internet and social movement action repertoires. *Information, Communication & Society*, 13 (8), 1146–1171. doi: 10.1080/13691181003628307.

Vasi, I. B. (2011). *Winds of change: The environmental movement and the global development of the wind energy industry*. Oxford: Oxford University Press.

Von Storch, L., Ley, L. & Sun, J., 2021. New climate change activism: before and after the Covid-19 pandemic. *Social Anthropology*, 29 (1), 205–209. doi: 10.1111/1469-8676.13005.

Von Zabern, L. & Tulloch, C.D., 2021. Rebel with a cause: The framing of climate change and intergenerational justice in the German press treatment of the Fridays for Future protests. *Media, Culture & Society*, 43 (1), 23–47.

Wahlström, M., Sommer, M., Kocyba, P., De Vydt, M., De Moor, J., Davies, S., Wouters, R., Wennerhag, M., Van Stekelenburg, J., Uba, K. & Buzogány, A., 2019. Protest for a future: Composition, mobilization and motives of the participants in Fridays For Future climate protests on 15 March, 2019 in 13 European cities.

Wallis, H. & Loy, L.S., 2021. What drives pro-environmental activism of young people? A survey study on the Fridays For Future movement. *Journal of Environmental Psychology*, 74, 101581.

Westwell, E. & Bunting, J., 2020. The regenerative culture of Extinction Rebellion: Self-care, people care, planet care. *Environmental Politics*, 29 (3), 546–551. doi: 10.1080/09644016.2020.1747136.

Willis, R., Curato, N. & Smith, G., 2022. Deliberative democracy and the climate crisis. *Wiley Interdisciplinary Reviews: Climate Change*, 13 (2), e759.

Winkelmann, R., Donges, J.F., Smith, E.K., Milkoreit, M., Eder, C., Heitzig, J., Katsanidou, A., Wiedermann, M., Wunderling, N. & Lenton, T.M., 2022. Social tipping processes towards climate action: A conceptual framework. *Ecological Economics*, 192, 107242.

Wong, W.H., 2012. Internal affairs. *Internal Affairs*. Cornell University Press.

Wouters, R., 2018. The persuasive power of protest. How protest wins public support. *Social Forces*, 98 (1), 403–426. doi: 10.1093/sf/soy110.

Wouters, R. & Walgrave, S., 2017. Demonstrating power: How protest persuades political representatives. *American Sociological Review*, 82 (2), 361–383.

Wozniak, A., Wessler, H., Chan, C.-h., & Lück, J. (2021). *The Event-Centered Nature of Global Public Spheres: The UN Climate Change Conferences, Fridays for Future, and the (Limited) Transnationalization of Media Debates* (15, 688–714).

Wright, E.O., 2010. *Envisioning real utopias* London: Verso.

Wuppertal Institute, 2020. *CO2-neutral bis 2035: Eckpunkte eines deutschen Beitrags zur Einhaltung der 1, 5-° C-Grenze; Diskussionsbeitrag für Fridays for Future Deutschland* Wuppertal.

Yates, L., 2021. Prefigurative politics and social movement strategy: The roles of prefiguration in the reproduction, mobilisation and coordination of movements. *Political Studies*, 69 (4), 1033–1052.

Zamponi, L., Baukloh, A.C., Bertuzzi, N., Chironi, D., Della Porta, D. & Portos, M., 2022. (Water) bottles and (street) barricades: the politicisation of lifestyle-centred action in youth climate strike participation. *Journal of Youth Studies*, 1–22. doi: 10.1080/13676261.2022.2060730.

Zantvoort, F., 2021. Movement pedagogies in pandemic times: Extinction Rebellion Netherlands and (un) learning from the margins. *Globalizations*, 1–14. doi: 10.1080/14747731.2021.2009319.

Zulianello, M. & Ceccobelli, D., 2020. Don't call it climate populism: on Greta Thunberg's technocratic ecocentrism. *The Political Quarterly*, 91 (3), 623–631.

27

GEOENGINEERING AND PUBLIC POLICY

Framing, Research, and Deployment

Ina Möller

27.1 Introduction

In the general landscape of fear, hope, anxiety, and optimism surrounding climate change, the concept of 'geoengineering' is making its way into public policy. Geoengineering is a term that has been widely used to describe a collection of large-scale interventions that aim to deliberately alter the Earth's climate. The concept usually invokes techno-scientific imaginaries like space mirrors, ocean current engineering, stratospheric aerosol injection, or large-scale afforestation. In their anticipatory nature, retroactive function, and unprecedented scale, these imaginaries can be considered different from the conventional mitigation strategies that are commonly used to avoid or prevent the release of greenhouse gases.

The techniques that have been imagined under the concept of geoengineering are often divided into two categories: the large-scale, *ex post* removal and storage of atmospheric carbon dioxide (carbon dioxide removal, or CDR), and the increase of planetary reflectivity (solar radiation management, or SRM). Of these two approaches, CDR has become the most widely discussed and accepted category, featuring prominently in the scenarios of the Intergovernmental Panel on Climate Change (IPCC) as 'negative emissions technologies'. CDR approaches often rely on enhancing or imitating the capacities of ecosystems to absorb and store greenhouse gases, for example by increasing the amount of trees that can absorb carbon dioxide, or by creating 'artificial trees' – machines that filter greenhouse gases from the ambient air. By contrast, SRM describes a group of interventions that cool the planet by increasing the Earth's reflectivity. SRM techniques are less socially accepted, and remain most strongly associated with the geoengineering term. One frequently discussed option in this group is the dispersal of reflective aerosols in the Earth's stratosphere, imitating the cooling effect of large volcanic eruptions. But also approaches like increasing the reflectivity of glaciers, sea ice, or marine clouds fall under this category.

More importantly, geoengineering is a term that evokes contestation and debate. Often users of the term employ it to highlight how a given technique is unusual or different from conventional climate policy. By contrast, employing the terms 'mitigation' or 'adaptation' to describe any of the above techniques usually emphasizes commonalities and familiarity with

DOI: 10.4324/9781003043843-31

what is already known. The exact collection of techniques and approaches that fall under the concept of geoengineering therefore depends on the perspective of the term's user, and the context and purpose of the conversation.

In general, the use of geoengineering techniques to address climate change is still speculative. Large-scale afforestation in China comes closest to what can be considered a 'real-life' case of geoengineering, although this policy has been implemented under the banner of ecological restoration (Cao *et al.*, 2011). Nascent examples of policies explicitly relating to the reversal of global warming are emerging mainly in relation to net-zero strategies. The United Kingdom is situating itself as a front-runner in acknowledging the need for large-scale greenhouse gas removal and supporting targeted research on technologies like bioenergy with carbon capture and storage (BEIS, 2020). In the United States, policy is moving ahead to address atmospheric carbon dioxide removal by providing tax incentives for activities like direct air capture (Congressional Research Service, 2021). In the European Union, the Commission has launched a policy package to encourage its agricultural sector into 'carbon farming' at a large scale (European Commission, 2021). Yet in general, research shows that countries are still facing much uncertainty about what role such techniques can and should play in a wider policy portfolio (Schenuit *et al.*, 2021). Similar to other speculative technologies like hydrogen and nuclear fusion, public policy around geoengineering is still very much at the beginning of things.

This chapter provides an overview of key issues that have been studied with respect to the governance of geoengineering and provides directions for future areas of inquiry. It is divided into three sections, each discussing a distinct question that is relevant to public policy making on this subject. Section 27.2 addresses the fundamental question of how geoengineering is defined and how this impacts public policy making. Section 27.3 addresses expectations about how public policy should govern geoengineering research and development. Section 27.4 addresses imaginaries of where and when geoengineering might take place, sketching out the key debates, anticipated actors, and the geopolitics involved in geoengineering deployment.

27.2 Defining Geoengineering: A Basis for Public Policy?

What one means by 'geoengineering' and whether one chooses to use the term at all is often subject to extensive debate, as well as a matter of political aim, personal conviction, and professional background. The idea of 'geoengineering' the climate has passed through a century-long definitional evolution, being part of scientific discourse ever since anthropogenic climate change was discovered. In this evolution, the concept has been associated with different political aims and used (or shunned) by different actors. This section gives a short outline that highlights how the evolution of definitions is relevant to public policy making.

27.2.1 The Definitional Evolution of Geoengineering

Our contemporary understanding of geoengineering is strongly shaped by a widely cited Royal Society report that defined geoengineering as "the deliberate large-scale intervention in the Earth's climate system, in order to moderate global warming", dividing the various imagined techniques into SRM and CDR (Shepherd *et al.*, 2009, p. ix). Yet discussions of the idea can be traced back to the beginning of the 20th century. In the wake of disastrous global cooling events caused by volcanic eruptions, the Swedish climatologist Svante Arrhenius speculated how burning coal might lead to a virtuous cycle of global warming. He argued that in this cycle, "we may hope to enjoy ages with more equable and better climates ... when the earth will bring forth more abundant crops than at present" (Arrhenius, 1908, p. 63). Decades later, as climate science recognized the detrimental effect of burning fossil fuels, scientists began to think the other way around and search for ways to reduce global temperatures. In 1965, the White House published a document in which it described possibilities to bring about "countervailing climatic changes" such as increasing the reflectivity of the Earth, in order to address excessive global warming (President's Science Advisory Committee, 1965, p. 127). Nine years later, in the aftermath of the global oil crisis and the Watergate Scandal, American and Russian atmospheric scientists discussed the notion of purposeful climate influence and stabilization, and the problems that humanity might face when trying to use climate control (Budyko, 1974; Kellogg and Schneider, 1974).

The idea to purposefully intervene in the climatic system was continuously engaged with over many years, regularly appearing in policy relevant reports such as those published by the United States National Academy of Sciences (1983, 1992, 2015a, 2015b). Geoengineering encompasses grand narratives of space mirrors and the control of ocean currents, as well as highly technical discussions around chemical reactions and intellectual property rights. It is also an area subject to moral debates and heated discussions about what is right and what is wrong (Oomen, 2021). And while the conventional story is that geoengineering was subject to a 'taboo' lifted by atmospheric chemist Paul Crutzen in 2006 (Lawrence and Crutzen, 2017), the increase of scientific and political engagement that provides the justification for writing this chapter is also due to long-term changes in the standing of climate science vis-à-vis the state, and short-term changes in the composition and dynamics of related knowledge networks (Schubert, 2021; Möller, 2022).

Since the publication of the IPCC's Special Report on 1.5 Degrees, the usefulness of geoengineering as an umbrella term that covers all forms of global climatic intervention has become questioned. In 2018, the IPCC declared that it would no longer use the term. Instead, it highlighted a need for more nuance and differentiation between individual approaches, pointing to the fundamental differences between technologies that remove carbon dioxide and technologies that increase the reflectivity of the Earth (de Coninck *et al.*, 2018). Within each of these categories, public discourse is now finding new umbrella headings. Thus CDR (also termed 'greenhouse gas removal' or 'negative emissions technologies') is being separated into 'nature-based' or 'natural climate solutions' on the one hand, and 'technological solutions' on the other (Bellamy and Osaka, 2019). Meanwhile SRM (also termed 'climate stabilization' or 'solar radiation modification') is becoming separated into the global approach of 'solar geoengineering' (mostly equated with 'stratospheric aerosol injection'), and smaller-scale approaches that use techniques like marine cloud brightening or ice preservation.

The categories that I will use in this chapter differ somewhat from the categories described above. Given that all geoengineering techniques aim to change the global climate, they would all need to be implemented at a globally relevant scale – regardless of how they physically function with respect to the Earth's radiation balance, or whether they use 'technological' or 'natural' approaches. Implementation at this scale is always, fundamentally, a question of political organization. I thus distinguish geoengineering techniques by their political mode of operation, based on how many actors it takes for the technique to cause change at a climatically relevant scale. Depending on the answer to this question, each technique can be conceptualized as a centralized, an industrial, or an emergent intervention (see Box 27.1). Distinguishing techniques according to this dimension serves as a helpful tool when thinking about cross-cutting public policy issues in the rest of the chapter.

Box 27.1 Geoengineering Techniques and Their Political Mode of Operation

Geoengineering refers to large-scale, deliberate, and mostly speculative interventions to change the Earth's climate. Approaches are commonly grouped by their physical mode of operation, separating between techniques that remove greenhouse gases from the atmosphere (CDR) and techniques that increase the reflectivity of the Earth's surface (SRM). All aim to reverse or counterbalance the effect of global warming *after* greenhouse gases have been emitted and require large-scale implementation in order to affect planetary systems.

For a discussion on public policy, I find it helpful to group these technologies by their (potential) political mode of operation.

Centralized interventions are techniques with concentrated implementation power, requiring only a few actors in a few places to affect change at the planetary scale. Ideas might include engineering the movement of ocean currents, placing reflective mirrors in space, or spraying reflective aerosols in the atmosphere.

Industrial interventions are techniques that require the alignment of a large but limited number of actors (notably governments and industry) in a limited number of places to affect change at the planetary scale. Ideas might include planting large amounts of biomass, harvesting and storing the resulting carbon in liquid or solid form, artificially changing the pH level of the oceans, or increasing the reflectivity of polar ice sheets.

Emergent interventions are techniques that require many actors in many places to affect change at the planetary level. Ideas might include adding restoring degraded ecosystems, painting infrastructure white, or growing light-coloured crops.

Concentration of implementation power ↑	Carbon Dioxide Removal	Solar Radiation Management
centralized	Ocean current engineering	Stratospheric aerosol injection Space mirrors
industrial	Ocean iron fertilisation / ocean liming Large-scale afforestation Bioenergy with carbon capture and storage (BECCS) Direct air capture and storage (DACS)	Marine cloud brightening Arctic/Antarctic ice sheet preservation
emergent	Biochar/soil carbon enhancement Ecosystem restoration	Painting cities and rooftops white Growing light-coloured crops

Figure 27.1 An overview of geoengineering techniques that distinguishes their physical mode of operation as well as their political mode of operation.

Note: The graph sorts techniques according to the anticipated concentration of implementation power, referring to how many actors are likely needed to affect global change using one technique.

27.2.2 Geoengineering as Public Policy

The definitional evolution that geoengineering has experienced over the years is reflected in the way that public policy making has interacted with it. At the height of the scientific engagement with the term (accompanied by a row of outdoor experiments with ocean iron fertilization, a hype around their potential to generate carbon credits, as well as an emerging anti-geoengineering discourse among civil society organizations), geoengineering was also addressed in international policy making fora. In 2008, parties to the Convention on Biological Diversity (CBD) as well as parties to the London Convention and London Protocol on marine dumping (LC-LP) adopted decisions to discourage private enterprises from conducting ocean iron fertilization. Both conventions later followed up with additional decisions and amendments that would regulate geoengineering and marine geoengineering activities more widely, though not imposing a ban on scientific research (Ginzky and Frost, 2014; Talberg *et al.*, 2017).

A decade later, the term was again introduced to an international forum. In 2019, Switzerland and a diverse coalition of countries tabled a resolution at the United Nations Environment Assembly (UNEA), calling on the United Nations Environment Programme to write a report that would elucidate scientific and governance questions related to geoengineering. In the ensuing discussions, a small group of countries engaged with the proposal over several days. While the United States and Saudi Arabia argued against introducing international regulation on geoengineering that might affect their discretion, the European Union and Bolivia feared that initiating regulatory activities under UNEA might weaken existing regulation under the Biodiversity Convention and the London Protocol (Jinnah and Nicholson, 2019). As is often the case in negotiations, interests were defended on the basis of procedure. The key arguments that were brought forth in public revolved around timing, forum, and definition, with concerns expressed about the institutional mandate and the choice of terminology in defining what the resolution should be about. Despite extensive efforts to accommodate everyone's preferences, the final draft was eventually withdrawn from the negotiation table (McLaren and Corry, 2021a).

The contrast between the engagement of the parties at the Biodiversity Convention/London Protocol and at the UNEA points toward the importance of scientific discourse in shaping the foundations of public policy and governance. In what can be described as a kind of 'de facto governance', authoritative assessments like those of the Royal Society, the National Academy of Sciences, and the IPCC set the scene for what public institutions can do and how negotiators can argue (Gupta and Möller, 2019). Thus, an assessment that highlights the need to engage with the governance of geoengineering provides the basis for passing a resolution, while an assessment that dismisses the usefulness of the entire concept puts the need of such a resolution in question.

'Public policy' around geoengineering therefore starts not on the desks of policy makers, but much earlier – in the editorial boardrooms and coordination meetings of authors contributing to authoritative scientific reports, and in the discursive structures that they build on (Boettcher, 2020). These discourses provide the starting point for any public policy discussion. At the same time, these discourses often work with concepts and problem definitions that need to be adjusted to match the political and institutional context of decision makers (Möller, 2020; Boettcher and Kim, 2022). Whether geoengineering is then conceptualized as a political project distinct from other forms of climate policy making, or whether it is unravelled into a multitude of technicalities and accounting procedures, affects the content and scope of public policies. For both scholars and practitioners, keeping an eye on these definitional struggles is therefore key to understanding where, when, and how 'geoengineering' enters the realm of climate change policy making.

27.3 Geoengineering Research: An Issue for Public Participation?

Accompanying the imaginary of geoengineering has always been the advancement of scientific research. As a techno-scientific project, methods of intentionally controlling the climate have been depicted in experimental designs and engineering ideas that range from plans to fertilize the Earth's oceans, to imitating the cooling effect of large volcanic eruptions (Martin, 1990;

MacMartin, Caldeira and Keith, 2014). In the scholarly literature as well as in public debate, any kind of outdoor experimentation linked to these plans is often considered the frontier – and the bone of contention – of geoengineering research.

27.3.1 Outdoor Experiments and the Social Licence to Operate

In 2021, geoengineering made the front cover of newspapers around the world due to a confrontation between Harvard University researchers and an international network of civil society organizations. The Harvard research group had planned to launch a small balloon experiment with the help of a private space company based in Kiruna, northern Sweden. They were met by much public attention across the Swedish media and were requested to cancel their experiment in a letter signed by representatives of the indigenous Sami Council. In the end, the researchers withdrew, stating that they would make efforts to improve their public consultation and participation process before attempting any further experiments of the kind (Goering, 2021).

What happened in Sweden is essentially a repetition of similar attempts to do outdoor experimentation with geoengineering techniques in the past (also see Low, Baum and Sovacool, 2022). A series of tests involving ocean iron fertilization between 1993 and 2009 met with increasing resistance from civil society organizations who gathered around an anti-geoengineering rhetoric (Strong *et al.*, 2009). They eventually brought their concerns to the attention of international policy makers, resulting in the earlier discussed resolutions and decisions on geoengineering regulation under the Biodiversity Convention and the London Protocol (Fuentes-George, 2017). Shortly after, a British team of scientists planned to conduct an outdoor experiment linked to the delivery mechanisms that would be needed to conduct stratospheric aerosol injection. Their plan to hoist a one-kilometre-long water hose into the sky was again met with protest from anti-geoengineering civil society groups, and eventually cancelled (although the reason given by the scientists was a problem of patents and intellectual property rights) (Pidgeon *et al.*, 2013).

Within this setting of scientific experimentation and civil society protest, experts often highlight the need for public participation in the governance of geoengineering research. Underlying the discussion around public participation is often a quest to obtain consent, or a 'social licence to operate' in the research and deployment of geoengineering technologies (Lenton *et al.*, 2019). A large amount of literature thus engages with scoping public opinions and organizing workshops in which to discuss geoengineering scenarios. A 2021 report by the US National Academy of Sciences stated that public engagement in solar geoengineering research is necessary for building trust and understanding what 'the public' considers permissible and what not (NASEM, 2021, p. 178). Also in the realm of CDR, public participation is coming to be seen as an important part of ensuring the social legitimacy of greenhouse gas removal technologies (O'Beirne *et al.*, 2020).

Yet given the global reach of geoengineering, it is often uncertain who this public might be and what the implications are of introducing the idea of climate control to audiences who were previously unaware of this possibility (Bellamy and Lezaun 2015). Furthermore, contemporary societal arrangements often work with models of representative democracy. If officials who represent the interests of their voters are in a position to make informed decisions, lack of explicit public engagement on the research and use of geoengineering would not necessarily be undemocratic (Wong 2016). The crux of this discussion – similar to many other issue areas of policy making – would then boil down to how legitimate these decision makers, and the systems in which they operate, are perceived to be. Given the less-than-impressive results that governments have delivered on climate change policy so far, key issues in the coming debate around incentivizing and governing both CDR and SRM technologies will likely revolve around climate communication and trust in the actors involved (Colvin *et al.*, 2019; Raimi, 2021).

27.3.2 Designing Governance for Research

The assortment of proposals for how to design governance for geoengineering research is as proliferous as proposals of how to design research itself. One recent review in the field of solar geoengineering maps out the different formats of governance that have been proposed, covering state and non-state actor engagement, possible moratoria on outdoor research and their problems, the legitimation of decision making procedures, the regulation of commercial actors and intellectual property, as well as compensation mechanisms. It concludes by highlighting a need for more research on the role that non-state actors could take in governing geoengineering, as well as the need for the IPCC to dedicate an explicit focus on SRM (Reynolds, 2019). Another review maps the landscape of principles, frameworks, procedures, and institutions that have been suggested and critically assesses their shortcomings in terms of Western norm reproduction, instrumentalist conceptions of public engagement, and the problematic separation of indoor and outdoor research. This review points to the need for a top-down, international research governance regime that can explicitly reflect on the emerging social, ethical, political, and technical implications of geoengineering research, as well as regulating knowledge production on geoengineering more generally (McLaren and Corry, 2021b).

These different conclusions are examples of different understandings of how, and more particularly why, geoengineering should be governed. In their analysis of governance rationales for solar geoengineering, Aarti Gupta et al. (2020) depict a continuum of reasons for why scholars propose governance mechanisms in the first place. Simply put, it ranges from the desire to enable research to the desire to restrict it, with more nuanced rationales in between that focus on ensuring adequate oversight or safeguarding the interests of under-represented populations. These variations in governance rationales somewhat explain the wide variety of governance proposals that exist in the geoengineering literature, and point to the underlying values that necessarily shape discussions around geoengineering research and its regulation.

The attention given to outdoor experiments, public participation, and governance design obscures some of the more inconspicuous, but more potent, aspects of geoengineering research. Climate modelling groups across the world routinely calculate what would happen if the Earth were engulfed in a layer of sulphur particles, if its oceans were mixed with hundreds of millions of tonnes of lime, or if it were covered in half a billion hectares of forest or biofuel crops. Such global modelling experiments have become authoritative sources of scientific knowledge that inform policy making at the highest levels, most notably through the regular assessment reports of the IPCC (Beck and Mahony, 2017; Hansson *et al.*, 2021). In graphs and scenarios, they show the outcome of a whole range of what one might call 'Earth Experiments', evoking (but rarely engaging with) important questions about responsibility and justice (Stilgoe, 2016; Rubiano Rivadeneira and Carton, 2022).

The way in which digital 'Earth experiments' are communicated differs, depending on which modelling group is engaging with which type of model, and how much access they have to authoritative authorship positions. What sort of digital experiment is considered 'feasible' or 'realistic' and therefore deemed 'policy relevant' is essentially down to the value judgement of the individual modelling group. Thus the modelling culture of influential groups like the Integrated Assessment Modelling community, which has prime access to writing the IPCC's Working Group III report on mitigation, has had a substantial impact on contemporary climate policy (Cointe, Cassen and Nadaï, 2019). Led by the International Institute for Applied Systems Analysis (IIASA) in Austria, this community has focused primarily on the role of terrestrial carbon removal, with some of its leading figures identifying a combination of biomass production with carbon capture and storage (BECCS) as an 'overlooked' but highly promising solution to the climate change problem early on (Kraxner, Nilsson and Obersteiner, 2003). The resulting focus of the IPCC's AR5 report on afforestation and BECCS as key elements of future climate trajectories has influenced climate policy around the world, leading to a flurry of net-zero targets that rely on unprecedented amounts of afforestation and biomass production (Peters and Geden, 2017; Rogelj *et al.*, 2021). The political implications of these targets remain to be seen, but they do raise an important question about the power that expert-driven climate modelling has on contemporary climate policy (Beck and Oomen, 2021).

27.4 Geoengineering Deployment: A Task for Policy Cooperation?

Despite frequent efforts to separate research from deployment, the potential use of geoengineering is a constant companion in any discussions around its governance. Particularly among the approaches labelled SRM, the global nature of interventions like stratospheric aerosol injection raises questions as to whether there can be a real separation between research and deployment, and whether the two would not need to be governed hand in hand. Meanwhile in the area of CDR, discussions revolve around creating incentives for upscaling techniques like afforestation or carbon capture and storage in order to meet the tall order of going 'net-negative' by the middle of this century, and how such an enormous effort would be coordinated and financed.

27.4.1 Common Concerns: Slippery Slope, Moral Hazard, and the Question of Justice

One core theme that has shaped this discussion is the idea of a 'slippery slope'. By talking about a slippery slope, geoengineering critics refer to the possibility that setting up mechanisms that allow or incentivize research may lead to the creation of vested interests or a socio-technical lock-in, resulting in no other option than deployment (Anshelm and Hansson, 2014; Callies, 2019). The slippery-slope argument is commonly refuted by pointing to examples where investment in a technology did result in failure and abandonment, and some even positing that the investment in geoengineering is actually more of an 'uphill struggle', as there is still very little political and societal interest in these methods (Bellamy and Healey, 2018). Nevertheless, the question of systemic path dependency is worth taking into account. Particularly in the case of CDR, large-scale investment in removing greenhouse gases from the atmosphere could shift a heavy-weight industrial system onto a pathway that is not so easy to reverse, and the systemic consequences of this shift should be taken into account at early stages in the political process.

A second core theme is the idea of 'moral hazard' or 'mitigation deterrence' caused by geoengineering. This concern relates to the possibility that including speculative geoengineering technologies in the climate policy portfolio will result in less efforts to reduce emissions through conventional mitigation (McLaren, 2016; Markusson, McLaren and Tyfield, 2018). Whether this is actually the case is very hard to measure, given that there is no parallel world to compare ours with. Yet learning from other cases in which speculative technologies have been named as key solutions to climate change (notably multi-decade-long discussions about the potential of carbon capture and storage, hydrogen, or nuclear fusion), we may assume that adding CDR or even solar geoengineering to the climate policy agenda is unlikely to accelerate efforts in reducing greenhouse gases emissions. The recent wave of net-zero pledges by companies who continue to invest in fossil fuels seems to support this point (Bhargava *et al.*, 2022). One key thing to note here is that while some studies find evidence against the relevance of moral hazard (e.g. Austin and Converse, 2021), this research rests on the responses of individuals and does not tell us much about the dynamics that take place at a collective level. Collective climate change policy, with its decade long debates about who should be responsible, who should pay, and what kind of information would be needed to ensure action, is subject to a tangle of interests and historical political relations. How these will be affected by the introduction of geoengineering policies is difficult to test or predict based on the responses of individual citizens. Arguably, the largest risk in this realm is the delay of *any kind* of climate policy, be it emissions reduction, carbon removal, or climate adaptation.

A third common theme is the question of intra- and intergenerational justice (see Box 27.2). The potential deployment of a centralized intervention like stratospheric aerosol injection evidently raises questions about who will decide what to do, how much of it, and where and when this will take place. Given stark differences in power and influence between different communities and nations, observers highlight the need to account for the interests of peoples who are not heard or represented in the halls of decision making power. This can usually go either way: for example, while advocates use justice as a way to argue *for* using solar geoengineering (stating that those who are most vulnerable to climate change would also benefit most from a technology like stratospheric aerosol injection), critics highlight the neo-colonial taste of this argument and use it to argue *against* solar geoengineering, given the concentration of political and economic power that such a technique would reinforce (Horton and Keith, 2016; Stephens and Surprise, 2020).

In the area of CDR, concerns around justice relate mainly to issues of land use (particularly land and water grabbing) and what large-scale carbon removal might do to key commodities. Models suggest that industrial-scale use of BECCS or afforestation could raise staple food prices in the Global South by fivefold compared to 2010, and may exacerbate water stress beyond what climate change itself would cause (Fuhrman *et al.*, 2020; Stenzel *et al.*, 2021). Direct air

Box 27.2 Justice and Geoengineering

Centralized interventions like stratospheric aerosol injection are most prominent in the justice discussion, as these raise thorny questions about who can make decisions on behalf of whom, who can be held responsible for side effects, and who should compensate for any harm experienced.

Industrial interventions like bioenergy with carbon capture and storage raise important questions about economic path dependencies and what these might mean for future generations, as well as possible effects on other socio-environmental systems including food and energy. As these industrial interventions require substantial resource use, key concerns revolve around the distribution and respect for land and water rights.

Emergent interventions like biochar or reflective infrastructure are less prominent in the justice debate, but may nevertheless raise questions around how incentives for implementation are distributed. Many of these concerns are not exclusive to geoengineering techniques, but recognizable from other kinds of public policy. It is thus advisable to consider the implications of geoengineering techniques in relation to a wider portfolio of climate-relevant public policy.

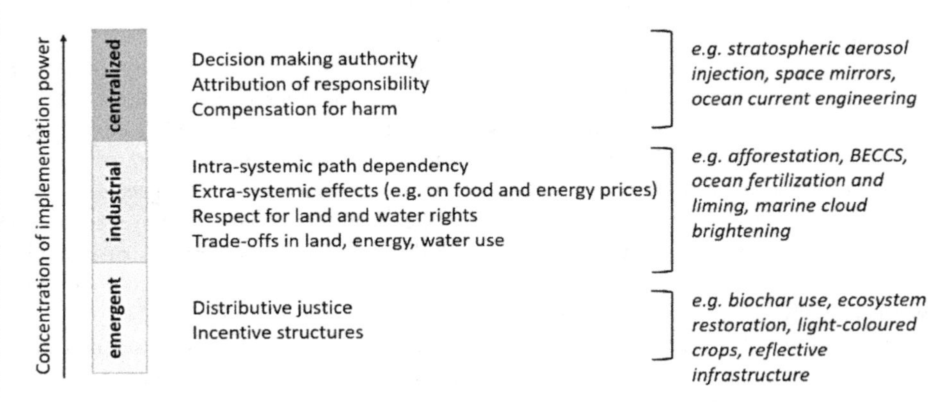

Figure 27.2 An overview of the most pertinent justice concerns associated with different geoengineering techniques.

Note: The listed concerns are not exclusive to the techniques mentioned on the right-hand side of the graph. Rather, the graph highlights the most pertinent justice concerns associated with each type of intervention.

capture would likewise require 12–20 percent of the global energy supply if it is to be globally relevant (Chatterjee and Huang, 2020). Given that any industrial intervention to absorb CO_2 would have substantial effects on global markets, those who are already most vulnerable to fluctuations in prices would likely be the first to feel the effects. On the other hand, given the substantial (and increasing) climate footprint of emerging powers that host a large percentage of the world's poorest people, such techniques are seen as a suitable climate policy for countries like China and India, with the potential to minimize damage to economic growth while supporting environmental restoration efforts (Chen *et al.*, 2019; Weng, Cai and Wang, 2021).

Discussions about justice and geoengineering are always held with the caveat that very few of the people who are spoken about actually participate in the conversation. Future generations are perhaps the most difficult to include as they lack political presence and as it remains notoriously hard to know what the future will bring. Although models can tell us something about the possible (physical) effects of both climate change and geoengineering, they provide a necessarily simplified view of the world. Risks related to economic dependencies, disinformation campaigns, perceived security hazards, and pure bad luck are difficult to account for, and the social instabilities that may or may not be caused by various geoengineering techniques (or indeed, a world without geoengineering in which climate change alone affects these questions) add a significant layer of uncertainty. Yet even when models give reasonable estimates about the future, Western liberal philosophy tends to prioritize the interests of contemporary (human) generations over others. Suggestions to overcome this challenge highlight the need for a more deliberative style of decision making in which the plurality of values is taken seriously (Hourdequin, 2019). In addition, a valuable source of wisdom can be found in the cyclical or spiralling understanding of time held by many first nations communities, an ontology that would enable a more equal evaluation of the interests held by past present and future generations (Winter, 2021).

At the same time, it is worth noting that the expertise produced on geoengineering and climate change science more generally lies firmly in the hand of scientists based in highly industrialized countries (Corbera *et al.*, 2016; Biermann and Möller, 2019). The few studies that do include voices from commonly under-represented communities paint a picture of diverse positions, but also highlight the need to place the various geoengineering techniques in a much larger context than is commonly done. Studies that have tried to gauge perceptions of geoengineering interventions among non-Western publics highlight the variation in viewpoints and philosophical ideas that can be found in these settings, but also mention a commonality that could be characterized as a kind of 'conditional acceptance' within communities that are heavily affected by climate change. Here, geoengineering is seen as a last-resort option in the face of devastating environmental change, but concerns remain around the enhanced economic dependencies, unequal distribution of power, and marginalization or exploitation that these technologies might cause (Winickoff, Flegal and Asrat, 2015; Carr and Yung, 2018; Gannon and Hulme, 2018). More fundamentally, Kyle Powys Whyte (2019) points out that most geoengineering discourses are not set up in a way that would allow marginalized communities like indigenous peoples to properly express their concerns about risk, research, and power. Judging by the perspectives that many of these peoples have on other topics, he argues that geoengineering would hardly be considered a discreet topic in relation to climate justice. Instead, it would form part of a much wider discussion around colonialism and how this has defined and continues to shape indigenous vulnerability to climate change.

27.4.2 Imaginaries of Deploying Solar Radiation Management

Further questions that shape speculation and inquiry around geoengineering deployment revolve around who will develop and use the technology, what their intentions might be, and whether they have the necessary legitimacy to be successful in their endeavour (see Box 27.3). One key imaginary that shapes this discussion in the field of stratospheric aerosol injection is the notion of 'rogue actors'. Because stratospheric aerosol injection is touted as a comparatively simple and cheap technology (involving a few dozen aircraft and a few hundred tonnes of sulphur powder), scientists often point out that basically anyone with a reasonably decent budget, including wealthy individuals, could run a stratospheric aerosol injection operation. This has led governance scholars to speculate about the role of 'green-finger' billionaires, desperate small island states, and environmentally minded non-governmental organizations as potential unilateral deployers of stratospheric aerosol injection (SAI) (Victor, 2008; Millard-Ball, 2012; Reynolds and Wagner, 2019; Schenuit, Gilligan and Viswamohanan, 2021). Although the initiation of an SAI operation through such actors is imaginable, one could also argue that any unsanctioned efforts at manipulating the climate would soon be put to an end by other, more powerful actors (Rabitz, 2016). Furthermore, the upkeep of such an operation would quickly dry out the resources of any individual actor, save large and militarily powerful nations (Smith, 2020). Yet even the incentive of these large powers to unilaterally deploy SAI is questionable, with the social and political costs (including possible trade sanctions, diplomatic isolation, and reprisals in other issue areas) deemed higher than the benefits that any one nation could gain from a non-sanctioned use of the technology (Horton, 2011).

While one could say that unilateral deployment of SAI is possible (though unlikely), it is also unlikely that a multilateral approach to managing this technology will emerge without an urgent reason to do so. Inquiries into the perceptions of policy makers confirm that expectations and social norms play an important role in reasoning about geoengineering. While there is no scientific consensus around the desirability of and need for (research on) centralized forms of SRM, policy makers are cautious about taking a public position on the subject. Rich countries mainly fear that their reputation might be damaged by associating with the topic while poor countries lack the capacity to engage with yet another complex scientific topic

Box 27.3 Anticipated Actors in Geoengineering Implementation

Centralized interventions like stratospheric aerosol injection or ocean current engineering are likely to be implemented by economically and militarily powerful actors, given the long-term upkeep and political controversy associated with these techniques. A coalition of smaller, heavily affected nations might also embark on such an endeavour if they can secure the legitimacy not to face retribution.

Industrial interventions like bioenergy with carbon capture and storage or marine cloud brightening are more comparable to the activities of the oil and gas, timber, agribusiness, and shipping industries. Their expertise and infrastructure is highly relevant if these techniques are to be implemented at a large scale, but would require reorientation through the policies of national governments and the guidelines of finance, insurance, and standard setting organizations.

Emergent interventions like biochar or reflective infrastructure are activities more likely to be led by municipalities, public and private infrastructure owners, and (consortia of) individual land owners. Accompanying all three types of interventions are actors engaging in research and/or advocacy of the various techniques.

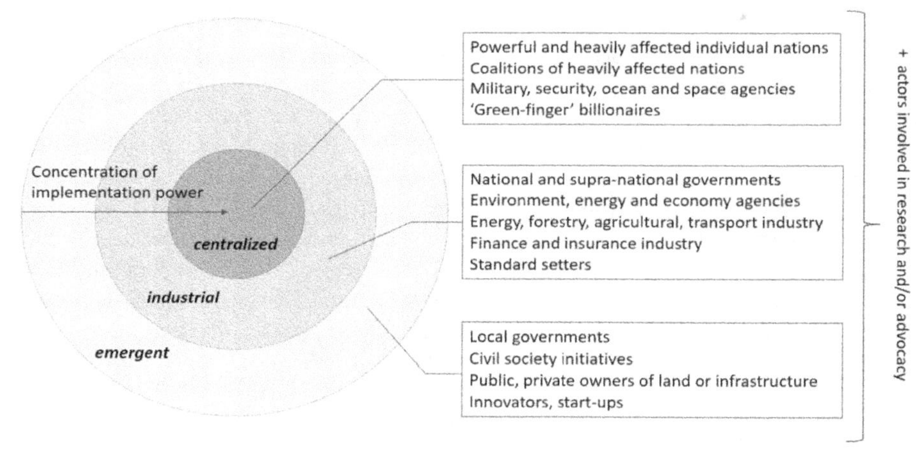

Figure 27.3 Key actors that might be involved in implementing different geoengineering techniques, grouped according to each intervention's political mode of operation.

on their already overburdened climate change policy agendas (Möller, 2020). This caution to engage means that negotiations at the multilateral level are difficult to initiate, and an international agreement on how to govern solar geoengineering is not yet visible on the horizon of international climate negotiations.

In the absence of a multilateral agreement for global-scale solar radiation management, scenarios in which we might still see SRM techniques being deployed are in industrial or emergent form. Climate modellers are increasingly engaging with scenarios of regional cooling, and examples of initiatives are starting to take shape on the ground. Most prominently, a consortium of local government actors, civil society organizations, and scientific institutes has recently embarked on a quest to save the Great Barrier Reef, and within its portfolio of techniques it aims to use marine cloud brightening to provide a cooling umbrella for the region (Tollefson, 2021). Other ongoing endeavours in the use of SRM include plans to save glaciers and ice sheets, such as the one described by the 'Arctic Ice Project' (formerly called 'Ice911'). The principal idea of this non-profit organization is to scatter glass beads on the surface of Arctic sea ice in order to increase its reflectivity and reduce the speed of melting (Arctic Ice Project, 2021). The incentivization of such reflectivity enhancement procedures has also been introduced to the International Standards Organization (ISO), where environmental certification companies from the USA tried to introduce a new standard for measuring organizations' climate footprints. This standard, based on the concept of radiative forcing, would have allowed the inclusion of 'climate coolants' in the calculation of an organizations footprint, enabling the generation of tradeable credits for such projects. Due to concerns about the unintended effects that such a standard might have, the initiative was eventually downgraded to a technical document with no guiding power (Möller, 2021).

27.4.3 Realities of Deploying Carbon Dioxide Removal

In the field of CDR, the discussion around who will deploy negative emission technologies is less about ensuring control and more about creating incentives for investment and upscaling. Removing carbon from the atmosphere at an industrial scale for the sake of permanent storage is not yet financially viable, mainly due to low carbon prices. When looking at carbon removal through land-use change, incentives for afforestation and reforestation are still lower than for deforestation. Although many developed countries (as well as China and India) are reporting net gains in forest cover, the amount of embodied deforestation in their imports has risen, contributing to the overall trend of global forest loss (Hoang and Kanemoto, 2021). In addition, changes in the climate are affecting the frequency of wildfires and forest dieback, further contributing to the release of carbon emissions. More 'technological' solutions to carbon removal are also facing severe challenges. The most viable economic approach envisioned for companies that filter CO_2 from the atmosphere is to turn the carbon into a commodity, selling it on to other companies who use it for the cultivation of algae or (ironically) for enhanced oil recovery (Wilcox, Psarras and Liguori, 2017). The permanence of this absorbed carbon is therefore not guaranteed, and the scale at which these companies are operating is still minimal. In addition, scaling carbon removal and permanent storage faces significant challenges in terms of energy sourcing, the land needed to generate that (ideally renewable) energy, and the infrastructure necessary for CO_2 transportation.

The core actors projected to play a role in the industrial development of CDR are the forest and agricultural sectors, but also the oil and gas industry. The capture of carbon dioxide and its storage underground is a key element of many of the 'technological' solutions that have been put forward. Meanwhile, the primary use of captured carbon is currently in enhanced oil recovery, where underground oil is pumped out and replaced with liquid carbon dioxide. This means that oil and gas companies have both the necessary skill set and the infrastructure to deal with captured carbon dioxide. On the one hand, this is an opportunity because this industry is highly influential and a globally powerful player that – given the right incentives – could contribute substantially to the acceleration of the net-zero transition (Garcia Freites and Jones, 2021). On the other hand, it is a problem because the fossil fuel industry contributes substantially to climate change, and has used the promise of carbon removal for many years in order to avoid any reduction of emissions (Carton *et al.*, 2020). Meanwhile, the forest and agricultural sectors are already well known as actors from previous debates around carbon offsetting, land-use change, and biodiversity conservation. The challenge of reaching net negative emissions by 2050 only increases the demands on these sectors to adapt their practices with environmental goals in mind (von Hedemann *et al.*, 2020). At the same time, they face increasing pressure to produce timber and food for a growing population and – more problematically – highly wasteful consumption patterns among the world's wealthy citizens.

How then do scholars and practitioners propose to overcome these enormous hurdles to CDR deployment? Discussions in this field are akin to classic policy dilemmas already familiar from climate change policy more generally. The availability of finance, for example through an increase in the carbon price, is most often mentioned as an important necessity for improving incentives to invest in carbon removal and storage, and ramping up (voluntary) carbon markets to facilitate the financing and trading of carbon credits is a common solution discussed in climate policy circles (Honneger *et al.*, 2021). But there are also words of caution. Too much focus on carbon is liable to create perverse incentives, negatively affecting communities and livelihoods, heightening the cost of food, and contributing to the further depletion of biodiversity. Therefore scholars argue for separating emission reductions from emission removal targets, providing financing for other environmentally and societally relevant goals, and ensuring that the generation of carbon credits is subject to regulation and monitoring (McLaren *et al.*, 2019; WBGU, 2021). Some say that solutions that generate co-benefits, for example carbon removal techniques that also help to ameliorate soils, should be prioritized (Cox and Edwards, 2019). Most importantly, the upscaling of carbon markets should not justify a continued reliance on fossil fuels, as early climate action would lessen the negative impact of future negative emission technologies (Hasegawa *et al.*, 2021). As this discussion moves from the theoretical to the practical, actors that will be relevant to shaping this trajectory are (re)-insurance companies and standard setting organizations. Together with the targets and goals defined by policy makers, these actors will have an important hand in guiding industry towards large-scale CDR.

27.5 Conclusion and Future Research

Despite the controversy of the concept, geoengineering in various shapes and guises has entered national and international policy agendas. As countries face increasing pressure to act on climate change, large-scale carbon removal and perhaps also large-scale solar radiation management are becoming unavoidable topics. This chapter has laid out how questions around the definition, research, and deployment of geoengineering are affecting public policy agendas. In doing so, it has introduced a political categorization to better discuss the policy implications of different types of geoengineering techniques, distinguishing between centralized, industrial, and emergent interventions. One important thing to keep in mind during this discussion is that all types of geoengineering are still highly speculative, and that the idea of being able to actively engineer the climate is intrinsically linked to the global perspective of climate science. Yet any form of climate policy will always take place in specific local settings, with local actors and local interests involved. Bridging this gap will be a key challenge for policy making. It will also require the support of analysts who can critically reflect on the global solutions offered by climate science, as well as find ways to make these match the needs and limitations of regional, national, and local settings.

Box 27.4 Chapter Summary

- Geoengineering is a contested concept that is usually associated with techno-scientific imaginaries of halting or reversing global warming.
- Public policy needs to be aware of the different meanings and intentions with which the term is used.
- Geoengineering techniques are inherently anticipatory; they shape contemporary policy despite large uncertainties about whether or not they will ever exist.
- To facilitate governance, it is helpful to think about how geoengineering techniques might differ in terms of the political organization of their implementation.

References

Anshelm, J. and Hansson, A. (2014) 'Battling Promethean dreams and Trojan horses: Revealing the critical discourses of geoengineering', *Energy Research & Social Science*, 2(2014), pp. 135–144.

Arctic Ice Project (2021) *Technology Focus Areas*. Available at: https://www.arcticiceproject.org/technology-focus-areas/#materials (Accessed: 11 October 2021).

Arrhenius, S. (1908) *Worlds in the making: The evolution of the universe*. New York and London: Harper & Brothers.

Austin, M. M. K. and Converse, B. A. (2021) 'In search of weakened resolve: Does climate-engineering awareness decrease individuals' commitment to mitigation?', *Journal of Environmental Psychology*, p. 101690. doi: 10.1016/J.JENVP.2021.101690.

Beck, S. and Mahony, M. (2017) 'The IPCC and the politics of anticipation', *Nature Publishing Group*, 7(5), pp. 311–313. doi: 10.1038/nclimate3264.

Beck, S. and Oomen, J. (2021) 'Imagining the corridor of climate mitigation – What is at stake in IPCC's politics of anticipation?', *Environmental Science & Policy*, 123, pp. 169–178. doi: 10.1016/J.ENVSCI.2021.05.011.

BEIS (2020) *UK government's view on greenhouse gas removal technologies and solar radiation management - GOV.UK, Department for Business, Energy and Industrial Strategy*. Available at: https://www.gov.uk/government/publications/geo-engineering-research-the-government-s-view/uk-governments-view-on-greenhouse-gas-removal-technologies-and-solar-radiation-management (Accessed: 12 January 2022).

Bellamy, R. and Healey, P. (2018) '"Slippery slope" or "uphill struggle"? Broadening out expert scenarios of climate engineering research and development', *Environmental Science & Policy*, 83(October 2017), pp. 1–10. doi: https://doi.org/10.1016/j.envsci.2018.01.021.

Bellamy, R. and Lezaun, J. (2015) 'Crafting a public for geoengineering', *Public Understanding of Science*, (2014). doi: 10.1177/0963662515600965.

Bellamy, R. and Osaka, S. (2019) 'Unnatural climate solutions?', *Nature Climate Change 2019 10:2*, 10(2), pp. 98–99. doi: 10.1038/s41558-019-0661-z.

Bhargava, A. et al. (2022) 'Climate-washing litigation: Legal liability for misleading climate communications', *CSSN Research Report*, pp. 1–27.

Biermann, F. and Möller, I. (2019) 'Rich man's solution? Climate engineering discourses and the marginalization of the Global South', *International Environmental Agreements: Politics, Law and Economics*, 19(2), pp. 151–167. doi: 10.1007/s10784-019-09431-0.

Boettcher, M. (2020) 'Cracking the code: How discursive structures shape climate engineering research governance', *Environmental Politics*, 29(5), pp. 890–916. doi: 10.1080/09644016.2019.1670987.

Boettcher, M. and Kim, R. E. (2022) 'Arguments and architectures: Discursive and institutional structures shaping global climate engineering governance', *Environmental Science and Policy*, 128, pp. 121–131. doi: 10.1016/j.envsci.2021.11.015.

Budyko, M. I. (1974) 'Metod Vozdeystviya Na Klimat (Method of Influencing the Climate)', *Meteorologiia i Gidrologiia*, 2, pp. 91–97.

Callies, D. E. (2019) 'The slippery slope argument against geoengineering research', *Journal of Applied Philosophy*, 36(4), pp. 675–687. doi: 10.1111/JAPP.12345.

Cao, S. et al. (2011) 'Excessive reliance on afforestation in China's arid and semi-arid regions: Lessons in ecological restoration', *Earth-Science Reviews*, 104(4), pp. 240–245. doi: 10.1016/J. EARSCIREV.2010.11.002.

Carr, W. A. and Yung, L. (2018) 'Perceptions of climate engineering in the South Pacific, Sub-Saharan Africa, and North American Arctic', *Climatic Change*. 147, pp. 119–132. doi: 10.1007/s10584-018-2138-x.

Carton, W. et al. (2020) 'Negative emissions and the long history of carbon removal', *Wiley Interdisciplinary Reviews: Climate Change*, 11(e671), pp. 1–25. doi: 10.1002/WCC.671.

Chatterjee, S. and Huang, K. W. (2020) 'Unrealistic energy and materials requirement for direct air capture in deep mitigation pathways', *Nature Communications*, 11(1), pp. 1–3. doi: 10.1038/s41467-020-17203-7.

Chen, C. et al. (2019) 'China and India lead in greening of the world through land-use management', *Nature Sustainability*, 2(2), pp. 122–129. doi: 10.1038/s41893-019-0220-7.

Cointe, B., Cassen, C. and Nadaï, A. (2019) 'Organising policy-relevant knowledge for climate action: Integrated assessment modelling, the IPCC, and the emergence of a collective expertise on socioeconomic emission scenarios', *Science and Technology Studies*, 32(4), pp. 36–57. doi: 10.23987/sts. 65031.

Colvin, R. M. et al. (2019) 'Learning from the climate change debate to avoid polarisation on negative emissions', *Environmental Communication*, 14(1), pp. 23–35. doi: 10.1080/17524032.2019.1630463.

Congressional Research Service (2021) *The Tax Credit for Carbon Sequestration (Section 45Q)*. Available at: https://crsreports.congress.gov (Accessed: 12 January 2022).

de Coninck, H. et al. (2018) 'Strengthening and Implementing the Global Response', in Masson-Delmotte, V. et al. (eds) *Global Warming of 1.5°C. An IPCC Special Report on the Impacts of Global Warming of 1.5°C Above Pre-Industrial Levels and Related Global Greenhouse Gas Emission Pathways*. Intergovernmental Panel on Climate Change.

Corbera, E. et al. (2016) 'Patterns of authorship in the IPCC Working Group III report', *Nature Climate Change*, 6(1), pp. 94–99. doi: 10.1038/nclimate2782.

Cox, E. and Edwards, N. R. (2019) 'Beyond carbon pricing: Policy levers for negative emissions technologies', *Climate Policy*, 19(9), pp. 1144–1156. doi: 10.1080/14693062.2019.1634509.

European Commission (2021) *Carbon Farming, Climate Action*. Available at: https://ec.europa.eu/clima/eu-action/forests-and-agriculture/sustainable-carbon-cycles/carbon-farming_en (Accessed: 14 September 2022).

Fuentes-George, K. (2017) 'Consensus, certainty, and catastrophe: Discourse, governance, and ocean iron fertilization', *Global Environmental Politics*, 17(2), pp. 125–143. doi: 10.1162/GLEP.

Fuhrman, J. et al. (2020) 'Food–energy–water implications of negative emissions technologies in a +1.5 °C future', *Nature Climate Change*, 10(10), pp. 920–927. doi: 10.1038/s41558-020-0876-z.

Gannon, K. E. and Hulme, M. (2018) 'Geoengineering at the "Edge of the World": Exploring perceptions of ocean fertilisation through the Haida Salmon Restoration Corporation', *Geo: Geography and Environment*, 5(1), pp. 1–21. doi: 10.1002/GEO2.54.

Garcia Freites, S. and Jones, C. (2021) *A Review of the Role of Fossil Fuel-Based Carbon Capture and Storage in the Energy System*. Manchester: Tyndall Centre.

Ginzky, H. and Frost, R. (2014) 'Marine geo-engineering: Legally binding regulation under the London protocol', *Carbon & Climate Law Review*, 8(2), pp. 82–96.

Goering, L. (2021) *Sweden rejects pioneering test of solar geoengineering tech, Reuters*. Available at: https://www.reuters.com/article/us-climate-change-geoengineering-sweden-idUSKBN2BN35X (Accessed: 1 October 2021).

Gupta, A. et al. (2020) 'Anticipatory governance of solar geoengineering: Conflicting visions of the future and their links to governance proposals', *Current Opinion in Environmental Sustainability*, 45, pp. 10–19. doi: 10.1016/j.cosust.2020.06.004.

Gupta, A. and Möller, I. (2019) 'De facto governance: How authoritative assessments construct climate engineering as an object of governance', *Environmental Politics*, 28(3), pp. 480–501. doi: 10.1080/09644016.2018.1452373.

Hansson, A. et al. (2021) 'Boundary work and interpretations in the IPCC review process of the role of bioenergy with carbon capture and storage (BECCS) in limiting global warming to 1.5°C', *Frontiers in Climate*, 3(643224), pp. 1–14. doi: 10.3389/FCLIM.2021.643224/XML/NLM.

Hasegawa, T. et al. (2021) 'Land-based implications of early climate actions without global net-negative emissions', *Nature Sustainability* 2021(4), pp. 1052–1059. doi: 10.1038/s41893-021-00772-w.

von Hedemann, N. et al. (2020) 'Forest policy and management approaches for carbon dioxide removal', *Interface Focus*, 10(5), pp. 1–16. doi: 10.1098/RSFS.2020.0001.

Hoang, N. T. and Kanemoto, K. (2021) 'Mapping the deforestation footprint of nations reveals growing threat to tropical forests', *Nature Ecology & Evolution*, 5(6), pp. 845–853. doi: 10.1038/s41559-021-01417-z.

Honneger, M. et al. (2021) 'Who is paying for carbon dioxide removal? Designing policy instruments for mobilizing negative emissions technologies', *Frontiers in Climate*, 3(672996), pp. 1–15. doi: 10.3389/fclim.2021.672996.

Horton, J. B. (2011) 'Geoengineering and the myth of unilateralism: Pressures and prospects for international cooperation', *Stanford Journal of Law, Science and Policy*, IV, pp. 56–69.

Horton, J. B. and Keith, D. (2016) 'Solar Geoengineering and Obligations to the Global Poor', in Preston, C. (ed.) *Climate Justice and Geoengineering: Ethics and Policy in the Atmospheric Anthropocenein the Atmospheric*. London: Rowman & Littlefield International, pp. 79–92.

Hourdequin, M. (2019) 'Geoengineering justice: The role of recognition', *Technology, & Human Values*, 44(3), pp. 448–477. doi: 10.1177/0162243918802893.

Jinnah, S. and Nicholson, S. (2019) 'The hidden politics of climate engineering', *Nature Geoscience*, 12(11), pp. 876–879. doi: 10.1038/s41561-019-0483-7.

Kellogg, W. W. and Schneider, S. H. (1974) 'Climate stabilization: For better or for worse?', *Science*, 186(4170), pp. 1163–1172. doi: 10.1126/SCIENCE.186.4170.1163.

Kraxner, F., Nilsson, S. and Obersteiner, M. (2003) 'Negative emissions from BioEnergy use, carbon capture and sequestration (BECS)— the case of biomass production by sustainable forest management from semi-natural temperate forests', *Biomass and Bioenergy*, 24(2003), pp. 285–296. doi: 10.1016/S0961-9534(02)00172-1.

Lawrence, M. G. and Crutzen, P. J. (2017) 'Was breaking the taboo on research on climate engineering via albedo modification a moral hazard, or a moral imperative?', *Earth's Future*, 5, pp. 136–143. doi: 10.1002/eft2.172.

Lenton, A. et al. (2019) 'Foresight must guide geoengineering research and development', *Nature Climate Change 2019 9:5*, 9(5), pp. 342–342. doi: 10.1038/s41558-019-0467-z.

Low, S., Baum, C. M. and Sovacool, B. K. (2022) 'Taking it outside: Exploring social opposition to 21 early-stage experiments in radical climate interventions', *Energy Research & Social Science*, 90(102594), pp. 1–21. doi: 10.1016/J.ERSS.2022.102594.

MacMartin, D. G., Caldeira, K. and Keith, D. W. (2014) 'Solar geoengineering to limit the rate of temperature change', *Philosophical Transactions. Series A, Mathematical, Physical, and Engineering Sciences*, 372(20140134), pp. 1–13. doi: 10.1098/rsta.2014.0134.

Markusson, N., McLaren, D. and Tyfield, D. (2018) 'Towards a cultural political economy of mitigation deterrence by negative emissions technologies (NETs)', *Global Sustainability*, 1(e10), pp. 1–9. doi: 10.1017/SUS.2018.10.

Martin, J. H. (1990) 'Glacial-interglacial CO2 Change: The iron hypothesis', *Paleoceanography*, 5(1), pp. 1–13. doi: 10.1029/PA005i001p00001.

McLaren, D. (2016) 'Mitigation deterrence and the "moral hazard" of solar radiation management', *Earth's Future*, 4(12), pp. 596–602. doi: 10.1002/2016EF000445.

McLaren, D. et al. (2019) 'Beyond "Net-Zero": A case for separate targets for emissions reduction and negative emissions', *Frontiers in Climate*, 1(4), pp. 1–5. doi: 10.3389/fclim.2019.00004.

McLaren, D. and Corry, O. (2021a) 'Clash of geofutures and the remaking of planetary order: Faultlines underlying conflicts over geoengineering governance', *Global Policy*, 12(S1), pp. 1758–5899.12863. doi: 10.1111/1758-5899.12863.

McLaren, D. and Corry, O. (2021b) 'The politics and governance of research into solar geoengineering', *Wiley Interdisciplinary Reviews: Climate Change*, 12(e707), pp. 1–20. doi: 10.1002/WCC.707.

McNutt, M. K. et al. (2015a) *Climate Intervention: Carbon Dioxide Removal and Reliable Sequestration*. Washington DC: The National Academies Press.

McNutt, M. K. et al. (2015b) *Climate Intervention: Reflecting Sunlight to Cool Earth*. Washington DC: The National Academies Press.

Millard-Ball, A. (2012) 'The tuvalu syndrome: Can geoengineering solve climate's collective action problem?', *Climatic Change*, 110(3–4), pp. 1047–1066. doi: 10.1007/s10584-011-0102-0.

Möller, I. (2020) 'Political perspectives on geoengineering: Navigating problem definition and institutional fit', *Global Environmental Politics*, 20(2), pp. 57–82. doi: 10.1162/glep_a_00547.

Möller, I. (2021) *CSSN Briefing CSSN Position Paper 2021:1: Potential obstruction of climate change mitigation through ISO standard on radiative forcing management*. Available at: https://www.cssn.org/wp-content/uploads/2021/01/Copy-of-CSSN-Briefing-v3.pdf (Accessed: 5 February 2021).

Möller, I. (2022) *The Emergence of Geoengineering: How Knowledge Networks form Governance Objects*. Cambridge University Press.

NASEM (2021) *Reflecting Sunlight, Reflecting Sunlight*. National Academies Press. doi: 10.17226/25762.

National Academy of Sciences (1992) *Policy Implications of Greenhouse Warming: Mitigation, Adaptation, and the Science Base, Public Policy*. doi: 10.17226/1605.

Nierenberg, W. A. et al. (1983) *Changing Climate: Report of the Carbon Dioxide Assessment Committee*. Washington DC: The National Academies Press.

O'Beirne, P. et al. (2020) 'The UK net-zero target: Insights into procedural justice for greenhouse gas removal', *Environmental Science and Policy*, 112, pp. 264–274. doi: 10.1016/j.envsci.2020.06.013.

Oomen, J. (2021) *Imagining Climate Engineering, Imagining Climate Engineering*. Routledge. doi: 10.4324/9781003043553.

Peters, G. P. and Geden, O. (2017) 'Catalysing a political shift from low to negative carbon', *Nature Climate Change*, 7(9), pp. 619–621. doi: 10.1038/nclimate3369.

Pidgeon, N. et al. (2013) 'Deliberating stratospheric aerosols for climate geoengineering and the SPICE project', *Nature Climate Change*, 3(5), pp. 451–457. doi: 10.1038/nclimate1807.

President's Science Advisory Committee (1965) *Restoring the Quality of Our Environment*. Washington DC.

Rabitz, F. (2016) 'Going rogue? Scenarios for unilateral geoengineering', *Futures*, 84(Part A), pp. 98–207. doi: 10.1016/j.cogdev.2004.05.003.

Raimi, K. T. (2021) 'Public perceptions of geoengineering', *Current Opinion in Psychology*, 42, pp. 66–70. doi: 10.1016/J.COPSYC.2021.03.012.

Reynolds, J. L. (2019) 'Solar geoengineering to reduce climate change: A review of governance proposals', *Proceedings of the Royal Society A*, 475(20190255). doi: 10.1098/rspa.2019.0255.

Reynolds, J. L. and Wagner, G. (2019) 'Highly decentralized solar geoengineering', *Environmental Politics*, 29(5), pp. 917–933. doi: 10.1080/09644016.2019.1648169.

Rogelj, J. et al. (2021) 'Three ways to improve net-zero targets', *Nature*, 591(7850), pp. 365–368. doi: 10.1038/d41586-021-00662-3.

Rubiano Rivadeneira, N. and Carton, W. (2022) '(In)justice in modelled climate futures: A review of integrated assessment modelling critiques through a justice lens', *Energy Research & Social Science*, 92(102781), pp. 1–11. doi: 10.1016/J.ERSS.2022.102781.

Schenuit, F. et al. (2021) 'Carbon dioxide removal policy in the making: Assessing developments in 9 OECD cases', *Frontiers in Climate*, 3(638805), pp. 1–22. doi: 10.3389/FCLIM.2021.638805/XML/NLM.

Schenuit, F., Gilligan, J. and Viswamohanan, A. (2021) 'A scenario of solar geoengineering governance: Vulnerable states demand, and act', *Futures*, 132, p. 102809. doi: 10.1016/J.FUTURES.2021.102809.

Schubert, J. (2021) *Engineering the Climate: Science, Politics, and Visions of Control*. Mattering Press.

Shepherd, J. et al. (2009) *Geoengineering the Climate: Science, Governance and Uncertainty*. London: The Royal Society.

Smith, W. (2020) 'The cost of stratospheric aerosol injection through 2100', *Environmental Research Letters*, 15(114004), pp. 1–15. doi: 10.1088/1748-9326/aba7e7.

Stenzel, F. et al. (2021) 'Irrigation of biomass plantations may globally increase water stress more than climate change', *Nature Communications*, 12(1512), pp. 1–9. doi: 10.1038/s41467-021-21640-3.

Stephens, J. C. and Surprise, K. (2020) 'The hidden injustices of advancing solar geoengineering research', *Global Sustainability*, 3(e2), pp. 1–6. doi: 10.1017/sus.2019.28.

Stilgoe, J. (2016) *Experiment Earth: Responsible Innovation in Geoengineering*. Routledge.

Strong, A. et al. (2009) 'Ocean fertilization: Time to move on', *Nature*, 461(September 2009), pp. 347–348. doi: 10.1038/461347a.

Talberg, A. et al. (2017) 'Geoengineering governance-by-default: An earth system governance perspective', *International Environmental Agreements: Politics, Law and Economics*, 18, pp. 229–253. doi: 10.1007/s10784-017-9374-9.

Tollefson, J. (2021) 'Can artificially altered clouds save the Great Barrier Reef?', *Nature*, 596(7873), pp. 476–478. doi: 10.1038/D41586-021-02290-3.

Victor, D. G. (2008) 'On the regulation of geoengineering', *Oxford Review of Economic Policy*, 24(2), pp. 322–336. doi: 10.1093/oxrep/grn018.

WBGU (2021) *Beyond Climate Neutrality*. Berlin.

Weng, Y., Cai, W. and Wang, C. (2021) 'Evaluating the use of BECCS and afforestation under China's carbon-neutral target for 2060', *Applied Energy*, 299(117263), pp. 1–13. doi: 10.1016/J.APENERGY.2021.117263.

Whyte, K. P. (2019) 'Indigeneity in geoengineering discourses: Some considerations', *Ethics, Policy & Environment*, 21(3), pp. 289–307. doi: 10.1080/21550085.2018.1562529.
Wilcox, J., Psarras, P. C. and Liguori, S. (2017) 'Assessment of reasonable opportunities for direct air capture', *Environmental Research Letters*, 12(065001), pp. 1–7. doi: 10.1088/1748-9326/aa6de5.
Winickoff, D. E., Flegal, J. A. and Asrat, A. (2015) 'Engaging the global South on climate engineering research', *Nature Climate Change*, 5(7), pp. 627–634. doi: 10.1038/nclimate2632.
Winter, C. J. (2021) *Subjects of Intergenerational Justice: Indigenous Philosophy, the Environment and Relationships*. Routledge. doi: 10.4324/9781003097457.
Wong, P.-H. (2016) 'Consenting to geoengineering', *Philosophy & Technology*, 29(2), pp. 173–188. doi: 10.1007/S13347-015-0203-1.

28

ENVIRONMENTAL POLICYMAKING IN AUTHORITARIAN COUNTRIES

The Cases of Singapore and Russia

Catherine Chen and Christian Aschenbrenner

28.1 Introduction

The challenge of confronting climate change has increasingly drawn attention to environmental policymaking in non-democracies. On the one hand, China, the largest non-democracy in the world, has invested more in renewable energy than any other country (Beeson, 2018). Therefore, some argue that authoritarian countries might be better placed to take rapid and decisive environmental actions (von Stein, 2022). On the other hand, the global effort to combat climate change would be incomplete without significant input from non-democracies and their alternative to Western-style environmental politics: authoritarian environmentalism (Beeson, 2018).

Much scholarly literature on environmental policymaking in non-democracies has been dedicated to China, the world's biggest fossil-fuel carbon dioxide emitter (World Resources Institute, 2022). To paint a fuller picture, we first review the propositions of the authoritarian environmentalism literature, using China as the leading example. Next, we focus on the specific cases of Singapore and Russia. We argue that Singapore's environmental policymaking demonstrates a clear pattern of authoritarian environmentalism. In the meantime, Russia's trajectory is quite different due to its regime changes, power fragmentation, and lack of consistency in framing the environment's role in economic growth and state development. Future research is needed to understand the factors contributing to the distinctive patterns of environmental policymaking in different non-democracies.

28.2 Environmental Authoritarianism

Authoritarian environmentalism is defined as "a public policy model that concentrates authority in a few executive agencies manned by capable and uncorrupt elites seeking to improve environmental outcomes" (Gilley, 2012, p. 288). In this model, barring a selected group of scientific and technocratic elites, public participation is limited to state-led mobilization for implementation purposes. The characteristics of authoritarian environmentalism shape the

DOI: 10.4324/9781003043843-32

corresponding policy outputs: a rapid and comprehensive response to the issue, on the one hand, and limits on individual freedom, on the other.

The main causes of authoritarian environmentalism are twofold (Gilley, 2012). On the issue-characteristic side, the public can be ignorant or irrational about environmental issues. They may also be incentivized to free ride, urge immediate action, or have diverging interests (Stone, 2009). Three factors contribute to authoritarian environmentalism on the side of traditions and structures of state domination. First, the environmental threat is perceived by policymakers to have authoritarian issue characteristics. Second, the existing structures of state domination of policymaking are strong. Third, political elites are united about the need to act and be capable of providing effective leadership (Beeson, 2010; Shearman & Smith, 2007).

Democratic environmentalism is described as a public policy model where authority is spread across several levels and agencies of government. In contrast to authoritarian environmentalism, direct public participation from a broad cross-section of society is encouraged (Holden, 2002; Humphrey, 2009). Gilley (2012) emphasizes that public participation involves two dimensions: the *stage* in the policy process where participation takes place and the *level* of participation. Although environmental policy models across regime types are likely to be mixtures of democratic and authoritarian features, environmental policy outputs in authoritarian countries have features distinctive from democratic countries.

China's response to climate change has been used as a typical case to illustrate the meaning, causes, and consequences of authoritarian environmentalism (Gilley, 2012). Lawrence (2013) argues that because of China's exploitative economic development, the widespread environmental degradation necessitates an authoritarian role played by the state in environmental protection. Such a role has indeed been assumed and reinforced by the scale and severity of China's environmental problems, lending legitimacy to and winning public support for the state's stronghold over environmental policymaking and implementation. As a result, authoritarian political elites manage to consolidate the existing regime and maintain domestic stability (Beeson, 2010).

An important feature of authoritarian environmentalism is the state's ability to co-opt environmental concerns to pursue its political objectives. Moore (2014) uses the South-North Water Transfer Project (SNWTP) to demonstrate how Beijing mounts large-scale, technocratic, and top-down solutions to environmental policy problems. SNWTP's origin lies in Mao's suggestion that "since the north has very little water, and the south a great deal, perhaps we should borrow some of it". In execution, SNWTP consists of three canals – the Eastern, Middle, and Western – with serious planning initiated as early as 1978 and construction beginning in 2002. The Eastern (1,156 km) and Middle routes (1,241 km), with a budget of $9 billion and $10 billion, respectively, were finished in the 2010s.

The SNWTP reflects the Chinese Communist Party's long-term fascination with hydraulic engineering rather than careful planning in environmental policymaking. The project has been marketed as a solution to water scarcity and other environmental problems in China's north. However, pollution, ecological disruption, and population displacement were ever present in the construction process. Moore (2014) argues that SNWTP is, above all, a political project where the authoritarian government applied persuasion and coercion to deter opposition. While the authoritarian state is enabled by its resourcefulness to take on mammoth projects, the long-term environmental and ecological consequences are unforeseeable and can be disastrous.

In democracies, there is a gap between environmental policy outputs and outcomes (Adam et al., 2019; Bättig & Bernauer, 2009). Governments must be responsive to citizen demands to uphold their legitimacy, leading to an accumulation of policies that are not necessarily followed

Table 28.1 Central characteristics of environmental authoritarianism in East Asia

Authoritarian path dependency	*Environmental degradation*
• Rapid modernization • State control of the economy • Export-driven economic growth • Compromised social and political rights	• Need for large-scale development projects • Environmental stress caused by rapid urbanization • Necessity for authoritarian measures curbing environmental degradation

by effective implementation and intended outcomes (Adam et al., 2019). A similar output–outcome gap exists in non-democracies (Gilley, 2012), albeit for different reasons. In China, the central government delegates the implementation and monitoring of policies to provincial and local governments. However, local pushback against central directives has been widespread, creating local implementation gaps. Because of this, Beijing has aimed to reconfigure central–local power relations by turning environmental governance into a highly politicized task. Enforcing party discipline and holding local officials accountable leads to long-term uncertainties due to possible backlash from the local level (Shen & Jiang, 2021).

Zooming out from China, some Asian countries also demonstrate the pattern of authoritarian environmentalism. With 30% of the world's land area and 60% of the world's current population (United Nations, 2022), Asia's outlook on economic development and environmental protection significantly impacts the world's future. In East Asia, authoritarianism has played a central role in the region's economic development (Beeson, 2010). The rapid industrialization typical of the area has led to the escalation of environmental degradation. Faced with such imperative challenges, many parts of the region have resorted to an authoritarian approach toward environmental policymaking and implementation (Beeson, 2010; Bruun, 2020; Choi, 2021).

Two main factors contribute to the emergence of authoritarian environmentalism in East Asia. First, because of the region's historical trajectory of economic and political development, some countries in this region demonstrate a path-dependent propensity toward authoritarianism. Specifically, 'soft authoritarianism' describes many Asian societies where fundamental social and political rights are compromised due to rapid modernization, state control of the economy, and an emphasis on export-driven growth (Nasir & Turner, 2013). Beyond East Asia, authoritarian regimes such as Iran also use environmentalism to consolidate existing power structures (Doyle & Simpson, 2006).

Second, major environmental failures necessitate authoritarian measures to address this salient issue with a strong hand. Environmental issues in fast-developing countries reflect the socio-economic features existing in such societies (Gupta & Asher, 1998). First, industrialization necessitates the rapid implementation of many development projects related to water, agriculture, power generation, and population distribution and control. Second, urbanization accelerates as cities rapidly expand, putting acute stress on the environment (Gupta & Asher, 1998). Many countries, faced with the major challenge of protecting their environment, may resort to different degrees of authoritarian measures to curb environmental degradation.

28.3 The Case of Singapore

Located in Southeast Asia, Singapore is a highly developed country with its Human Development Index consistently amongst the top of all countries in the world (United Nations, 2020). Along with Singapore, South Korea and Taiwan are rare examples of developmental states. Doner et al. (2005) define developmental states as organizational complexes in which expert and coherent bureaucratic agencies collaborate with organized private sectors to spur national economic transformation. Such institutional features play a determining role in building a strong capacity for promoting economic development unseen in other Asian countries such as Malaysia and Thailand.

Doner et al. (2005) argue that developmental states only emerge when political leaders confront extraordinarily constrained political environments, termed 'systematic vulnerability'. Specifically, such environments are characterized by the threats of potential civil unrest caused by deterioration in living standards, the need for foreign exchange induced by national insecurity, and the budget constraint imposed by the lack of easy revenue sources. The institutional features of developmental states, emerging from such systematic vulnerabilities, enhance information flows within and between the public and private sectors such that bureaucrats are allowed more leverage to help firms compete with global rivals.

While developing its economy quickly, Singapore has also built itself into the world-renowned 'Garden City' along explicit developmental lines (Han, 2017). Singapore's environmental policymaking dates to the 1960s, when a state-led effort was undertaken to tackle air and water pollution caused by rapid industrialization. Later, the transition to a Garden City began in the 1990s (Chia & Chionh, 1987; Han, 2017). The project featured tree planting measures and was followed by other environmental reforms: the establishment of an anti-pollution unit in 1970, the passage of the Clean Air Act in 1971, the launching of the Ministry of the Environment in 1972, the formation of the Garden City Action Committee in 1973, the establishment of environmental infrastructure in the 1980s, and the international demonstration of its environmental commitment in the 1990s.

The impacts of Singapore's developmental state tradition on its environmental governance are significant (Han, 2017). The Garden City project embodies a technocratic approach to resolving negative environmental externalities from rapid economic modernization. The impressive environmental output, rather than being the democratic response to public demand, results from the developmental state's deliberate planning and management based on a utilitarian view toward nature. The environmental policymaking in Singapore remains inherently top-down and non-participatory and is understood as a subset of national economic goals; bottom-up input is only welcome when co-opted by the state (Han, 2017; Ortmann, 2012).

In line with the argument attributing Singapore's environmental authoritarianism to its developmental state's features, Anwar and Sam (2012) observe that economic interests drive the measures taken to address environmental issues and that the introduction of emission control measures is conditional on their economic feasibility. Goh (2001) also emphasizes the legitimacy of the Singapore state from continual economic development. If the government failed at being environmentally sound and cleaning up after industrialization, its legitimacy base would deteriorate due to the perception that environmental degradation is the consequence of state-engineered economic development.

In the meantime, concerns that do not generate immediate utilitarian values, such as ecosystem diversity and climate change, have not received enough attention from the state. Under state-led ecological modernization, the framing of the environment left out critical aspects. Wong (2012) notes that Singapore's environmental framings are narrowly defined by state and industrial interests, leading to a focus on carbon emissions, energy security, and the impact on gross domestic product. Specifically, the state-driven assumption of modernization and efficiency being the solution to the environmental crisis falls short when technological advances cause other environmental and social problems (Weida, 2009). Indeed, the Environmental Performance Index places Singapore low in climate change and the protection of natural habitats category (Wolf et al., 2022).

Other scholars have argued that top-down authoritarian environmentalism does not capture the full picture of Singapore's environmental governance landscape. Hobson (2005) observes that environmental NGOs in Singapore create a space for volunteers to relate to the land, partaking in practical activities and programs that do not fall into the state-controlled political space. On a national level, though, the power of environmental policymaking is solidly held by the state; NGOs are invited to provide input on predetermined issues as part of the 'consultation' process and are expected to stay silent once consulted and respect the state's final say on the matter. More generally, despite moving toward liberalization, central government and planning departments dominate the policymaking sphere and leave little space for external policy advocacy (Ortmann, 2012).

Empirically, we find evidence that Singapore's environmental policy output triangulates the patterns described in the existing literature. Through the ACCUPOL project (Adam et al., 2019; Aschenbrenner et al., 2022; Knill et al., 2020), we collected major environmental legislation passed in Singapore from 1970 to 2020 and manually coded the policy targets and the instruments used in these policies. Policy targets summarize the aims of the policies. Policy instruments describe how and what instrument is used for achieving the target. Throughout the period we examine, air pollution issues make up most of the policy targets, followed by nature conservation and water protection policies (Figure 28.1).

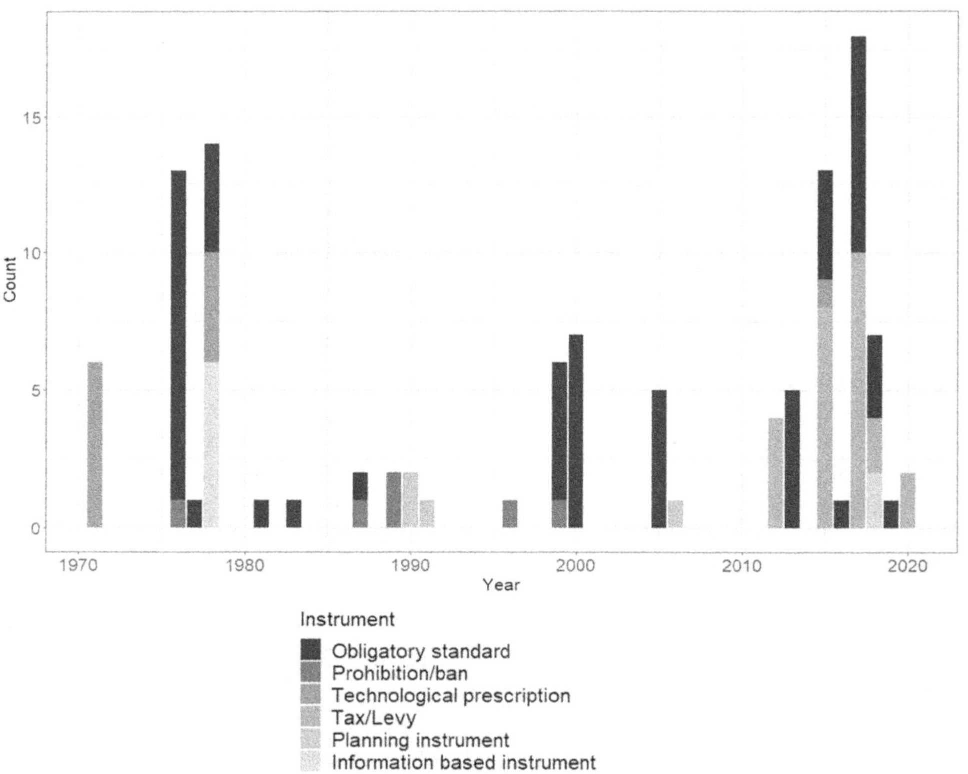

Figure 28.1 Number of Singapore's instruments in environmental policies, 1970–2020.

In terms of the clean air policy, the 1970s heavily feature limits on emissions from combustion plants with some policy outputs on emission limits from other sources (Clean Air Act 1971; Clean Air Regulation 1978). This focus on air pollution legislation and emission discharges (also seen in early water pollution control) was the result of the Singaporean government reacting to the wide-ranging pollution in the city-state resulting from its fast industrialization and rampant urbanization (Chia & Chionh, 1987; Tortajada & Joshi, 2014). Starting from around 2000, limits on emissions from passenger vehicles and heavy vehicles were introduced (Environmental Pollution Control Regulations 1999), with the amount of regulatory activity on passenger vehicle emission limits peaking in 2017 (Environmental Protection and Management Regulations 2017; Road Traffic Rules 2017). Such a shift in emphasis highlights the nature of environmental issues changing from being driven by industrial development to being driven by consumer behavior. The shift also went hand-in-hand with the ordered transfer of power to a softer leadership and an evolution toward a more competitive authoritarian state (Leong, 2000; Ortmann, 2011).

The policy targets for nature conservation produced in the late 1980s, 1990s, and 2000s highlight the continued effort to build Singapore as a Garden City. These acts include the Wild Animals and Birds Act (1987), Endangered Species Act (1989), National Parks Act (1990), and Parks and Trees Order (1991, 2006). In particular, the bottom-up input seems to have been the most extensive with nature conservation, with the regime co-opting and channeling civic endeavors to protect mangroves and natural/historical landmarks. The problematic pressure of nature conservation spurned this unusual civic engagement in a small, industrialized city-state with only a few green areas left (Neo, 2007; Ooi, 1994, 2002).

Overall, a distinctive pattern in Singapore's policy targets is that once they are addressed in one year's policy output, they are not likely to appear in the policy output again for at least a few years (except for emission limits on passenger vehicles around 2000 and in the 2010s). Such a pattern reflects the state's capacity to implement environmental policies to the intended effect. For example, in 1975, the Water Pollution and Drainage Act were passed, addressing emission limits in water protection. Since then, water protection was not mentioned again until the 2000s (Environmental Pollution Control Regulations 2005). The 1976 bill was likely comprehensive in scope and effective in implementation, and no further water protection act was needed for a long time. Such effectiveness is an indication of Singapore's state capacity. Hanson and Sigman (2019) rate the city-state among the most capable states in the world, making it the most capable autocratic system (Alizada et al., 2021). Government effectiveness is consistently ranked very high and trumps even the most democratic regimes (Kaufmann et al., 2010) (Figure 28.2).

The types of instruments present in acts have also changed over the years. Technological prescription, obligatory standards, and information-based instruments dominated the 1970s in the Clean Air Act and Water Pollution and Drainage Act. Around 1990, planning instruments, prohibitions, and bans were introduced, followed by the dominance of obligatory standards again around 2000. Here, one must note that the policy instruments pertaining to vehicular emissions are directly taken from the EU/EC legislation, following mechanisms of policy diffusion (Crippa et al., 2016). In the 2010s, tax and levy instruments became highly prevalent along with obligatory standards, especially once Singapore joined the general global trend in enacting and implementing carbon pricing policies for industrial and motor vehicle emissions (Steinebach et al., 2020) with the introduction of its own carbon taxation policy.

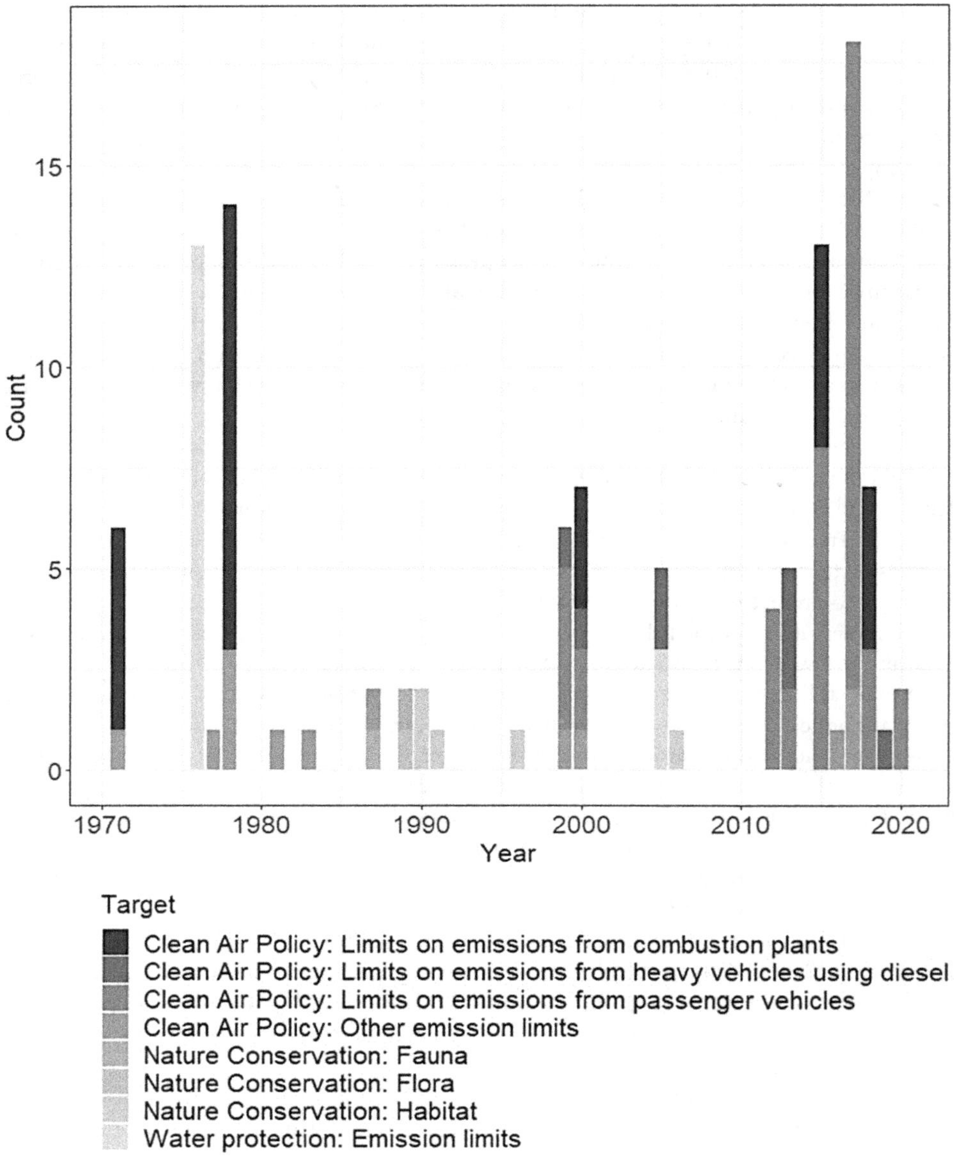

Figure 28.2 Number of Singapore's targets in environmental policies, 1970–2020.

28.4 The Case of Russia

The Russian Federation is the largest country in the world, with a high level of development that places it between Singapore, a highly advanced state, and China, a powerful developing country (United Nations, 2022). Unlike China or Singapore, Russia has experienced stark regime changes, ranging from the closed authoritarian one-party state of the Soviet Union to competitive authoritarianism during the Yeltsin presidency in the 1990s, and a turn towards further autocratization since the Putin presidencies (Alizada et al., 2021; Pleines & Schröder, 2010).

Today's Russia is an authoritarian state often characterized as a highly corrupt, oligarchic, and personalist regime (Baturo & Elkink, 2021). This personalist regime is marked by de-institutionalization, patronage networks, media control, and office permanency. In the environmental protection realm, the personalist traits manifest in the deinstitutionalization of the Environment Ministry and its incorporation and subjugation into the Ministry of Natural Resources (Mol, 2009).

A major feature of the current regime in Russia is its inherent conservatism and focus on resource extraction following a petrostate model (Fish, 2018). Beyond simple rent-seeking behavior from resource extraction, there are only sporadic modernization attempts that are hampered by the need to preclude possible political changes (Wilson, 2015). Russia's resource extraction is heavily concentrated in a few large oligarchic and state-owned corporations that can easily be used for patronage. Politically active social groups are also rewarded by social spending, contributing to stabilizing the regime (Knutsen & Rasmussen, 2018; Logvinenko, 2020). The overarching goal of policymaking is political stability and retaining the socio-economic status quo (Easter, 2008; Fish, 2018).

Some of the distinctive environmental-policymaking patterns emerged in the Soviet period. Soviet environmental policies were largely symbolic and declaratory due to their focus on heavy industry, the underpricing of energy and resources, and the inefficiencies of the Soviet state administration, especially in the environmental sector (Henry & Douhovnikoff, 2008; Höhmann et al., 1973; Kelley, 1976; OECD, 1999). Soviet environmental policy goals could be more stringent than Western countries and technically infeasible in implementation (Höhmann, et al., 1973).

In the 1960s and 1970s, air and water pollution caused by industrialization hatched civic environmentalism (Höhmann et al., 1973; Yanitsky, 2012; Ziegler, 1982). Unfortunately, environmentally oriented interest groups had to contend with bureaucratic and industrial interests within the interlocking and fragmented decision-making structure of the Soviet state with little success (Kelley, 1976). Environmental policies were often bogged down in the conflicts between the various levels of government. Until the late Gorbachev era, environmental implementation remained decentralized across at least 15 different Soviet ministries, specialized agencies or committees, and the bureaucracies of the 16 Union republics (Henry & Douhovnikoff, 2008; Höhmann et al., 1973; Kelley, 1976). This fragmentation also explains the limitation of legislative activity at the national level, which remained within the realms of nature conservation and air pollution control, for example, maximum allowable concentrations for air emissions and the Forest Code of 1976.

After the Chernobyl disaster, environmental policies became a more serious topic for Soviet/Russian policymaking, leading to the creation of the State Committee on Environmental Protection (Goskompriroda) as the first agency to centralize environmental protection policy (Henry & Douhovnikoff, 2008). The Soviet collapse saw the furthering of environmental policymaking in the Russian successor state despite difficult conditions during the economic and political crises of the 1990s (Henry & Douhovnikoff, 2008; OECD, 2006). The former Goskompriroda was elevated into a line ministry, and major environmental legislation was passed with new regulations on industrial plants enacted following the polluter-pays-principle and an introduction of a general water levy for water discharges and air emissions. The adding of these market-based instruments also marked a logic of neoliberal modernization, which was imported from the 'West' (Kotov & Nikitina, 2002; OECD, 1999) and signalled a turn away from strict state instruments, like obligatory standards or prohibitions that were common in the Soviet Union.

One of these market instruments was the 1991 introduction of pollution charges for enterprises. The federal government set the basic fees (adjusted by inflation) for each specified pollutant. At the same time, regional and local authorities licensed limits of allowable emissions for each particular firm or branch of a firm (Kotov & Nikitina, 2002). As Kotov and Nikitina point out (2002), this system was prone to corruption and 'modification' detrimental to the environmental goals. The fact that the government also ignored the repeated bending and violation of its environmental laws shows the general changing attitude of the Yeltsin regime by 1995, with a turn towards economics, regime stability, and survival as well as the beginning of a turn away from democratization (Henry & Douhovnikoff, 2008; Mommsen, 2010). This turn also saw the demotion of the Environment Ministry and later abolition under Putin in a process of environmental deinstitutionalization (Mol, 2009).

During this time, the Soviet legacy and the notion of neoliberal developmentalism both held sway in the policy outputs, reflected by the introduction of obligatory standards for water quality in 1999 and the polluter-pays-principle into the water code in 2002. The introduction of these policies can largely be considered a catch-up to similar Soviet policies on air pollution and an attempt by the state to adopt market-based techniques (OECD, 1999, 2006). The turn towards market-based instruments continued under the early Putin presidency in the clean air and water protection policy sub-sectors concerning industrial discharges. On motor vehicles, Russian policymaking followed in the footsteps of the Euro standards, similar to Singaporean practices.

The turnover of the presidency toward Medvedev in 2008 sparked a new discourse on modernization, including ecological topics, in Russia. Unlike Putin, Medvedev was more invested in policy content than vague rhetoric concerning environmental matters (Martus, 2021). Regarding environmental issues, the Medvedev presidency saw the most significant increase in the policy portfolio, with the large-scale introduction of a permits scheme, best available technology instruments (BATs), and the launch of a monitoring and information gathering regime to secure compliance and enforcement.

This policymaking package was initiated during the Medvedev presidency and concluded in the early years of Putin's second presidency. In this case, the Ministry of Natural Resources, responsible for environmental policymaking, overcame the industrial and bureaucratic interests against the environmental reform package by convincing the president to step in their favor. This uncharacteristic weighing in by the president allowed for a large increase in the environmental policy stocks and the introduction of a wide array of policies, ranging from BATs to permitting schemes and information-based instruments for better control and enforcement. Hampered by widespread corruption and poor state capacities, environmental enforcement has always remained an inadequacy of Russian environmental policy (Newell & Henry, 2016). The strengthening of compliance and enforcement instruments underlined the seriousness of the reform attempt.

Contrary to the logic of authoritarian environmentalism, in the Russian regime, environmental policy is not made by a centralized group in conjunction with a government unified in its environmental goals. Instead, the president only weighs in as an initiator or arbiter of the different interests of autonomous bureaucrats and industries. At the same time, many ministries pursue their own agenda against other state and industry actors (Martus, 2017). In a word, Russian environmental policymaking is dependent on the attention of the largely disinterested political leadership over the fragmented and infighting bureaucrats, industrial interests, and peripheral environmental movements (Yanitsky, 2012), making it distinct from the authoritarian environmentalism observed in China or Singapore.

The volatility in the regime, leadership interest, and bureaucratic interest also has consequences on the type and frequency of policy that is pursued. The pattern in policy items tackled in the 1970 to 2020 period suggests the essential inactivity of the Soviet Union on national environmental matters. Following the decline of the Soviet system, a catch-up can be witnessed, especially in the policy areas where the problem pressure due to industrialization was most significant – air and water pollution. After the 1990s and early 2000s, we also see a switch in policy targets from tackling pollution by industrialization to pollution due to consumption. What is striking is the massive number of items in the environmental reform package of 2014, which once more tackled problems related to air and water pollution by industry, indicating that previous efforts had not been sufficient despite much previous policy output (Figure 28.3).

The high-output, low-efficacy observation is confirmed when we look at the policy instruments utilized by Russian legislators and regulators. The Soviet period saw a clear preference for state-centric instruments such as obligatory standards and the issuance of permits (Ziegler, 1982). The shift towards a neoliberal development paradigm in the Russian Federation under Yeltsin saw the introduction of many market-based policy instruments, chiefly levies and liability schemes, that proved largely ineffective in tackling the country's environmental issues. Charges and fines for pollution were often too low, while monitoring and data collection were lacking. In short, the state did not and, at times, could not effectively build and maintain the surrounding regulatory enforcement for such market-based instruments, leading to distortion of the policies (Kotov & Nikitina, 2002; World Bank, 2014).

A return to more common regulatory instruments concerns motor vehicles, with emission standards enacted following the EU's lead (Crippa et al., 2016). However, unlike Singapore or China (Heggelund et al., 2019), the Russian Federation has not followed the global trend of enacting carbon pricing policies. Enforcement and compliance mechanisms such as monitoring and data collection were introduced during the tempered Medvedev 'modernization'. This reform package also saw the further expansion of the state with the introduction of BATs and permit schemes (Figure 28.4).

28.5 Conclusion and Future Research

Comparing Singapore and Russia, one can see that the 'ideal' of authoritarian environmentalism is better realized in the city-state of Singapore. We offer three propositions to explain this. First, the regime in Singapore has been continuously in charge of the country since 1959 with a relatively stable ruling elite, in contrast to Russia, which experienced a mammoth regime change and a long period of political chaos and adjustment.

Second, the regimes in Singapore and Russia follow different ideological tenets. In Singapore, the ruling ideology was based on a developmental authoritarian state that integrated environmental governance as a strategic piece of its pursuit of economic growth, rapid industrialization, and urbanization. On the contrary, the Soviet ruling ideology framed environmental pollution

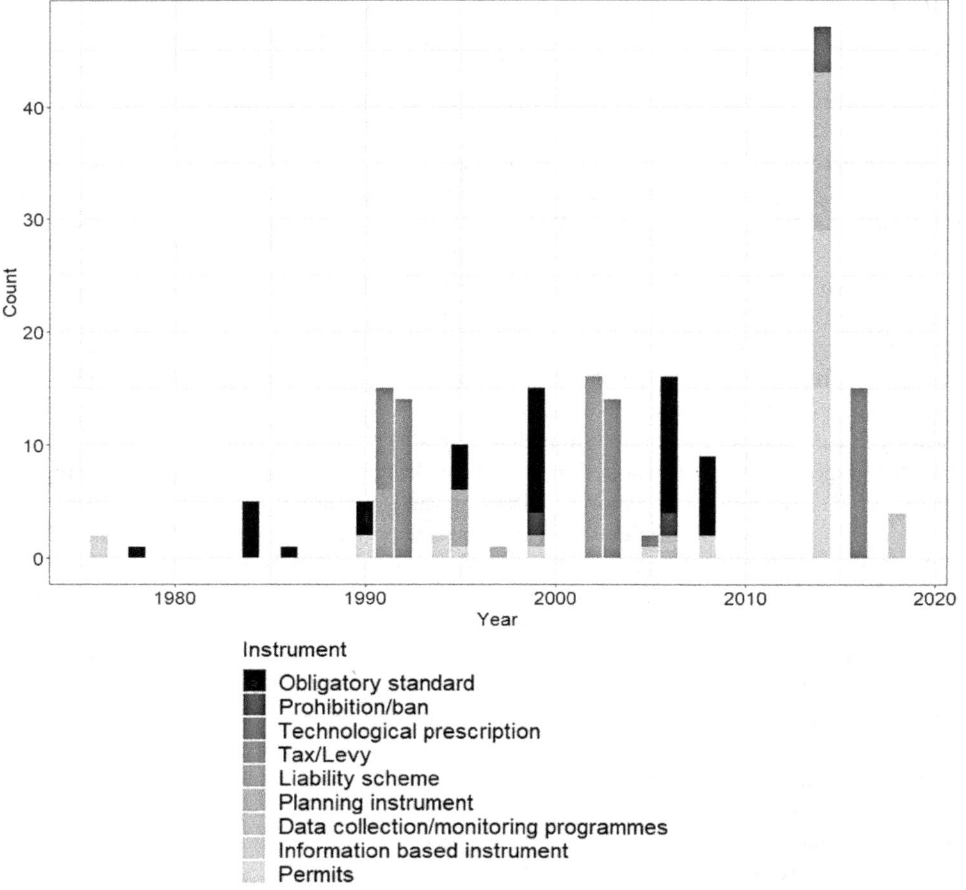

Figure 28.3 Number of Russia's instruments in environmental policies, 1970–2020.

as a 'capitalist externality' and proved incapable of effectively tackling environmental issues caused by rapid industrialization and urbanization. Following the collapse of the Soviet Union, the regime in Russia – after an initial acceptance of a neoliberal developmental paradigm – shifted more and more toward authoritarianism based on a petrostate model, hindering effective environmental policymaking and implementation.

Third, Singapore quickly overcame bureaucratic and elite fragmentation and became the second East Asian country (after Japan) to establish a dedicated Environment Ministry. The state succeeded in bundling together elite, bureaucratic, and industry/co-opted civic interests. In Russia, the Soviet-era fragmentation only ended at the point of regime collapse. Later, the Russian Environment Ministry was demoted and then abolished, with its functions integrated into the Ministry of Natural Resources. The fragmentation diluted the responsibilities for environmental policymaking in the process of deinstitutionalization.

However, there remain some commonalities in environmental policymaking between the two countries. Regulations on vehicular emissions follow the guidance of EU/EC regulations in both countries. Another common feature is the relative weakness of independent

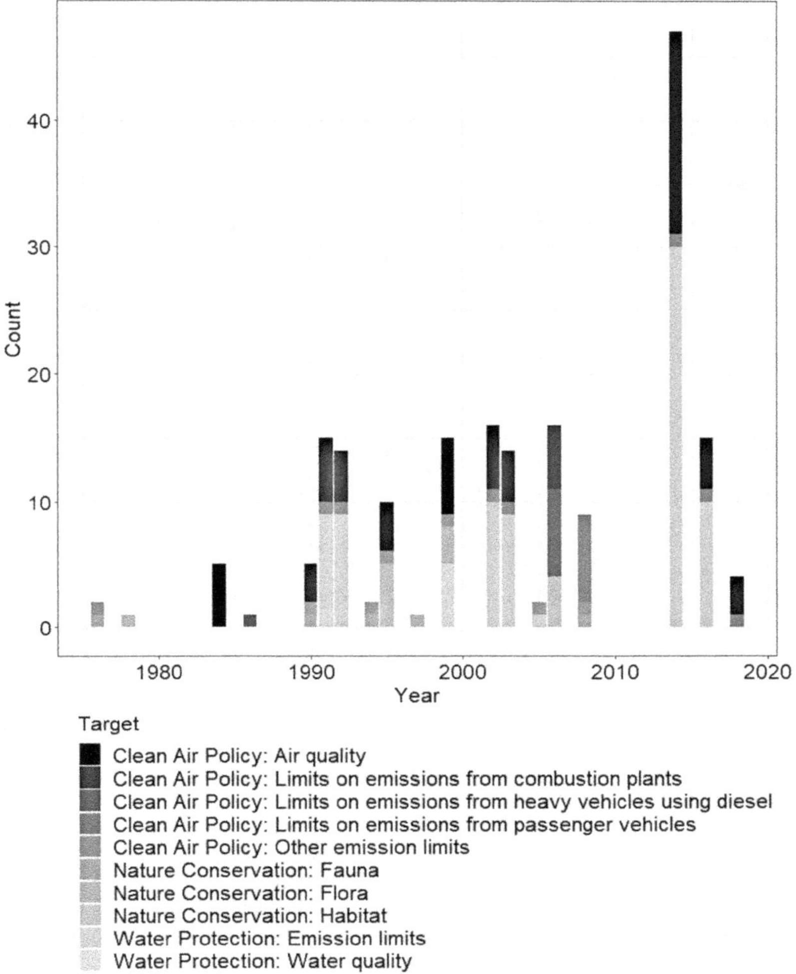

Figure 28.4 Number of Russia's targets in environmental policies, 1970–2020.

environmental activism, especially when compared to public participation on environmental issues in democracies. In Russia, environmental activism is either irrelevant or repressed, while in Singapore, the state has learned to co-opt such initiatives.

Avenues for future research on authoritarian environmentalism remain open. As the Russian case shows, not all authoritarian counties favor the developmental state model that is well suited to authoritarian environmentalism. However, neither are they more inclined toward democratic environmentalism. Here, researchers could concentrate on cases where factors such as regime stability, ideology, state capacity, or resource richness (fossil fuels) cause deviation from the typical cases of authoritarian environmentalism. Additionally, scholars can take a comparative lens to specify the conditions for effective or ineffective authoritarian environmental policymaking, coupled with comprehensive data collection on policy implementation and environmental outcomes.

> **Box 28.1 Chapter Summary**
>
> - Understanding environmental policymaking in non-democracies.
> - Investigating the applicability of authoritarian environmentalism to different non-democracies.
> - Uncovering and explaining distinctive patterns of environmental policymaking in Singapore and Russia.
> - Empirically summarizing characteristics of environmental policies passed in Singapore and Russia during the period 1970–2020.

References

Adam, C., Hurka, S., Knill, C., & Steinebach, Y. (2019). *Policy Accumulation and the Democratic Responsiveness Trap*. Cambridge University Press.

Alizada, N., Cole, R., Gastaldi, L., Grahn, S., Hellmeier, S., Kolvani, P., Lachapelle, J., Lührmann, A., Maerz, S. F., Pillai, S., & Lindberg, S. I. (2021). *Autocratization Turns Viral*. Varieties of Democracy Institute (V-Dem).

Anwar, S., & Sam, C.-Y. (2012). Is Economic Nationalism Good for the Environment? A Case Study of Singapore. *Asian Studies Review, 36*(1), 39–58.

Aschenbrenner, C., Knill, C., & Steinebach, Y. (2022). *Autocracies and Policy Accumulation: The Case of Singapore* [Manuscript].

Bättig, M. B., & Bernauer, T. (2009). National Institutions and Global Public Goods: Are Democracies More Cooperative in Climate Change Policy? *International Organization, 63*(2), 281–308. https://doi.org/10.1017/S0020818309090092

Baturo, A., & Elkink, J. (2021). *New Kremlinology. Understanding Regime Personalization in Russia*. Oxford University Press.

Beeson, M. (2010). The Coming of Environmental Authoritarianism. *Environmental Politics, 19*(2), 276–294. https://doi.org/10.1080/09644010903576918

Beeson, M. (2018). Coming to Terms with the Authoritarian Alternative: The Implications and Motivations of China's Environmental Policies: Environmental authoritarianism in China. *Asia & the Pacific Policy Studies, 5*(1), 34–46. https://doi.org/10.1002/app5.217

Bruun, O. (2020). Environmental Protection in the Hands of the State: Authoritarian Environmentalism and Popular Perceptions in Vietnam. *The Journal of Environment & Development, 29*(2), 171–195. https://doi.org/10.1177/1070496520905625

Chia, L. S., & Chionh, Y. H. (1987). Singapore. In L. S. Chia (Ed.), *Environmental Management in Southeast Asia: Directions and Current Status* (pp. 109–168). Faculty of Science, National University of Singapore.

Choi, S. Y. (2021). Resilient Peripheralisation Through Authoritarian Communication Against Energy Democracy in South Korea. *Environmental Politics, 30*(6), 1002–1023. https://doi.org/10.1080/09644016.2020.1843884

Crippa, M., Janssens-Maenhout, G., Guizzardi, D., & Galmarini, S. (2016). EU Effect: Exporting Emission Standards for Vehicles Through the Global Market Economy. *Journal of Environmental Management, 183*, 959–971. https://doi.org/10.1016/j.jenvman.2016.09.068

Doner, R. F., Ritchie, B. K., & Slater, D. (2005). Systemic Vulnerability and the Origins of Developmental States: Northeast and Southeast Asia in Comparative Perspective. *International Organization, 59*(2), 327–361.

Doyle, T., & Simpson, A. (2006). Traversing More Than Speed Bumps: Green politics under authoritarian regimes in Burma and Iran. *Environmental Politics, 15*(5), 750–767. https://doi.org/10.1080/09644010600937199

Easter, G. M. (2008). The Russian State in the Time of Putin. *Post-Soviet Affairs, 24*(3), 199–230. https://doi.org/10.2747/1060-586X.24.3.199

Fish, M. S. (2018). What Has Russia Become? *Comparative Politics, 50*(3), 327–346. https://doi.org/10.5129/001041518822704872

Gilley, B. (2012). Authoritarian Environmentalism and China's Response to Climate Change. *Environmental Politics, 21*(2), 287–307. https://doi.org/10.1080/09644016.2012.651904

Goh, D. P. S. (2001). The Politics of the Environment in Singapore? Lessons from a 'Strange' Case. *Asian Journal of Social Science*, *29*(1), 9–34. https://doi.org/10.1163/156853101X00307

Gupta, A., & Asher, M. G. (1998). *Environment and the Developing World: Principles, Policies, and Management*. Wiley.

Han, H. (2017). Singapore, a Garden City: Authoritarian Environmentalism in a Developmental State. *The Journal of Environment & Development*, *26*(1), 3–24. https://doi.org/10.1177/1070496516677365

Hanson, J. K., & Sigman, R. (2019). Leviathan's Latent Dimensions: Measuring State Capacity for Comparative Political Research. *Journal of Politics*, *83*(4), 1495–1510.

Heggelund, G., Stensdal, I., Duan, M., & Wettestad, J. (2019). China's Development of ETS as a GHG Mitigating Policy Tool: A Case of Policy Diffusion or Domestic Drivers? *Review of Policy Research*, *36*(2), 168–194. https://doi.org/10.1111/ropr.12328

Henry, L. A., & Douhovnikoff, V. (2008). Environmental Issues in Russia. *Annual Review of Environment and Resources*, *33*, 437–460. https://doi.org/10.1146/annurev.environ.33.051007.082437

Hobson, K. (2005). 'Green' Practices: NGOs and Singapore's Emergent Environmental-Political Landscape *Sojourn: Journal of Social Issues in Southeast Asia*, *20*(2), 155–176.

Höhmann, H.-H., Seidenstecher, G., & Vajna, T. (1973). *Umweltschutz und ökonomisches System in Osteuropa. Drei Beispiele: Sowjetunion, DDR, Ungarn*. Kohlhammer.

Holden, B. (2002). *Democracy and global warming*. Continuum.

Humphrey, M. (2009). *Ecological Politics and Democratic Theory: The Challenge to the Deliberative Ideal* (Transferred to digital print). Routledge.

Kaufmann, D., Kraay, A., & Mastruzzi, M. (2010). *The Worldwide Governance Indicators: Methodology and Analytical Issues* (Working Paper (5430)).

Kelley, D. R. (1976). Environmental Policy-Making in the USSR: The Role of Industrial and Environmental Interest Groups. *Soviet Studies*, *28*(4), 570–589.

Knill, C., Steinbacher, C., & Steinebach, Y. (2020). Balancing Trade-offs Between Policy Responsiveness and Effectiveness: The Impact of Vertical Policy-Process Integration on Policy Accumulation. *Public Administration Review*, puar.13274. https://doi.org/10.1111/puar.13274

Knutsen, C. H., & Rasmussen, M. (2018). The Autocratic Welfare State: Old-Age Pensions, Credible Commitments, and Regime Survival. *Comparative Political Studies*, *51*(5), 659–695. https://doi.org/10.1177/0010414017710265

Kotov, V., & Nikitina, E. (2002). Reorganisation of Environmental Policy in Russia: The Decade of Success and Failures in Implementation and Perspective Quests. *Nota Di Lavoro*, *57*.

Lawrence, D. P. (2013). *Impact Assessment: Practical Solutions to Recurrent Problems and Contemporary* (Second edition). John Wiley & Sons, Inc.

Leong, H. K. (2000). Prime Ministerial Leadership and Policy-Making Style in Singapore: Lee Kuan Yew and Goh Chok Tong Compared. *Asian Journal of Political Science*, *8*(1), 91–123. https://doi.org/10.1080/02185370008434161

Logvinenko, I. (2020). Authoritarian Welfare State, Regime Stability, and the 2018 Pension Reform in Russia. *Communist and Post-Communist Studies*, *53*(1), 100–116. https://doi.org/10.1525/cpcs.2020.53.1.100

Martus, E. (2017). Contested Policymaking in Russia: Industry, Environment, and the "best available technology" Debate. *Post-Soviet Affairs*, *33*(4), 276–297. https://doi.org/10.1080/1060586X.2016.1209315

Martus, E. (2021). Policymaking and Policy Framing: Russian Environmental Politics under Putin. *Europe-Asia Studies*, *73*(5), 869–889. https://doi.org/10.1080/09668136.2020.1865275

Mol, A. P. J. (2009). Environmental Deinstitutionalization in Russia. *Journal of Environmental Policy & Planning*, *11*(3), 223–241. https://doi.org/10.1080/15239080903033812

Mommsen, M. (2010). Das politische System unter Jelzin—Ein Mix aus Demokratie, Oligarchie, Autokratie und Anarchie. In H. Pleines & H.-H. Schröder (Eds.), *Länderbericht Russland* (pp. 55–70). Bundeszentrale für Politische Bildung.

Moore, S. M. (2014). Modernisation, Authoritarianism, and the Environment: The Politics of China's South–North Water Transfer Project. *Environmental Politics*, *23*(6), 947–964. https://doi.org/10.1080/09644016.2014.943544

Nasir, K. M., & Turner, B. S. (2013). Governing as Gardening: Reflections on Soft Authoritarianism in Singapore. *Citizenship Studies*, *17*(3–4), 339–352. https://doi.org/10.1080/13621025.2012.707005

Neo, H. (2007). Challenging the Developmental State: Nature Conservation in Singapore. *Asia Pacific Viewpoint*, *48*(2), 186–199. https://doi.org/10.1111/j.1467-8373.2007.00340.x

Newell, J. P., & Henry, L. A. (2016). The State of Environmental Protection in the Russian Federation: A Review of the Post-Soviet Era. *Eurasian Geography and Economics*, *57*(6), 779–801. https://doi.org/10.1080/15387216.2017.1289851

OECD. (1999). *OECD Environmental Performance Reviews: Russian Federation 1999*. OECD. https://doi.org/10.1787/9789264180116-en

OECD. (2006). *Environmental Policy and Regulation in Russia. The Implementation Challenge*. OECD.

Ooi, G. L. (1994). A Centralised Approach to Environmental Management: The Case of Singapore. *Sustainable Development, 2*(1), 17–22. https://doi.org/10.1002/sd.3460020104

Ooi, G. L. (2002). The Role of the State in Nature Conservation in Singapore. *Society & Natural Resources, 15*(5), 455–460. https://doi.org/10.1080/08941920252866800

Ortmann, S. (2011). Singapore: Authoritarian but Newly Competitive. *Journal of Democracy, 22*(4), 153–164. https://doi.org/10.1353/jod.2011.0066

Ortmann, S. (2012). Policy Advocacy in a Competitive Authoritarian Regime: The Growth of Civil Society and Agenda Setting in Singapore. *Administration & Society, 44*(6_suppl), 13S-25S. https://doi.org/10.1177/0095399712460080

Shearman, D. J. C., & Smith, J. W. (2007). *The Climate Change Challenge and the Failure of Democracy*. Praeger Publishers.

Shen, W., & Jiang, D. (2021). Making Authoritarian Environmentalism Accountable? Understanding China's New Reforms on Environmental Governance. *The Journal of Environment & Development, 30*(1), 41–67. https://doi.org/10.1177/1070496520961136

Steinebach, Y., Fernández-i-Marín, X., & Aschenbrenner, C. (2020). Who Puts a Price on Carbon, Why and How? A Global Empirical Analysis of Carbon Pricing Policies. *Climate Policy*, 1–13. https://doi.org/10.1080/14693062.2020.1824890

Tortajada, C., & Joshi, Y. K. (2014). Water Quality Management in Singapore: The Role of Institutions, Laws and Regulations. *Hydrological Sciences Journal, 59*(9), 1763–1774. https://doi.org/10.1080/02626667.2014.942664

United Nations. (2020). Human Development Report 2020. In *Human Development Reports*. United Nations. https://hdr.undp.org/content/human-development-report-2020

United Nations. (2022). *World Population Prospects 2022*. https://population.un.org/wpp/

von Stein, J. (2022). Democracy, Autocracy, and Everything in Between: How Domestic Institutions Affect Environmental Protection. *British Journal of Political Science, 52*(1), 339–357. https://doi.org/10.1017/S000712342000054X

Weida, L. (2009). Climate Change Policies in Singapore: Whose "Environments" Are We Talking About? *Environmental Justice, 2*(2), 79–83. https://doi.org/10.1089/env.2009.0017

Wilson, K. (2015). Modernization or More of the Same in Russia: Was There a "Thaw" Under Medvedev? *Problems of Post-Communism, 62*(3), 145–158. https://doi.org/10.1080/10758216.2015.1019803

Wolf, M. J., Emerson, J. W., Esty, D. C., de Sherbinin, A., Wendling, Z. A., & et al. (2022). *2022 Environmental Performance Index*. Yale Center for Environmental Law & Policy. https://epi.yale.edu/about-epi

Wong, C. M. L. (2012). The Developmental State in Ecological Modernization and the Politics of Environmental Framings: The Case of Singapore and Implications for East Asia. *Nature and Culture, 7*(1), 95–119. https://doi.org/10.3167/nc.2012.070106

World Bank. (2014). *Environmental Perspective of Russia's Accession to the World Trade Organization*. World Bank. https://openknowledge.worldbank.org/handle/10986/17799

World Resources Institute. (2022). *Climate Watch*. https://www.climatewatchdata.org

Yanitsky, O. N. (2012). From Nature Protection to Politics: The Russian Environmental Movement 1960–2010. *Environmental Politics, 21*(6), 922–940. https://doi.org/10.1080/09644016.2012.724216

Ziegler, C. E. (1982). Economic Alternatives and Administrative Solutions in Soviet Environmental Protection. *Policy Studies Journal, 11*(1), 175–187.

29
ENVIRONMENTAL POLICY IN FAST-GROWING ECONOMIES
The Case of India

Kirsten Jörgensen

29.1 Introduction

Since the 1990s, India, the world's second most populous country, has seen rapid economic and industrial growth, becoming the third-largest energy consumer in the world. At the same time, despite advances in poverty eradication, India has not been able to leapfrog to a significantly better ranking in the Human Development Index, and, owing to severe environmental pollution and degradation, the country also ranks consistently low in the Environmental Performance Index. In this sense, India demonstrates the explanatory limits of the Environmental Kuznets Curve (EKC). The EKC assumes that in the early stages of industrial development, economic growth damages the environment, but that over time, as income grows, environmental pollution will decline because of increased attention to environmental policy (Stern, 2004). As the trajectory of ecologically unsustainable development in India demonstrates, growth is not a sufficient condition on its own to initiate and successfully implement environmental policy. In fact, rapid economic growth has caused environmental quality in India to decrease over time (Managi and Jena, 2008).

This chapter deals first with India's development trajectory, including the emergence of the growth-first paradigm after the country's independence in 1947, and the role of the state. Second, it examines India's emerging environmental policy in the context of the advent of international environmental governance. Third, it focuses on India's environmental politics, describing the actors and structures which influenced its domestic environmental policy. The fourth subsection introduces India's role as a global climate power and sheds light on domestic climate policy and politics. Finally, the fifth subsection discusses the linkages between democracy and the environment in India.

29.2 Development: A Post-Colonial State Catching Up

After independence in 1947, India was a developing state challenged by a lagging economy with a largely impoverished population that included many people regularly facing famine. Poverty and social disparity were closely linked to the underdeveloped and unsustainable economy that the former colonial power, Britain, had left behind. India's inheritance of environmental

DOI: 10.4324/9781003043843-33

colonialism included the exploitation and degradation of its forests (Das, 2020; Kashwan, 2017; Gadgil und Guha, 1997). Additionally, India's domestic caste system has persistently constrained development and equality (Drèze und Sen, 2014). In 1951, 82% of the population belonged to the 'poor and vulnerable' group below the poverty line. Economic inequality across rural and urban households was significant (Drèze and Sen, 2014, S. 6; Joseph et al., 2014, S. 153).

The high-growth development paradigm emerged as an answer. India's Industrial Policy Resolution of 1948 argued: "meagre redistribution of existing wealth would make no difference, and a dynamic policy must therefore be directed to continuous increase in production" (Joseph et al., 2014, S. 162). India chose the 'third way' of development, to be a non-Western democracy, committed to social justice, combining 'Soviet-style' Five-Year Planning with private entrepreneurship (Sil, 2014, S. 345). The state became the central actor in India's economic and social development as a planner and regulator, exerting strict control over industrial development via the License Raj, a heritage from British colonial rule. Beginning in the 1950s, led by the Five-Year Plans, India's historical path-dependence on a fossil-fuelled economy emerged and its state-controlled coal production accelerated. Related to that, the 1950s also saw the emergence of the growth-first policy paradigm, which was later reinforced in the 1980s. The growth-first paradigm and the carbon lock-in, which denotes the technological and political-administrative lock-in of industrial economies into fossil fuel-based energy systems and the obstacles this path-dependence poses to policy change and the development of new low-carbon technologies (Unruh, 2000), have significant explanatory power in understanding India's environmental and climate policy.

In the 1970s slow development and modest growth rates revealed a "mismatch between the limited capacities of the Indian state and the highly ambitious statist model of development" (Kohli, 2006, S. 1251). Both India's economic inward-looking self-reliance policy paradigm and Indira Gandhi's populist '*garibi hatao*' (remove poverty) slogan faltered. Following an authoritarian interlude through Gandhi's Emergency rule (1975–1977) India's development strategy and economic policy transformed gradually in the 1980s during her last term in office (1980–1984). This process continued after her assassination under her successor, Rajiv Gandhi. This included a limited opening of the economy for foreign investors, limiting investments in public sector enterprises, and a pro-business orientation that was different from the neo-liberal paradigm in that it was still supportive of an active role for the state (Kohli, 2006, S. 1256). New state–business actor constellations were developed to pursue India's growth cooperatively with a pro-business strategy, including restrictions on labour activism and "a more sharply capitalist path of development" (Kohli, 2006, S. 1255). Less emphasis was now placed on distributive justice and more on economic growth. Policies focused more on the "creation of the cake rather than its distribution" (Joseph et al., 2014, S. 168).

India's development policy paradigm had shifted and policies were set more rigorously on the 'growth first' trajectory. Growth rates increased to annually 5.8% of the GDP in the 1980s preceding India's double-digit growth rates which materialized after the economic liberalization reforms in 1991. Social inequality on the other hand remained high (Kohli, 2006, S. 1253). India's rapid economic development was accompanied by severe pollution and environmental degradation and the country ranked continuously low in the Yale Environmental Performance Index. It was ranked 155th out of 178 countries in 2014, and 168th out of 180 countries in 2020 (Hsu et al., 2014; Wendling et al., 2020). Yet, the domestic, political, and academic controversies about India's development were not focused on environmental issues. In their book, *Why Growth Matters: How Economic Growth in India Reduced Poverty and the Lessons for Other Developing Countries*, Bhagwati and Panagariya, two internationally renowned, influential economists, mention 'environment' only four times (Bhagwati and Panagariya, 2013), two times criticizing the Environmental Ministry for support of NGOs concerned with the approval of genetically modified seeds and agricultural crops (p. 176) and for blocking clearances for major industrial projects (p. 208). Climate change is not mentioned at all.

29.3 India and the Global Rise of Environmental Policy

India's environmental crisis has accelerated dramatically since the 1950s, reinforced by state-planned industrialization and increasing urbanization. The crisis manifests itself in soil erosion, decline of forests, scarce renewable freshwater, overexploitation of groundwater, and river and air pollution (Agrawal and Yokozuka, 2002; Gupta and Tyagi, 2020; Singhal and Jain, 2020). Development has also been accompanied by social deprivation: 20 to 40 million people are believed to have been displaced as 'development refugees', with Adivasis, India's large indigenous ethnic groups, being particularly affected (Kashwan, 2017, S. 68).

The beginning of environmental policy in India is associated with Indira Gandhi's often cited speech at the United Nations Conference on Human Environment in Stockholm in 1972: "We do not wish to impoverish the environment any further and yet we cannot for a moment forget the grim poverty of large numbers of people. Are poverty and need not the greatest polluters?" (cited in Dwivedi, 2008, S. 114). At the beginning of the 1970s, not later than in other industrialized and newly industrializing countries worldwide, environmental policy emerged in India and the political agenda for this was to a great extent set from the outside by the Stockholm Conference in 1972 (Agrawal and Yokozuka, 2002). While legal authorization to file national environmental law was not yet anchored in India's constitutional distribution of tasks between the states and the centre, its policy makers at the federal level drew this authorization from the obligation to implement international agreements.

In addition to the Insecticides Act (1968), the Wild Life Act (1972), and the Water Act (1974) – India's first pollution control law (which was passed in a controversial policy process) – a Department of Environment (DOE) was instituted in 1980. This was followed by the Air Act (1981) (Reich and Bowonder, 1992, S. 645). India was amongst the first countries in the world to introduce a Constitutional Amendment (48 A) in 1976, establishing the state's task to "protect and improve the environment and to safeguard the forests and wildlife of the country" (cited in Dwivedi, 2008, S. 121). A coherent national environmental policy coalescing environmental regulation and setting up political-administrative capacities failed to develop, however, and environmental performance remained weak (Dwivedi, 2008; Chopra, 2017; Agrawal und Yokozuka, 2002).

29.4 Environmental Politics in India

Despite severe environmental pollution driven by India's increased growth rates and new types of environmental problems involving toxic chemicals and toxic waste, the pressure of environmental degradation was not sufficient to set the political agenda for a more efficacious environmental policy (Agrawal and Yokozuka, 2002; Reich and Bowonder, 1992). A focusing event led to the expansion of environmental legislation and capacity building (Agrawal and Yokozuka, 2002; Reich and Bowonder, 1992). In 1984 a gas explosion at the plant of Union Carbide India Limited (UCIL) caused a catastrophe in Bhopal that was estimated to have killed 10,000 people in the days after the accident and 15,000 to 20,000 in the following two decades (Broughton, 2005). The chemical plant produced pesticides and was 51% owned by the Union Carbide and Carbon Corporation (UCC) USA, consisting of foreign investment and Indian financing from the central and local governmental levels. In response to the disaster, India passed the Environment Protection Act in 1986, which integrated fragmented environmental rules and implemented international obligations arising from environmental treaties and established procedures for environmental policy in India's federal system (Swenden and Saxena, 2020). The Ministry of Environment and Forests (MOEF) was founded and authorized to administer and enforce environmental laws and policies.

Chronic weaknesses of India's environmental policy capacity, the lack of political will by governmental actors, and implementation failure by the administration resulted in weak environmental performance. Yet, India's judiciary and civil society have played an important role in triggering environmental policy (Ciecierska Holmes and Jörgensen, 2020). India's courts and their jurisprudence have stimulated environmental institution building. Drawing on international law, the courts promoted the introduction of environmental policy principles, such as the polluter pays principle, the precautionary principle, the principle of intergenerational equity, the concept of sustainable development, and "the notion of the state as a trustee of all natural resources" (Rajamani, 2007, S. 295). Public interest litigation has become an approach by which judges sometimes fill the void when policy is lacking and thereby design "innovative solutions" (Rajamani, 2007, S. 294). In some outstanding cases the courts have passed orders to enforce environmental performance regarding the corrosive air pollution of the Taj Mahal, air pollution in India's megacities, for example the Delhi Vehicular Pollution Case, as well as river pollution and waste management (Rajamani, 2007). Water problems were often addressed; the right to pollution-free water corroborated; and local powers were admonished to enable self-governance on drinking water and other issues related to local water management (Gupta and Tyagi, 2020).

India's civil society exerted much influence on environmental policy, demanding environmental governance and triggering paradigmatic and policy change in various domains, as is particularly evident in forestry policy. Grievances against the practices of India's forest departments had emerged already in 1958 (Guha, 1989, S. 153) and new actor networks including civil society actors triggered the reform of India's National Forest Policy of 1988 (Das, 2020). The introduction of the Joint Forest Management Scheme in 1990 and the Forest Rights Act (2006) are, despite implementation problems, regarded as policy innovations that link social inclusion and forest protection (Kashwan, 2017).

The environmental crisis led to social mobilization and environmental grassroots movements emerged – a phenomenon which was termed 'environmentalism of the poor' and understood as environmental protection 'offered by ecosystem people' (Guha and Martinez-Alier, 2006, S. 12). Environmental injustice against peasants, tribal peoples, and pastoralists caused several conflicts, exemplified by the highly contested Sardar Sarovar dam construction (1961) on the Narmada River in Gujarat which submerged large numbers of villages and displaced 200,000 people. An impressive body of scholarly literature shows the role of grassroots movements, particularly involving indigenous tribes, in protecting environmental resources and livelihoods against state planned industrial projects. A very prominent one was the eco-feminist tree-hugging Chipko movement which started in the 1970s and became a forerunner and model for other movements (Dwivedi, 2008; Ganguly, 2020). Grassroots movements have been mobilized against a number of industrial projects including timber, mining, steel, the chemical industry, and hydropower (Agrawal and Yokozuka, 2002). The modern, centralized, Nehruvian bureaucratic state has clashed in different ways with these vibrant local initiatives and "age-old custom and practice", revealing "social divisions about appropriate models of development and governance " (Jasanoff, 1993, S. 49).

Environmental non-governmental organizations (ENGOs) also contribute heavily to environmental performance. Public and international funding for ENGO involvement in environmental assessment, outreach, and education programmes has grown since the 1980s (Ganguly, 2020). Several think tanks have been particularly visible and influential in the environmental and climate debate, such as the Energy and Resources Institute (TERI), and the Centre for Science and Environment (CSE), which has since the 1980s regularly published 'State of the Environment' reports and contributed to getting air-pollution on the governmental agenda. However, the relation between the state and ENGOs differs from the much more intense state–business relations. ENGO influence and involvement has remained modest in environmental and climate policy processes (Jasanoff, 1993; Ciecierska Holmes and Jörgensen, 2020; Fernandes et al., 2020).

Another condition in favour of ecologically sustainable development is India's innovation policy, which started quite early with Prime Minister Nehru's emphasis on technology and science. It was related to India's focus on domestic production, substitution of imports, and investment in domestic science and technology development and boosted environmental capacity building (Jasanoff, 1993; Joseph et al., 2014). Institutional arrangements and capacity building emerged which still today manifests itself in national and subnational technical-political missions (Joseph et al., 2014). Prominent examples are the Smart City Mission (2015) and the Solar Mission (2008) (Singhal and Jain, 2020; Bandyopadhyay et al., 2020). Energy policy, a high priority policy domain in a developing country with significant import dependence on coal and millions of people still lacking access to electricity, has since the late 1970s been a policy domain where experimentation and technology development are important. Both at the centre as well as in India's states, various policy frameworks have had a significant impact on renewable energy development and approaches to environmental leapfrogging (Schmid, 2012; Chaudhary et al., 2015; Jörgensen et al., 2015).

29.5 India and the Global Climate Governance System

Climate policy first appeared on India's governmental agenda in the run-up to the United Nations Conference on Environment and Development (UNCED) 1992 and has since then intensively interacted with the global climate governance system. Processed in the context of foreign policy while only marginally involving the environmental ministry, the subject was framed as a purely international issue, not a domestic concern (Michaelowa and Michaelowa 2012). Throughout its development, climate policy in India has been torn between the priorities of rapid economic growth, poverty eradication, and moves towards a low-carbon economy (Jörgensen, 2017; Pillai and Dubash 2021).

At the beginning of the 1990s, the global call for climate action hit India at a time of political instability and in the middle of its economic transition. India's situation of rapidly changing governments after Rajiv Gandhi's assassination and "a new sense of isolation on the world stage", after the collapse of the Soviet Union, along with the related "loss of that primary ally" (Jasanoff, 1993, S. 32), added to the challenges posed to the country, which hosted 30% of the world's poorest people and had more than 300 million people lacking electricity or a reliable supply (Shukla et al., 2015). Climate change was not denied in India (Jasanoff, 1993), yet a powerful and persistent narrative emerged and became a core belief widely shared across governmental and civil society actors which shielded India's growth-first paradigm and kept it from developing a domestic climate mitigation policy. Originating from the publication of two well-known environmental activists, the policy paradigm of 'climate equity' developed, which would guide India's foreign climate policy (Dubash, 2013; Fisher, 2012). In the report, "Global Warming in an Unequal World" (1982), Agarwal and Sunita Narain, the leaders of the Centre for Science and Environment in Delhi, called for a national climate budget which should reflect "each nation's just and fair share" on a per capita basis (cited in Jasanoff, 1993, S. 35). The climate equity frame had not only guided India's negotiating position but also provided cognitive leadership, furthering the introduction of the 'Common but Differentiated Responsibility' (CBDR) principle into the climate treaties in 1992 (Jörgensen, 2017). India had made a point and both India and China became leaders in the negotiating group of developing countries (G77) in the 1990s by insisting on the differentiated architecture of the treaties, thus leaving developing countries free from obligations (Gupta et al., 2015). As a domestic climate policy paradigm widely shared across governmental, civil society actors, and the business sector, it sealed India's domestic debate from the climate mitigation issue for at least two decades until 2007 (Dubash, 2013).

29.6 India's Climate Politics

In the 2010s India's global role as a rapidly growing economy was changing. Future emission scenarios suggested that India was to become a global climate power and a potential climate policy veto player. Expectations from the developing country coalition and the European Union, the international climate leader, were now for India to join the group of countries which committed to CO_2 reduction scenarios (Torney, 2020). After 2007, as a result of the growing scientific consensus about India's vulnerability and the impacts of climate change on the country, its water resources, agriculture, coastal areas, and cities, policy change slowly surfaced under the Congress-led government. The emergence of a domestic climate advocacy coalition and a slightly more open policy process which allowed for participation of stakeholders from science and think tanks furthered the domestic debate (Fernandes et al., 2020; Aamodt and Stensdal, 2017; Dubash, 2013). Due to the linking of economic opportunities offered by finance and technology transfer in the context of the Kyoto Protocol, business interests showed enthusiasm for it (Jörgensen, 2017).

Two years prior to the COP 15 Climate Summit in Copenhagen (2009), incumbent Prime Minister Manmohan Singh addressed the issue of global climate mitigation and announced that per capita greenhouse gas emissions (GHGEs) in India would never exceed those of the Organisation for Economic Co-operation and Development (OECD) nations. He established a National Advisory Panel mandated to develop a national climate strategy, and India's first National Action Plan on Climate Change (NAPCC) was launched in 2008. This included national objectives such as the promotion of renewable energies ('Solar Mission') and energy efficiency ('Enhanced Energy Efficiency') (Government of India, Prime Minister's Council on Climate Change, 2008). The NAPCC 2008 indicated a shift in India's domestic climate policy and a reconsideration of the concerns over threats that climate mitigation policy poses to development. The introduction of the 'co-benefit' principle in the NAPCC in combination with emphasis on energy security concerns was expected to potentially unlock India's domestic climate policy blockade and create an opportunity structure for low-carbon development that would not clash with its climate equity paradigm (Raghunandan, 2020; Dubash, 2013). Initially introduced by the Intergovernmental Panel on Climate Change (IPCC), the co-benefit idea focuses economic, social, and environmental advantages inherent in climate mitigation measures. The NAPCC focuses on the climate mitigation potential inherent in economic development (Mayrhofer and Gupta, 2016); and India's Twelfth Five-Year Plan (2012–2017) proposed "faster, more inclusive and sustainable growth", outlining the need for a low-carbon strategy for inclusive growth "in order to improve the sustainability of its growth process, [with] carbon mitigation [as] an important co-benefit" (Planning Commission of India, 2013, S. 117).

At the 2009 Copenhagen Summit, driven by increasing demands from the international community and developing country coalitions, incumbent environmental minister Jairam Ramesh slightly modified India's negotiating position and pledged alignment with the BASIC group, which included Brazil, South Africa, and China, to voluntarily decrease India's 2005 levels of emission intensity by 20–25% by 2020 (Fernandes et al., 2020, S. 164).

After the 2014 change from the coalition government led by the Congress Party to a new coalition government led by Prime Minister Modi and his Hindu-nationalist Bharatiya Janata Party (BJP), the new policymakers were in charge of formulating India's approach to the 2015 United Nations Climate Change Conference in Paris including a Nationally Determined Contribution (NDC) from India's side. Climate policy was at this juncture not highly institutionalized, coordination capacities were weak, but "a largely stable framing of India's role in mitigation efforts" existed (Pillai and Dubash, 2021, S. 11). India's NDC included a rather moderate new goal for a reduction in the emissions intensity of GDP by 33 to 35 percent by 2030 (compared to 2005), an ambitious goal for renewable energy, in which non-fossil fuels would make up a 40 percentage of the installed electricity mix by 2030, and additional carbon reductions of 2.5 to 3 billion tons through an increase of forest cover (Government of India, 2015). The goal to enhance solar power capacity to 100 GW by 2022 particularly stood out. In cooperation with France, India launched an International Solar Alliance involving more than 120 countries, including several African nations.

Despite ambitious and successful technological missions in renewable energy and energy efficiency, which suggest that the time for the idea of low-carbon development has come in India (Jörgensen, 2021), Pillai and Dubash argue convincingly that the country's political-administrative system and framing of the climate problem is still caught in the carbon lock-in (Pillai und Dubash, 2021). A phase out of coal before 2040 is not in sight. The BJP government has rather shown support for new private coal mining projects, ignoring nature protection issues. Coal production is assessed to rise from 200 to 300 GW (Climate Action Tracker, 2020). Political accountability, a coherent climate and energy policy strategy, political coordination of climate action across governmental ministries and agencies, and a pluralistic climate governance structure are all missing (Pillai and Dubash, 2021). Under Prime Minister Modi, Policy formulation and decision-making were centralized in the office of the Prime Minister (Pillai and Dubash, 2021), while researchers and NGOs enjoyed less access to the policy process (Aamodt and Stensdal, 2017). And there are no signs that a post-COVID recovery strategy will be more considerate of a greening of the economy.

29.7 Democracy and Environment

Overall the outcome of India's environmental policy is unsatisfying. Poor efficacy has characterized environmental performance and India has been ranking continuously low among the weakest group of countries in the Environmental Performance Index up until today. Various conditions hamper environmental policy success in India: corruption (Dwivedi, 2008), state–business relations that focus on economic growth (Jaffrelot, 2019), over-centralization in India's federal system (Sharma and Swenden, 2018; Swenden and Saxena, 2020; Beermann et al., 2016; Jörgensen, 2020), and a lack of accountability and weak administrative capacities at the subnational levels (Dwivedi, 2008; Jörgensen, 2020).

One interesting question in this context is how democracy affects environmental policy (Hochstetler 2012). Agrawal and Yokozuka (2002) argued that India's democratic system and the mixture of bureaucratic organization and participatory politics provides an opportunity structure for environmental policy. Apart from a few success stories there is considerable evidence, however, that India's centralized and heavily bureaucratized state, which reserves decision-making to elite groups, does not offer opportunity structures in many issue areas for more vibrant policy processes, processes which could otherwise catalyse intense controversies about the direction of economic and development paths (Ciecierska Holmes and Jörgensen, 2020). India's democracy has seen a curtailment of democratic institutions in different phases, allowing for an environmental race-to-the bottom and impeding climate policy (Ganguly, 2020; Jaffrelot, 2019). A power-concentrating policy style, a push back on democratic rights, and restrictions of civil society movements have been significant elements of government policy since Prime Minister Modi and the BJP took office in 2014 (Sharma and Swenden, 2018).

29.8 Conclusion and Future Research

Environmental politics in India is a fascinating field in the study of comparative politics. It is multifaceted, contradictory, and diverse, and various factors have shaped environmental and climate policy outcomes. International factors including environmental and climate governance, the North–South debate, and the equity paradigm have influenced environmental policy development. The institutional structures of India's political system, such as federalism and centralization, and its democracy, including its vibrant civil society and judiciary, have each played a role.

During the course of the past four decades India has transformed from a poor developing country to a rapidly growing economy and a regional power. At the same time, the economic growth trajectory came at the disastrous expense of India's environment and natural resources and has destroyed traditional livelihoods for large numbers of people. Since independence, confronted with tremendous poverty, India's policymakers followed the "growth first" development paradigm of the 20th century, which can be traced back to the Kuznets Curve in development economics and left environmental issues mostly unattended. Despite the remarkable growth in the number of environmental policy institutions since the 1980s, environmental concerns are still today predominantly pitted against growth issues. It is quite difficult to reflect on India's idiosyncratic environmental development trajectory from a comparative politics perspective. Systematic comparative research about the environmental policy paths taken in rapidly developing countries is still rudimentary. In any comparison it is important to acknowledge that India started from and remains on a much lower base of GNI per capita relative to China, South Africa, Brazil, or Russia, and the pressure to develop infrastructures and employment for its growing population is huge. In India's Twelfth Five-Year Plan (2012–2017) "Faster, More Inclusive and Sustainable Growth" had its main emphasis on growth and inclusion. Overall, compared to other rapidly growing countries, India's environmental situation has worsened. The Environmental Performance Index 2022 assigned India the lowest rank, downgrading it from 169 in the EPI 2020 to 180 (Wolf et al., 2022; Wendling et al., 2020). India's deteriorating air quality, unsafe drinking water and sanitation, threats to biodiversity and habitat, and rapidly rising greenhouse gas emissions were highlighted as particular challenges. In the EPA peer group ranking of the 27 'emerging markets', India ranked 26 in environmental health, 27 in ecosystem vitality, and 23 in climate change (Wolf et al., 2022, pp. 30–37). The growth-first thinking has also informed India's domestic climate policy debate, coalescing with the call for inter- and intragenerational climate equity, introducing the common but differentiated responsibility principle (CBDR) into the international climate regime in 1992. Climate equity is a highly significant and legitimate aspect of climate policy that considers both the historical emissions of the industrialized countries and the pressing need to develop. Yet, it has also obstructed the view and restricted policy learning by drawing a red line between forward looking environmental policy and the co-benefits lying in low-carbon development.

India's state has always played an active role, centrally planning economic development, controlling major economic sectors – including energy production – and paving the way for its entry in the fossil fuel economy during the first three decades after independence. By relaxing state control and gradually liberalizing the economy and collaborating more closely with India's business community since the 1980s, the state maintained an active role that continuously focused on economic growth. The outside initiative of the first United Nations Conference on the Human Environment in Stockholm 1972 led to India's initial placement of environmental policy on the political agenda. The dramatic industrial catastrophe in Bhopal brought a focus on environmental policy. Environmental legislative frameworks developed, but within a context of limited political-administrative capacities, particularly at the subnational levels. These frameworks lacked empowerment and resources, as well as being over-centralized and rife with corruption. As a result, environmental performance in India was hampered, leaving implementation gaps in many environmental domains.

India's vibrant civil society organizations (CSOs) and grassroots movements have won significant influence mobilizing and campaigning for environmental protection, agenda setting, and implementation. Although by far still not on an equal footing with the business sector, since they are not part of the state–business alliances which have developed since the 1980s, CSOs have developed large research and outreach capacities. India's courts and their jurisprudence have stimulated environmental institution building, promoted the introduction of international environmental policy principles, and sometimes filled the void when policy was lacking.

India's climate policy emerged, intensively interacting with the global climate governance system pushing for climate justice. Domestic climate policy surfaced after 2007 under the Congress-led government, associated with a growing scientific consensus about India's environmental vulnerability. The NAPCC 2008 indicated a shift in India's domestic climate policy and a reconsideration of the concerns about the threats climate mitigation policy posed to development. In the run-up to the UN climate conference in Paris 2015 the coalition government led by Prime Minister Modi and his Hindu-nationalist BJP came up with an INDC which included a rather moderate new goal for a reduction in emissions intensity, although an ambitious goal for renewable energy and the goal to enhance solar power capacity particularly stood out. India's quite ambitious and successful renewable energy goals are connected to a path chosen at the end of the 1970s when the first renewable energy projects were developed and reinforced by policy frameworks and technological missions that were developed at the national and subnational levels. Yet, approaches to the expansion of renewable energy do not indicate a phase out of coal. As reported in the EPI 2022, India will belong to the four countries including the USA, China, and Russia that will account for over 50% of residual global greenhouse gas emissions in 2050.

Overall the outcomes of India's environmental policy are dissatisfying. On the one hand India's democratic system offers an opportunity structure for environmental actors; on the other hand, its centralized and heavily bureaucratized state impedes environmental action. India's democracy has seen a curtailment of democratic institutions in different phases, allowing for an environmental race-to-the bottom and impeding climate policy, and this is also the case in the most recent phase since 2014 under the BJP government, which pursues a power-concentrating policy style which pushes back on democratic rights and restricts civil society movements.

Box 29.1 Chapter Summary

- This chapter has dealt with India's development trajectory and examines its emerging environmental policy and climate policy in the context of international environmental governance and domestic politics and institutions.
- The growth-first paradigm and the carbon lock-in, international environmental and climate governance, the North–South debate, and the equity paradigm have shaped environmental policy in India. The institutional structures of India's political system, the strong role of the state, federalism and centralization, the judiciary, and India's democracy have each played a role.
- India's vibrant civil society organizations and grassroots movements have won significant influence in mobilizing and campaigning for environmental protection, agenda-setting, and implementation.
- On the one hand, India's democratic system offers an opportunity structure for environmental actors; on the other hand, its centralized and heavily bureaucratized state impedes environmental action.

References

Aamodt, Solveig; Stensdal, Iselin (2017): Seizing policy windows. Policy Influence of climate advocacy coalitions in Brazil, China, and India, 2000–2015. In: *Global Environmental Change Part A: Human & Policy Dimensions* 46, S. 114–125. DOI: 10.1016/j.gloenvcha.2017.08.006.

Agrawal, Arun; Yokozuka, Noriko (2002): Environmental capacity-building: India's democratic politics and environmental management. In: Helmut Weidner, Martin Jänicke und Helge Jörgens (Hg.): *Capacity building in national environmental policy. A comparative study of 17 countries*. Berlin, New York: Springer.

Bandyopadhyay, Kaushik Ranjan; Joshi, Madhura; Quitzow, Rainer (2020): Sustainable Energy: Prospects and Challenges. In: Natalia Ciecierska-Holmes, Kirsten Jörgensen, Lana Laura Ollier und D. Raghunandan (Hg.): *Environmental policy in India*. London: Routledge.

Beermann, Jan; Damodaran, Appukuttan; Jörgensen, Kirsten; Schreurs, Miranda A. (2016): Climate action in Indian cities: An emerging new research area. In: *Journal of Integrative Environmental Sciences*, S. 1–12. DOI: 10.1080/1943815X.2015.1130723.

Bhagwati, Jagdish N.; Panagariya, Arvind (2013): *Why growth matters. How economic growth in India reduced poverty and the lessons for other developing countries*. 1. ed. New York, NY: Public Affairs (A Council on Foreign Relations book).

Broughton, Edward (2005): The Bhopal disaster and its aftermath: A review. In: *Environmental Health* 4 (1), S. 6. DOI: 10.1186/1476-069X-4-6.

Chaudhary, Ankur; Krishna, Chetan; Sagar, Ambuj (2015): Policy making for renewable energy in India: lessons from wind and solar power sectors. In: *Climate Policy* 15 (1), S. 58–87. DOI: 10.1080/14693062.2014.941318.

Chopra, Kanchan (Hg.) (2017): *Development and environmental policy in India. The last few decades*. Singapore: Springer Singapore (SpringerBriefs in Economics).

Ciecierska Holmes, Natalia; Jörgensen, Kirsten (2020): Environmental politics in India. Institutions, actors and environmental governance. In: Natalia Ciecierska-Holmes, Kirsten Jörgensen, Lana Laura Ollier und D. Raghunandan (Hg.): *Environmental policy in India*. London: Routledge, S. 241–258.

Climate Action Tracker (Hg.) (2020): India. Online verfügbar unter https://climateactiontracker.org/countries/india/zuletztgeprüftam 13.09.2021.

Das, Smriti (2020): Forest governance in India: Achieving balance within a complex policy subsystem. In: Natalia Ciecierska-Holmes, Kirsten Jörgensen, Lana Laura Ollier und D. Raghunandan (Hg.): *Environmental policy in India*. London: Routledge, S. 89–110.

Drèze, Jean; Sen, Amartya (2014): *An uncertain glory. India and its contradictions*. London: Penguin Books.

Dubash, Navroz K. (2013): The politics of climate change in India: Narratives of equity and cobenefits. In: *WIREs Climate Change* 4 (3), S. 191–201. DOI: 10.1002/wcc.210.

Dwivedi, O P (2008): Environmental challenges facing India O.P. Dwivedi. In: O. P. Dwivedi und Jordi Díez (Hg.): *Global environmental challenges. Perspectives from the South. Peterborough, Ont.*, Buffalo, N.Y: Broadview Press, S. 113–138.

Fernandes, D.; Jörgensen, K.; Narayanan, N.C. (2020): Factors shaping the climate policy process in India. In: Natalia Ciecierska-Holmes, Kirsten Jörgensen, Lana Laura Ollier und D. Raghunandan (Hg.): *Environmental policy in India.* London: Routledge, S. 158–173.

Fisher, Susannah (2012): Policy storylines in Indian climate politics: opening new political spaces? In: *Environmental Planning C* 30 (1), S. 109–127. DOI: 10.1068/c10186.

Gadgil, Madhav; Guha, Ramachandra (1997): *This fissured land. An ecological history of India. 4th impression.* Delhi: Oxford Univ. Press (Oxford India paperbacks).

Ganguly, Sunayana (2020): Civil society and state interaction in environment policy in India. In: Natalia Ciecierska-Holmes, Kirsten Jörgensen, Lana Laura Ollier und D. Raghunandan (Hg.): *Environmental policy in India.* London: Routledge, S. 60–85.

Government of India, Prime Minister's Council on Climate Change (2008): National Action Plan on Climate Change. Online verfügbar unter http://pmindia.nic.in/Pg01-52.pdf,zuletztgeprüftam 29.11.2011.

Government of India (GOI) (2015): India's Intended Nationally Determined Contribution: Working Towards Climate Justice. Hg. v. United Nations Climate Change (UNFCC). Bonn. Online verfügbar unter https://www4.unfccc.int/sites/submissions/INDC/Published%20Documents/India/1/INDIA%20INDC%20TO%20UNFCCC.pdf,zuletztgeprüftam 05.01.2019.

Guha, Ramachandra (1989): *The unquiet woods.* Delhi: Oxford University Press.

Guha, Ramachandra; Martinez-Alier, Juan (2006): *Varieties of environmentalism. Essays north and south. Repr.* London: Earthscan.

Gupta, Joyeeta; Tyagi, Richa (2020): Dilemmas of water governance India: Dilemmas of water governance. In: Natalia Ciecierska-Holmes, Kirsten Jörgensen, Lana Laura Ollier und D. Raghunandan (Hg.): *Environmental policy in India.* London: Routledge, S. 111–132.

Gupta, Neha; Yadav, Krishna Kumar; Kumar, Vinit (2015): A review on current status of municipal solid waste management in India. In: *Journal of environmental sciences (China)* 37, S. 206–217. DOI: 10.1016/j.jes.2015.01.034.

Hochstetler, Kathryn (2012): Democracy and the Environmental in Latin America and Eastern Europe. In: Paul F. Steinberg und Stacy D. VanDeveer (Hg.): *Comparative environmental politics.* Cambridge, Mass.: MIT Press (American and comparative environmental policy), S. 199–229.

Hsu, A.; Emerson, J.; Levy, M.; de Sherbinin, A.; Johnson, L.; Malik, O. et al. (2014): *The 2014 environmental performance index.* New Haven, CT: Yale Center for Environmental Law and Policy. Online verfügbar unter https://www.epi.yale.edu.

Jaffrelot, Christophe (2019): Business-friendly Gujarat under Narendra Modi: The implications of a new political Economy. In: Christophe Jaffrelot, Atul Kohli und Kanta Murali (Hg.): *Business and politics in India.* New York: Oxford University Press (Modern South Asia), S. 211–233.

Jasanoff, Sheila (1993): India at the crossroads in global environmental policy. In: *Global Environmental Change Part A: Human & Policy Dimensions* 3 (1), S. 32–52. DOI: 10.1016/0959-3780(93)90013-B.

Jörgensen, Kirsten (2017): India: The global climate power torn between 'growth-first' and 'green growth'. In: Wurzel, Rudiger K. W, James Connelly und Duncan Liefferink (Hg.): *The European Union in international climate change politics. Still taking a lead?* London, New York: Routledge, Taylor & Francis Group, S. 270–283.

Jörgensen, Kirsten (2020): The role India's states play in environmental policymaking. In: Natalia Ciecierska-Holmes, Kirsten Jörgensen, Lana Laura Ollier und D. Raghunandan (Hg.): *Environmental policy in India.* London: Routledge, S. 39–59.

Jörgensen, Kirsten (2021): Low-carbon development: An idea whose time has come. Unlocking climate cooperation between India and the EU. In: Philipp Gieg, Timo Lowinger, Manuel Pietzko, Anja Zürn, Umma Salma Bava und Gisela Müller-Brandeck-Bocquet (Hg.): *EU-INDIA relations. The strategic partnership in the light of the European Union Global Strategy. [S.l.]*: Springer International Publishing, S. 185–199.

Jörgensen, Kirsten; Mishra, Arabinda; Sarangi, Gopal K. (2015): Multi-level climate governance in India: the role of the states in climate action planning and renewable energies. In: *Journal of Integrative Environmental Sciences* 12 (4), S. 267–283. DOI: 10.1080/1943815X.2015.1093507.

Joseph, K J; Singh, Lakhwinder; Abraham, Vinoj (2014): Dealing with the innovation - inequality - the conundrum: The Indian Experience. In: Maria Clara Couto Soares, Mario Scerri und Rasigan Maharajh (Hg.): *Inequality and development challenges*. New Delhi: Routledge, Taylor & Francis Group, S. 149–189.

Kashwan, Prakash (2017): *Democracy in the woods. Environmental conservation and social justice in India, Tanzania, and Mexico*. New York, NY: Oxford University Press (Studies in comparative energy and environmental politics).

Kohli, Atul (2006): Politics of economic growth in India. 1980-2005. Part 1: The 1980s. In: *Economic & Political Weekly* 41 (13), S. 1251–1259.

Managi, Shunsuke; Jena, Pradyot Ranjan (2008): Environmental productivity and Kuznets curve in India. In: *Ecological Economics* 65 (2), S. 432–440. DOI: 10.1016/j.ecolecon.2007.07.011.

Mayrhofer, Jan P.; Gupta, Joyeeta (2016): The science and politics of co-benefits in climate policy. In: *Environmental Science & Policy* 57, S. 22–30. DOI: 10.1016/j.envsci.2015.11.005.

Michaelowa, K., & Michaelowa, A. (2012). India as an emerging power in international climate negotiations. *Climate Policy*, 12(5), 575–590.

Pillai, Aditya Valiathan; Dubash, Navroz K. (2021): The limits of opportunism: the uneven emergence of climate institutions in India. In: *Environmental Politics*, S. 1–25. DOI: 10.1080/09644016.2021.1933800.

Planning Commission of India (2013): Twelfth Five Year Plan (2012–2017) Faster, More Inclusive and Sustainable Growth. Hg. v. Government of India. New Delhi [u.a.]. Online verfügbar unter http://planningcommission.gov.in/plans/planrel/12thplan/pdf/12fyp_vol1.pdf,zuletztgeprüftam 06.03.2018.

Raghunandan, D (2020): Factors shaping India's International Climate Policy. In: Natalia Ciecierska-Holmes, Kirsten Jörgensen, Lana Laura Ollier und D. Raghunandan (Hg.): *Environmental policy in India*. London: Routledge.

Rajamani, L. (2007): Public interest environmental litigation in India: Exploring issues of access, participation, equity, effectiveness and sustainability. In: *Journal of Environmental Law* 19 (3), S. 293–321. DOI: 10.1093/jel/eqm020.

Reich, Michael R.; Bowonder, B. (1992): Environmental policy in India. Strategies for better implementation. In: *Policy Studies Journal* 20 (4), S. 643–661. DOI: 10.1111/j.1541-0072.1992.tb00188.x.

Schmid, Gisèle (2012): The development of renewable energy power in India: Which policies have been effective? In: *Energy Policy* (45), S. 317–326.

Sharma, Chanchal Kumar; Swenden, Wilfred (2018): Modi-fying Indian Federalism? Center-state relations under Modi's tenure as Prime Minister. In: *INPP* 1 (1). DOI: 10.18278/inpp.1.1.4.

Shukla, P. R.; Garg, A.; Dholakia, H. H. (2015): Energy-emissions trends and policy landscape for India: Allied publishers. Online verfügbar unter https://books.google.de/books?id=u6nACwAAQBAJ.

Sil, Rudra (2014): India. In: Jeffrey Kopstein, Mark Lichbach und Stephen E. Hanson (Hg.): *Comparative politics*: Cambridge University Press, S. 339–390.

Singhal, S.; Jain, S. (2020): Smart Sustainable Cities. In: Natalia Ciecierska-Holmes, Kirsten Jörgensen, Lana Laura Ollier und D. Raghunandan (Hg.): *Environmental policy in India*. London: Routledge, S. 174–200.

Stern, David I. (2004): The rise and fall of the environmental kuznets curve. In: *World Development* 32 (8), S. 1419–1439. DOI: 10.1016/j.worlddev.2004.03.004.

Swenden, W; Saxena, R (2020): Environmental competencies in India's federal system. In: Natalia Ciecierska-Holmes, Kirsten Jörgensen, Lana Laura Ollier und D. Raghunandan (Hg.): *Environmental policy in India*. London: Routledge, S. 17–38.

Torney, D. (2020): India's relations with the EU on environmental policy. In: Natalia Ciecierska-Holmes, Kirsten Jörgensen, Lana Laura Ollier und D. Raghunandan (Hg.): *Environmental policy in India*. London: Routledge.

Unruh, Gregory C. (2000): Understanding carbon lock-in. In: *Energy Policy* 28 (12), S. 817–830. DOI: 10.1016/S0301-4215(00)00070-7.

Wendling, Z A; Emerson, J W; de Sherninin, A; Esty, D C (2020): *Environmental performance index*. New Haven, CT. Online verfügbar unter https://epi.yale.edu/epi-country-report/IND.

Wolf, M.J. et al., 2022. *Environmental performance index*. New Haven: CT: Yale Center for Environmental Law & Policy. epi.yale.edu.

30

CONCLUSIONS

Past Achievements and Future Directions for Environmental Policy Research

Helge Jörgens, Christoph Knill, and Yves Steinebach

30.1 Introduction

The contributions in this book provide a thorough review of the state-of-the-art in the study of environmental policy. They study environmental policy from different analytical angles and perspectives, comparing empirical developments across countries and different world regions. The chapters also provide clear indications for future research. This chapter will summarize these claims and identify general needs and avenues for future research. Before addressing this point, however, we will shed light on another, often neglected aspect associated with the study of environmental policy. More specifically, as will be shown in the following, the study of environmental policy has never merely been an end in itself but has inspired various subdisciplines of political science in manifold ways. In other words, environmental policy studies have been a catalyst for political science research *in general*.

30.2 Environmental Policy Analysis as a Stimulus for Political Science

Environmental policy is a relatively young policy domain (Chapter 2). Nevertheless, environmental policy studies have triggered many important research impulses in the political science subfields of policy analysis, comparative public policy, comparative politics, and global governance. In the following, we present these impacts.

30.2.1 Environmental Policy as a Stimulus for Analytical Frameworks for Public Policy Analysis

In the field of public policy analysis, scholars have developed a series of theoretical concepts and analytical frameworks that today guide empirical research in various public policy domains. Several of these analytical frameworks have been inspired or empirically underpinned by research on environmental policy. One example is the Institutional Analysis and Development (IAD) framework developed by Ostrom and colleagues to better understand the factors that influence the ecological functioning of common pool resources, such as forests, fisheries, or agricultural land (Ostrom, 1990, 2007). Another prominent case is the Advocacy Coalition Framework

DOI: 10.4324/9781003043843-34

(ACF) developed by Sabatier and colleagues (Sabatier, 1998; Weible et al., 2009). The ACF was largely based on and inspired by Sabatier's previous work on environmental regulation and environmental policy implementation (Sabatier, 1988). Whereas, initially, "applications of the ACF were on environmental or energy issues, … today the framework is applied almost evenly across topical issues, including health, finance/economic, social welfare, disaster and crisis management, and education" (Weible and Jenkins-Smith, 2016, pp. 16–17). Another analytical model, the capacity-building framework, was developed by Jänicke and colleagues to explain variations in the environmental policy performance of countries by accounting for multiple factors, including problem structures, actors and their strategies, as well as issue-specific cognitive, institutional, economic, and technological framework conditions (Jänicke, 1997; Jänicke and Weidner, 1997a; Weidner and Jänicke, 2002; Jänicke and Jörgens, 2009). Finally, Maarten Hajer (Hajer, 1995; Hajer and Versteeg, 2005) developed a discourse-centered model of the policy process based on his research on the ecological modernization discourse of the late 1980s and early 1990s (see also Chapter 6), which added a focus on the social construction of policy problems to the predominant emphasis on the role of actors, interests, and institutions in the policy process.

Environmental policy research has also been a constant source of conceptual innovation (Meadowcroft and Fiorino, 2017). Concepts initially developed in the field of environmental policy research now play a role far beyond the boundaries of this policy domain. One example is Jänicke's concept of ecological modernization (Jänicke, 1985; Jänicke, 2008, see also Chapter 6), which was decisive for recognizing that economic growth and environmental protection do not need to be mutually exclusive (see also Mol et al., 2009). Without the concept of ecological modernization, the Brundtland Commission's notion of sustainable development (World Commission on Environment and Development, 1987) would certainly not have been the same (Barry, 2007). Today, the concept of sustainability is an integral element of the self-description of nearly all policy domains as the United Nations' Sustainable Development Goals clearly show (Kanie and Biermann, 2017; Browne, 2017). But the stimuli from the field of environmental policy to rethink traditional notions of markets and economic growth go even further. The recognition that ecological modernization as a political strategy is reaching its limits has brought concepts of sufficiency (Schneidewind and Zahrnt, 2014; Lorek and Spangenberg, 2019; Barry and Doran, 2006; Princen, 2003, see Chapter 25) and degrowth (Stuart et al., 2021; Kallis et al., 2020; Alexander, 2012) into the political debate.

Another concept that has significantly benefited from environmental policy research is that of policy integration (Tosun and Lang, 2017; Trein et al., 2019). Environmental policy is a cross-cutting policy whose success depends heavily on integrating environmental concerns into the goals, measures, and programs of other policy domains (Chapter 9). As a result, environmental policy research has been a pioneer in exploring the chances and barriers of integrating the concerns of one policy area into the goals, policies, and measures of other domains (Lenschow, 2002; Lafferty and Hovden, 2003; Jacob and Volkery, 2004; Jordan and Lenschow, 2010). But environmental policy integration is not only a horizontal challenge of taking environmental concerns into account in other policy domains. The fact that there are often several decades between environmentally damaging actions and their effects makes it necessary to treat environmental policy as a long-term policy and to take into account the problems this raises for processes of policymaking that normally have a shorter time horizon (Sprinz, 2012; Leschine, 2007; see also Chapter 22). While problems of making the policy process more suitable for long-term challenges are also inherent in other policy domains, such as pension policies or policies dealing with demographic changes, these challenges have been discussed most intensively in the field of environmental governance.

Environmental policy is also the field where the concept of policy diffusion, which had originally been developed to describe interdependent policymaking in the US states (Walker, 1969; Gray, 1994; Berry and Berry, 2007), started to be systematically applied to the comparative analysis of public policies at the national level (Hoberg, 1991; Jörgens, 1996, 2001; Kern, 1997; Jänicke and Weidner, 1997b; Kern et al., 2001; Tews et al., 2003; see Chapter 18). Today, the policy diffusion approach is successfully applied in a wide range of policy domains to explain the interplay of international and domestic factors for adopting policies at the national level (Simmons and Elkins, 2003; Holzinger et al., 2007; Gilardi, 2012).

30.2.2 Environmental Policy as a Research Stimulus for Comparative Public Policy

In many ways, the study of environmental policy, which took off in the early 1970s, provided a fruitful ground for the emergence and advancement of the field of comparative public policy. Many concepts that are now part of the standard vocabulary of comparative policy analysis were developed with a view to the environmental field. This can be illustrated by various examples. First, the entire debate on policy design (Chapter 19), choice of policy instruments, and classification of different types of policy instruments has been driven by work focusing on environmental policy. This includes differentiation between old and new tools, procedural and substantive instruments, and distinctions between hierarchical, economic, and communicative instruments (Knill and Lenschow, 2000; Bemelmans-Videc et al., 1998; Héritier et al., 1996; Jordan et al., 2003).

Second, and related to this discussion, public policy scholars discovered that for many policy problems, it is rarely sufficient to rely on single instrument types. Instead, effective problem-solving requires the combination of different instruments or instrument mixes (Chapter 24). It also became clear that different policy instruments might reinforce but also weaken each other; that is, designing public policies is a challenging endeavor (Howlett and Rayner, 2013; Gunningham and Grabosky, 1998; Gunningham and Sinclair, 1999).

In this context, the preference for certain instrument types or combinations of instruments has inspired the development of concepts like policy styles, styles of regulations, or administrative styles that capture variation in policymaking approaches across countries. In this context, particular emphasis is on the role of dominant policy paradigms that structure policy and regulatory approaches in different policy areas and that display high stability over time due to their manifestations in actors' beliefs and ideas (Knill, 2001; Bayerlein et al., 2020; Vogel, 1986; Howlett and Tosun, 2021).

Third, the study of environmental policy inspired researchers to develop concepts that capture aggregate development in policy outputs, that is, features of environmental policy stocks rather than individual environmental policies (Chapter 16). By focusing on environmental policy, many studies identified growing policy complexity as an essential feature that characterizes policymaking in modern democracies. As policy stocks are growing in size, the number of mutual interactions between different policy elements is rising exponentially. The emergence of complex policy mixes comes with fundamental challenges for designing but also for evaluating new policies (Limberg et al., 2022; Adam et al., 2018; Hurka et al., 2022).

In addition to policy complexity, research has developed the related concepts of policy growth, policy accumulation, and rule growth to analytically capture the constant increase in the size of policy stocks. As a relatively young policy area, environmental policy is characterized by a pronounced dynamic of policy growth. But comparative studies show that impressive growth rates also characterize older and more mature policy areas. In other words, the study of environmental policy provided an important impetus for the general study of governmental activity and the resulting dynamics of policy accumulation (Adam et al., 2019).

Fourth, the study of environmental policy – particularly EU environmental policy – revitalized the study of policy implementation (Chapter 10). Implementation research started in the early 1970s but lost momentum by the early 1980s. It was only in the context of the growing academic and political interest in the implementation effectiveness of EU (environmental) policy that implementation research took off again. The fact that the EU turned out as an ideal laboratory to compare member state compliance with EU requirements inspired a large amount of scientific work. These studies significantly advanced our general understanding of potential causes of implementation failures, such as the role of instrument choice and the design of administrative structures for policy implementation (Knill and Lenschow, 2000; Jordan, 1999; Knill and Liefferink, 2007; Steinebach, 2022b; Börzel, 2021).

Important insights into problems of policy implementation also come from recent research, which highlights that there is a widening gap between policy production and the administrative capacities available for policy implementation. While governments constantly produce new policies, they have limited incentives to simultaneously expand the necessary implementation capacities. As a result, implementers are increasingly forced to set priorities and sacrifice specific implementation tasks at the expense of others. This phenomenon of "policy triage" (Knill et al., forthcoming) constitutes an essential feature of environmental policy implementation that is also of relevance for other policy areas. Moreover, the concept of policy triage complements existing analytical perspectives in implementation research. The latter has typically focused on the implementation of individual policies, but largely lacks a perspective that takes into account the fact that there might be trade-offs in implementation effectiveness across different policies in the policy stock, that is, the need to implement a new policy A might lead to insufficient implementation of policies B or C due to resource constraints (Knill et al., 2021a; Limberg et al., 2021; Knill et al., 2021b).

Fifth, research on environmental policy has brought about essential insights into the cross-national dynamics of policymaking. Studies of environmental policy convergence and policy diffusion have improved our understanding of causal mechanisms that drive environmental policymaking beyond merely domestic factors (Busch and Jörgens, 2005; Chapter 18). Research has shown, for instance, that international institutions constitute an essential source of domestic policymaking (Holzinger et al., 2008a; Holzinger et al., 2008b). Apart from the global harmonization of policies, this effect merely emerges from the stimulation of processes of transnational policy learning and policy diffusion (Jörgens, 2004; Busch et al., 2005). Research has also shown that economic globalization does not necessarily come with often-feared races to the bottom or policy dumping (Vogel, 1995, 1997). On the contrary, market incentives and the establishment of level playing fields through international harmonization can create regulatory dynamics that point in the opposite direction of ever more ambitious policies (Holzinger et al., 2008a; Jörgens et al., 2014).

With regard to cross-national dynamics, a further fruitful debate centred around the role of environmental pioneer and leader countries and the extent to which such first movers and their advocacy for similar policy adoptions at the international level might have a stimulating effect on laggard and follower countries. The conditions under which countries might benefit from adopting innovative and ambitious policies ahead of others are still subject to long-standing academic debate (see Chapter 17). The fact that policy leadership by individual countries may stimulate cross-national policy change is of high importance, especially in areas where there are high hurdles to adopting internationally harmonized policies. These findings have also informed the study of other policy areas, such as, for instance, competition and tax policies (Thisted and Thisted, 2020; Wurzel et al., 2019).

Sixth, especially research on EU environmental policy has shown that policymaking in international settings may be driven by policy spillovers from other policy areas. In the case of the EU, the overarching goal of market integration fueled the development of environmental policy as a side product of economic integration. Yet, these spillover dynamics provided the ground for the emergence of environmental policy as a genuine field of EU responsibility. In a similar vein, the presence of spillovers and policy interaction across policy sectors increased the scholarly attention to study the challenges of policy integration (see above, see also Chapter 9). In fact, this research topic has largely been developed with a view to environmental policy (Trein et al., 2019; Jordan and Lenschow, 2010).

Finally, environmental policy has provided an important impetus to study policy performance (Chapter 7). Researchers have developed a variety of valuable indicators and indexes to capture changes in the state of the environment across countries and over time. Based on these achievements, scholars have been increasingly able to study the effects of policy outputs on policy performance (Fernández-i-Marin et al., 2021).

30.2.3 Environmental Policy as a Research Stimulus for Comparative Politics

The study of environmental policy has not only inspired public policy but also turned out to be a fruitful ground for studying the role of the institutional and procedural characteristics of policymaking and hence touches upon important research topics of comparative politics.

First, the study of environmental policy paved the way to advance our understanding of political parties in policymaking (Chapter 7). Whether parties matter for policy is a core topic of comparative politics. In this context, the classical cleavage has traditionally been on the left–right dimension, that is, between parties representing labour versus parties representing employer interests. Left parties generally emphasize the need to provide social protection through the control of market mechanisms and the provision of social transfers. Right parties, by contrast, display more liberal and conservative ideologies, focusing on creating a free and prosperous society through individual political rights, property rights, and a limited state. Yet, it soon became obvious that focusing on this cleavage might not be the most appropriate approach to capture partisan influence in policy areas beyond social and economic policies. This insight became manifest in particular with regard to environmental policy (Knill et al., 2010; Schulze, 2014). Environmental issues cut across the left–right dimension when looking at specific party families (Gallagher et al., 2006). For policy preferences of the centre-right-wing parties, such as those of the Christian-democratic party family or other parties that stress religious aspects, protecting the environment might be important in terms of saving "God's creation". At the same time, socialist or social-democratic parties as political actors from the left-wing ideological spectrum might prefer economic growth over environmental protection to secure employment for their core voter clientele (Neumayer, 2003). Thus, the left–right dimension should be of minor importance for explaining the variance in environmental policy developments over time and across countries.

Instead, a stronger emphasis has been placed on the cleavage between parties representing post-material and material interests. With national boundaries becoming more and more permeable, an open–closed society cleavage has emerged in which social groups are either in favor of or against the opening of markets. Those who feel disadvantaged by economic globalization favor trade barriers to protect local manufacturing and 'locals first' policies in the labour market. The economically defensive attitude of these groups is reinforced by anti-immigration stances, which stress religious and national values against a multi-ethnic society (Kriesi et al., 2012; Walter, 2017). Due to these developments, Marks et al. (2021) argue that conventional parties on the left–right have become much less socially structured. However, parties on the socio-cultural transnational divide – GAL (green, alternative, libertarian) and TAN (traditionalist, authoritarian, nationalist) – have a sharply divergent social base.

Second, major works in this subfield studied the role of institutional factors in shaping policy outputs and policy impacts. The environmental field has been an important area for studying the nexus between institutions and policies. This holds not only with regard to Lijphart's (2012) distinction between consensual and majoritarian systems and their impact on the problem-solving capacity of political systems. It also applies to systematic investigations on the effects of institutional veto points and institutional veto players on environmental policy outputs and policy impacts (Scruggs, 2003; Jahn, 2016; see also Chapter 7).

Third, while for long the study of interest groups and their influence on policymaking had been centered on social policy and the cleavage between trade unions and business associations, the rise of environmental policy from the 1970s onwards offered new ground for studying and applying existing concepts and theories of interest group politics. Particular emphasis in this regard has been placed on the role of neo-corporatism for environmental policymaking (Jahn, 1998, 2016; Scruggs, 2001). In line with the findings for the social sector, many studies found a strong positive impact of neo-corporatism on environmental performance. Neo-corporatism not only improves the chances for public interest groups to effectively participate in policymaking but also enhances the implementation effectiveness of the adopted policies through the involvement of different societal interests (Steinebach, 2019; Steinebach, 2022a).

Fourth, the study of environmental policy is essentially connected to the analysis of social movements in comparative politics. In fact, claims have been made for the centrality of the environmental movement to processes of macrosocial and political change towards a post-industrial society (Touraine, 1981; Rootes, 2004). In line with existing conceptions of social movements, an environmental movement

> may be defined as a loose, noninstitutionalized network of informal interactions that may include, as well as individuals and groups who have no organizational affiliation, organizations of varying degrees of formality, that are engaged in collective action motivated by shared identity or concern about environmental issues.
>
> *(Rootes, 2004, p. 610)*

There exists an extensive literature in comparative politics that covers a variety of aspects of environmental movements: public opinion, attitudes, and political values that form the potential for the rise and mobilization of the movement; environmental movement organizations and networks; protest and other kinds of activities carried out by the movements; discourses and framings put forward by environmental actors in the public domain (Giugni and Grasso, 2015). The study of environmental movements has been highly important to study not only the variety, structural heterogeneity, and transformation of social movements. It has also advanced our understanding of the processes of the institutionalization of social movements, manifesting itself in the emergence of international environmental organizations and green parties. At the same time, it can be observed that institutionalization only refers to some parts of the environmental movement. Overall, the study of environmental movement constitutes an important showcase to understand the dynamics and impacts of social movements more generally (see Chapters 13 and 26).

30.2.4 *Environmental Policy as a Research Stimulus in the Field of Global Governance*

Also, at the international level, environmental policy research has been influential in developing analytical frameworks and new research agendas. First, analytical models for understanding and comparing the effectiveness of international regimes have drawn heavily on examples of international environmental governance (Young, 1999; Miles et al., 2002). Early attempts to empirically assess the regime effectiveness built on qualitative case studies of different international institutions in the environmental domain (Haas et al., 1993; Levy et al., 1993; Keohane, 1996). Subsequent efforts to develop quantitative measures of regime effectiveness also used empirical data on international environmental agreements and treaties (Helm and Sprinz, 2000; Hovi et al., 2003). Second, research on variations in national compliance with multilateral agreements had its predominant empirical focus on international environmental agreements (Weiss and Jacobson, 1998; Chayes and Chayes, 1993; Mitchell, 1993). Based on the analysis of environmental agreements, scholars developed a managerial model of compliance that focused less on sanctioning non-compliance and more on enabling countries to comply with their international commitments by building up domestic capacities (Chayes et al., 1998).

Third, Peter Haas's concept of epistemic communities was developed against the empirical background of cases of global environmental governance (Haas, 1992; Adler and Haas, 1992; Haas, 1989). Further research on environmental issues helped refine the concept (Haas, 2016; Dunlop, 2016). Whereas, in the beginning, the concept of epistemic communities has been used primarily to explain variations in the success of international environmental policies, over the years, it has spilled over to other policy domains (see, for example, Howorth, 2004; King, 2005) and also to studies of policymaking at the national level (Löblová, 2018).

Finally, a fourth and more recent research agenda in international relations focuses on the role and influence of international public administrations, such as treaty secretariats or the secretariats of international organizations, in international policymaking (Chapter 12). An influential publication that set the stage for subsequent systematic research on international secretariats was Biermann and Siebenhüner's edited volume *Managers of Global Change* (Biermann and Siebenhüner, 2009), which compared the influence of several international environmental secretariats and bureaucracies on global environmental policy outputs. This publication was followed by a series of studies that focused on individual or groups of international public administrations to assess the preconditions and causal mechanisms of their influence on processes of global environmental governance (Jinnah, 2014; Jörgens et al., 2016; Saerbeck et al., 2020; Kolleck et al., 2017).

Overall, the study of environmental policy, despite its comparatively short history, has continuously provided theoretical stimuli and solid empirical bases for advancing the broader fields of policy analysis, comparative public policy, comparative politics, and global governance. In the next section, and based on the chapters in this volume, we identify a comprehensive and detailed research agenda for future environmental policy analysis. To facilitate reading and orientation for the reader, we present the research agenda in tables, each oriented to the topics of one of the four parts of the book.

30.3 A Future Research Agenda for Environmental Policy Analysis

The first part of this Handbook focused on analytical concepts and paradigms in environmental policy analysis. Important topics addressed in this section are the formation and development of policy domains (see for example Burstein, 1991; Knoke and Laumann, 1982; Haunss and

Hofmann, 2015), the distinction between policy outputs, outcomes, and impacts and how these dimensions relate to environmental performance (Knill et al., 2012; see also Fiorino, 2011; Esty and Porter, 2005; Jahn, 2016; Scruggs, 2003), the transformations of the state in the light of global environmental change (Dingwerth and Jörgens, 2015; Barry and Eckersley, 2005; Duit et al., 2016; Duit, 2014), the emerging concept of polycentric governance (Ostrom, 2010; Jordan et al., 2018; Dorsch and Flachsland, 2017), and the transformations of ecological modernization as a strategic approach to environmental policymaking (Jänicke, 2008; Mol et al., 2009). Table 30.1 summarizes the main topics for future research identified in Chapters 2–6.

Table 30.1 Analytical concepts and paradigms in environmental policy analysis

Chapter	*Topics for future research*
Emergence and Development of the Environmental Policy Field (Chapter 2, Böcher)	• Explore the relationship between climate change and other environmental issues within the environmental policy domain. Will other environmental topics such as nature conservation or air pollution control be subsumed under the umbrella of climate policy or will they retain their own status? • Study the potential emergence of climate policy as a distinct and autonomous policy domain. • Explore processes of polarization and the potential emergence of new conflicts in the environmental policy field. These conflicts may arise due to a radicalization of climate activism, populist parties attempting to roll back environmental policy decisions, or between the Global South and the Global North along questions of global environmental justice.
Environmental Policy Outputs, Outcomes, and Impacts (Chapter 3, Steinebach)	• Regarding environmental *policy outputs*, research should move beyond the study of advanced industrialized democracies and include economically less developed or autocratic countries. There is a need to assess whether the knowledge gained on policy outputs also holds in other spatial and cultural contexts. • With respect to *policy outcomes*, research has not yet fully taken into account the dynamics of policy accumulation and the growing complexity of environmental policy mixes. Instrument interaction needs to be studied in order to identify the most effective environmental policy mixes, including multiple instrument combinations. • Regarding *policy impacts*, future research needs to address the relationship between accumulation and implementation capacities in order to identify the right balance between the environmental policy stock which is up for implementation and the administrative capacities available as well as potential 'tipping points'.

(*Continued*)

Table 30.1 (Continued)

Chapter	*Topics for future research*
The Environmental State (Chapter 4, Duit)	• With a few exceptions, the scholarly debate on the environmental state has been theoretical and conceptual in nature. What is lacking are empirically grounded accounts of what the state can and cannot be expected to do with regard to the environmental crisis. • A main task for future research is to develop an empirically informed program which can then serve as the foundation for a second wave of theorizing about the state. • Future research should develop a better understanding of the driving forces of state involvement in environmental matters and its problem-solving capacity. • A more detailed idea of the political economy of the environmental state will allow for theorizing the environmental state *sui generis*, as something distinct from the welfare state, and governed by its own political economy.
Polycentric Governance (Chapter 5, Jordan and Huitema)	• One important priority in future research on polycentric governance is to understand better the role that governments and thus public policy should play. • There is need to further explore how the state and the structure of national political systems affect the political opportunity structures encountered by sub-national and non-state actors. Similarly, it is important to understand how state structures affect how new ideas circulate and become transplanted in national policy systems. • Although some scholarly work exists, more attention needs to be paid to the role of political power in the functioning of polycentric governance systems. • How long does polycentric governance take to emerge and how and why does it change over time? • Finally, research has to focus on the effectiveness of polycentric governance systems. Do polycentric systems make a material contribution to the resolution of environmental problems?
Ecological Modernization and Beyond (Chapter 6, Jänicke and Jörgens)	• There is a need for systematic and comparative analyses of the limits of ecological modernization as an environmental policy strategy and how these limits can be extended. • There is a lack of research on the potential, feasibility, and effectiveness of ecological modernization policies in developing countries. A mapping of best practices of ecological modernization policies in countries of the Global South could contribute to a better understanding of the factors that facilitate or hinder the adoption and implementation of these policies.

(Continued)

Table 30.1 (Continued)

Chapter	*Topics for future research*
	• Forward-looking typologies of environmental policy paradigms in combination with empirical assessments of actual environmental policy change in industrialized and developing countries could help to assess ongoing processes of paradigmatic policy change in the environmental and climate policy domain. • Environmental policy analysis should focus on the feasibility of different types of policies. However, political feasibility is still a relatively vague and under-researched concept. In order to improve its usefulness for environmental policy research, it must be operationalized and empirically applied.

The second part of the Handbook dealt with environmental policy performance and its key determinants. Topics addressed include the role and impact of environmental bureaucracies, both nationally and internationally, the integration of environmental concerns into other policy areas (environmental policy integration), the implementation of environmental policies, and the different approaches and methods to evaluating environmental policies. Further topics include the role of litigation by environmental NGOs, the consideration of indigenous and local knowledge in environmental decision-making, and the complex relationship between science and policy in evidence-based policymaking. Table 30.2 summarizes the main topics for future research identified in Chapters 7–15.

Table 30.2 Determinants of environmental policy performance

Chapter	*Topics for future research*
Determinants of Performance in National Environmental Policies (Chapter 7, Jahn and Klagges)	• Previous and current research has placed too much emphasis on what factors can improve environmental performance. More attention should be paid those factors that hinder environmental improvement or even promote pollution. • A major challenge for comparative environmental research in analyzing the determinants of environmental performance is the lack of reliable data and relevant concepts for many countries and over longer periods of time. • Macro-comparative environmental studies have begun to shift policy analysis from the consideration of individual key factors to the study of interactions. However, there are countless interactions and a unified concept of which interactions are most relevant has not yet been developed. • A challenge for future research is to model and to analyze the interconnectedness between domestic and international politics in environmental policymaking. There is a need for developing models for multilevel process analysis and an integrated theory which is suitable for this type of macro-comparative research.

(*Continued*)

Table 30.2 (Continued)

Chapter	*Topics for future research*
Bureaucracy and Environmental Policy (Chapter 8, Knill and Steinebach)	• Future research should systematically examine whether and how the institutionalization of environmental and climate issues in the state bureaucracy affects later policy outputs. • More and context-specific data on the capacities of the environmental administrations that are in charge of implementing individual environmental policies is needed. This data will help to assess which capacities different policy types require and whether the bureaucracies in charge of implementation possess these capacities. • Research on the dismantling of political programs has mainly focused on environmental policies. There is a need to explore processes of bureaucratic dismantling. Future research might thus examine whether and how successful pro-growth and less environmentally friendly actors use attacks on the administration as a distinct form of policy dismantling.
Analytical Perspectives on Environmental Policy Integration (Chapter 9, Steinbacher)	• Following the distinction between (1) horizontal and vertical environmental policy integration (EPI) and (2) policy substance versus policy process orientations, EPI initiatives should not be evaluated based on their overall environmental performance, but rather on their intermediate outcomes in each of the four dimensions. • Research should focus on prioritization (or political commitment) as a contextual factor that may affect the likelihood of EPI success or failure.
Environmental Policy Implementation (Chapter 10, Tosun and Schaub)	• Future research on environmental policy implementation should systematically link the degree of political (dis)agreement at the time when a given environmental policy instrument was adopted to its impact on the behavior of the target groups and its broader policy outcomes. • So far, street-level bureaucrats have received only limited attention in the literature on environmental policy implementation. By focusing on this type of actor, research on environmental policy implementation could benefit from insights provided by the existent literature on regulatory inspections. • Research should continue to focus on how participatory arrangements affect target group behavior and policy outcomes and connect it in a systematic fashion to other subdisciplines in political science, such as Political Sociology or Political Psychology.
Environmental Policy Evaluation (Chapter 11, Schoenefeld)	• Cumulative and especially international databases of environmental policy evaluations do not (yet) exist, but are a key requisite for the advancement of environmental policy knowledge. • Studies should explore how to better integrate evaluation knowledge into policymaking and explore the current and future role of evaluation in complex policy environments that do not follow the sequential logic described by the policy cycle model. • Questions related to the governance of evaluation have been little addressed in the field of (environmental policy) evaluation and merit greater attention in order to improve and further develop evaluation practices.

(*Continued*)

Table 30.2 (Continued)

Chapter	*Topics for future research*
	• Research should also address the challenge of linking and integrating different evaluative practices, such as impact assessment, policy monitoring, and policy evaluation. • The formulation of environmental policy evaluation standards or guidelines is needed in order to ensure a minimum level of evaluation quality as well as comparability and complementarity between instances of evaluation.
International Public Administrations in Environmental Governance (Chapter 12, Jörgens et al.)	• The observation that unelected bureaucracies beyond the nation state may actively seek to influence multilateral policies raises questions of democratic legitimacy. A comprehensive and systematic analysis of multiple cases of IPA influence could identify best practices in increasing the transparency, accountability, and participation of instances of IPA influence in environmental governance. • Future research could focus on the emergence of global administrative spaces in different fields of environmental policy and explore to what extent these transnational administrative structures have an impact on environmental governance. Different administrative spaces can be compared with regard to parameters such as network density, the centrality of different types of actors, or the role of IPAs in establishing such structures. • The study of IPAs is confronted with unique methodological challenges which require innovative methods and research designs. New methods for assessing the influence of IPAs should go beyond the traditional combination of interviews and document analysis. Mixed-method designs which combine quantitative and qualitative approaches and use longitudinal data are promising. • Research designs that systematically compare the role and influence of IPAs across different policy domains could be developed in order to relate IPA influence to variations in problem structures and features of the policy domain.
The role of Litigation of Environmental Non-Governmental Organizations in Environmental Politics and Policy (Chapter 13, Töller et al.)	• More research is needed on how ENGOs' right to take legal action affects environmental policymaking. • While there is a considerable body of research on how ENGOs use their right of legal action under the Aarhus Convention, more countries need to be covered, and in particular a finer differentiation of the phenomenon to be explained is needed. • Research could focus on the relative importance suing has for individual ENGOs in the portfolio of possible strategies and how these portfolios change over time. How are lawsuits distributed across different environmental associations operating in a country? Can a specialization of some associations in suing be observed? • Another important question relates to whether the right to sue leads to a shift of resources within associations, so that more resources go into enforcing existing law but fewer into influencing policymaking.

(*Continued*)

Table 30.2 (Continued)

Chapter	*Topics for future research*
	• There is need to assess the effect of suits on environmental quality (policy impacts). Future studies should seek to establish a causal relationship between lawsuits and environmental improvements. • Future research should also investigate the conditions that lead to the filing or non-filing of lawsuits and overcome the current selection bias towards cases in which lawsuits are filed.
Indigenous and Local Knowledge in Environmental Decision Making: The Case of Climate Change (Chapter 14, Solorio et al.)	• Research on climate justice needs to be strengthened in national and local contexts in order to overcome the state-centered approach that is dominant in climate policy analysis. • More attention should be paid to self-organization experiences on the part of indigenous groups. While the implementation of participative processes such as indigenous consultations have received much attention, less attention has been paid to the ways in which indigenous groups have tried to overcome the constraints posed by the state. • More generally, environmental research needs to connect with debates on decolonizing knowledge. Otherwise, proposals for integrating indigenous knowledge do nothing but to perpetuate current injustices.
The Science–Policy Interface and Evidence-Based Policymaking in Environmental Policy (Chapter 15, Wellstead et al.)	• So far, only little attention has been paid to how scientists and experts engage in policymaking and political advocacy. • Research should focus more on the causal mechanisms underpinning processes of evidence-based policymaking by applying qualitative methods such as process tracing. • Future research could focus on the emergence of policy innovation labs in the environmental sector and their potential to enhance evidence-based policymaking through their emphasis on experimental approaches and stakeholder engagement. • A research area that has received relatively little scholarly attention in the environmental policy field but will have a significant impact on evidence-based policymaking is algorithmic policymaking.

The third part addressed the drivers, dynamics, and selected examples of policy change in the environmental domain. The first three chapters in this part of the Handbook take a comparative perspective, focusing on processes of policy accumulation and the growth of environmental policy portfolios, cross-national leader-pioneer-follower dynamics in environmental governance, and the diffusion and convergence of environmental policies. The remaining chapters explore the interplay of procedural and substantive policy instruments in policy design for sustainable energy, the transformation of climate change and energy into security issues, and the dynamics of environmental policy change in Southeast Asia. Table 30.3 summarizes the main topics for future research identified in Chapters 16–21.

Table 30.3 Environmental policy change

Chapter	*Topics for future research*
Policy Change and Policy Accumulation in the Environmental Domain (Chapter 16, Knill)	• There is a need to further investigate trade-offs between the adoption of new policies and potential disadvantages emerging from policy growth with regard to policy complexity and bureaucratic overburdening. • Research indicates that institutional arrangements facilitating the vertical integration of policy production and policy implementation might help to avoid these problems associated with policy accumulation. These arguments need to be explored more systematically in future research.
Leaders, Pioneers, and Followers in Environmental Governance (Chapter 17, Liefferink et al.)	• While there is a considerable body of research on leaders and pioneers, the work on followership has only just started. Especially the dynamics between leaders and followers deserve attention: when, why, and under which circumstances do actors follow or refrain from doing so even in the case of active leadership efforts? • Research should focus more strongly on causal mechanisms, linking to scholarly work on the logics of consequences and appropriateness and the policy diffusion and transfer literature. • Leadership by non-state actors and, to a lesser degree, sub-national actors (such as cities) is still under-researched; their roles need to be more systematically investigated. The notion of polycentricity may provide a fruitful basis for understanding local action, experimentation, learning, and diffusion. • Studies of environmental leadership tend to focus on rich, industrialized countries, although there are notable exceptions. Widening the scope to the Global South should be an urgent priority.
Convergence and Diffusion of Environmental Policies (Chapter 18, Knill et al.)	• While there is the overall finding that diffusion and convergence contribute to the overall strengthening of environmental policies, more research is needed on the conditions under which such scenarios are more or less likely. • In this regard, particular emphasis should be placed on the temporal sequence between different drivers of policy diffusion and policy convergence, in particular with regard to economic drivers and drivers linked to international harmonization and cooperation.
Policy Design for Sustainable Energy and the Interplay of Procedural and Substantive Policy Instruments (Chapter 19, Mukherjee)	• Ongoing and future research work in the area of policy design is shifting the focus towards understanding not only how substantive and procedural instruments work individually, but how they can be deliberately planned to interact and work together in policy mixes to address complex policy problems. • Theorization about the particular relationship that procedural means share within the design of policy mixes – whether with each other or with more substantive tools – remains a promising area of future research. • In a time where sustainability will not automatically result from effective environmental policy blueprints, it is important to examine how good design can bring about the sound application of policies in sectors such as renewable energy and energy efficiency. • Further important avenues for future inquiry have to do with the temporality and sequencing of procedural means during policy design.

(*Continued*)

Table 30.3 (Continued)

Chapter	*Topics for future research*
Securitization, Climate Change, and Energy (Chapter 20, Trombetta)	• Future research on climate security discourses needs to explore how different discourses are assembled together, which elements are prioritized, and which are excluded, and what are the implications for security provision and for security logics and practices. • As new security practices and ways of governing through security like risk or resilience are gaining relevance, it becomes more urgent to explore their potential and limits. • A contextual approach that is open to transformation and more empirical analysis is necessary.
Environmental Policy Dynamics in Southeast Asia: Two Steps Forward, One Step Back (Chapter 21, Nair et al.)	• Additional data and indicators are needed to overcome uncertainty and improve environmental management in the region. There is a need for 'sustainability disclosures' which, although accepted in principle by all Southeast Asian countries, have only been incorporated into environmental and social risk management systems in Singapore and the Philippines. • The fact that there are still many environmental problems in low-capacity states (Cambodia, Laos) and backsliders (Myanmar, Philippines, Thailand) and continuing ongoing problems in Malaysia and Indonesia that have never been satisfactorily addressed, calls for additional research on the potential and prerequisites of environmental capacity-building in the region. • Overall, systematic comparative case studies are needed to identify the determinants of and obstacles to environmental policymaking in the region.

Finally, the fourth part addresses key paradigmatic challenges that sustainability transitions are confronted with and discusses how environmental policies and politics are transforming to meet these challenges. A first set of chapters in this section focuses on the challenge of long-term environmental policymaking, the role of cities and urban transformations in multi-level climate governance, and the turn towards complex policy mixes for addressing environmental challenges. A second set of chapters addresses the concept of sufficiency and whether and how it can be translated into concrete environmental policies, the emergence of a new climate movement characterized by more radical forms of protest and a refusal to make compromises, and the potential and risks of geoengineering as a technology-based strategy to alleviate the impacts of climate change. The last two chapters in this part explore processes of environmental policymaking in two distinct groups of countries: authoritarian states, with a focus on Singapore and Russia, and fast-growing economies, exemplified by the case of India. Table 30.4 summarizes the main topics for future research identified in Chapters 22–29.

In sum, this concluding chapter – and thereby this entire Handbook – has demonstrated that the study of environmental policy has made huge developments over the course of the last decades and has also triggered major advancements in adjacent disciplines. Yet, this Handbook has also demonstrated that this is not the end of the 'journey' but there is still much to learn and that there are new and pressing challenges requiring scholarly attention. We hope that the contributions in this Handbook have helped those who wanted to learn more about environmental policy and inspired those already experienced to think about 'old' issues in new ways. Moreover, we hope that the identified avenues for future research will be part of and addressed by future research projects.

Table 30.4 Transformation of environmental policies: paradigmatic challenges

Chapter	*Topics for future research*
The Challenge of Long-Term Environmental Policy (Chapter 22, Sprinz)	• More research should be directed at how long-term constituencies for Long-term Environmental Policy (LoPo) can be created. The influence that Fridays for Future is trying to exert on politicians and other stakeholders is one prominent example thereof. • An important, yet under-explored, question is whether environmental LoPo policymaking can be predicted. The use of policy prediction models should be explored in depth, e.g., on the implementation of net zero greenhouse gas emission goals, their timing, intermediate goals and the revisions thereof, but also on whether and when over-fished areas will (not) be reopened for harvest, and by which time stringent policies to build net carbon sinks will be pursued so as to add credibility to so-called net zero climate emissions goals. • Assessments of the effectiveness of LoPo, i.e., the causal effect of policy decisions on LoPo, are needed, although causal attribution over time will prove challenging. • Building on Putnam's dictum of two-level games about the nexus of international and domestic (environmental) policies, it would be fruitful to undertake systematic research on LoPo two-level challenges. • Research is also needed on which policy tools prove most successful in avoiding delayed starts in long-term environmental policymaking.
Cities and Urban Transformations in Multi-Level Climate Governance (Chapter 23, Kern)	• Research on local climate action tends to concentrate on leading cities, while 'ordinary' cities are neglected. Thus, there is still a lack of knowledge on smaller cities and towns which heavily depend on national funding due to a lack of own capacities. • A similar bias exists in research on city networks. Researchers prefer examining transnational city networks, such as ICLEI and C40, but overlook national and regional networks. • Research on local sustainability and climate strategies needs to be complemented by studies on actual implementation of such strategies on the ground. • There is still a lack of knowledge on the relationship between climate mitigation and adaptation at local level, in particular with respect to strategies, organizational solutions, and project integration. • The relations between cities and their 'hinterland' have hardly been studied, although sustainable, climate-neutral, and resilient cities depend on their periphery. • Research on new governance arrangements and emerging climate policies, such as climate budgets and climate contracts between cities and national agencies, is still missing. • Innovative methods which go beyond the state-of-the-art are needed.

(*Continued*)

Table 30.4 (Continued)

Chapter	*Topics for future research*
'Policy Mixes for Addressing Environmental Challenges (Chapter 24, Rogge and Song)	• Future research may benefit from drawing on advanced analytical models of the functional role of policy mixes in achieving transitions and extending these approaches in order to deepen our understanding of the role of transformative policy mixes in accelerating desired system changes. • Regarding the elements of multi-level policy mixes, an inclusion of policy strategies and instruments from different governance levels is necessary. However, transformative policy mixes addressing global or cross-sectoral environmental challenges may easily become too complex to map and harmonize all policy strategies and instrument mixes across administrative levels and policy departments. • There is a need for evaluating the impact of multi-level policy mixes. However, such a multi-level evaluation further complicates the already substantial challenge of evaluating policy mixes rather than single policy instruments. • Strengthening the consistency and coherence of policy mixes across policy fields is one of the core issues in future research, as possible conflicts between different policy fields might lead to ineffective policy mixes. • Future research and practice around policy mixes addressing environmental challenges should pay greater attention to governance, coordination, and capacity challenges such as the strategic coordination between governing.
Fifty Shades of Sufficiency: Semantic Confusion and No Policy (Chapter 25, Fuchs et al.)	• The development of methodologies for designing and evaluating sufficiency policies remains a pivotal task for scholars and practitioners alike. • It will be important to accompany relevant processes aiming to turn recent decisions by policy actors to integrate sufficiency in their sustainability strategies into practice. • Most fundamentally, a broad societal dialogue about the question of "what is enough" is needed.
The New Climate Movement: Organization, Strategy, and Consequences (Chapter 26, Buzogány and Scherhaufer)	• The scholarly literature is separated into three strands focusing on either on the characteristics, the strategies, or the consequences of new climate protest movements. A closer integration of these separate literatures is needed. • The new climate movement is studied and depicted as a predominantly Western and Northern phenomenon. There is a lack of research on transformations of environmental and climate movements and forms of protest in the Global South. • More research on the relationship between environmental movements and political parties and the influence of social movements on different forms of political participation, including voting, is needed. • Future research could focus more strongly on how the climate movement can use discursive opportunity structures and produce new imaginaries by coupling emotions with highly complex, and thus insecure, scientific knowledge. • Research on new climate movements could be linked to the rich literature on democratic innovations.

(*Continued*)

Table 30.4 (Continued)

Chapter	*Topics for future research*
Geoengineering and Public Policy: Framing, Research, and Deployment (Chapter 27, Möller)	• As countries face increasing pressure to act on climate change, there is a growing need to explore the potential, the risks, and the political regulation of technologies of large-scale carbon removal and large-scale solar radiation management. • Research has to consider the still highly speculative nature of geoengineering. • So far, geoengineering is discussed mostly in abstract terms and as a global response to climate change. Yet climate policy always takes place in specific local settings, with local actors and local interests involved. Bridging the gap between a global strategy and its local implementation will be a key challenge for researchers and policymakers.
Environmental Policymaking in Authoritarian Countries (Chapter 28: Chen and Aschenbrenner)	• Future research could concentrate on cases where factors such as regime stability, ideology, state capacity, or resource richness (fossil fuels) cause deviation from the typical cases of authoritarian environmentalism. • Scholars might take a comparative lens to specify the conditions for effective or ineffective authoritarian environmental policymaking, coupled with comprehensive data collection on policy implementation and environmental outcomes.
Environmental Policy in Fast-Growing Economies: The Case of India (Chapter 29, Jörgensen)	• Systematic comparative research about the environmental policy paths taken in rapidly developing countries is still rudimentary. • India's comparatively poor environmental performance, despite active and influential civil society organizations, needs to be studied more systematically. • Environmental policy in India is a unique case that is difficult to compare with other cases. But not least because of its outstanding importance as a polluter and emitter of greenhouse gases, India's environmental policy is an important topic for future environmental policy research.

References

Adam, Christian; Hurka, Steffen; Knill, Christoph; Steinebach, Yves (2019): *Policy Accumulation and the Democratic Responsiveness Trap*. Cambridge: Cambridge University Press.

Adam, Christian; Steinebach, Yves; Knill, Christoph (2018): Neglected Challenges to Evidence-Based Policy-Making: The Problem of Policy Accumulation. In *Policy Sciences* 51 (3), pp. 269–290.

Adler, Emanuel; Haas, Peter M. (1992): Conclusion: Epistemic Communities, World Order, and the Creation of a Reflective Research Program. In *International Organization* 46 (1), pp. 367–390.

Alexander, Samuel (2012): Planned Economic Contraction: The Emerging Case for Degrowth. In *Environmental Politics* 21 (3), pp. 349–368.

Barry, John (2007): Towards a Model of Green Political Economy: From Ecological Modernisation to Economic Security. In *International Journal of Green Economics* 1 (3/4), pp. 446–464.

Barry, John; Doran, Peter (2006): Refining Green Political Economy: From Ecological Modernisation to Economic Security and Sufficiency. In *Analyse & Kritik* 28 (2), pp. 250–275.

Barry, John; Eckersley, Robyn (Eds.) (2005): *The State and the Global Ecological Crisis*. Cambridge, MA: MIT Press.

Bayerlein, Louisa; Knill, Christoph; Steinebach, Yves (2020): *A Matter of Style: Organizational Agency in Global Public Policy*. Cambridge: Cambridge University Press.

Bemelmans-Videc, Marie-Louise; Rist, Ray C.; Vedung, Evert (Eds.) (1998): *Carrots, Sticks, and Sermons: Policy Instruments and Their Evaluation*. New Brunswick: Transaction.

Berry, Frances Stokes; Berry, William D. (2007): Innovation and Diffusion Models in Policy Research. In Paul A. Sabatier (Ed.) *Theories of the Policy Process*. 2nd ed. Boulder: Westview Press, pp. 223–260.

Biermann, Frank; Siebenhüner, Bernd (Eds.) (2009): *Managers of Global Change: The Influence of International Environmental Bureaucracies*. Cambridge, Mass: MIT Press.

Börzel, Tanja A. (2021): *Why Noncompliance: The Politics of Law in the European Union*. Ithaca: Cornell University Press.

Browne, Stephen (2017): *Sustainable Development Goals and UN Goal-Setting*. London: Routledge.

Burstein, Paul (1991): Policy Domains: Organization, Culture, and Policy Outcomes. In *Annual Review of Sociology* 17 (1), pp. 327–350.

Busch, Per-Olof; Jörgens, Helge (2005): The International Sources of Policy Convergence: Explaining the Spread of Environmental Policy Innovations. In *Journal of European Public Policy* 12 (5), pp. 860–884.

Busch, Per-Olof; Jörgens, Helge; Tews, Kerstin (2005): The Global Diffusion of Regulatory Instruments: The Making of a New International Environmental Regime. In *ANNALS of the American Academy of Political and Social Science* 598 (1), pp. 146–167.

Chayes, Abram; Chayes, Antonia Handler (1993): On Compliance. In *International Organization* 47 (2), pp. 175–205.

Chayes, Abram; Chayes, Antonia Handler; Mitchell, Ronald B. (1998): Managing Compliance: A Comparative Perspective. In Edith Brown Weiss, Harold K. Jacobson (Eds.) *Engaging Countries. Strengthening Compliance with International Environmental Accords*. Cambridge, MA: MIT Press, pp. 39–62.

Dingwerth, Klaus; Jörgens, Helge (2015): Environmental Risks and the Changing Interface of Domestic and International Governance. In Stephan Leibfried, Frank Nullmeier, Evelyne Huber, Matthew Lange, Jonah Levy, John Stephens (Eds.) *The Oxford Handbook of Transformations of the State*. Oxford: Oxford University Press, pp. 338–354.

Dorsch, Marcel J.; Flachsland, Christian (2017): A Polycentric Approach to Global Climate Governance. In *Global Environmental Politics* 17 (2), pp. 45–64.

Duit, Andreas (Ed.) (2014): *State and Environment: The Comparative Study of Environmental Governance*. Cambridge, MA: MIT Press.

Duit, Andreas; Feindt, Peter H.; Meadowcroft, James (2016): Greening Leviathan: The Rise of the Environmental State? In *Environmental Politics* 25 (1), pp. 1–23.

Dunlop, Claire A. (2016): Knowledge, Epistemic Communities, and Agenda Setting. In Nikolaos Zahariadis (Ed.): *Handbook of Public Policy Agenda Setting*. Cheltenham: Edward Elgar, pp. 273–294.

Esty, Daniel C.; Porter, Michael E. (2005): National Environmental Performance: An Empirical Analysis of Policy Results and Determinants. In *Environment and Development Economics* 10 (4), pp. 391–434.

Fernandéz-i-Marín, Xavier; Knill, Christoph; Steinebach, Yves (2021): Studying Policy Design Quality in Comparative Perspective. In *American Political Science Review* 115 (3), pp. 931–947.

Fiorino, Daniel J. (2011): Explaining National Environmental Performance: Approaches, Evidence, and Implications. In *Policy Sciences* 44 (4), pp. 367–389.

Gallagher, Michael; Laver, Michael; Mair, Peter (2006): *Representative Government in Modern Europe: Institutions, Parties and Governments*. 4th ed. Boston: McGraw-Hill.

Gilardi, Fabrizio (2012): Transnational Diffusion: Norms, Ideas, and Policies. In Walter Carlsnaes, Thomas Risse, Beth A. Simmons (Eds.): *Handbook of International Relations*. 2nd ed. London: Sage, pp. 453–477.

Giugni, Marco; Grasso, Maria T. (2015): Environmental Movements in Advanced Industrial Democracies: Heterogeneity, Transformation, and Institutionalization. In *Annual Review of Environment and Resources* 40 (1), pp. 337–361.

Gray, Virginia (1994): Competition, Emulation, and Policy Innovation. In Lawrence C. Dodd, Calvin Jillson (Eds.): *New Perspectives on American Politics*. Washington, D.C.: CQ Press, pp. 230–248.

Gunningham, Neil; Grabosky, Peter (1998): *Smart Regulation: Designing Environmental Policy*. Oxford: Clarendon Press.

Gunningham, Neil A.; Sinclair, Darren (1999): Regulatory Pluralism: Designing Policy Mixes for Environmental Protection. In *Law & Policy* 21 (1), pp. 49–76.

Haas, Peter M. (1989): Do Regimes Matter? Epistemic Communities and Mediterranean Pollution Control. In *International Organization* 43 (3), pp. 377–403.

Haas, Peter M. (1992): Introduction: Epistemic Communities and International Policy Coordination. In *International Organization* 46 (1), pp. 1–35.

Haas, Peter M. (2016): *Epistemic Communities, Constructivism, and International Environmental Politics*. London: Routledge.

Haas, Peter M.; Keohane, Robert O.; Levy, Marc A. (Eds.) (1993): *Institutions for the Earth: Sources of Effective International Environmental Protection*. Cambridge, Mass: MIT Press.

Hajer, Maarten; Versteeg, Wytske (2005): A Decade of Discourse Analysis of Environmental Politics: Achievements, Challenges, Perspectives. In *Journal of Environmental Policy & Planning* 7 (3), pp. 175–184. DOI: 10.1080/15239080500339646.

Hajer, Maarten A. (1995): *The Politics of Environmental Discourse: Ecological Modernization, and the Policy Process*. Oxford: Clarendon Press.

Haunss, Sebastian; Hofmann, Jeanette (2015): Entstehung von Politikfeldern – Bedingungen einer Anomalie. In *Der moderne Staat* 8 (1), pp. 29–49.

Helm, Carsten; Sprinz, Detlef F. (2000): Measuring the Effectiveness of International Environmental Regimes. In *Journal of Conflict Resolution* 44 (5), pp. 630–652.

Héritier, Adrienne; Knill, Christoph; Mingers, Susanne (1996): *Ringing the Changes in Europe: Regulatory Competition and the Transformation of the State*. Britain, France, Germany. Berlin: de Gruyter.

Hoberg, George (1991): Sleeping with an Elephant: The American Influence on Canadian Environmental Regulation. In *Journal of Public Policy* 11 (1), pp. 107–131.

Holzinger, Katharina; Jörgens, Helge; Knill, Christoph (Eds.) (2007): *Transfer, Diffusion und Konvergenz von Politiken*. Wiesbaden: VS Verlag für Sozialwissenschaften (Politische Vierteljahresschrift, Sonderheft 38).

Holzinger, Katharina; Knill, Christoph; Arts, Bas (Eds.) (2008a): *Environmental Policy Convergence in Europe. The Impact of International Institutions and Trade*. Cambridge: Cambridge University Press.

Holzinger, Katharina; Knill, Christoph; Sommerer, Thomas (2008b): Environmental Policy Convergence: The Impact of International Harmonization, Transnational Communication, and Regulatory Competition. In *International Organization* 62 (4), pp. 553–587.

Hovi, Jon; Sprinz, Detlef F.; Underdal, Arild (2003): The Oslo-Potsdam Solution to Measuring Regime Effectiveness: Critique, Response, and the Road Ahead. In *Global Environmental Politics* 3 (3), pp. 74–96.

Howlett, Michael; Rayner, Jeremy (2013): Patching vs Packaging in Policy Formulation: Assessing Policy Portfolio Design. In *Politics and Governance* 1 (2), pp. 170–182.

Howlett, Michael; Tosun, Jale (Eds.) (2021): *The Routledge Handbook of Policy Styles*. London: Routledge.

Howorth, Jolyon (2004): Discourse, Ideas, and Epistemic Communities in European Security and Defence Policy. In *West European Politics* 27 (2), pp. 211–234.

Hurka, Steffen; Haag, Maximilian; Kaplaner, Constantin (2022): Policy Complexity in the European Union, 1993-today: Introducing the EUPLEX Dataset. In *Journal of European Public Policy* 29 (9), pp. 1512–1527.

Jacob, Klaus; Volkery, Axel (2004): Institutions and Instruments for Government Self-Regulation: Environmental Policy Integration in a Cross-Country Perspective. In *Journal of Comparative Policy Analysis: Research and Practice* 6 (3), pp. 291–309.

Jahn, Detlef (1998): Environmental Performance and Policy Regimes: Explaining Variations in 18 OECD-Countries. In *Policy Sciences* 31 (2), pp. 107–131.

Jahn, Detlef (2016): *The Politics of Environmental Performance: Institutions and Preferences in Industrialized Democracies*. Cambridge: Cambridge University Press.

Jänicke, Martin (1985): *Preventive Environmental Policy as Ecological Modernisation and Structural Policy*. Discussion Paper 85/2. Berlin: Social Science Research Center (WZB).

Jänicke, Martin (1997): The Political System's Capacity for Environmental Policy. In Martin Jänicke, Helmut Weidner (Eds.): *National Environmental Policies. A Comparative Study of Capacity-Building. With assistance of Helge Jörgens*. Berlin: Springer, pp. 1–24.

Jänicke, Martin (2008): Ecological Modernisation: New Perspectives. In *Journal of Cleaner Production* 16 (5), pp. 557–565.

Jänicke, Martin; Jörgens, Helge (2009): New Approaches to Environmental Governance. In Arthur P. J. Mol, David A. Sonnenfeld, Gert Spaargaren (Eds.): *The Ecological Modernisation Reader: Environmental Reform in Theory and Practice*. London: Routledge, pp. 159–189.

Jänicke, Martin; Weidner, Helmut (Eds.) (1997a): *National Environmental Policies. A Comparative Study of Capacity-Building. With assistance of Helge Jörgens*. Berlin: Springer.

Jänicke, Martin; Weidner, Helmut (1997b): Summary: Global Environmental Policy Learning. In Martin Jänicke, Helmut Weidner (Eds.): *National Environmental Policies. A Comparative Study of Capacity-Building. With assistance of Helge Jörgens*. Berlin: Springer, pp. 299–313.

Jinnah, Sikina (2014): *Post-Treaty Politics: Secretariat Influence in Global Environmental Governance*. Cambridge, Mass.: MIT Press.

Jordan, Andrew (1999): The Implementation of EU Environmental Policy: A Policy Problem without a Political Solution? In *Environment and Planning C: Government and Policy* 17 (1), pp. 69–90.

Jordan, Andrew; Huitema, Dave; van Asselt, Harro; Forster, Johanna (Eds.) (2018): *Governing Climate Change: Polycentricity in Action?* Cambridge: Cambridge University Press.

Jordan, Andrew; Lenschow, Andrea (2010): Environmental Policy Integration: A State of the Art Review. In *Environmental Policy and Governance* 20 (3), pp. 147–158.

Jordan, Andrew; Wurzel, Rüdiger K. W.; Zito, Anthony R. (Eds.) (2003): *New Instruments of Environmental Governance? National Experiences and Prospects*. London: Frank Cass.

Jörgens, Helge (1996): Die Institutionalisierung von Umweltpolitik im internationalen Vergleich. In Martin Jänicke (Ed.): *Umweltpolitik der Industrieländer. Entwicklung - Bilanz - Erfolgsbedingungen*. Berlin: Edition Sigma, pp. 59–112.

Jörgens, Helge (2001): The Diffusion of Environmental Policy Innovations - Findings from an International Workshop. In *Environmental Politics* 10 (2), pp. 122–127.

Jörgens, Helge (2004): Governance by Diffusion: Implementing Global Norms Through Cross-National Imitation and Learning. In William M. Lafferty (Ed.): *Governance for Sustainable Development. The Challenge of Adapting Form to Function*. Cheltenham: Edward Elgar, pp. 246–283.

Jörgens, Helge; Kolleck, Nina; Saerbeck, Barbara (2016): Exploring the Hidden Influence of International Treaty Secretariats: Using Social Network Analysis to Analyse the Twitter Debate on the 'Lima Work Programme on Gender'. In *Journal of European Public Policy* 23 (7), pp. 979–998.

Jörgens, Helge; Lenschow, Andrea; Liefferink, Duncan (Eds.) (2014): *Understanding Environmental Policy Convergence: The Power of Words, Rules and Money*. Cambridge: Cambridge University Press.

Kallis, Giorgos; Paulson, Susan; D'Alisa, Giacomo; Demaria, Federico (2020): The Case for Degrowth. Cambridge: Polity.

Kanie, Norichika; Biermann, Frank (Eds.) (2017): *Governing Through Goals: Sustainable Development Goals As Governance Innovation*. Cambridge, Mass.: MIT Press.

Keohane, Robert O. (1996): Analyzing the Effectiveness of International Environmental Institutions. In Robert O. Keohane, Marc A. Levy (Eds.): *Institutions for Environmental Aid: Pitfalls and Promise*. Cambridge, Mass: MIT Press, pp. 3–27.

Kern, Kristine (1997): Politikkonvergenz durch Politikdiffusion - Überlegungen zu einer vernachlässigten Dimension der vergleichenden Politikanalyse. In Lutz Mez, Helmut Weidner (Eds.): *Umweltpolitik und Staatsversagen. Perspektiven und Grenzen der Umweltpolitikanalyse. Festschrift für Martin Jänicke zum 60. Geburtstag*. Berlin: Edition Sigma, pp. 270–279.

Kern, Kristine; Jörgens, Helge; Jänicke, Martin (2001): The Diffusion of Environmental Policy Innovations: A Contribution to the Globalisation of Environmental Policy. Berlin (WZB-Discussion Paper, FS II 01-302).

King, Michael (2005): Epistemic Communities and the Diffusion of Ideas: Central Bank Reform in the United Kingdom. In *West European Politics* 28 (1), pp. 94–123.

Knill, Christoph (2001): *The Europeanisation of National Administrations: Patterns of Institutional Change and Persistence*. Cambridge: Cambridge University Press.

Knill, Christoph; Debus, Marc; Heichel, Stephan (2010): Do Parties Matter in Internationalised Policy Areas? The Impact of Political Parties on Environmental Policy Outputs in 18 OECD Countries, 1970-2000. In *European Journal of Political Research* 49 (3), pp. 301–336.

Knill, Christoph; Lenschow, Andrea (Eds.) (2000): *Implementing EU Environmental Policy: New Directions and Old Problems*. Manchester: Manchester University Press.

Knill, Christoph; Liefferink, Duncan (2007): *Environmental Politics in the European Union: Policy-making, Implementation and Patterns of Multi-level Governance*. Manchester: Manchester University Press.

Knill, Christoph; Schulze, Kai; Tosun, Jale (2012): Regulatory Policy Outputs and Impacts: Exploring a Complex Relationship. In *Regulation & Governance* 6 (4), pp. 427–444.

Knill, Christoph; Steinbacher, Christina; Steinebach, Yves (2021a): Balancing Trade-Offs between Policy Responsiveness and Effectiveness: The Impact of Vertical Policy-Process Integration on Policy Accumulation. In *Public Administration Review* 81 (1), pp. 157–160.

Knill, Christoph; Steinbacher, Christina; Steinebach, Yves (2021b): Sustaining Statehood: A Comparative Analysis of Vertical Policy-Process Integration in Denmark and Italy. In *Public Administration* 99 (4), pp. 758–774.

Knill, Christoph; Steinebach, Yves; Zink, Dionys: How Policy Growth Affects Policy Implementation: Bureaucratic Overload and Policy Triage. In *Journal of European Public Policy* forthcoming.

Knoke, David; Laumann, Edward O. (1982): The Social Organization of National Policy Domains: An Exploration of Some Structural Hypotheses. In Peter V. Marsden, Nan Lin (Eds.): *Social Structure and Network Analysis*. Beverly Hills: Sage, pp. 255–270.

Kolleck, Nina; Well, Mareike; Sperzel, Severin; Jörgens, Helge (2017): The Power of Social Networks: How the UNFCCC Secretariat Creates Momentum for Climate Education. In *Global Environmental Politics* 17 (4), pp. 106–126.

Kriesi, Hanspeter; Grande, Edgar; Dolezal, Martin; Helbling, Marc; Höglinger, Dominic; Hutter, Sven; Wüest, Bruno (Eds.) (2012): *Political Conflict in Western Europe*. Cambridge: Cambridge University Press.

Lafferty, William M.; Hovden, Eivind (2003): Environmental Policy Integration: Towards an Analytical Framework. In *Environmental Politics* 12 (3), pp. 1–22.

Lenschow, Andrea (Ed.) (2002): *Environmental Policy Integration: Greening Sectoral Policies in Europe*. London: Earthscan.

Leschine, Thomas M. (Ed.) (2007): *Long-Term Management of Contaminated Sites*. Amsterdam: Elsevier.

Levy, Marc A.; Keohane, Robert O.; Haas, Peter M. (1993): Improving the Effectiveness of International Environmental Institutions. In Peter M. Haas, Robert O. Keohane, Marc A. Levy (Eds.): *Institutions for the Earth: Sources of Effective International Environmental Protection*. Cambridge, Mass: MIT Press, pp. 397–426.

Lijphart, Arend (2012): *Patterns of Democracy: Government Forms and Performance in Thirty-Six Countries*. 2nd ed. New Haven: Yale University Press.

Limberg, Julian; Knill, Christoph; Steinebach, Yves (2022): Condemned to Complexity? Growing State Activity and Complex Policy Systems. In *Governance* online first. Available online at https://onlinelibrary.wiley.com/doi/full/10.1111/gove.12684.

Limberg, Julian; Steinebach, Yves; Bayerlein, Louisa; Knill, Christoph (2021): The More the Better? Rule Growth and Policy Impact from a Macro Perspective. In *European Journal of Political Research* 60 (2), pp. 438–454.

Löblová, Olga (2018): When Epistemic Communities Fail: Exploring the Mechanism of Policy Influence. In *Policy Studies Journal* 46 (1), pp. 160–189.

Lorek, Sylvia; Spangenberg, Joachim H. (2019): Identification of Promising Instruments and Instrument Mixes to Promote Energy Sufficiency. *EUFORIE - European Futures for Energy Efficiency*. Deliverable 5.5 [Available online: https://ec.europa.eu/research/participants/documents/downloadPublic?documentIds=080166e5c39c2b51&appId=PPGMS].

Marks, Gary; Attewell, David; Rovny, Jan; Hooghe, Liesbet (2021): Cleavage Theory. In Marianne Riddervold, Jarle Trondal, Akasemi Newsome (Eds.): *The Palgrave Handbook of EU Crises*. Cham: Palgrave Macmillan, pp. 173–193.

Meadowcroft, James; Fiorino, Daniel J. (Eds.) (2017): *Conceptual Innovation in Environmental Policy*. Cambridge, MA: MIT Press.

Miles, Edward L.; Underdal, Arild; Andresen, Steinar; Wettestad, Jørgen; Skjærseth, Jon Birger; Carlin, Elaine M. (Eds.) (2002): *Environmental Regime Effectiveness. Confronting Theory with Evidence*. Cambridge, Mass: MIT Press.

Mitchell, Ronald B. (1993): Compliance Theory: A Synthesis. In *Review of European Community & International Environmental Law* 2 (4), pp. 327–334.

Mol, Arthur P. J.; Sonnenfeld, David A.; Spaargaren, Gert (Eds.) (2009): *The Ecological Modernisation Reader: Environmental Reform in Theory and Practice*. London: Routledge.

Neumayer, Eric (2003): Are Left-wing Party Strength and Corporatism Good for the Environment? Evidence from Panel Analysis of Air Pollution in OECD Countries. In *Ecological Economics* 45 (2), pp. 203–220.

Ostrom, Elinor (1990): *Governing the Commons: The Evolution of Institutions for Collective Action*. Cambridge: Cambridge University Press.

Ostrom, Elinor (2007): Institutional Rational Choice: An Assessment of the Institutional Analysis and Development Framework. In Paul A. Sabatier (Ed.): *Theories of the Policy Process*. 2nd ed. Boulder: Westview Press, pp. 21–64.

Ostrom, Elinor (2010): Beyond Markets and States: Polycentric Governance of Complex Economic Systems. In *American Economic Review* 100 (3), pp. 641–672.

Princen, Thomas (2003): Principles for Sustainability: From Cooperation and Efficiency to Sufficiency. In *Global Environmental Politics* 3 (1), pp. 33–50.

Rootes, Christopher (2004): Environmental Movements. In David A. Snow, Sarah A. Soule, Hanspeter Kriesi (Eds.): *The Blackwell Companion to Social Movements*. Malden, MA: Blackwell, pp. 608–640.

Sabatier, Paul (1998): The Advocacy Coalition Framework: Revisions and Relevance for Europe. In *Journal of European Public Policy* 5 (1), pp. 98–130.

Sabatier, Paul A. (1988): An Advocacy Coalition Framework of Policy Change and the Role of Policy-Oriented Learning Therein. In *Policy Sciences* 21 (2/3), pp. 129–168.

Saerbeck, Barbara; Well, Mareike; Jörgens, Helge; Goritz, Alexandra; Kolleck, Nina (2020): Brokering Climate Action: The UNFCCC Secretariat Between Parties and Nonparty Stakeholders. In *Global Environmental Politics* 20 (2), pp. 105–127.

Schneidewind, Uwe; Zahrnt, Angelika (2014): *The Politics of Sufficiency: Making it easier to live the Good Life*. München: oekom.

Schulze, Kai (2014): Do Parties Matter for International Environmental Cooperation? An Analysis of Environmental Treaty Participation by Advanced Industrialised Democracies. In *Environmental Politics* 23 (1), pp. 115–139. doi: 10.1080/09644016.2012.740938.

Scruggs, Lyle (2001): Is There Really a Link Between Neo-Corporatism and Environmental Performance? Updated Evidence and New Data for the 1980s and 1990s. In *British Journal of Political Science* 31 (4), pp. 686–692.

Scruggs, Lyle (2003): *Sustaining Abundance: Environmental Performance in Industrial Democracies*. Cambridge: Cambridge University Press.

Simmons, Beth A.; Elkins, Zachary (2003): Globalization and Policy Diffusion: Explaining Three Decades of Liberalization. In Miles Kahler, David A. Lake (Eds.): Governance in a Global Economy: Political Authority in Transition. Princeton: Princeton University Press, pp. 275–304.

Sprinz, Detlef F. (2012): Long-Term Environmental Policy: Definition–Origin–Response Options. In Peter Dauvergne (Ed.): *Handbook of Global Environmental Politics*. 2nd ed. Cheltenham: Edward Elgar, pp. 183–193.

Steinebach (2019): Water Quality and the Effectiveness of European Union Policies. In *Water* 11 (11), p. 2244.

Steinebach, Yves (2022a): Administrative Traditions and the Effectiveness of Regulation. In *Journal of European Public Policy*, online first, 1–20.

Steinebach, Yves (2022b): Instrument Choice, Implementation Structures, and the Effectiveness of Environmental Policies: A Cross-national Analysis. In *Regulation & Governance* 16 (1), pp. 225–242.

Stuart, Diana; Gunderson, Ryan; Petersen, Brian (2021): *The Degrowth Alternative: A Path to Address our Environmental Crisis?* London: Routledge.

Tews, Kerstin; Busch, Per-Olof; Jörgens, Helge (2003): The Diffusion of New Environmental Policy Instruments. In *European Journal of Political Research* 42 (4), pp. 569–600.

Thisted, Ebbe V.; Thisted, Rune V. (2020): The Diffusion of Carbon Taxes and Emission Trading Schemes: The Emerging Norm of Carbon Pricing. In *Environmental Politics* 29 (5), pp. 804–824.

Tosun, Jale; Lang, Achim (2017): Policy Integration: Mapping the Different Concepts. In *Policy Studies* 38 (6), pp. 553–570.

Touraine, Alain (1981): *The Voice and the Eye: An Analysis of Social Movements*. Cambridge: Cambridge University Press.

Trein, Philipp; Meyer, Iris; Maggetti, Martino (2019): The Integration and Coordination of Public Policies: A Systematic Comparative Review. In *Journal of Comparative Policy Analysis: Research and Practice* 21 (4), pp. 332–349.

Vogel, David (1986): *National Styles of Regulation: Environmental Policy in Great Britain and the United States*. Ithaca, NY: Cornell University Press.

Vogel, David (1995): *Trading Up. Consumer and Environmental Regulation in a Global Economy*. Cambridge, Mass: Harvard University Press.

Vogel, David (1997): Trading Up and Governing Across: Transnational Governance and Environmental Protection. In *Journal of European Public Policy* 4 (4), pp. 556–571.

Walker, Jack L. (1969): The Diffusion of Innovations Among the American States. In *American Political Science Review* 63 (3), pp. 880–899.

Walter, Stefanie (2017): Globalization and the Demand-Side of Politics: How Globalization Shapes Labor Market Risk Perceptions and Policy Preferences. In *Political Science Research and Methods* 5 (1), pp. 55–80.

Weible, Christopher M.; Jenkins-Smith, Hank C. (2016): The Advocacy Coalition Framework: An Approach for the Comparative Analysis of Contentious Policy Issues. In B. Guy Peters, Philippe Zittoun (Eds.): *Contemporary Approaches to Public Policy: Theories, Controversies and Perspectives*. London: Palgrave Macmillan, pp. 15–34.

Weible, Christopher M.; Sabatier, Paul A.; McQueen, Kelly (2009): Themes and Variations: Taking Stock of the Advocacy Coalition Framework. In *Policy Studies Journal* 37 (1), pp. 121–140.

Weidner, Helmut; Jänicke, Martin (Eds.) (2002): *Capacity Building in National Environmental Policy: A Comparative Study of 17 Countries*. Berlin: Springer.

Weiss, Edith Brown; Jacobson, Harold K. (Eds.) (1998): *Engaging Countries. Strengthening Compliance with International Environmental Accords*. Cambridge, Mass: MIT Press.

World Commission on Environment and Development (1987): *Our Common Future*. New York: Oxford University Press.

Wurzel, Rüdiger K.W.; Liefferink, Duncan; Torney, Diarmuid (2019): Pioneers, Leaders and Followers in Multilevel and Polycentric Climate Governance. In *Environmental Politics* 28 (1), pp. 1–21.

Young, Oran R. (Ed.) (1999): *The Effectiveness of International Environmental Regimes. Causal Connections and Behavioral Mechanisms*. Cambridge, Mass: MIT Press.

INDEX

Pages in *italics* refer to figures, pages in **bold** refer to tables, and pages followed by n refer to notes.